# Selected Papers from SDEWES 2017: The 12th Conference on Sustainable Development of Energy, Water and Environment Systems

# Selected Papers from SDEWES 2017: The 12th Conference on Sustainable Development of Energy, Water and Environment Systems

Special Issue Editors

**Francesco Calise**
**Mário Costa**
**Qiuwang Wang**
**Zhang Xiliang**
**Neven Duic**

MDPI • Basel • Beijing • Wuhan • Barcelona • Belgrade

*Special Issue Editors*

Francesco Calise
University of Naples Federico II
Italy

Mário Costa
Instituto Superior Técnico Mechanical
Engineering Department
Portugal

Qiuwang Wang
Xi'an Jiaotong University
China

Zhang Xiliang
Tsinghua University
China

Neven Duic
University of Zagreb
Croatia

*Editorial Office*
MDPI
St. Alban-Anlage 66
4052 Basel, Switzerland

This is a reprint of articles from the Special Issue published online in the open access journal *Energies* (ISSN 1996-1073) in 2018 (available at: https://www.mdpi.com/journal/energies/special_issues/ SDEWES_2017)

For citation purposes, cite each article independently as indicated on the article page online and as indicated below:

LastName, A.A.; LastName, B.B.; LastName, C.C. Article Title. *Journal Name* **Year**, *Article Number, Page Range.*

ISBN 978-3-03897-396-6 (Pbk)
ISBN 978-3-03897-397-3 (PDF)

Cover image courtesy of shutterstock.com user cge2010.

# Contents

# About the Special Issue Editors

**Francesco Calise** was born in 1978 and graduated cum laude in mechanical engineering from the University of Naples Federico II, Italy in 2002. He obtained the Ph.D. degree in Mechanical and Thermal Engineering in 2006. From 2006 to 2014, he was Researcher and Assistant Professor of Applied Thermodynamics at the University of Naples Federico II. In 2014 he became Associate Professor at the University of Naples Federico II. His research activity has been mainly focused on the following topics: fuel cells, advanced optimization techniques, solar thermal systems, concentrating photovoltaic/thermal photovoltaic systems, energy saving in buildings, solar heating and cooling, organic Rankine cycles, geothermal energy, dynamic simulations of energy systems, and renewable polygeneration systems. He was invited as a lecturer for some courses or conferences (UK and Finland). He teaches several courses on energy management and applied thermodynamics at the University of Naples Federico II for BsC, MS, and PhD students. He was a supervisor of several Ph.D. degree theses. He is a reviewer of about 30 international journals. He was involved in several research projects funded by EU and the Italian Government. He is member of the Editorial Board of 10 international journals. He was a conference chair and/or a member of the scientific committee in several sessions of international conferences.

**Neven Duić**, Prof. Dr. Sc., is a Full Professor of Power Engineering at the Department of Energy, Power, Engineering, and Environment, Faculty of Mechanical Engineering and Naval Architecture. He has been a member of the Postgraduate Study Committee since 2008 and Chair for the PhD study of the process energy field. He was Chair of Faculty committee for international projects from 2010 until 2014. Prof. Duić has published 108 original scientific papers and 7 review papers in scientific journals referenced in SCI and CC publications; 82 of these scientific papers are Q1 papers, of which 39 are in the top 5% of the journal category (referring to ISI Web of knowledge data base). His papers were cited 2173 times according to Scopus data base and 1779 times according to Web of knowledge data base. His H-index according to Scopus is 28, and according to the Web of Knowledge data base it is 24. According to CROSBI (Croatian Scientific Bibliography), Prof. Duić has published the most CC scientific papers in technical science since 2007. He received a National Award for significant scientific achievement in 2016 in the field of engineering sciences. Prof. Duić has successfully supervised eight PhD students, and he is currently mentoring five FMENA PhD students. He was member of the PhD thesis Juries for 11 PhD students, of which seven were at foreign universities (Macquarie University, Aalborg University, University of Limerick, Instituto Superior Tecnico, University of Sarajevo, University of Beograd, and University of Santander). He is editor of Q1 journal *Energy Conversion Management*, subject editor for Q1 journal *Energy*, and a member of the editorial board of the following journals: Q1 journal *Applied Energy*; Q2 journal Clean Technologies and Environmental Policy"; Q3 journal "Thermal Science"; and since 2013, he has been Editor-in-Chief of *Journal of Sustainable Development of Energy, Water, and Environment Systems—JSDEWES*, which has been indexed in Scopus data base at very short notice and from the first attempt. He has organised several of the series of conferences on Sustainable Development of Energy, Water, and Environment Systems and was a member of organising, scientific, and programming committees of more than 57 research conferences. Prof. Duić is coordinator of Croatian participation in two projects of the Croatian Foundation for Science and over 40 international scientific research projects. He is national representative of Horizon 2020 projects for ERC/MSCA/FET. His main areas of interests are energy policy and planning, energy economics, sustainable development policy and resource planning, climate change mitigation, combustion engineering and modelling, research, and innovation policy.

Full CV: http://powerlab.fsb.hr/nduic/

**Mário Costa** is a Full Professor in the area of Environment and Energy at the Mechanical Engineering Department of Instituto Superior Técnico. He graduated in Chemical Engineering at University of Coimbra in 1984, obtained his PhD in Mechanical Engineering at Imperial College London in 1992, and obtained his Habilitation in Mechanical Engineering at Technical University of Lisbon in 2009. Currently, he teaches courses on thermodynamics, combustion, renewable energies, and integrated energy systems. He has supervised 8 Postdoc, 18 PhD, 75 MSc, and 35 Diploma students. He has participated in more than 50 national and international projects in the area of energy and environment and has (co-)authored one book, more than 130 papers in international peer-reviewed journals, and more than 170 papers in international conferences, and given more than 30 invited lectures at other universities and international symposia. Currently, he serves as Associate Editor of the Proceedings of the Combustion Institute and belongs to the Editorial Board of the *Aerospace, Combustion and Flame, Energy Conversion and Management, Energies, and Energy and Fuels*. He was the recipient of the Caleb Brett Award of the Institute of Energy in 1991, of the Sugden Award of the British Section of the Combustion Institute in 1991, of the Prémio Científico Universidade Técnica de Lisboa/Santander Totta in 2010, of a Menção Honrosa Universidade de Lisboa/Santander Universidades in 2016, and of Prémio Científico Universidade de Lisboa/Santander Universidades in 2017.

**Qiuwang Wang** is a Full Professor of School of Energy and Power Engineering, Xi'an Jiaotong University. He received a B.Sc. in Fluid Machinery and a Ph.D. degree in Engineering Thermophysics from Xi'an Jiaotong University in 1991 and 1996, respectively. He was a visiting scholar at City University of Hong Kong (1998–1999) and a Guest Professor at Kyushu University of Japan (2003), and has been Senior Visiting Scholar at Rutgers, The State University of New Jersey, USA from 2013. He is currently teaching the course "Heat Transfer" for undergraduate students and the course of "Advanced Heat Transfer" for graduate students, respectively. His research interests include heat transfer enhancement and its applications to engineering problems, high-temperature/high-pressure heat transfer and fluid flow, transport phenomena in porous media, numerical simulation, prediction and optimization, etc. He is a recipient of National Funds for Distinguished Young Scientists by NSF of China (2010) and Changjiang Scholarship Chair Professor by Ministry of Education of China (2013), the leader of the Innovation Team in Key Areas of Ministry of Science and Technology (2016), and the leader of the People Plan of Science and Technology Innovation Leading Talents (2017). His research team obtained the 2nd Grade National Award for Technological Invention of China (2015) and the National Science and Technology Progress Innovation Team Award of China (2017). He is the Chinese Delegate of Assembly for International Heat Transfer Conferences (AIHTC) (2015-), a member of the Scientific Council of the International Centre for Heat and Mass Transfer (ICHMT) (2009-), Vice President of Chinese Society of Engineering Thermophysics in Heat and Mass Transfer (2010-), has been an Associate Editor of *Heat Transfer Engineering* since 2011, and has been an Editorial Board Member for several international journals such as *Energy Conversion and Management, Applied Thermal Engineering, Energies, Frontiers in Energy*, etc. He has been the Initiator and Chairman of the International Workshop on Heat Transfer Advances for Energy Conservation and Pollution Control (IWHT) (every two year since 2011). He has also delivered more than 40 invited/keynote lectures at international conferences or foreign universities. He has also authored or co-authored four books and more than 180 international journal papers, and obtained 25 Chinese Invention Patents and two US Patents.

**Zhang Xiliang**, Dr., is Professor and Director of the Institute for Energy, Environment, and Economy, Tsinghua University. Prof. Zhang is also Director of the Tsinghua–MIT Chinese Energy and Climate Project and Deputy Director of the Low-Carbon Energy Laboratory, Tsinghua University. His current research interests include low-carbon energy economy transformation, the integrated assessment of energy and climate policies, renewable energy, and automotive energy. He leads the expert group for China's national carbon market design under the Department of Climate Change of Ministry of Ecology and Environment. He also heads a major research initiative, the "Green and Low-Carbon Economy Transformation Management and Policy of the National Natural Science Foundation of China". He co-led the expert group for drafting "China's Renewable Energy Law" in 2004–2005 under the Environment and Resource Committee of the People's Congress. Prof. Zhang has been a lead author of the 4th and 5th IPCC Climate Change Assessment Report. He is Chair of the Energy Systems Engineering Committee of China Energy Research Society, and Vice President of the Chinese Renewable Energy Industry Association. Prof. Zhang holds a PhD in Systems Engineering from Tsinghua University.

# Preface to "Selected Papers from SDEWES 2017: The 12th Conference on Sustainable Development of Energy, Water and Environment Systems"

EU energy policy is increasingly promoting a resilient, efficient, and sustainable energy system. Several agreements have been signed in the last few months that set ambitious goals in terms of energy efficiency and emission reductions, to reduce the energy consumption in buildings. These actions are expected to fulfill the goals negotiated at the Paris Agreement in 2015. The successful development of this ambitious energy policy needs to be supported by scientific knowledge: a huge effort must be made in order to develop more efficient energy conversion technologies based both on renewables and fossil fuels. Similarly, researchers are also expected to work on the integration of conventional and novel systems and take into account the requirements for the management of novel energy systems in terms of energy storage and device management. Therefore, a multi-disciplinary approach is required in order to achieve these goals. To ensure that scientists belonging to the different disciplines are aware of the scientific progress in the other research areas, specific conferences are periodically organized. One of the most popular conferences in this area is the Sustainable Development of Energy, Water, and Environment Systems (SDEWES) Series Conference. The 12th Sustainable Development of Energy, Water, and Environment Systems Conference was recently held in Dubrovnik, Croatia. The present Special Issue of *Energies*, specifically dedicated to the 12th SDEWES Conference, is focused on four main fields: energy policy and energy efficiency in smart energy systems, polygeneration and district heating, advanced combustion techniques and fuels, and biomass and building efficiency.

**Francesco Calise, Mário Costa, Qiuwang Wang, Zhang Xiliang, Neven Duic**
*Special Issue Editors*

*Review*

# Recent Advances in the Analysis of Sustainable Energy Systems

Francesco Calise [1,*] , Mário Costa [2] , Qiuwang Wang [3] , Xiliang Zhang [4] and Neven Duić [5]

1   Department of Industrial Engineering, University of Naples Federico II, P.le Tecchio 80, 80125 Naples, Italy
2   IDMEC, Mechanical Engineering Department, Instituto Superior Técnico, Universidade de Lisboa,
    1049-001 Lisboa, Portugal; mcosta@tecnico.ulisboa.pt
3   Key Laboratory of Thermo-Fluid Science and Engineering (Ministry of Education), Xi'an Jiaotong University,
    Xi'an 710049, China; wangqw@mail.xjtu.edu.cn
4   Institute of Energy, Environment & Economy, China Automotive Energy Research Center,
    Tsinghua University, Beijing 100084, China; zhang_xl@tsinghua.edu.cn
5   Faculty of Mechanical Engineering and Naval Architecture, University of Zagreb, Ivana Lučića 5,
    10000 Zagreb, Croatia; neven.duic@fsb.hr
*   Correspondence: frcalise@unina.it; Tel.: +39-0817682301

Received: 3 August 2018; Accepted: 17 September 2018; Published: 21 September 2018

**Abstract:** EU energy policy is more and more promoting a resilient, efficient and sustainable energy system. Several agreements have been signed in the last few months that set ambitious goals in terms of energy efficiency and emission reductions and to reduce the energy consumption in buildings. These actions are expected to fulfill the goals negotiated at the Paris Agreement in 2015. The successful development of this ambitious energy policy needs to be supported by scientific knowledge: a huge effort must be made in order to develop more efficient energy conversion technologies based both on renewables and fossil fuels. Similarly, researchers are also expected to work on the integration of conventional and novel systems, also taking into account the needs for the management of the novel energy systems in terms of energy storage and devices management. Therefore, a multi-disciplinary approach is required in order to achieve these goals. To ensure that the scientists belonging to the different disciplines are aware of the scientific progress in the other research areas, specific Conferences are periodically organized. One of the most popular conferences in this area is the Sustainable Development of Energy, Water and Environment Systems (SDEWES) Series Conference. The 12th Sustainable Development of Energy, Water and Environment Systems Conference was recently held in Dubrovnik, Croatia. The present Special Issue of Energies, specifically dedicated to the 12th SDEWES Conference, is focused on five main fields: energy policy and energy efficiency in smart energy systems, polygeneration and district heating, advanced combustion techniques and fuels, biomass and building efficiency.

**Keywords:** renewable energy; smart cities; district heating and cooling; sustainable development

---

## 1. Introduction

On 18 June 2018 a new agreement, dealing with EU energy policy, was reached between negotiators from the EU Parliament, Commission and Council. This agreement will promote the development of a reliable and environmental friendly energy system for the countries of the European Union. This legislative proposal is included in the Clean Energy for All Europeans package presented by the European Commission on November 2016. The previous proposals were negotiated on the revised Renewable Energy Directive and on the Energy Performance in Buildings Directive. This new agreement includes extremely ambitious targets in terms of energy efficiency: 32.5% for 2030 with an upwards revision clause by 2023. Similarly, a target of 40% in terms of reduction of emissions was also agreed. These targets, combined with 32% renewable energy target for the EU for 2030 (see STATEMENT/18/4155) and the revision of Energy Building Performance Directive, will allow Europe to achieve the goals established by the Paris Agreement in order to complete the transition towards a clean energy system. In addition, EU citizens may benefit from a more efficient energy market, where a substantial reduction in energy bills will be achieved, along with an improved security of the energy supply systems and a more comfortable and healthy environment [1].

These ambitious goals require a huge effort by industry and academia in order to design and analyze novel energy conversion systems and the integration of renewables in conventional energy systems [2]. In particular, several researchers have been involved in this area [3], analyzing energy, environmental and economic aspects of sustainable development initiatives [4], promoting and disseminating the results of their research [5] and developing novel solutions for specific sectors [6–9]. Sustainable development is a highly interdisciplinary concept, involving a number of different disciplines (energy, water, renewable, electrical engineering, control engineering, etc.). To analyze these topics, a couple of decades ago, the Sustainable Development of Energy, Water and Environment Systems (SDEWES) Conference series was initiated.

In 2017, the 12th SDEWES Conference (SDEWES 2017) was held in Dubrovnik, Croatia. The Conference was attended by 530 researchers coming from around 60 countries. The Conference also included 12 special sessions, two special events, four invited lectures and two panels by some of the most eminent experts in the field. A total of 450 papers and 100 posters were presented.

The papers in this Special Issue (SI) are selected among the works presented at the SDEWES 2017 Conference. The conference covered a plurality of research fields and skills. Papers included technical, economic, environmental and social analyses aiming to develop sustainable energy, transport and water systems. From among the 450 accepted manuscripts, 17 were selected for this special issue of *Energies*. The present paper aims to provide an overview of the papers included in the abovementioned special issue. In addition, this paper also includes a comprehensive literature review presenting the works previously published in past SDEWES special Issues, dealing with the same topics addressed in the present SI. This literature review is intended to provide an overview of the development of the research performed by the scientists participating to the SDEWES conferences in order to highlight the recent progress of the research in these fields.

This is the first cooperation between *Energies* and SDEWES which will be certainly continued due to the success of the present special issue. The papers within the present special issue can be classified into five main research fields. These research fields are: energy policy and energy efficiency in smart energy systems (three papers), polygeneration and district heating (three papers), advanced combustion techniques and fuels (two papers), biomass (two papers) and building efficiency (four papers). Three additional papers deal with other miscellaneous topics related to this special issue.

## 2. Background

This section presents a review of the papers previously published in journals' special issues dedicated to past SDEWES conferences. The papers here analyzed can be classified into the five research fields mentioned before.

### 2.1. Energy Policy and Energy Efficiency in Smart Energy Systems

The topic of the development of a novel and sustainable energy system, especially at the urban level, was widely investigated during previous SDEWES conferences. Dozens of papers are available in the various previous SDEWES special issues regarding this topic, using a plurality of different approaches [10,11]. The following Table 1 summarizes the papers analyzed in this subsection, showing the methodologies, topics and main findings.

**Table 1.** Main topics, methodologies and outcomes of the previous SDEWES papers dealing with energy policy and energy efficiency in smart energy systems.

| Reference | Topic | Methodology | Main Outcomes |
|---|---|---|---|
| [12] | Energy security | Numerical model | Energy Security indicators |
| [13] | Energy security | Geo-economic model | Geo-economic Index of Energy Security |
| [14] | Energy poverty | Analytical Hierarchy Process | Multi-Criteria Decision Analysis |
| [15] | Energy poverty | Analytical Model | Multidimensional energy poverty index (MEPI), |
| [16] | Socio-technical optimality | Conceptual model | Conceptual model |
| [17] | Sustainable urban mobility plans | mutual learning workshop | Workshops |
| [18] | GHG emission reduction | ALTER-MOTIVE model | Policy scenario |
| [19] | Electric Vehicles | Numerical model | Promotional policies |
| [20] | Sustainable Transportation | Numerical model | Alternatives to the current fossil fuel systems |
| [21] | Hydrogen transportation | Numerical model | Design of the hydrogen infrastructure in Croatia |
| [22] | Sustainable mobility | Environmental Impact Study | A case study for Macchia Lucchese |
| [23] | Energy efficiency in the EU | Review analysis | Comparison of existing studies |
| [24] | Energy efficiency obligations schemes | Numerical model | Calculation of real savings |
| [25] | South-East European power system in 2050 | Cost-optimal analysis | Decarbonization scenarios |
| [26] | Heat savings scenarios | Least Cost Tool method | Cost-optimal mix of heat savings, district heating and individual heating |
| [27] | Smart energy grids | HOMER | Economic indexes |
| [28] | Zero emission city | ENERGYplan, TRNSYS | Energy and economic indexes |
| [29] | Energy transition | ENERGYplan | Transition strategies |
| [30,31] | Universities as models for sustainability | Numerical model | Case studies |

Many papers focused on the problem of energy security [12]. Radovanic et al. [13] presented an analysis of the energy security measurement based on a geo-economic approach. The authors proposed a new approach to measure the energy security in a quantitative way. The new technique is based on a new geo-economic concept of energy security. The authors used the conventional indicators, combined with the sovereign credit rating in order to measure also the economic, financial and political stability. Using this technique, authors measured this newly proposed Geo-economic Index of Energy Security which showed significant deviations with respect to the conventional approach based on simple indicators. The authors concluded that the least impact on energy security is due to energy dependence and renewable energy production. Thus, the sovereign credit rating must be further investigated and the reliability of the energy dependence indicator must be verified in order to consider this index as a measure of energy security. März [14] focused on the fuel poverty vulnerability of urban neighborhoods using a spatial multi-criteria decision technique. They presented a case study for the German city of Oberhausen. Fuel poverty is becoming a critical issue in several EU countries. Therefore, several institutions are promoting specific policies in order to reduce households' fuel poverty vulnerability. However, such programs are not able to reach people really needing help since it is not easy to identify poor people needing such support. In this framework, the author developed a new approach based on GIS- Multi-Criteria Decision Analysis (MCDA), using an Analytical Hierarchy Process (AHP). The analysis was performed considering three vulnerability dimensions: heating burden, socio-economic and building vulnerability. Then, they proposed an overall Fuel Poverty Index in order to investigate the relative fuel poverty vulnerability of 168 urban neighbourhoods. The authors of this study concluded future policies must consider a trade off between ecological and social targets. This problem was also investigated by Okushima [15], who proposed a new multidimensional energy poverty index (MEPI), to be used to evaluate energy poverty. MEPI includes three dimensions, to be used only for the case of developed countries: income, energy costs and efficiency of housing. This methodology is applied to the area involved in the Fukushima accident. Results show that since the 2000s, Japan has suffered a remarkable aggravation of its energy poverty. Vulnerable households are in a serious energy poverty situation. In addition, the increase of energy prices, caused by the Fukushima accident, dramatically affected the energy poverty of vulnerable households and the elderly. In all these studies, authors remarked that societal aspects must be carefully addressed in order to achieve the goal of a sustainable energy transition [16].

Several papers focused on a clean transition of the mobility sector [17]. Ajonovic and Hass [18] compared the different emission reduction policies (e.g., incentives for biofuels, threshold for specific $CO_2$ emissions, fuel and registration taxes, etc.) in Europe in the car mobility sector. Using the ALTER-MOTIVE model, they found that greenhouse gas (GHG) emissions can be reduced by 33% in 2030 compared to the Business as Usual (BAU) scenario. They also concluded that several different measures and alternative technologies and fuels must be simultaneously used in order to achieve a significant reduction of GHG emissions. A similar study was also performed by Knez et al. [19], regarding the development of electrical vehicles. Dominkovic et al. [20] investigated the role of the transportation sector for a sustainable clean energy transition. This sector accounts for about 30% of the total energy consumed in EU and a huge effort must be performed in order to reduce energy demand and emissions. Obviously, this circumstance is mainly caused by heavy-weight vehicles and by long-range vehicles. A dramatic energy consumption is also caused by airplanes. The authors performed a comprehensive literature review in order to detect possible solutions. The energy transition could be obtained by implementing four different actions: biofuels, hydrogen, synthetic fuels (electrofuels) and electricity. Results showed that the possibility of using electric vehicles has the largest impact on the overall energy and emission balance. The authors estimated that in EU 72.3% of the transport energy demand could be directly electrified by the technology existing today. For the remaining part, 3069 TWh of additional biomass was needed for biofuels. In addition, 2775 TWh of electricity and 925 TWh of heat were also needed for renewable electrofuels. Firak et al. [21] analyzed a future scenario where the Croatian transportation sector is dominated by fuel-cell vehicles and hydrogen infrastructure. In particular, authors calculated the volume of hydrogen required to supply the tourist's hydrogen fuel cell vehicles for three phases up to the year of 2030. The authors assumed that hydrogen will be produced via water electrolysis driven by PV fields located at suitable sites. Hydrogen refueling stations sites are proposed on the basis of traffic volumes on selected road directions in Croatia (mostly approaching the Adriatic coast). This approach will determine a dramatic reduction of GHGs emissions. In addition, the proposed system will allow one to solve the issues related to energy storage and clean transportation. The problem of the sustainable development of the transportation sector was also investigated by Briggs et al. [11], focusing on the simulation of non-automotive and off-highway vehicles. This work analyzes the simulation techniques for such vehicles comparing the approach based on drive cycle testing and experimental validations. The study considers two case studies: an urban hybrid diesel-electric bus and a forklift truck powered by an Internal Combustion Engine (ICE). A novel sustainable mobility system for campsites was investigated by Del Moretto et al. [22]. The authors evaluated a sustainable mobility connection among three campsites and the coastal area of Tuscany, Italy. They considered two alternatives, namely: a diesel tourist train and an electric tourist train. The two alternatives were compared considering energy, environmental and socio-economic aspects. They concluded both solutions can be viable when suitable funding policies are implemented.

Knoop analyzed the energy efficiency targets for the EU [23]. In 2014, the EU planned a minimum 27% energy efficiency improvement by 2030. These targets must be achieved by voluntary actions set by each Member State, which may also set more restrictive national objectives. However, there is still much debate regarding the potential improvements for each Member State. Therefore, this paper aims to fill this knowledge gap, providing a review of scientific works investigating the possible improvements in energy by each EU Member States by 2030. The analyzed papers detect a significant potential for energy efficiency, showing very different outcomes, depending on the analyzed country. In the worst scenario, 10–28% energy savings could be achieved by 2030 with respect to the BAU scenario. Conversely, in the best case, 7–44% can be achieved. Energy efficiency potentials range between 14% and 52%, depending on the selected EU Member State. Moser [24] analyzed the energy efficiency obligation schemes. Such measures are used in order to implement energy saving actions. Such schemes were adopted as a consequence of the EU Energy Efficiency Directive and EU Institutions claim that determined a significant increase of the energy savings at relatively low costs. However, this paper criticizes these optimistic results since in author's opinion, the energy savings are dramatically overestimated. This idea is based on the fact that bargaining processes determine an improvement of the accredited savings of each considered measure. In addition, the author pointed out that non-standardized methods of measurement may determine a remarkable overestimation of real savings.

Pleßmann and Blechinger [25] focused on the decarbonization pathway for South-East Europe (SEE) in order to achieve 2050 EU mitigation goals. The authors implemented a multi-regional power system model to investigate an economically viable decarbonization pathway for SEE countries. The authors discuss the optimal strategies to be implemented in the power sector. On the other hand, they neglect cross-sectoral demand shifts due to the heat pumps and electrical vehicles. Results show that a huge effort must be performed to meet the decarbonization targets. SEE must implement important actions in order to achieve the GHG emission reduction targets: PV and wind capacities must be increased by 120.7 GW and 92.4 GW by 2050, respectively; transmission capacities to near countries must be increased by 32.7 GW until 2050. High investments are expected to achieve these goals. As a consequence, the levelized cost of power supply will be 12.1 ctEUR/kWh.

Many papers focused on the clean energy transition related to urban areas. Amer-Allam et al. [26] focused on the Danish municipality of Helsingør, trying to evaluate possible reductions in energy consumption and emissions. The authors analyzed the heating system of Helsingør developing future scenarios in order to detect the combination of individual heating, district heating and heat savings, maximizing system economic profitability. Results show that in 2030: (i) the heating demand can be reduced by 20–39% by implementing heat savings; (ii) 32–41% of the overall heat will be supplied by district heating systems; heating-related $CO_2$ emissions will be reduced by up to 95%. In 2050, in the optimal thermo-economic scenario, the share of district heating in Helsingør will increase by up to 44%. The topic of energy management in municipalities was also investigated by Batas-Bjelic et al. [27]. The authors used HOMER to simulate a smart grid proving heat and electricity to a municipal area. The final goal was the reduction of yearly energy costs. A number of different technologies were considered, namely: wind power plants, PV plants, combined heat and power (CHP) plants. Such a plurality of devices is crucial in order to manage fluctuating renewable energy production and the excess heat from cogeneration plants. The results of the simulations proved the economic and environmental benefits of the proposed smart municipal energy grids. From the economic point of view, the internal rate of return is ranged between 6.87% and 15.3%; $CO_2$ emissions varied from −4885 to 5166 t/year. The calculated number of CHP operating hours varied from 2410 to 7849 h/year. Another study aiming to analyze the possible transition to a clean energy system at urban level was presented by De Luca et al. [28]. The goal of this study is to convert an Italian city to a zero greenhouse gas city by 2030. They proposed to use a number of efficient technologies, namely: wind turbines, photovoltaic panels, biogas cogeneration, thermal solar panels, cogeneration and heat pumps. They combined both ENERGYPlan and TRNSYS software to perform their analysis. The results show that calculated thermal and electric energy prices are very promising: 0.11 €/kWhe and 0.12 €/kWht, respectively. Therefore, the whole system is also economically profitable. ENERGYPlan was also used by Vidal-Amaro et al. [29] to design an electrical renewable energy system in Mexico. In this paper several scenarios for the development of renewables for the Mexican electricity system were analyzed, aiming to meet the target of a 100% renewable system. Presently, the Mexican electricity system generates 260.4 TWh/year (85% based on fossil fuels) of electricity. The authors evaluated the impact of a higher utilization of several renewable technologies (PV, wind, geothermal, biomass, hydro and concentrating solar power) on the system capacity to match user demand. Several other studies are available in this area regarding universities [30,31]. In all the cases, universities are taken as models for the development of sustainability and energy efficiency actions and the implementation of the climate strategies.

Similarly, this topic was analyzed in detail in previous SDEWES special issues paying attention to a number of different aspects, namely: the role of energy prices as a driving force towards a sustainable development [32], renewable energy integration in an EU northern power market [33], impact of electric vehicles on the Croatian transportation system [34], carbon emission reduction targets for New Zeland in 2050 [35], guidelines for power utilities to reach specific decarbonization targets [36], the role of PV and concentrating solar power for the decarbonization targets [37], policies and subsidies to support renewables [38,39] and many others [40–48].

### 2.2. Polygeneration and District Heating

The topic of polygeneration, with special focus on district heating and cooling systems, was widely discussed during the previous SDEWES conferences. As a consequence, dozens of papers dealing with this topic are included in previous journal special issues dedicated to the SDEWES conferences. The following Table 2 presents the main features of the main papers analyzed in this subsection.

**Table 2.** Main topics, methodologies and outcomes of the previous SDEWES papers dealing with polygeneration and district heating.

| Reference | Topic | Methodology | Main Outcomes |
|---|---|---|---|
| [49] | District heating systems | Measured data | Comparison between two systems |
| [50] | District heating systems | Modelica® library | Simulation tool |
| [51] | Power and water distribution networks | Multi-objective model | Design procedure for power and water distribution networks |
| [52] | District heating systems | Numerical model | A model for planning and scheduling of district heating systems |
| [53] | Renewable district heating | TRNSYS | Energy and economic indexes |
| [54] | Biomass trigeneration system | Optimization model | Energy and economic indexes |
| [55] | Cogeneration | Review and market analysis | Evaluation of the profitability |
| [56–63] | Polygeneration | TRNSYS | Energy and economic indexes |
| [64] | Polygeneration including desalination | Exergy and exergoeconomic analysis | Exergy, energy and economic indexes |
| [65] | Polygeneration including desalination | Energy analysis | Energy indexes |
| [66] | Trigeneration | Energy and economic analysis | Feasibility for small islands |
| [67] | Polygeneration | Review Analysis | Feasibility indexes |
| [68] | Biomass Cogeneration | Numerical Model | Feasibility analysis |
| [69,70] | Cogeneration and wastewater treatment | Aspen | Energy and economic indexes |
| [71] | Cogeneration | Energy and economic model | Simulation tool |

Several papers specifically focused on district heating and cooling networks, paying special attention to the novelties in terms of research and development [72,73]. A comparison between district heating systems in Zagreb and Aalborg was presented by Culig-Tolic et al. [49]. They aimed to analyze similarities and differences in order to improve the systems. The authors concluded that the Aalborg district heating system is better than the Zagreb one. This district heating system can be improved by replacing pipes in order to decrease water and heat losses. In addition, lowering hot stream temperature is crucial in order to promote the integration with renewable energy sources. Arce et al. [50] presented the results of the project funded in the framework of the FP7 programme. They developed a modeling platform of heat transport in district heating systems. The platform includes several models of the components of the district heating systems. A code-to-code validation procedure is presented in order to test the accuracy of the proposed simulation platform. The developed model showed a good agreement in steady state. On the other hand, some significant deviations are detected in the case of highly dynamic phenomena. Gonzales-Bravo et al. [51] also focused on district networks (energy and water) presenting a novel approach considering simultaneously economic, environmental and social aspects. In particular, the novel method presented in this paper also considers the variety of criteria for the stakeholders involved in the design of operational strategies and new facilities. A multi-objective model was used in order to consider this plurality of different objective functions. This method selects system layout and design parameters as a tradeoff between the multiple stakeholders. A case study was presented by Hermosillo, Guaymas and Obregon for the problem of the water scarcity in Mexico. The results of this analysis show that this approach may determine significant economic, environmental and social benefits for inhabitants and a good profitability for the investors. The optimal size of district heating and cooling systems was also investigated by Pavicevic et al. [52]. The authors argued that the optimization of district heating and cooling systems is typically a very hard task, due to the large number independent variables and the large time horizons (at least one year), requiring a huge computational effort. This paper aims to develop an optimization model capable to calculate both design and operational parameters. The proposed district heating system includes the following components: solar thermal collectors, boilers, electric heaters, heat pumps and thermal storage units. The authors simultaneously consider building refurbishment. Nine different scenarios were considered in the analyses. The results show that the proposed renewable system is economically profitable with respect to a conventional systems based on boilers. A novel district heating and cooling system was investigated by Carotenuto et al. [53]. The system was fed by a plurality of renewable sources (geothermal, biomass and solar) and it was operated at low temperature. In particular, geothermal wells, evacuated solar thermal collectors and auxiliary wood-chip boilers are used in winter. In summer, cooling energy is provided by an adsorption chiller. The system produces domestic hot water all year long. A suitable dynamic simulation model, developed in TRNSYS, is used to perform the calculations. A case study is presented for the town of Monterusciello, near the city of Pozzuoli, in the South of Italy where a geothermal source is available at about 55 °C. Buildings rehabilitation is also considered. The results show that during winter geothermal and solar energy are only used for thermal purpose. In summer the auxiliary biomass boiler must be mandatory activated to meet the space cooling demand. The energy performance of the system is satisfactory since solar collector efficiency is above 40% and the Coefficient of Performance of the adsorption chiller is about 0.5. However, the system is far from an economic feasibility.

The idea of polygeneration consists in the simultaneous production of different energy vectors (electricity, cool and heat) and byproducts using both fossil fuels and renewables. A plurality of renewable energy sources may be used in polygeneration systems [54]. This topic is extremely attractive since this technology is expected to dramatically reduce energy consumption and emissions. Cogeneration is the simplest case of polygeneration systems [55]. As a consequence, many papers were published in this area in the previous SDEWES Special Issues. Calise et al. [56] presented a novel layout of a polygeneration system for the island of Pantelleria. The system supplies energy to a district heating and cooling network and it simultaneously produce: electricity, thermal energy, cooling energy and desalinated water. The system layout includes: geothermal wells, Parabolic Through Collectors (PTC), a Multi-Effect Desalination (MED) and an Organic Rankine Cycle (ORC). A detailed dynamic simulation model was developed and special control strategies were proposed to manage the system. A thermo-economic analysis was also implemented and the results showed that the payback period was 8.5 years. Several other works dealing with polygeneration were presented by the same research group, analyzing solar assisted heat pumps system coupled with photovoltaic/thermal collectors [57], optimal control strategies for trigeneration systems [58], novel layouts integrating geothermal and solar energy, ORC and MED technologies [59], energy and exergy analyses of a novel system producing electricity, heat and cool and desalinized water based on Concentrating PVT and MED technologies [60,61], small-scale polygeneration systems based on building integrated PVT collectors [62,63]. The integration of desalination technologies with cogeneration systems was also investigated by Catrini et al. [64]. The authors analyzed a Combined Heat and Power steam cycle coupled with Multi Effect Distillation-Thermal Vapour Compression, from both economic and exergy points of view. Here, a detailed exergy and economic model is presented. Different scenarios are evaluated. In a first scenario, concentrated brine is assumed to be discharged to sea, wasting its physical and chemical exergy. In a second scenario, a certain amount of the outlet brine is supplied to a Reverse Electrodialysis unit, producing electricity. Unfortunately, unit costs resulted high for both cases. The highest exergy destruction is due to freshwater in the first configuration and Reverse Electrodialysis electric output in the second one. A similar analysis was also presented by Tamburini et al. [65], focusing on retrofitting existing CHP systems using multiple effect distillation (MED) along with thermal vapour compression (TVC) technology. Polygeneration was also investigated by Beccali et al. [66] investigating the feasibility of the installation of trigeneration systems for a number of Italian islands. In particular, they focused on retrofitting existing power plants, evaluating the feasibility of heat recovery from the existing diesel engines and the related installation of a suitable district heating network. Six different islands were analyzed. For the analyzed case studies, different boundary conditions are considered, since several parameters (number of inhabitants, climatic conditions and touristic fluxes, etc.) are significantly different. The calculations are performed by detailed dynamic simulations and a number of scenarios were analyzed. The economic analysis showed that the proposed system is beneficial from an energy point of view but it is extremely far from an economic profitability, even when suitable funding policies are considered. Better results could be achieved only in case of a high number of permanent inhabitants. The results of the Energy Agency Annex 54 project ("Integration of Micro-Generation and Related Energy Technologies in Buildings") was presented Angrisani et al. [67]. They presented a review of the available methodologies and indexes for the calculation of polygeneration systems performance. A novel index is also introduced and discussed in order to evaluate the system economic profitability. The reviewed indexes were calculated with respect to two alternative systems, using three commercially available cogenerators. The authors evaluated the feasibility of combined production with respect to the separated one as a function of the energy prices and users. In all the cases, thermoeconomic indexes showed good reliability and robustness. Cogeneration systems can be also supplied by biomass, as investigated by Pfeifer et al. [68]. They focused on biomass from unused agricultural land in Croatia. This type of biomass could be used to supply cogeneration systems up to 15 MWe. Their calculations showed that the novel systems including the combined heating and cooling plants with seasonal storage are not profitable. However, more conventional CHP systems would be feasible. Di Fraia et al. [69,70] analyzed the integration of CHP system in wastewater treatment plant. They proposed a CHP system fueled by the biogas produced by the treatment plant. CHP waste heat is used to dry outlet sludge. This arrangement allows one to dramatically reduce the costs for sludge disposal. The authors implemented a suitable energy and economic model and they calculated a payback period slightly lower than seven years. Piacentino et al. [71] presented a simulation tool for cogeneration systems, which optimizes the plant layout, the design parameters and their operation strategy. A case study is presented for a building in the hotel sector. The authors identified the most promising configurations as a function of: tax exemption and operating map of the engine.

Several other papers were published in this area in the previous SDEWES special issues, dealing with: trigeneration systems including fuel cells and supplied by municipal waste [74], building integrated trigeneration systems [75], cogeneration for wood industry in Serbia [76], simulation of a cogenerative micro-gas turbine operating at partial load conditions [77], cogeneration systems supplied by waste forest biomass [78], interaction between cogeneration systems and electrical vehicles [77,79], integration of solar energy in district heating systems [80], incentivizing mechanisms for trigeneration plants in Italy [81].

## 2.3. Advanced Combustion Techniques and Fuels

In order to improve heat conversion of various types of fuels several novel techniques were investigated during the previous SDEWES conferences. The main features of the papers analyzed in this subsection are summarized in the following Table 3.

**Table 3.** Main topics, methodologies and outcomes of the previous SDEWES papers dealing with advanced combustion techniques and fuels.

| Reference | Topic | Methodology | Main Outcomes |
|---|---|---|---|
| [82] | Particulate filter for diesel boilers | Experimental analysis | New prototype based on Biomorphic Silicon Carbide (bioSiC) |
| [83] | Purification process of dry syngas | lab-scale test facility | High bed conversion of the impurity removal reaction |
| [84] | Catalytic decarboxylation of rubber seed oil for diesel fuel production | Aspen HYSYS V8.0 | Energy and economic indexes |
| [85] | Bio-oil pyrolysis | Experiments in a fixed bed tubular reactor | Biochar yields decrease in case of lower pyrolysis temperatures and heating rates |
| [86] | Model of pollutant emissions in diesel engines | Numerical model in FIRE | Biodiesel blends release lower nitrogen oxide emissions |
| [87] | Thermoelectric generator fed by natural gas. | ANSYS/FLUENT | Optimal set of design parameters |
| [88] | Slab heating characteristics in a reheating furnace. | ANSYS/FLUENT | Optimal set of design parameters |
| [89] | Aqueous bioethanol combustion | Experimental test rig | The emissions of all the tested alcohols are compliant with the present Hungarian standard |
| [90] | Grate biomass furnaces with water boilers and condensing economizers | Experimental | New method for moisture evaluation |
| [91] | Energy storage system based on sewage sludge gasification | Numerical | Syngas suitable to be injected into a natural gas distribution network |
| [92] | Spark ignition engines fed by biogas | Experimental and numerical | Validates simulation tool |
| [93] | Pure tire pyrolysis oil | Experimental | Experimental data |

Orihuela et al. [82] analyzed the possibility to use Biomorphic Silicon Carbide (bioSiC) as a particulate filter for diesel boilers. This material is a novel ceramic material which exhibits excellent thermal and mechanical properties and it is very attractive as article filter media of exhaust gases of diesel boilers. In this paper, authors designed and constructed an experimental setup to extract a sample of the boiler exhaust gas in order to filter it under controlled conditions. Several types of filters were experimentally analyzed, measuring the number and size of particles upstream and downstream. The results of the tests show that, for filters made from natural precursors, the efficiency dramatically depends on the cutting direction and associated microstructure. For samples derived from radially cut wood, a 95% initial efficiency of the filter is achieved. For samples obtained by an axial cut of the wood, the initial efficiency ranges between 70% and 90%. Then, due to the particles accumulation, the efficiency increases around 95%. Kobayashi et al. [83] analyzed the purification process of dry syngas. This syngas is produced by coal gas using oxy-fuel gasification and it is used in a combined cycle equipped with carbon dioxide capturing. This technology is extremely promising since it is expected to achieve thermal efficiency around 44%, simultaneously providing compressed $CO_2$ (93 vol %). The authors designed the dry syngas cleaning process and a lab-scale test facility to prove the feasibility pf the process. In particular, two types of sorbents were tested. The results, for both cases, showed enough removal within the satisfactory short depth of sorbent bed and a higher bed conversion of the impurity removal reaction. Cheah et al. [84] analyzed the catalytic decarboxylation of rubber seed oil for diesel fuel production in Malaysia. The process was simulated using Aspen HYSYS V8.0 and a suitable thermoeconomic model was also developed. The authors calculated the minimum fuel selling prices in order to achieve a reasonable profitability. The simulated systems can process 65 kL/day of inedible oil producing 20 ML/year of renewable diesel. For this system, authors calculated a return of investment equal to 12.1%. Pehilivan et al. [85] analyzed the pyrolysis for bio-oil, char and gases production. They used cherry pulp which was tested in a fixed bed tubular reactor at different temperatures and heating rates. Several analyses were performed in order to analyze chemical alterations after the pyrolysis process. Experimental results showed that biochar yields decrease in case of lower pyrolysis temperatures and heating rates. Petranovic et al. [86] modelled pollutant emissions in diesel engines, analyzing the case of biofuels. They started from a literature review showing that there is no consensus regarding the influence of biodiesel in emission concentration. Therefore, they developed a suitable numerical model in FIRE environment in order to calculate such emissions. The model takes into account spray particles, combustion, fuel evaporation and disintegration and the chemical process for pollutant formation. The developed model was also validated using experimental data. Then, they compared the emissions of biodiesel vs the ones of conventional diesel engines. They found that biodiesel blends release lower nitrogen oxide emissions than the engines powered with the regular diesel. Another simulation model was presented by Bargiel et al. [87]. The authors developed a numerical model of a thermoelectric generator fed by natural gas. The rated capacity is around 50–100 W and the device is designed in order to supply electricity to remote objects of the natural gas infrastructure, where the electrical grid is not available. The numerical model was implemented in the ANSYS/FLUENT platform. An optimization procedure was implemented in order to optimize temperature difference across the thermoelectric module. An optimum response surface was calculated leading to a selection of the design parameters which will be subsequently used as a basis for the experimental analysis of the prototype. Wang et al. [88] analyzed slab heating characteristics in a reheating furnace. For this process, the performance is strictly related to the combustion process and to the fluid dynamics. The heating efficiency depends on a number of different parameters, such as: locations of both slabs and burners, type of fuel, geometry of slab supporting systems and thermal properties of slabs. A suitable finite-volume simulation model was implemented in ANSYS/FLUENT in order to perform the calculations. The results of the simulations showed that the best configuration was obtained in case of six side burners. Kun-Balog et al. [89] analyzed the process of aqueous bioethanol combustion in order to calculate the related pollutant emissions. A 12% ethanol-water solution was used to analyze the distillation process. Then, a suitable experimental test rig was used for the calculation of the combustion performance. During the tests, a 15 kW combustion power was used at an air-to-fuel ratio of 1.17. In addition, 96–50% ethanol-water solutions were used in the tests (both liquid and gaseous). The emissions of all the tested alcohols are compliant with the present Hungarian standard. Striugas et al. [90] focused on grate biomass furnaces equipped with water boilers and condensing economisers. These systems are widely used in Lithuania in district heating systems. Unfortunately, such systems are very sensitive to the inlet biomass composition which may determine an unstable system operation. Therefore, authors developed an indirect method to evaluate the moisture included in the fuel stream and this method was implemented in a suitable controller. The method was validated using the data of a 6 MW grate-fired furnace fueled by biomass, showing deviations below 3%. Kokalj et al. [91] developed a novel energy storage system based on sewage sludge gasification. A suitable thermodynamic model of the gasification process was used to analyze the performance of the system. The model returns the calculated syngas amount and composition. The aim of the study is to produce syngas suitable to be injected into a natural gas distribution network. To this scope, the best configuration is achieved gasifying sewage sludge with 35–40 wt % moisture. Conversely, the best performance is obtained with SS dried to 20 wt % of moisture content, when syngas is used onsite. Nunes de Faria et al. [92] developed experimental analyses in order to predict the performance of suitably modified spark ignition engines using biogas produced in a sewage plant in Brazil. The tests showed that the reduction in engine efficiency was higher than the advantages in terms of emission reduction. The authors also developed a zero-dimensional thermodynamic simulation model of the system, based on a system of differential equations. The model allows one to calculate the pressure inside the cylinder, indicated power and the mean effective pressure as a function of the crank angle is several different engine operating conditions. The results of the simulations were consistent with the experimental ones, showing deviations below 5%. Vihar et al. [93] investigated the possibility to use pyrolysis oil produced from waste tires in an automotive Diesel engine. The authors aim to further extend the operating range towards lower loads by implementing a novel arrangement based on the exhaust gas recirculation and tailored main injection strategy. This study also provides a detailed experimental analysis of the particulate emissions of the tire pyrolysis oil. The authors found excellent results both in terms of performance and emissions. Gas recirculation in diesel engines was also investigated in several other papers [94,95].

Other papers were published in the previous SDEWES special issues in this area, investigating similar topics, namely: analysis of the fuel injection timing and ignition position in a direct-injection natural gas engine [96], effects of engine cooling water temperature on performance and emission biofuel engines [97], biofuel pellet torrefaction [98], bio-oil production [99] and thermal treatment [100], advanced fuels for gas turbines [101], coal and biomass cofiring [102], optimization of the furnaces of the aluminum industry [103].

*2.4. Biomass*

Biomass is often considered crucial to achieve the goals in terms of sustainable development [104]. This is clearly reflected by a literature review of the previous SDEWES special issues which include a large number of papers investigating this topic, summarized in the following Table 4.

**Table 4.** Main topics, methodologies and outcomes of the previous SDEWES papers dealing with biomass.

| Reference | Topic | Methodology | Main Outcomes |
|---|---|---|---|
| [105] | Substrate feed control in anaerobic digesters | Review analyis | Closed-loop feed control are often missing |
| [106] | Energy crops for biogas production | Life Cycle Assessment | System performance indexes |
| [107] | Biomass gasification | Dynamic neural network model | Simulation tool |
| [108] | Biomass from *Miscanthus giganteus* | Field experiments performed on a cultivation | Experimental correlations |
| [109] | Bioethanol production | Experimental | Experimental data |
| [110] | Phytoextraction by pyrolysis | Thermo-economic model | Economic indexes |
| [111] | Transesterification of rapeseed oil by butanol | Experimental | Experimental data |
| [112] | Bioethanol production from *Chenopodium formosanum* | Experimental | Experimental data |
| [113] | Bioethanol microalgae | Experimental | Experimental data |
| [114] | Biorefinery for lignocellulosic biomass | Numerical | Novel system layout |
| [115] | Sustainability of algal-based biorefineries | Review Analysis | Energy, economic and environmental indexes |

In this field, biogas and gasified biomass are probably the most promising bio-fuels for their potential to decarbonize energy systems [116–119]. Gaida et al. [105] presented a detailed review of different strategies to control substrate feed in anaerobic digesters. Their analysis shows that the majority of full-scale biogas plants are not equipped with a closed-loop feed control. The control strategy is a tradeoff between economic profitability, ecological footprint and reliability. Such controls are sometimes used for anaerobic wastewater treatment, but it is not in use in agricultural or industrial biogas plants due to the lack of robust and reliable process monitoring. Lijo et al. [106] studied the effect of substituting energy crops for food waste as feedstock for biogas production. The authors aimed to evaluate the environmental consequences of feedstock selection in biogas production. To this scope two different plants were analyzed using Life Cycle Assessment approach. Plant A performs the co-digestion of energy crops (78%) and animal waste (22%) while Plant B consumes energy crops (4%), food waste (29%) and animal manure (67%). According to the authors' calculations, producing electricity from biomass is better than the existing electric mix from the environmental point of view. Maize silage (650 $Nm^3$/TVSfed) and food waste (660 $Nm^3$/TVSfed) were identified as the most promising sources of bioenergy. The authors also concluded that specific guideless should be established in order to promote bioenergy environmental sustainability. Mikulandrić et al. [107] presented a numerical model of a biomass gasification process in a co-current fixed bed gasifier. In particular, they implemented a dynamic neural network model for biomass gasification in various operating conditions. The model is based on the data extracted from a co-current, fixed bed gasifier operated by TU Dresden. Results showed that the accuracy of the dynamic neural network model was higher than the one achieved by multiple linear regression models. In fact, it can predict process temperature and syngas quality with average errors lower than 10% and 30%, respectively.

Another interesting option is the cultivation of specific plants for energy purposes [120]. Szulczewski et al. [108] presented a new technique for the calculation of biomass yield of *Miscanthus giganteus* in the course of vegetation. The authors implemented a simplified approach where the biomass increase was simply modelled considering simple biometric measurements. The analysis is based on field experiments performed on a cultivation. On the basis of these data, an experimental correlation was determined between shoot volume index and shoot mass. The accuracy of estimation of is strictly dependent on the number of shoots. The authors concluded that the best tradeoff is obtained for 10 shoots of miscanthus. The results of the statistical estimation are satisfactory, showing relative errors below 17%. Ko et al. [109] analyzed bioethanol production from recovered Napier grass with heavy metals. These plants are used to absorb heavy metals from polluted soils. However, the management of the recovered explants may be extremely complex. Therefore, authors of this study proposed to convert it into bio-ethanol which is a very promising bio-fuel. The plants were used for soils contaminated by Zn, Cd and Cr. Unfortunately, such heavy metals also inhibit biomass production which is significantly lower (from 4% to 21% with respect to the case of uncontaminated soil, depending on heavy metal concentration). In addition, bacteria fermentation was enhanced by the presence of heavy metals. On the other hand, the fermentation efficiency was lower. The authors concluded that the overall effect of the utilization of Napier grass phytoremediation for bioethanol production is significantly positive for the sustainability of environmental resources. A similar study was also performed by Kuppens et al. [110] focusing on fast pyrolysis as a phytoextraction methodology. A suitable technoeconomic analysis was performed to this scope. Kejek et al. [111] focused on the process of transesterification of rapeseed oil by butanol and separation of butyl ester. They present in detail the chemical processes occurring in this system, analyzing in detail the effects of the variations of the main independent variables: the amount of catalyst, the reaction temperature and time, the method of oil addition to butanol and the molar ratio of butanol. On the basis of the measured performance data, specific statistical correlations were developed. They concluded that using a strong acid significantly improves the separation process and it determines a zero content of potassium and free glycerol. The separation is enhanced also by the addition of a small amount of water and by the removal of butanol. In this field another interesting study was also presented by Yang et al. [112] presenting bioethanol production from *Chenopodium formosanum*. Cheng et al. [113] investigated the production of bioethanol using lipid-extracted biomass from a specific microalgae. They presented a new process where the lipid-extracted biomass was directly subjected to simultaneous saccharification and fermentation. The proposed system does not need any expensive pretreatment, also decreasing the contamination risk and complication of high sugar content. The results of their analysis showed the optimum configuration was found for temperature at 36 °C, pH 5, 60 units/mL enzyme concentration, and yeast loading of 3 g/L. In such configuration, an overall conversion of more than 90% of the theoretical yield was achieved with maximum bioethanol yield of 0.26 g bioethanol/g lipid-extracted biomass.

Biorefineries are also often studied as a viable option for a sustainable energy production. Özdenkçi et al. [114] presented a novel concept of integrated biorefinery for lignocellulosic biomass. The novel conversion technology is based on partial wet oxidation coupled with lignin recovery with acidification. The process is performed in reactor including hydrothermal liquefaction and supercritical water gasification. Thomassen et al. [115] presented a review of the sustainability of algal-based biorefineries. Their study aims to evaluate the data available in literature regarding both economic and environmental performances of such plants. In fact, they noted that algal-based bioenergy products have faced multiple economic and environmental problem. However, there is no consensus in literature about the sources of such problems. The performed literature review identified four main challenges: (1) the use of harmonized assumptions; (2) the adaptation of the methodology to all stages of technological maturity; (3) the use of a clear framework; (4) the integration of the technological process. The authors also proposed a specific methodology integrating techno-economic analyses and life cycle assessments. This technique can be iteratively performed for each stage of technology development. This methodology allows one to identify the crucial technological parameters and it can be useful in order to promote a commercial development of the proposed systems.

Other papers were published in the previous SDEWES special issues in this area, investigating similar topics, namely: biodiesel production from cooking oil [121], multiple biomass corridor [122], biomass for power solutions [123], hybrid solar biomass systems [124], biofuels for marine vehicles [125], production of biosolid fuels from municipal sewage sludge [126], analysis of active solid catalysts for esterification of tall oil fatty acids with methanol [127], analysis of woody biomass in Japan [128], biodiesel production using injection of superheated methanol technology [129].

*2.5. Building Efficiency*

Buildings account for about 40% of the overall energy consumption for the majority of developed countries [130–139]. Therefore, the reduction of building energy consumption is crucial in order to achieve the goals in terms of decarbonization recently established in Paris Agreement. To this scope, several actions are required, such as: building envelope refurbishments [140–142], optimization of the HVAC systems [143–149], utilization of renewable energy sources [150–156]. This topic was initially marginally investigated during the first SDEWES conferences. Then, due to the growing interest in this area, more and more papers were included in SDEWES special issues dealing with building energy efficiency. Several papers focused on passive buildings, as summarized in the Table 5 below reporting the features of the main papers involved in this topic.

**Table 5.** Main topics, methodologies and outcomes of the previous SDEWES papers dealing with building efficiency.

| Reference | Topic | Methodology | Main Outcomes |
|---|---|---|---|
| [157] | Passive buildings | Simulation model in EnergyPlus | Optimal envelope features |
| [158] | Energy required for the manufacturing processes materials for buildings | Life Cycle Assessment | Overall impact of materials |
| [159] | Summer overheating in industrial buildings | Simulation model | Proposal of passive measures |
| [160] | Indoor confort in bioclimatic architecture | Review analysis | Methodologies for evaluating thermal comfort |
| [161] | Insulating materials for buildings | Review analysis | Methodologies for evaluating thermal properties |
| [162] | Building energy performance | Numerical model | New methodology for the calculation of building energy performance |

Chen et al. [157] presented a simulation model developed in order to optimize passive buildings. They started from a complete literature review of the papers available in literature, investigating simulation-based techniques for the optimization of passive buildings. Then, they proposed a simulation method developed in EnergyPlus environment. In particular, suitable control algorithms are applied to a local green building assessment scheme to control ventilation and lighting dimming. In addition, a NSGA-II genetic algorithm was used in order to calculate the Pareto Frontier and the final optimum as a function of: building layout and geometry, envelope thermophysical properties and infiltration and air-tightness. This approach allows one to design optimal passive buildings in different operating conditions. Passive buildings were also investigated by Kovacic et al. [158] focusing on the energy required for the manufacturing processes of several materials used for the construction of such buildings. The problem was analyzed also considering environmental aspects. A case study for a passive housing block in Austria was investigated. The authors implemented the Life Cycle Assessment (LCA) technique, using the real data available by an energy monitoring campaign. LCA returns the environmental impacts of the building materials, HVAC systems, and the operational energy for time scenarios of 20, 50 and 80 years. Results show that distribution pipes accounts for 10% of the Global Warming Potential. In addition, authors concluded that the passive house performs only slightly better in terms of environmental impacts, with respect to a reference case. The same research group also presented an overview of the strategies for the renovation of building stock for ageing society [163] and a life cycle optimization tools for building early design phases [164]. Gourlis et al. [159] analyzed passive measures for preventing summer overheating in industrial buildings. In industrial buildings space cooling demand is dominated by the internal gains due to the production process which may be very fluctuating. The authors analyze different retrofit scenarios for a case study in Austria. Specific measurements were performed in order to analyze the indoor climate conditions for the reference scenario. Then, using a suitable simulation model, energy efficiency actions were analyzed. The authors found that several actions are available for reducing overheating risk. In another paper, authors implemented Building Information Modelling (BIM), for the analysis of energy efficiency actions in industrial buildings [165].

Beccali et al. [160] presented a literature review of the implications of indoor thermal comfort and bioclimatic architecture in hot-humid climates. The paper analyzed in detail the methodologies commonly used in literature to evaluate thermohygrometric comfort in buildings featured by natural ventilation, based on adaptive approaches. The authors focus on the Mozambican buildings, analyzing a new healthcare facility. The same research group also implemented a specific tool (based on an artificial neural network) for assessment of the energy performance and the refurbishment actions for the non-residential building stock in Southern Italy [166]. Ricciu et al. [161] presented a review of the methodologies for the calculation of insulating materials for buildings. The authors analyze the available techniques for the calculation of the heat capacities of materials, paying special attention to the thermal behavior of lightweight insulation materials. For all the cases, the relationship between these properties and energy efficiency and indoor comfort were discussed. The paper also includes a new simplified approach for the estimation of the specific heat capacity, showing errors around 5%. Horvat et al. [162] presented a novel methodology for the calculation of building energy performance. Their approach is based on a detailed mathematical model for the calculation of indoor temperatures, space cooling and heating and domestic hot water demands. A case study is presented for a family house. The results are compared with the ones of EN ISO 13790 and EN 15316, showing significant differences in determination of the energy demands.

Further papers dealing with this topic can be found in the previous SDEWES special issues, investigating other aspects, such as: retrofitting actions for historical buildings in Italy [167], building water and energy regulations in England [168] and in Indian Hill Towns [169], energy monitoring of an apartment [170], definitions of suitable typologies representing residential building stocks [171], natural ventilation strategies for near zero energies school buildings [172], Energy Performance

Certification for faculty buildings in Zaragoza (Spain) [173], building energy demand for hotels, hospitals, and offices in Korea [174].

## 3. Research Topics Represented in This Special Issue

This special issue includes 17 papers from the 12th SDEWES Conference selected for this SI. This section briefly summarizes the methodologies and the main findings of those papers.

### 3.1. Energy Policy and Energy Efficiency in Smart Energy Systems

Van de Dobbelsteen et al. [175] presented the results of the EU City-zen project which aims to develop an urban energy transition methodology to be used by the cities in order to achieve the goals of sustainability and carbon neutrality. In order to disseminate the results of the project and to implement the proposed methodology a number of Roadshows are organized in cities aiming to achieve sustainable lifestyle e carbon neutrality. In those Roadshows, experts from across Europe cooperate in order to propose strategies and timelines with the local stakeholders to achieve the above mentioned goals. One of these Roadshows was related to the district of Gruz, in the city of Dubrovnik. A detailed analysis of the peculiarities of this district was performed in order to detect possible actions to be implemented to improve sustainability and carbon neutrality. The analysis showed that the majority of the problems were related to the cruise ships and to the related tourists' activity. An energy master planning was implemented in order to define the actions to be implemented. A number of strategies are implemented, such as energy renovation of the buildings of the district and a complete renovation of the port based on renewables and algae arrays for cruise ships waste water treatment. These actions were negotiated with the local population and stakeholders and they may significantly improve the environmental impact.

The promotion of efficiency was also investigated by Sousa et al. [176]. The authors focused on the Portuguese Electricity Demand-Side Efficiency Promotion Plan (PPEC), which is a voluntary financial tool used to submit energy efficiency proposals. This plan is directly related to the EU Energy Efficiency Directive (EED), introducing Energy Efficiency Obligations (EEO). According to EEO, each energy company must achieve an energy saving of 1.5% of the overall energy consumed by the customers. This paper aims to analyze the role of the PPEC mechanism Portugal in relation to EU commitments. The authors performed a review of the existing Market-Based Instruments for energy efficiency in the world, detecting 46 mandatory and six non-mandatory schemes. The majority of those schemes were related to USA and EU. In the case of Portugal, where no EEO were adopted, such efficiency actions may be adopted both by the Distribution System Operator (DSO) and by private companies under the PPEC mechanism. The results of the overview showed that during the previous six PPEC editions, the number of promoters increased more than 10 times. In the same period, 3.6 times additional measures were implemented. Such measures are mainly related to public lighting or traffic lighting, lighting in buildings. In the same period, the total investment determined by PPEC increased nearly four times. Thus, authors concluded that PPEC is an effective mechanism to promote energy saving and energy efficiency.

Another possibility to increase efficiency and sustainability for cities is the so called "Hydrogen Economy", investigated by Kilkis et al. [177]. In this work, authors start from the idea of smart energy system where hydrogen plays a key role in the interactions between renewables, fossil fuel power plant and user demand. In fact, surplus electricity can be converted in hydrogen by electrolysis and this hydrogen can be subsequently used for power production in fuel cells during peak demand hours. In particular, authors assume to produce hydrogen using renewables and to use the existing natural gas pipelines to distribute hydrogen which is supposed to be used by fuel cells. The authors design a novel "hydrogen city" with two cycles at building and district levels. In the first cycle, the electricity of wind turbines and third-generation PVT panels is used to produce hydrogen by electrolysis. In the second cycle, fuel cells are supplied by hydrogen and supply thermal, cooling and electrical energy and water to a low-exergy buildings. Cooling energy is produced by an absorption chillers supplied by the waste heat of the fuel cell. Suitable thermal energy storage systems (TES) are also included in the system. The system was also integrated with ground source geothermal heat pump and geothermal ORC in a "Circular Geothermal Option". The results for the hydrogen city model and Circular Geothermal option showed that the two hydrogen cycles at the district and building levels dramatically improve the possibilities to reach near-zero exergy targets. The proposed systems allow one to improve the utilization of renewables, also increasing the savings in terms of emissions with respect to the reference values, including avoidable $CO_2$ emissions. The authors also concluded that the integration energy-water nexus, renewables, hydrogen economy and net-zero targets is crucial in order to achieve a sustainable energy system.

*3.2. Polygeneration and District Heating*

Doracic et al. [178] focused on the utilization of the excess heat in district heating systems. The majority of EU countries consider cogeneration as a crucial technology in order to achieve the targets in terms of energy efficiency and $CO_2$ reduction. To this scope, district heating networks are often used in order to deliver heat to the final users. Unfortunately, we are far from a massive utilization of this technology, since in EU only 13% of the heat supply is covered by district heating networks. However, this share increases up to 50% for northern countries and it expected to achieve 70% by the next few years. The novel district heating networks will also include renewables and low-temperature heating systems (e.g., heat pumps), thermal storages, utilization of industrial thermal wastes and a suitable integration with the electricity network. These novel district heating systems (4th generation) will be also managed by advanced ICT technologies, fully integrated in the smart cities framework. In such novel networks, which include a plurality of production devices, the optimal management of heat sources is a crucial point in order to achieve good economic profitability. In particular, authors focused on a specific case study related to the city of Ozalj, implementing a suitable simulation model using both Matlab and QGIS software. The input data were obtained by a survey performed for 391 households, classified in eight different categories. The authors combined the data of this survey and the data regarding building gross area and localization in order to calculate the heat demand map. Then, they calculated the levelized cost of heat for a possible scenario considering a natural gas district heating system. In such calculations, both operating and capital costs were accurately considered. In a second scenario, a part of the heat provided by the natural gas was replaced by the excess heat provided by facilities, located outside the city. On the basis of the calculated heat demand map, authors identified the parts of the city where the installation of the district heating network was feasible. For these areas, they calculated a final heat demand equal to 75 MWh/year, obtaining a levelized cost of heat much lower than the one achieved by other conventional technologies. Their analysis also showed that the second scenario was feasible considering possible range of variation of excess heat supply, costs of pipes and heat price. From the environmental point of view, $CO_2$ emissions reductions up to 87% can be achieved by implementing the district heating network.

Taneczuk et al. [179] analyzed the technical possibilities to use district heating boiler slag in order to recover further thermal energy. In the framework of a development of sustainable, efficient and environmental friendly energy paradigm, it is crucial to recover energy by whatever source. The authors considered that a significant potential of energy recovery may be obtained by the utilization of the physical enthalpy of the combustion solid products, such as slag, due to their high temperature (250–450 °C). This energy accounts for about 3.5–25% of the total input primary energy. Thus, it seems extremely interesting the possibility to use the hot slag produced by the boilers fired with solid fuels. The authors describe a new process for the heat recovery from district heating grate-fired boiler slag. The proposed system includes two cogenerative engines (steam turbine and gas turbine), one coal fired water boiler, two hard coal stoker-fired water boilers. In order to perform the research, authors measured both volume and temperature of the slag leaving the boiler. They noticed that slag temperature was strictly related to the boiler load, varying from about 300 °C to 900 °C as a function of boiler output capacity. The authors assumed to use the slag heat in order to supply a high temperature heat pump transferring the heat from the water in the deslagger to the district heating network. The authors also proposed two modifications of the system in order to enhance its efficiency, namely: direct heat recovery in the existing slag trap extended by a heat exchanger; direct heat recovery in the existing slag trap. Results showed that the waste heat available is low (lower than 1% of boiler maximum thermal output). However, results also validated the technical feasibility of the two proposed modifications which are economically feasible only in certain circumstances.

Gimelli et al. [180] proposed a novel optimization algorithm for polygeneration systems. Polygeneration is a novel and efficient technology where multiple energy vectors (electricity, heat and cool) can be produced simultaneously along with different types of byproducts (hydrogen, desalinated water, glycerin, etc.) using both fossil fuels and renewable energy sources. Polygeneration systems are expected to dramatically contribute to the goals in terms of energy efficiency and reduction of emissions. However, such polygeneration systems are extremely complex systems which include a plurality of different devices. Therefore, such systems can be competitive from both energy and economic points of view only when a rigorous optimization is performed, in order to calculate the optimal set of design parameters. In this framework, authors propose a novel optimization algorithm which takes into account the main external conditions. In particular, authors implemented a vector optimization process which is able to determine system layout, fluid selection and design parameters using a Pareto dominance approach. The optimization is performed using a Genetic Algorithm. A case study is analyzed for a cogeneration system based on an Organic Rankine Cycle (ORC). The Pareto front was reported as a function of the Primary Energy Saving (PES) and Simple Pay Back period (SPB). Here, dominant solutions showed PES higher than 16.5% and SPB around 3–5 years.

*3.3. Advanced Combustion Techniques and Fuels*

Kazagic et al. [181] analyzed co-firing low-rank Bosnian coals using different types of biomass, focusing on ash-related problems and emissions. They presented the state of the art showing that conventional fossil-fuel based power stations are going to be converted into multi-fuel power plants. In such hybrid systems a plurality of different biomasses are co-fired with coal aiming at improving fuel mix diversity and at improving the security of energy supply system. In this framework, authors focused on an experimental analysis of co-firing Bosnian low-rank coal with different types of biomass. In particular, they focused on woody biomass and *Miscanthus*. In their system, a multi-fuel pulverized combustion concept was proposed. The tests were performed using a lab-scale furnace, aiming at determining the optimal values of: operating temperature, air distributions, fuel portions and reburning ratio. Six different fuel combinations were considered in the experiments whose composition was accurately measured before the tests. The lab-scale furnace includes an alumina-silicate ceramic tube combustor, surrounded by SiC electric heaters and a suitable insulation. The temperature of the reaction zone is controlled in the range from ambient to 1560 °C. The results of the experiments showed that the tested blends could be used in real systems. In all the cases major ash-related problems were not detected. For all the tested mixtures, a marginal or medium slagging propensity was detected. A minor increase of unburnt carbon content (UBC) in the ash deposits was detected when the co-firing rate was increased from 0.0 to 0.3. Conversely, in the slag collected at the bottom of the furnace, the UBC significantly increased for 0.15 and 0.2 biomass co-firing. From the emissions point of view, results showed that $SO_2$ emissions remained within the expected limits. NOx emissions slightly varied among the different considered fuel mixtures, showing a general decrease in case of higher biomass co-firing rates.

Eder et al. [182] presented a 3D CFD model of a Diesel ignited gas engine. The analysis was applied to large engines, typically used for marine application, where conventional diesel systems are going to be hybridized in dual-fuel (natural gas and Diesel) engines. In particular, authors focused on the Diesel ignited gas engines where a small amount of Diesel fuel is injected into the lean mixture. This technology allows one to enhance fuel flexibility and to decrease NOx and soot emissions. Conversely, a lower efficiency is achieved with respect to the case of gas fired engines. The simulation of such dual-fuel engines is a very hard task since several combustion mechanisms are included in system physics. As a consequence accurate 3D CFD models, including detailed calculations of the rates of reactions, are required in order to calculate and the 3D transient flame front propagation. To this scope authors implemented detailed models in order to take into account the ignition process and delay in a dual fuel operation. In order to perform a model validation, a suitable experimental setup was used. In particular, a Bosh Tube was used in order to measure the rate of injection. Results of the experiments showed a reasonable agreement between numerical and experimental data. Similarly, the combustion model was also validated against experimental data (provided by SCE in Graz). In this case, the calculated heat release related to the flame front propagation was higher than the corresponding measured data.

*3.4. Biomass*

The utilization of biomasses for energy purposes was also investigated by Guimaraes et al. [183], focusing on the development of an anaerobic digester supplied by food waste and sewage. In fact, in many countries the problem of the final disposal of municipal solid waste is extremely critical, causing environmental, social and economic issues. This is especially crucial for the organic matter which is dominant in the solid municipal waste of several countries like Brazil. However, a well-established technology (anaerobic digestion) allows one to stabilize such organic matter and to produce biogas which can be used for electricity and heat production. For a massive utilization of this technology a further research effort is required in order to reduce capital cost and implement optimized control and management strategies. This is the scope of this paper. The research is based on the samples of raw sewage collected by a wastewater treatment plant in Rio de Janeiro. Inoculum was collected from an Upflow Anaerobic Sludge Blanket at a local industry. Food waste was collected at meal time. The components were mixed and their parameters (COD, TS, PH) were properly measured. These samples were subsequently used in experiments in lab-scale bioreactors consisting in glass cylinders. Then a suitable control system was designed and installed on a PLC communicating with a supervisory software. During the experiments, different mixtures of the components were used in order to detect the optimal configuration. In this second part of the work, three biodigesters (B1, B2, and B3) were built. The three biodigesters contain different mixtures, namely: B1 food waste and Sewage; B2 Food Waste, Sewage and Anaerobic Sludge (inoculum); B3 Food Waste, Water and Anaerobic Sludge. Experiments were carried out for 60 days. B2 showed the best performance in terms of: biogas production (63 L), methane return (95%), TS, TVS and COD reductions. This result is due to the composition of substances present in sewage.

Tic et al. [184] proposed an innovative system for the sustainable utilization of sewage sludge. This waste mainly contains water and some organic matter, including microorganisms and various pollutants. Conventional management techniques are often inefficient and harmful for the environment and therefore EU promotes the utilization of novel techniques providing waste stabilization, energy recovery and recycling. In fact, due to the severe environmental restrictions, conventional utilization pathways (e.g., landfilling) are going to be replaced or combined with novel thermal processes. In this framework, authors focused on the drying and torrefaction of sewage sludge, combined with a gasifier and an engine for maximizing heat recovery. In particular, they started from a detailed analysis of the state of the art regarding dewatering and drying of sewage sludge, slagging gasifiers, inertization of solid residues and plasma gasification. Then in order to perform their analysis, a number of samples were obtained by the sewage treatment plant in Brzeg Dolny. These samples were used in two experimental setups. The first one, is a paddle dryer operated in batch mode which includes a number of meters required to measure all the relevant thermodynamic parameters. The second one is used to perform torrefaction tests which were carried out at a laboratory scale using isothermal rotary reactors. Here, gas analyzers, gas cromatographs and spectrometers were used in order to measure properties of inlet and outlet streams. Results of the tests showed that a partial drying dramatically increased electricity consumption. In fact, sludge is a non-Newtonian fluid and its viscosity dramatically increases in case of a lower water content. Torrefaction of the dried sewage sludge determined a marginal carbonization of the sludge and a relevant reduction of the moisture. GC-MS analysis showed remarkable amounts of different heterorganic compounds in the torrefaction gas. All these results are in line with the data available in literature.

*3.5. Building Efficiency*

The goal of a sustainable and efficient development can be also achieved reducing building energy consumption. In fact, buildings account for about 40% of the total final energy use. In this framework, Berardi et al. [1] presented a study dealing with dynamic modeling of building energy demand. In buildings, energy demand is mainly due to space heating and cooling which, in turn, depend on the peculiarities of the building envelope and on internal and external loads. A detailed time-dependent calculation of such demands is crucial in order to analyze the energy and economic feasibility of refurbishment actions. In addition, such calculations also affect the selection of the HVAC systems maximum capacity which dramatically affects the overall energy and economic performance of the buildings. In fact, when the selected capacity is higher than the required value, a higher capital cost is achieved and the production devices will operate in part load conditions where the efficiency is lower. Therefore, this paper analyzes temperature dependence of thermal conductivity. The authors implemented a suitable thermodynamic and hygrothermal model based on energy and mass balances in order to perform their calculation. Suitable measurement were performed in order to calculate the temperature-dependent functions of thermal conductivity. Such equations are subsequently implemented into the building simulation model in order to evaluate the variations of conductivity during the year. Results were compared to the conventional ones considering constant thermal conductivities. Results showed that for Turin the conventional techniques underestimate building thermal demand. Conversely, in Rome and in Palermo the difference between the conventional and the proposed technique is negligible, since winter temperature variation is limited. In summer conditions, the results depend on the sign of the average heat flux.

Obviously, the above mentioned goals may be also achieved by the utilization of conventional and innovative renewable energy technologies. In particular, for novel renewable technologies it is crucial to evaluate its economic profitability with respect to conventional ones. An example is reported by the work by da Klimes et al. [185] which compares an innovative renewable system (solar chimney) with a conventional one (photovoltaic, PV). A solar chimney converts the solar thermal energy of radiation into the kinetic energy. This kinetic energy may be converted in electricity by a wind turbine or used only for ventilation purposes. In fact, the solar chimney is beneficial for the building due to the increased ventilation rate which reduces the space cooling load. Several works are available in literature investigating solar chimneys from both numerical and experimental points of view. However, the profitability of those systems must still be proven. Therefore, in this study, authors compare a solar chimney with a PV-based fan providing the same ventilation rate of the solar chimney. The calculations were performed by a suitable thermodynamic model for the solar chimney and an electrical model for the PV-based fan. The comparison was performed fixing the solar incidence area. Results of the simulations show the case of the PV-powered DC fan supplies an air mass flow rate three times higher with respect to the case of the solar chimney. Thus, PV-powered DC fan overall efficiency is two orders of magnitude higher than the one of solar chimneys.

Building energy consumption can be reduced using a plurality of approaches, such as: increasing the utilization of renewables, improving building envelope performance, improving the efficiency of the heating and cooling systems. All these energy efficiency actions often require relevant investments and a major refurbishment of the buildings. However, the energy consumption may be also significantly reduced by optimizing the energy management system. In fact, in many cases up to 50% of the total input energy is dissipated by inefficient energy management system. Therefore, the development of efficient, robust and reliable control management system is crucial in order to achieve the goals in terms of buildings primary energy reduction. In this framework, Cottafava et al. [186]

proposed a novel tool to be used in the energy management of large building stocks, which is capable to detect anomalies due to abnormal energy use. Their approach is not based on complex thermodynamic dynamic simulations of the buildings and their HVAC systems. Conversely, they implemented that data visualization approach (DataViz) coupled with the Multidimensional detective approach, using he Scatter Plot Matrix and the Parallel Coordinates methods. Then, a clustering algorithm was implemented in order to analyze the data. A case study as discussed for the buildings of the University of Turin, Italy. This method allowed one to identify some outliers, detecting the buildings with abnormal energy consumptions. High specific energy consumption are due both to large IT centres and to electric chillers running 24/7. Then a clustering algorithm was used to test the initial hypothesis. This approach revealed that a number of buildings should be reorganized according to different clustering categories.

Oluleye et al. [187] investigated the utilization of thermal energy storage systems (TES) in dwellings. TES are becoming more and more crucial in the present energy system where the management of excess heat is a key point in order to achieve a good economic profitability. However, authors pointed out that no specific funding policy is available for TES to be used in buildings. Therefore, authors presented a novel optimization technique based on a multi-period MILP in order to integrate TES in an existing dwelling. The optimization procedure aimed to minimize the Equivalent Annual Cost (EAC) of a micro-CHP system including a suitable TES. The model considers, for each time step, the performance parameters of the devices and the present energy market prices. Results of the optimization significantly depend on the house type: highest advantages are calculated for a detached house. In this case, the integration of TES in a micro-CHP system reduces the TDE by 792 kWh/year, $CO_2$ by 146 kg/year and the homeowner makes 60 £/year. From the economic point of view, results showed that this technology needs to be supported by incentivization, also based on the achieved $CO_2$ reduction. In the non-incentivised case, TES showed a better economic performance with respect to the micro-CHP system.

*3.6. Other Topics*

Li et al. [188] performed a numerical thermo fluid dynamic simulation of packed beds with smooth or dimpled spheres featured by low channel to particle diameter ratio. Such devices are widely used in a plurality of applications, for example catalytic reactors and nuclear reactors. Several types of packed beds are under investigation, such as the composite structures packing which allows one to dramatically reduce pressure drops with respect to the randomly packed beds. In this framework, authors analyzed the influence of a series of dimpled spheres in wall bounded structured packed bed from the thermo fluid dynamic point of view. Two different low channel to particle diameter ratios ($N = 1.00$ and $N = 1.15$) were considered. The analysis is performed implementing a suitable computational fluid dynamics model, developed in ANSYS FLUENT environment, taking into account 3-D Navier-Stokes equations for steady state incompressible flow and assuming the RNG k-ε model. The simulations were performed considering two different low channel to particle diameter ratios ($N = 1.00$ and $N = 1.15$). The results of the simulations showed that, for $N = 1.00$, the packed bed with dimpled spheres should be used in real applications to enhance the performance, due to the lower pressure drops and higher heat transfer coefficient. In addition, for $N = 1.15$, pressure drop significantly increased Thus, the packed bed with dimpled spheres should be used only when the need to enhance heat transfer performance dominates the problem of pressure drop increase. Sun et al. [189] designed a path tracking steering controller for autonomous vehicles. Autonomous vehicle is a new challenging technology which aims to increase driving safety, reduce traffic congestion and emissions and to improve vehicle overall efficiency. In order to achieve these goals, the vehicle must accurately track the desired path. In particular, this paper presents a model predictive control (MPC) using a linearization method. The vehicle model is linearized by a sequence of supposed steering angles. This model was also coupled with linearized single-track 'bicycle' model and a brush tire model to simulate the high speed motion of the vehicle. A simulation has been performed to validate the accuracy of the method. Results indicates show that the steering controller with course-direction deviation reduces the average of absolute lateral deviation, compared to the controller with heading deviation, by nearly 20%.

Borjigin et al. [190] investigated longitudinal heat conduction in plate heat exchangers. A number of techniques are available in the open literature for the design and simulation of plat heat exchangers (LMTD, NTU, etc.). All these techniques are based on the assumption that longitudinal heat transfer can be considered negligible. However, in the present paper authors implemented a suitable numerical model in order to analyze the feasibility of such assumption for a small-scale plate heat exchanger. The model was implemented for a gas-to-gas plate heat exchanger, used in a cabinet cooling system. The system includes hot and cold fluid channels which are separated by two solid plates. The model is based on 3D steady state mass, energy and momentum balance equations. Both fluids are considered incompressible and turbulent flow is assumed. The developed numerical code was suitably validated using literature data. Results showed that longitudinal heat is not negligible and in small-scale counter-flow plate heat exchangers a more uniform temperature profile of is achieved with respect to the case of negligibility of longitudinal conduction. In balanced cross-flow and parallel-flow plate heat exchangers longitudinal conduction is negligible. In unbalanced flow, the small-scale longitudinal heat conduction is low for all types of heat exchangers. Finally, authors also concluded that the higher the thermal conductivity of the plate, the stronger the small-scale longitudinal heat conduction and the larger the thermal performance reduction.

## 4. Conclusions

This special issue of *Energies*, dedicated to the 12th Sustainable Development of Energy, Water and Environment Systems Conference, held in 2017 in Dubrovnik, Croatia, provided an insight of topics related to sustainable development. The guest editors of this special issue believe that the selected papers and addressed issues will be extremely interesting for the readers of *Energies*. The selected papers present recent advances in five main fields that are of strategic importance to the sustainable development: energy policy and energy efficiency in smart energy systems, polygeneration and district heating, advanced combustion techniques and fuels, biomass, building efficiency.

The papers included in the present special issue, and the ones previously published in past SDEWES SIs, clearly show that an integrated approach is required in order to achieve the goals in terms of sustainability and decarbonization, mentioned several times in the papers of this SI. First, a suitable energy planning is mandatory required in order to design the future energy scenarios. In this framework, it is also extremely important to design suitable funding policies to support the novel clean technologies, such as: fuel cells, electrical vehicles, hydrogen, etc. In addition, a special effort must be also performed in order to enhance energy efficiency and to promote system integration. It is extremely important to integrate different technologies in novel efficient polygeneration system, mainly based on renewables. Similarly, it is also crucial to recover all the energy flows which are presently wasted or dissipated. In terms of energy efficiency, a key role is due to the buildings which account for about 40% of the overall energy consumption. Thus, suitable energy saving actions must be performed for building envelope and its heating and cooling systems. Obviously, the goals of efficiency and sustainability can be achieved only in a future scenario where the use of renewables will become dominant over the fossil fuels. Solar, hydro and wind are very important in several countries. However, they suffer for unavoidable fluctuations and unpredictability of power production. Conversely, the use of biomass is much more efficient from this point of view. As a consequence, a huge research effort has been developed in order to select clean and sustainable novel biomass conversion techniques (algae, bioethanol, biorefineries, etc.).

Future SDEWES conferences will further contribute in disseminating new knowledge pursuing the goal of sustainabile development. Readers may refer to the International Centre for Sustainable Development of Energy, Water and Environment Systems (SDEWES Centre) for additional information regarding the conference series and the related activities.

**Author Contributions:** F.C. prepared an initial draft of the manuscript which was completed, corrected and reviewed by M.C., N.D., Q.W. and X.Z.

**Funding:** This research received no external funding.

**Acknowledgments:** The guest editors thank the authors of the papers that submitted their manuscripts to this Special Issue for the high quality of their work. We also thank all the reviewers who provided valuable and highly appreciated comments and advice, and the managing editors of Energies for their patience and excellent support.

**Conflicts of Interest:** The authors declare no conflict of interest.

## References

1. Berardi, U.; Tronchin, L.; Manfren, M.; Nastasi, B. On the Effects of Variation of Thermal Conductivity in Buildings in the Italian Construction Sector. *Energies* **2018**, *11*, 872. [CrossRef]
2. Urbaniec, K.; Mikulčić, H.; Wang, Y.; Duić, N. System integration is a necessity for sustainable development. *J. Clean. Prod.* **2018**, *195*, 122–132. [CrossRef]
3. Olawumi, T.O.; Chan, D.W.M. A scientometric review of global research on sustainability and sustainable development. *J. Clean. Prod.* **2018**, *183*, 231–250. [CrossRef]
4. Halati, A.; He, Y. Intersection of economic and environmental goals of sustainable development initiatives. *J. Clean. Prod.* **2018**, *189*, 813–829. [CrossRef]
5. Sinakou, E.; Boeve-de Pauw, J.; Goossens, M.; Van Petegem, P. Academics in the field of Education for Sustainable Development: Their conceptions of sustainable development. *J. Clean. Prod.* **2018**, *184*, 321–332. [CrossRef]
6. Kono, J.; Ostermeyer, Y.; Wallbaum, H. Investigation of regional conditions and sustainability indicators for sustainable product development of building materials. *J. Clean. Prod.* **2018**, *196*, 1356–1364. [CrossRef]
7. Scordato, L.; Klitkou, A.; Tartiu, V.E.; Coenen, L. Policy mixes for the sustainability transition of the pulp and paper industry in Sweden. *J. Clean. Prod.* **2018**, *183*, 1216–1227. [CrossRef]
8. Stoycheva, S.; Marchese, D.; Paul, C.; Padoan, S.; Juhmani, A.-S.; Linkov, I. Multi-criteria decision analysis framework for sustainable manufacturing in automotive industry. *J. Clean. Prod.* **2018**, *187*, 257–272. [CrossRef]
9. Xia, B.; Olanipekun, A.; Chen, Q.; Xie, L.; Liu, Y. Conceptualising the state of the art of corporate social responsibility (CSR) in the construction industry and its nexus to sustainable development. *J. Clean. Prod.* **2018**, *195*, 340–353. [CrossRef]
10. Medina-González, S.; Graells, M.; Guillén-Gosálbez, G.; Espuña, A.; Puigjaner, L. Systematic approach for the design of sustainable supply chains under quality uncertainty. *Energy Convers. Manag.* **2017**, *149*, 722–737. [CrossRef]
11. Briggs, I.; Murtagh, M.; Kee, R.; McCulloug, G.; Douglas, R. Sustainable non-automotive vehicles: The simulation challenges. *Renew. Sustain. Energy Rev.* **2017**, *68*, 840–851. [CrossRef]
12. Chung, W.-S.; Kim, S.-S.; Moon, K.-H.; Lim, C.-Y.; Yun, S.-W. A conceptual framework for energy security evaluation of power sources in South Korea. *Energy* **2017**, *137*, 1066–1074. [CrossRef]
13. Radovanović, M.; Filipović, S.; Golušin, V. Geo-economic approach to energy security measurement—principal component analysis. *Renew. Sustain. Energy Rev.* **2018**, *82*, 1691–1700. [CrossRef]
14. März, S. Assessing the fuel poverty vulnerability of urban neighbourhoods using a spatial multi-criteria decision analysis for the German city of Oberhausen. *Renew. Sustain. Energy Rev.* **2018**, *82*, 1701–1711. [CrossRef]
15. Okushima, S. Gauging energy poverty: A multidimensional approach. *Energy* **2017**, *137*, 1159–1166. [CrossRef]
16. Hinker, J.; Hemkendreis, C.; Drewing, E.; März, S.; Hidalgo Rodríguez, D.I.; Myrzik, J.M.A. A novel conceptual model facilitating the derivation of agent-based models for analyzing socio-technical optimality gaps in the energy domain. *Energy* **2017**, *137*, 1219–1230. [CrossRef]
17. Valero-Gil, J.; Allué-Poc, A.; Ortego, A.; Tomasi, F.; Scarpellini, S. What are the preferences in the development process of a sustainable urban mobility plan? New methodology for experts involvement. *Int. J. Innov. Sustain. Dev.* **2018**, *12*, 135–155.

18. Ajanovic, A.; Haas, R. The impact of energy policies in scenarios on GHG emission reduction in passenger car mobility in the EU-15. *Renew. Sustain. Energy Rev.* **2017**, *68*, 1088–1096. [CrossRef]

19. Knez, M.; Obrecht, M. Policies for promotion of electric vehicles and factors influencing consumers' purchasing decisions of low emission vehicles. *J. Sustain. Dev. Energy Water Environ. Syst.* **2017**, *5*, 151–162. [CrossRef]

20. Dominković, D.F.; Bačeković, I.; Pedersen, A.S.; Krajačić, G. The future of transportation in sustainable energy systems: Opportunities and barriers in a clean energy transition. *Renew. Sustain. Energy Rev.* **2018**, *82*, 1823–1838. [CrossRef]

21. Firak, M.; Đukić, A. Hydrogen transportation fuel in Croatia: Road map strategy. *Int. J. Hydrogen Energy* **2016**, *41*, 13820–13830. [CrossRef]

22. Del Moretto, D.; Colla, V.; Branca, T.A. Sustainable mobility for campsites: The case of Macchia Lucchese. *Renew. Sustain. Energy Rev.* **2017**, *68*, 1063–1075. [CrossRef]

23. Knoop, K.; Lechtenböhmer, S. The potential for energy efficiency in the EU Member States—A comparison of studies. *Renew. Sustain. Energy Rev.* **2017**, *68*, 1097–1105. [CrossRef]

24. Moser, S. Overestimation of savings in energy efficiency obligation schemes. *Energy* **2017**, *121*, 599–605. [CrossRef]

25. Pleßmann, G.; Blechinger, P. Outlook on South-East European power system until 2050: Least-cost decarbonization pathway meeting EU mitigation targets. *Energy* **2017**, *137*, 1041–1053. [CrossRef]

26. Ben Amer-Allam, S.; Münster, M.; Petrović, S. Scenarios for sustainable heat supply and heat savings in municipalities—The case of Helsingør, Denmark. *Energy* **2017**, *137*, 1252–1263. [CrossRef]

27. Batas-Bjelic, I.; Rajakovic, N.; Duic, N. Smart municipal energy grid within electricity market. *Energy* **2017**, *137*, 1277–1285. [CrossRef]

28. De Luca, G.; Fabozzi, S.; Massarotti, N.; Vanoli, L. A renewable energy system for a nearly zero greenhouse city: Case study of a small city in southern Italy. *Energy* **2018**, *143*, 347–362. [CrossRef]

29. Vidal-Amaro, J.J.; Sheinbaum-Pardo, C. A Transition Strategy from Fossil Fuels to Renewable Energy Sources in the Mexican Electricity System. *J. Sustain. Dev. Energy Water Environ. Syst.* **2018**, *6*, 47–66. [CrossRef]

30. Omrcen, E.; Lundgren, U.; Dalbro, M. Universities as role models for sustainability: A case study on implementation of University of Gothenburg climate strategy, results and experiences from 2011 to 2015. *Int. J. Innov. Sustain. Dev.* **2018**, *12*, 156–182. [CrossRef]

31. Opel, O.; Strodel, N.; Werner, K.F.; Geffken, J.; Tribel, A.; Ruck, W.K.L. Climate-neutral and sustainable campus Leuphana University of Lueneburg. *Energy* **2017**, *141*, 2628–2639. [CrossRef]

32. Kavvadias, K.C. Energy price spread as a driving force for combined generation investments: A view on Europe. *Energy* **2016**, *115*, 1632–1639. [CrossRef]

33. Zakeri, B.; Virasjoki, V.; Syri, S.; Connolly, D.; Mathiesen, B.V.; Welsch, M. Impact of Germany's energy transition on the Nordic power market—A market-based multi-region energy system model. *Energy* **2016**, *115*, 1640–1662. [CrossRef]

34. Novosel, T.; Perković, L.; Ban, M.; Keko, H.; Pukšec, T.; Krajačić, G.; Duić, N. Agent based modelling and energy planning—Utilization of MATSim for transport energy demand modelling. *Energy* **2015**, *92*, 466–475. [CrossRef]

35. Walmsley, M.R.W.; Walmsley, T.G.; Atkins, M.J.; Kamp, P.J.J.; Neale, J.R. Minimising carbon emissions and energy expended for electricity generation in New Zealand through to 2050. *Appl. Energy* **2014**, *135*, 656–665. [CrossRef]

36. Kazagic, A.; Merzic, A.; Redzic, E.; Music, M. Power utility generation portfolio optimization as function of specific RES and decarbonisation targets—EPBiH case study. *Appl. Energy* **2014**, *135*, 694–703. [CrossRef]

37. Pietzcker, R.C.; Stetter, D.; Manger, S.; Luderer, G. Using the sun to decarbonize the power sector: The economic potential of photovoltaics and concentrating solar power. *Appl. Energy* **2014**, *135*, 704–720. [CrossRef]

38. Moiseyev, A.; Solberg, B.; Kallio, A.M.I. The impact of subsidies and carbon pricing on the wood biomass use for energy in the EU. *Energy* **2014**, *76*, 161–167. [CrossRef]

39. Romagnoli, F.; Barisa, A.; Dzene, I.; Blumberga, A.; Blumberga, D. Implementation of different policy strategies promoting the use of wood fuel in the Latvian district heating system: Impact evaluation through a system dynamic model. *Energy* **2014**, *76*, 210–222. [CrossRef]

40. Pušnik, M.; Sučić, B. Integrated and realistic approach to energy planning—A case study of Slovenia. *Manag. Environ. Qual. Int. J.* **2014**, *25*, 30–51. [CrossRef]

41. Beke-Trivunac, J.; Jovanovic, L.; Radosavljevic, Ž.; Radosavljevic, M. An overview of environmental policies of local government organizations in the Republic of Serbia. *Manag. Environ. Qual. Int. J.* **2014**, *25*, 263–272. [CrossRef]

42. Kravanja, Z.; Čuček, L. Multi-objective optimisation for generating sustainable solutions considering total effects on the environment. *Appl. Energy* **2013**, *101*, 67–80. [CrossRef]

43. Schlör, H.; Fischer, W.; Hake, J.-F. Methods of measuring sustainable development of the German energy sector. *Appl. Energy* **2013**, *101*, 172–181. [CrossRef]

44. Golušin, M.; Munitlak Ivanović, O.; Redžepagić, S. Transition from traditional to sustainable energy development in the region of Western Balkans—Current level and requirements. *Appl. Energy* **2013**, *101*, 182–191. [CrossRef]

45. Metz, M.; Doetsch, C. Electric vehicles as flexible loads—A simulation approach using empirical mobility data. *Energy* **2012**, *48*, 369–374. [CrossRef]

46. Irsag, B.; Pukšec, T.; Duić, N. Long term energy demand projection and potential for energy savings of Croatian tourism-catering trade sector. *Energy* **2012**, *48*, 398–405. [CrossRef]

47. Peri, G.; Traverso, M.; Finkbeiner, M.; Rizzo, G. The cost of green roofs disposal in a life cycle perspective: Covering the gap. *Energy* **2012**, *48*, 406–414. [CrossRef]

48. Cantore, N. Sustainability of the energy sector in the Mediterranean region. *Energy* **2012**, *48*, 423–430. [CrossRef]

49. Čulig-Tokić, D.; Krajačić, G.; Doračić, B.; Mathiesen, B.V.; Krklec, R.; Larsen, J.M. Comparative analysis of the district heating systems of two towns in Croatia and Denmark. *Energy* **2015**, *92*, 435–443. [CrossRef]

50. del Hoyo Arce, I.; Herrero López, S.; López Perez, S.; Rämä, M.; Klobut, K.; Febres, J.A. Models for fast modelling of district heating and cooling networks. *Renew. Sustain. Energy Rev.* **2018**, *82*, 1863–1873. [CrossRef]

51. González-Bravo, R.; Fuentes-Cortés, L.F.; Ponce-Ortega, J.M. Defining priorities in the design of power and water distribution networks. *Energy* **2017**, *137*, 1026–1040. [CrossRef]

52. Pavičević, M.; Novosel, T.; Pukšec, T.; Duić, N. Hourly optimization and sizing of district heating systems considering building refurbishment—Case study for the city of Zagreb. *Energy* **2017**, *137*, 1264–1276. [CrossRef]

53. Carotenuto, A.; Figaj, R.D.; Vanoli, L. A novel solar-geothermal district heating, cooling and domestic hot water system: Dynamic simulation and energy-economic analysis. *Energy* **2017**, *141*, 2652–2669. [CrossRef]

54. Dominković, D.F.; Ćosić, B.; Bačelić Medić, Z.; Duić, N. A hybrid optimization model of biomass trigeneration system combined with pit thermal energy storage. *Energy Convers. Manag.* **2015**, *104*, 90–99. [CrossRef]

55. Radulovic, D.; Skok, S.; Kirincic, V. Cogeneration—Investment dilemma. *Energy* **2012**, *48*, 177–187. [CrossRef]

56. Calise, F.; Macaluso, A.; Piacentino, A.; Vanoli, L. A novel hybrid polygeneration system supplying energy and desalinated water by renewable sources in Pantelleria Island. *Energy* **2017**, *137*, 1086–1106. [CrossRef]

57. Calise, F.; Figaj, R.D.; Vanoli, L. A novel polygeneration system integrating photovoltaic/thermal collectors, solar assisted heat pump, adsorption chiller and electrical energy storage: Dynamic and energy-economic analysis. *Energy Convers. Manag.* **2017**, *149*, 798–814. [CrossRef]

58. Calise, F.; d'Accadia, M.D.; Libertini, L.; Quiriti, E.; Vanoli, R.; Vicidomini, M. Optimal operating strategies of combined cooling, heating and power systems: A case study for an engine manufacturing facility. *Energy Convers. Manag.* **2017**, *149*, 1066–1084. [CrossRef]

59. Calise, F.; d'Accadia, M.D.; Macaluso, A.; Vanoli, L.; Piacentino, A. A novel solar-geothermal trigeneration system integrating water desalination: Design, dynamic simulation and economic assessment. *Energy* **2016**, *115*, 1533–1547. [CrossRef]

60. Calise, F.; d'Accadia, M.D.; Piacentino, A. Exergetic and exergoeconomic analysis of a renewable polygeneration system and viability study for small isolated communities. *Energy* **2015**, *92*, 290–307. [CrossRef]

61. Calise, F.; Cipollina, A.; d'Accadia, M.D.; Piacentino, A. A novel renewable polygeneration system for a small Mediterranean volcanic island for the combined production of energy and water: Dynamic simulation and economic assessment. *Appl. Energy* **2014**, *135*, 675–693. [CrossRef]

62. Buonomano, A.; Calise, F.; Palombo, A.; Vicidomini, M. BIPVT systems for residential applications: An energy and economic analysis for European climates. *Appl. Energy* **2016**, *184*, 1411–1431. [CrossRef]

63. Buonomano, A.; Calise, F.; Palombo, A.; Vicidomini, M. Adsorption chiller operation by recovering low-temperature heat from building integrated photovoltaic thermal collectors: Modelling and simulation. *Energy Convers. Manag.* **2017**, *149*, 1019–1036. [CrossRef]

64. Catrini, P.; Cipollina, A.; Micale, G.; Piacentino, A.; Tamburini, A. Exergy analysis and thermoeconomic cost accounting of a Combined Heat and Power steam cycle integrated with a Multi Effect Distillation-Thermal Vapour Compression desalination plant. *Energy Convers. Manag.* **2017**, *149*, 950–965. [CrossRef]

65. Tamburini, A.; Cipollina, A.; Micale, G.; Piacentino, A. CHP (combined heat and power) retrofit for a large MED-TVC (multiple effect distillation along with thermal vapour compression) desalination plant: High efficiency assessment for different design options under the current legislative EU framework. *Energy* **2016**, *115*, 1548–1559. [CrossRef]

66. Beccali, M.; Ciulla, G.; Di Pietra, B.; Galatioto, A.; Leone, G.; Piacentino, A. Assessing the feasibility of cogeneration retrofit and district heating/cooling networks in small Italian islands. *Energy* **2017**, *141*, 2572–2586. [CrossRef]

67. Angrisani, G.; Akisawa, A.; Marrasso, E.; Roselli, C.; Sasso, M. Performance assessment of cogeneration and trigeneration systems for small scale applications. *Energy Convers. Manag.* **2016**, *125*, 194–208. [CrossRef]

68. Pfeifer, A.; Dominković, D.F.; Ćosić, B.; Duić, N. Economic feasibility of CHP facilities fueled by biomass from unused agriculture land: Case of Croatia. *Energy Convers. Manag.* **2016**, *125*, 222–229. [CrossRef]

69. Di Fraia, S.; Massarotti, N.; Vanoli, L.; Costa, M. Thermo-economic analysis of a novel cogeneration system for sewage sludge treatment. *Energy* **2016**, *115*, 1560–1571. [CrossRef]

70. Di Fraia, S.; Massarotti, N.; Vanoli, L. A novel energy assessment of urban wastewater treatment plants. *Energy Convers. Manag.* **2018**, *163*, 304–313. [CrossRef]

71. Piacentino, A.; Gallea, R.; Cardona, F.; Lo Brano, V.; Ciulla, G.; Catrini, P. Optimization of trigeneration systems by Mathematical Programming: Influence of plant scheme and boundary conditions. *Energy Convers. Manag.* **2015**, *104*, 100–114. [CrossRef]

72. Sayegh, M.A.; Danielewicz, J.; Nannou, T.; Miniewicz, M.; Jadwiszczak, P.; Piekarska, K.; Jouhara, H. Trends of European research and development in district heating technologies. *Renew. Sustain. Energy Rev.* **2017**, *68*, 1183–1192. [CrossRef]

73. Münster, M.; Morthorst, P.E.; Larsen, H.V.; Bregnbæk, L.; Werling, J.; Lindboe, H.H.; Ravn, H. The role of district heating in the future Danish energy system. *Energy* **2012**, *48*, 47–55. [CrossRef]

74. Katsaros, G.; Nguyen, T.-V.; Rokni, M. Tri-generation system based on municipal waste gasification, fuel cell and an absorption chiller. *J. Sustain. Dev. Energy Water Environ. Syst.* **2018**, *6*, 13–32. [CrossRef]

75. Sibilio, S.; Rosato, A.; Ciampi, G.; Scorpio, M.; Akisawa, A. Building-integrated trigeneration system: Energy, environmental and economic dynamic performance assessment for Italian residential applications. *Renew. Sustain. Energy Rev.* **2017**, *68*, 920–933. [CrossRef]

76. Danon, G.; Furtula, M.; Mandić, M. Possibilities of implementation of CHP (combined heat and power) in the wood industry in Serbia. *Energy* **2012**, *48*, 169–176. [CrossRef]

77. Dutra, J.C.; Gonzalez-Carmona, M.A.; Lazaro-Alvarado, A.F.; Coronas, A. Modeling of a Cogeneration System with a Micro Gas Turbine Operating at Partial Load Conditions. *J. Sustain. Dev. Energy Water Environ. Syst.* **2017**, *5*, 139–150. [CrossRef]

78. Borsukiewicz-Gozdur, A.; Klonowicz, P.; Król, D.; Wiśniewski, S.; Zwarycz-Makles, K. Techno-economic analysis of CHP system supplied by waste forest biomass. *Waste Manag. Res.* **2015**, *33*, 748–754. [CrossRef] [PubMed]

79. Vialetto, G.; Noro, M.; Rokni, M. Thermodynamic Investigation of a Shared Cogeneration System with Electrical Cars for Northern Europe Climate. *J. Sustain. Dev. Energy Water Environ. Syst.* **2017**, *5*, 590–607. [CrossRef]

80. Felipe, A.J.; Schneider, D.R.; Krajačić, G. Evaluation of integration of solar energy into the district heating system of the city of Velika Gorica. *Therm. Sci.* **2016**, *20*, 1049–1060. [CrossRef]

81. Di Palma, D.; Lucentini, M.; Rottenberg, F. Trigeneration plants in Italian large retail sector: A calculation model for the TPF projects with evaluation of all the incentivizing mechanisms. *J. Sustain. Dev. Energy Water Environ. Syst.* **2013**, *1*, 375–389. [CrossRef]

82. Orihuela, M.P.; Gómez-Martín, A.; Becerra, J.A.; Chacartegui, R.; Ramírez-Rico, J. Performance of biomorphic Silicon Carbide as particulate filter in diesel boilers. *J. Environ. Manag.* **2017**, *203*, 907–919. [CrossRef] [PubMed]

83. Kobayashi, M.; Akiho, H. Dry syngas purification process for coal gas produced in oxy-fuel type integrated gasification combined cycle power generation with carbon dioxide capturing feature. *J. Environ. Manag.* **2017**, *203*, 925–936. [CrossRef] [PubMed]

84. Cheah, K.W.; Yusup, S.; Gurdeep Singh, H.K.; Uemura, Y.; Lam, H.L. Process simulation and techno economic analysis of renewable diesel production via catalytic decarboxylation of rubber seed oil—A case study in Malaysia. *J. Environ. Manag.* **2017**, *203*, 950–961. [CrossRef] [PubMed]

85. Pehlivan, E.; Özbay, N.; Yargıç, A.S.; Şahin, R.Z. Production and characterization of chars from cherry pulp via pyrolysis. *J. Environ. Manag.* **2017**, *203*, 1017–1025. [CrossRef] [PubMed]

86. Petranović, Z.; Bešenić, T.; Vujanović, M.; Duić, N. Modelling pollutant emissions in diesel engines, influence of biofuel on pollutant formation. *J. Environ. Manag.* **2017**, *203*, 1038–1046. [CrossRef] [PubMed]

87. Bargiel, P.; Kostowski, W.; Klimanek, A.; Górny, K. Design and optimization of a natural gas-fired thermoelectric generator by computational fluid dynamics modeling. *Energy Convers. Manag.* **2017**, *149*, 1037–1047. [CrossRef]

88. Wang, J.; Liu, Y.; Sundén, B.; Yang, R.; Baleta, J.; Vujanović, M. Analysis of slab heating characteristics in a reheating furnace. *Energy Convers. Manag.* **2017**, *149*, 928–936. [CrossRef]

89. Kun-Balog, A.; Sztankó, K.; Józsa, V. Pollutant emission of gaseous and liquid aqueous bioethanol combustion in swirl burners. *Energy Convers. Manag.* **2017**, *149*, 896–903. [CrossRef]

90. Striūgas, N.; Vorotinskienė, L.; Paulauskas, R.; Navakas, R.; Džiugys, A.; Narbutas, L. Estimating the fuel moisture content to control the reciprocating grate furnace firing wet woody biomass. *Energy Convers. Manag.* **2017**, *149*, 937–949. [CrossRef]

91. Kokalj, F.; Arbiter, B.; Samec, N. Sewage sludge gasification as an alternative energy storage model. *Energy Convers. Manag.* **2017**, *149*, 738–747. [CrossRef]

92. De Faria, M.M.N.; Vargas Machuca Bueno, J.P.; Ayad, S.M.M.E.; Belchior, C.R.P. Thermodynamic simulation model for predicting the performance of spark ignition engines using biogas as fuel. *Energy Convers. Manag.* **2017**, *149*, 1096–1108. [CrossRef]

93. Vihar, R.; Žvar Baškovič, U.; Seljak, T.; Katrašnik, T. Combustion and emission formation phenomena of tire pyrolysis oil in a common rail Diesel engine. *Energy Convers. Manag.* **2017**, *149*, 706–721. [CrossRef]

94. Wang, S.; Zhu, X.; Somers, L.M.T.; de Goey, L.P.H. Effects of exhaust gas recirculation at various loads on diesel engine performance and exhaust particle size distribution using four blends with a research octane number of 70 and diesel. *Energy Convers. Manag.* **2017**, *149*, 918–927. [CrossRef]

95. Sjerić, M.; Taritaš, I.; Tomić, R.; Blažić, M.; Kozarac, D.; Lulić, Z. Efficiency improvement of a spark-ignition engine at full load conditions using exhaust gas recirculation and variable geometry turbocharger—Numerical study. *Energy Convers. Manag.* **2016**, *125*, 26–39. [CrossRef]

96. Wang, T.; Zhang, X.; Zhang, J.; Hou, X. Numerical analysis of the influence of the fuel injection timing and ignition position in a direct-injection natural gas engine. *Energy Convers. Manag.* **2017**, *149*, 748–759. [CrossRef]

97. Hossain, A.K.; Smith, D.I.; Davies, P.A. Effects of Engine Cooling Water Temperature on Performance and Emission Characteristics of a Compression Ignition Engine Operated with Biofuel Blend. *J. Sustain. Dev. Energy Water Environ. Syst.* **2017**, *5*, 46–57. [CrossRef]

98. Artiukhina, E.; Grammelis, P. Modeling of biofuel pellets torrefaction in a realistic geometry. *Therm. Sci.* **2016**, *156*. [CrossRef]

99. Chan, Y.H.; Yusup, S.; Quitain, A.T.; Tan, R.R.; Sasaki, M.; Lam, H.L.; Uemura, Y. Effect of process parameters on hydrothermal liquefaction of oil palm biomass for bio-oil production and its life cycle assessment. *Energy Convers. Manag.* **2015**, *104*, 180–188. [CrossRef]

100. Valle, B.; Remiro, A.; Aramburu, B.; Bilbao, J.; Gayubo, A.G. Strategies for maximizing the bio-oil valorization by catalytic transformation. *J. Clean. Prod.* **2015**, *88*, 345–348. [CrossRef]

101. Seljak, T.; Širok, B.; Katrašnik, T. Advanced fuels for gas turbines: Fuel system corrosion, hot path deposit formation and emissions. *Energy Convers. Manag.* **2016**, *125*, 40–50. [CrossRef]

102. Mikulčić, H.; Von Berg, E.; Vujanović, M.; Duić, N. Numerical study of co-firing pulverized coal and biomass inside a cement calciner. *Waste Manag. Res.* **2014**, *32*, 661–669. [CrossRef] [PubMed]

103. Royo, P.; Ferreira, V.J.; López-Sabirón, A.M.; García-Armingol, T.; Ferreira, G. Retrofitting strategies for improving the energy and environmental efficiency in industrial furnaces: A case study in the aluminium sector. *Renew. Sustain. Energy Rev.* **2018**, *82*, 1813–1822. [CrossRef]

104. Szarka, N.; Eichhorn, M.; Kittler, R.; Bezama, A.; Thrän, D. Interpreting long-term energy scenarios and the role of bioenergy in Germany. *Renew. Sustain. Energy Rev.* **2017**, *68*, 1222–1233. [CrossRef]

105. Gaida, D.; Wolf, C.; Bongards, M. Feed control of anaerobic digestion processes for renewable energy production: A review. *Renew. Sustain. Energy Rev.* **2017**, *68*, 869–875. [CrossRef]

106. Lijó, L.; González-García, S.; Bacenetti, J.; Moreira, M.T. The environmental effect of substituting energy crops for food waste as feedstock for biogas production. *Energy* **2017**, *137*, 1130–1143. [CrossRef]

107. Mikulandrić, R.; Böhning, D.; Böhme, R.; Helsen, L.; Beckmann, M.; Lončar, D. Dynamic modelling of biomass gasification in a co-current fixed bed gasifier. *Energy Convers. Manag.* **2016**, *125*, 264–276. [CrossRef]

108. Szulczewski, W.; Żyromski, A.; Jakubowski, W.; Biniak-Pieróg, M. A new method for the estimation of biomass yield of giant miscanthus (*Miscanthus giganteus*) in the course of vegetation. *Renew. Sustain. Energy Rev.* **2018**, *82*, 1787–1795. [CrossRef]

109. Ko, C.-H.; Yu, F.-C.; Chang, F.-C.; Yang, B.-Y.; Chen, W.-H.; Hwang, W.-S.; Tu, T.-C. Bioethanol production from recovered napier grass with heavy metals. *J. Environ. Manag.* **2017**, *203*, 1005–1010. [CrossRef] [PubMed]

110. Kuppens, T.; Van Dael, M.; Vanreppelen, K.; Thewys, T.; Yperman, J.; Carleer, R.; Schreurs, S.; Van Passel, S. Techno-economic assessment of fast pyrolysis for the valorization of short rotation coppice cultivated for phytoextraction. *J. Clean. Prod.* **2015**, *88*, 336–344. [CrossRef]

111. Hájek, M.; Skopal, F.; Vávra, A.; Kocík, J. Transesterification of rapeseed oil by butanol and separation of butyl ester. *J. Clean. Prod.* **2017**, *155*, 28–33. [CrossRef]

112. Yang, B.-Y.; Cheng, M.-H.; Ko, C.-H.; Wang, Y.-N.; Chen, W.-H.; Hwang, W.-S.; Yang, Y.-P.; Chen, H.-T.; Chang, F.-C. Potential bioethanol production from Taiwanese chenopods (*Chenopodium formosanum*). *Energy* **2014**, *76*, 59–65. [CrossRef]

113. Özdenkçi, K.; De Blasio, C.; Muddassar, H.R.; Melin, K.; Oinas, P.; Koskinen, J.; Sarwar, G.; Järvinen, M. A novel biorefinery integration concept for lignocellulosic biomass. *Energy Convers. Manag.* **2017**, *149*, 974–987. [CrossRef]

114. Thomassen, G.; Van Dael, M.; Lemmens, B.; Van Passel, S. A review of the sustainability of algal-based biorefineries: Towards an integrated assessment framework. *Renew. Sustain. Energy Rev.* **2017**, *68*, 876–887. [CrossRef]

115. Budzianowski, W.M.; Postawa, K. Renewable energy from biogas with reduced carbon dioxide footprint: Implications of applying different plant configurations and operating pressures. *Renew. Sustain. Energy Rev.* **2017**, *68*, 852–868. [CrossRef]

116. Chung, M.; Park, H.-C. Feasibility study for retrofitting biogas cogeneration systems to district heating in South Korea. *Waste Manag. Res.* **2015**, *33*, 755–766. [CrossRef] [PubMed]

117. Hublin, A.; Schneider, D.R.; Džodan, J. Utilization of biogas produced by anaerobic digestion of agro-industrial waste: Energy, economic and environmental effects. *Waste Manag. Res.* **2014**, *32*, 626–633. [CrossRef] [PubMed]

118. Sriwannawit, P.; Anisa, P.A.; Rony, A.M. Policy impact on economic viability of biomass gasification systems in Indonesia. *J. Sustain. Dev. Energy Water Environ. Syst.* **2016**, *4*, 56–68. [CrossRef]

119. Maier, S.; Szerencsits, M.; Narodoslawsky, M.; Ismail, I.M.I.; Shahzad, K. Current potential of more sustainable biomass production using eco-efficient farming practices in Austria. *J. Clean. Prod.* **2017**, *155*, 23–27. [CrossRef]

120. Chng, L.M.; Chan, D.J.C.; Lee, K.T. Sustainable production of bioethanol using lipid-extracted biomass from Scenedesmus dimorphus. *J. Clean. Prod.* **2016**, *130*, 68–73. [CrossRef]

121. Sheinbaum, C.; Balam, M.V.; Robles, G.; Lelo de Larrea, S.; Mendoza, R. Biodiesel from waste cooking oil in Mexico City. *Waste Manag. Res.* **2015**, *33*, 730–739. [CrossRef] [PubMed]

122. How, B.S.; Hong, B.H.; Lam, H.L.; Friedler, F. Synthesis of multiple biomass corridor via decomposition approach: A P-graph application. *J. Clean. Prod.* **2016**, *130*, 45–57. [CrossRef]

123. Kazagic, A.; Music, M.; Smajevic, I.; Ademovic, A.; Redzic, E. Possibilities and sustainability of "biomass for power" solutions in the case of a coal-based power utility. *Clean Technol. Environ. Policy* **2016**, *18*, 1675–1683. [CrossRef]

124. Hussain, C.M.I.; Norton, B.; Duffy, A. Technological assessment of different solar-biomass systems for hybrid power generation in Europe. *Renew. Sustain. Energy Rev.* **2017**, *68*, 1115–1129. [CrossRef]

125. Davis, G.W. Addressing Concerns Related to the Use of Ethanol-Blended Fuels in Marine Vehicles. *J. Sustain. Dev. Energy Water Environ. Syst.* **2017**, *5*, 546–559. [CrossRef]

126. Wzorek, M.; Tańczuk, M. Production of biosolid fuels from municipal sewage sludge: Technical and economic optimisation. *Waste Manag. Res.* **2015**, *33*, 704–714. [CrossRef] [PubMed]

127. Kocík, J.; Samikannu, A.; Bourajoini, H.; Pham, T.N.; Mikkola, J.-P.; Hájek, M.; Čapek, L. Screening of active solid catalysts for esterification of tall oil fatty acids with methanol. *J. Clean. Prod.* **2017**, *155*, 34–38. [CrossRef]

128. Ooba, M.; Hayashi, K.; Fujii, M.; Fujita, T.; Machimura, T.; Matsui, T. A long-term assessment of ecological-economic sustainability of woody biomass production in Japan. *J. Clean. Prod.* **2015**, *88*, 318–325. [CrossRef]

129. Ang, G.T.; Tan, K.T.; Lee, K.T.; Mohamed, A.R. Biodiesel production via injection of superheated methanol technology at atmospheric pressure. *Energy Convers. Manag.* **2014**, *87*, 1231–1238. [CrossRef]

130. Berardi, U. Building Energy Consumption in US, EU, and BRIC Countries. *Procedia Eng.* **2015**, *118*, 128–136. [CrossRef]

131. Berardi, U. A cross-country comparison of the building energy consumptions and their trends. *Resour. Conserv. Recycl.* **2017**, *123*, 230–241. [CrossRef]

132. Bourdeau, M.; Guo, X.; Nefzaoui, E. Buildings energy consumption generation gap: A post-occupancy assessment in a case study of three higher education buildings. *Energy Build.* **2018**, *159*, 600–611. [CrossRef]

133. Calero, M.; Alameda-Hernandez, E.; Fernández, M.; Ronda, A.; Martín-Lara, M.Á. Energy consumption reduction proposals for thermal systems in residential buildings. *Energy Build.* **2018**. [CrossRef]

134. Canale, L.; Dell'Isola, M.; Ficco, G.; Di Pietra, B.; Frattolillo, A. Estimating the impact of heat accounting on Italian residential energy consumption in different scenarios. *Energy Build.* **2018**, *168*, 385–398. [CrossRef]

135. Copiello, S.; Gabrielli, L. Analysis of building energy consumption through panel data: The role played by the economic drivers. *Energy Build.* **2017**, *145*, 130–143. [CrossRef]

136. D'Agostino, D.; Cuniberti, B.; Bertoldi, P. Energy consumption and efficiency technology measures in European non-residential buildings. *Energy Build.* **2017**, *153*, 72–86. [CrossRef]

137. Iturriaga, E.; Aldasoro, U.; Terés-Zubiaga, J.; Campos-Celador, A. Optimal renovation of buildings towards the nearly Zero Energy Building standard. *Energy* **2018**. [CrossRef]

138. Lu, Y.; Cui, P.; Li, D. Which activities contribute most to building energy consumption in China? A hybrid LMDI decomposition analysis from year 2007 to 2015. *Energy Build.* **2018**, *165*, 259–269. [CrossRef]

139. Mata, É.; Kalagasidis, A.S.; Johnsson, F. Contributions of building retrofitting in five member states to EU targets for energy savings. *Renew. Sustain. Energy Rev.* **2018**, *93*, 759–774. [CrossRef]

140. Fan, Y.; Xia, X. Energy-efficiency building retrofit planning for green building compliance. *Build. Environ.* **2018**, *136*, 312–321. [CrossRef]

141. Friess, W.A.; Rakhshan, K. A review of passive envelope measures for improved building energy efficiency in the UAE. *Renew. Sustain. Energy Rev.* **2017**, *72*, 485–496. [CrossRef]

142. Shi, L.; Zhang, H.; Li, Z.; Luo, Z.; Liu, J. Optimizing the thermal performance of building envelopes for energy saving in underground office buildings in various climates of China. *Tunn. Undergr. Space Technol.* **2018**, *77*, 26–35. [CrossRef]

143. Cho, J.; Kim, Y.; Koo, J.; Park, W. Energy-cost analysis of HVAC system for office buildings: Development of a multiple prediction methodology for HVAC system cost estimation. *Energy Build.* **2018**, *173*, 562–576. [CrossRef]

144. Granderson, J.; Lin, G.; Singla, R.; Fernandes, S.; Touzani, S. Field evaluation of performance of HVAC optimization system in commercial buildings. *Energy Build.* **2018**, *173*, 577–586. [CrossRef]

145. Manjarres, D.; Mera, A.; Perea, E.; Lejarazu, A.; Gil-Lopez, S. An energy-efficient predictive control for HVAC systems applied to tertiary buildings based on regression techniques. *Energy Build.* **2017**, *152*, 409–417. [CrossRef]

146. Razmara, M.; Maasoumy, M.; Shahbakhti, M.; Robinett, R.D. Optimal exergy control of building HVAC system. *Appl. Energy* **2015**, *156*, 555–565. [CrossRef]

147. Shi, J.; Yu, N.; Yao, W. Energy Efficient Building HVAC Control Algorithm with Real-time Occupancy Prediction. *Energy Procedia* **2017**, *111*, 267–276. [CrossRef]

148. Turner, W.J.N.; Staino, A.; Basu, B. Residential HVAC fault detection using a system identification approach. *Energy Build.* **2017**, *151*, 1–17. [CrossRef]

149. Yang, C.; Shen, W.; Chen, Q.; Gunay, B. A practical solution for HVAC prognostics: Failure mode and effects analysis in building maintenance. *J. Build. Eng.* **2018**, *15*, 26–32. [CrossRef]

150. Ascione, F.; Bianco, N.; De Masi, R.F.; De Stasio, C.; Mauro, G.M.; Vanoli, G.P. Multi-objective optimization of the renewable energy mix for a building. *Appl. Therm. Eng.* **2016**, *101*, 612–621. [CrossRef]

151. Guen, M.L.; Mosca, L.; Perera, A.T.D.; Coccolo, S.; Mohajeri, N.; Scartezzini, J.-L. Improving the energy sustainability of a Swiss village through building renovation and renewable energy integration. *Energy Build.* **2018**, *158*, 906–923. [CrossRef]

152. Li, D.; He, J.; Li, L. A review of renewable energy applications in buildings in the hot-summer and warm-winter region of China. *Renew. Sustain. Energy Rev.* **2016**, *57*, 327–336. [CrossRef]

153. Schaube, P.; Ortiz, W.; Recalde, M. Status and future dynamics of decentralised renewable energy niche building processes in Argentina. *Energy Res. Soc. Sci.* **2018**, *35*, 57–67. [CrossRef]

154. Wang, Z.; Zhao, J.; Li, M. Analysis and optimization of carbon trading mechanism for renewable energy application in buildings. *Renew. Sustain. Energy Rev.* **2017**, *73*, 435–451. [CrossRef]

155. Yuan, X.; Wang, X.; Zuo, J. Renewable energy in buildings in China—A review. *Renew. Sustain. Energy Rev.* **2013**, *24*, 1–8. [CrossRef]

156. Zhou, Y. Evaluation of renewable energy utilization efficiency in buildings with exergy analysis. *Appl. Therm. Eng.* **2018**, *137*, 430–439. [CrossRef]

157. Chen, X.; Yang, H.; Zhang, W. Simulation-based approach to optimize passively designed buildings: A case study on a typical architectural form in hot and humid climates. *Renew. Sustain. Energy Rev.* **2018**, *82*, 1712–1725. [CrossRef]

158. Kovacic, I.; Reisinger, J.; Honic, M. Life Cycle Assessment of embodied and operational energy for a passive housing block in Austria. *Renew. Sustain. Energy Rev.* **2018**, *82*, 1774–1786. [CrossRef]

159. Gourlis, G.; Kovacic, I. Passive measures for preventing summer overheating in industrial buildings under consideration of varying manufacturing process loads. *Energy* **2017**, *137*, 1175–1185. [CrossRef]

160. Beccali, M.; Strazzeri, V.; Germanà, M.L.; Melluso, V.; Galatioto, A. Vernacular and bioclimatic architecture and indoor thermal comfort implications in hot-humid climates: An overview. *Renew. Sustain. Energy Rev.* **2018**, *82*, 1726–1736. [CrossRef]

161. Ricciu, R.; Besalduch, L.A.; Galatioto, A.; Ciulla, G. Thermal characterization of insulating materials. *Renew. Sustain. Energy Rev.* **2018**, *82*, 1765–1773. [CrossRef]

162. Horvat, I.; Dović, D. Dynamic modeling approach for determining buildings technical system energy performance. *Energy Convers. Manag.* **2016**, *125*, 154–165. [CrossRef]

163. Kovacic, I.; Summer, M.; Achammer, C. Strategies of building stock renovation for ageing society. *J. Clean. Prod.* **2015**, *88*, 349–357. [CrossRef]

164. Kovacic, I.; Zoller, V. Building life cycle optimization tools for early design phases. *Energy* **2015**, *92*, 409–419. [CrossRef]

165. Gourlis, G.; Kovacic, I. Building Information Modelling for analysis of energy efficient industrial buildings—A case study. *Renew. Sustain. Energy Rev.* **2017**, *68*, 953–963. [CrossRef]

166. Beccali, M.; Ciulla, G.; Lo Brano, V.; Galatioto, A.; Bonomolo, M. Artificial neural network decision support tool for assessment of the energy performance and the refurbishment actions for the non-residential building stock in Southern Italy. *Energy* **2017**, *137*, 1201–1218. [CrossRef]

167. Galatioto, A.; Ciulla, G.; Ricciu, R. An overview of energy retrofit actions feasibility on Italian historical buildings. *Energy* **2017**, *137*, 991–1000. [CrossRef]

168. Garmston, H.; Pan, W. Non-Compliance with Building Energy Regulations: The Profile, Issues, and Implications on Practice and Policy in England and Wales. *J. Sustain. Dev. Energy Water Environ. Syst.* **2013**, *1*, 340–351. [CrossRef]

169. Kumar, A. Building Regulations Related to Energy and Water in Indian Hill Towns. *J. Sustain. Dev. Energy Water Environ. Syst.* **2017**, *5*, 496–508. [CrossRef]

170. Stutterecker, W.; Blümel, E. Energy plus standard in buildings constructed by housing associations? *Energy* **2012**, *48*, 56–65. [CrossRef]

171. Filogamo, L.; Peri, G.; Rizzo, G.; Giaccone, A. On the classification of large residential buildings stocks by sample typologies for energy planning purposes. *Appl. Energy* **2014**, *135*, 825–835. [CrossRef]

172. Gil-Baez, M.; Barrios-Padura, Á.; Molina-Huelva, M.; Chacartegui, R. Natural ventilation systems in 21st-century for near zero energy school buildings. *Energy* **2017**, *137*, 1186–1200. [CrossRef]

173. Herrando, M.; Cambra, D.; Navarro, M.; de la Cruz, L.; Millán, G.; Zabalza, I. Energy Performance Certification of Faculty Buildings in Spain: The gap between estimated and real energy consumption. *Energy Convers. Manag.* **2016**, *125*, 141–153. [CrossRef]

174. Chung, M.; Park, H.-C. Comparison of building energy demand for hotels, hospitals, and offices in Korea. *Energy* **2015**, *92*, 383–393. [CrossRef]

175. van den Dobbelsteen, A.; Martin, C.; Keeffe, G.; Pulselli, R.; Vandevyvere, H. From Problems to Potentials—The Urban Energy Transition of Gruž, Dubrovnik. *Energies* **2018**, *11*, 922. [CrossRef]

176. Sousa, J.; Martins, A. Portuguese Plan for Promoting Efficiency of Electricity End-Use: Policy, Methodology and Consumer Participation. *Energies* **2018**, *11*, 1137. [CrossRef]

177. Kılkış, B.; Kılkış, Ş. Hydrogen Economy Model for Nearly Net-Zero Cities with Exergy Rationale and Energy-Water Nexus. *Energies* **2018**, *11*, 1226. [CrossRef]

178. Doračić, B.; Novosel, T.; Pukšec, T.; Duić, N. Evaluation of Excess Heat Utilization in District Heating Systems by Implementing Levelized Cost of Excess Heat. *Energies* **2018**, *11*, 575. [CrossRef]

179. Tańczuk, M.; Masiukiewicz, M.; Anweiler, S.; Junga, R. Technical Aspects and Energy Effects of Waste Heat Recovery from District Heating Boiler Slag. *Energies* **2018**, *11*, 796. [CrossRef]

180. Gimelli, A.; Muccillo, M. The Key Role of the Vector Optimization Algorithm and Robust Design Approach for the Design of Polygeneration Systems. *Energies* **2018**, *11*, 821. [CrossRef]

181. Kazagic, A.; Hodzic, N.; Metovic, S. Co-Combustion of Low-Rank Coal with Woody Biomass and Miscanthus: An Experimental Study. *Energies* **2018**, *11*, 601. [CrossRef]

182. Eder, L.; Ban, M.; Pirker, G.; Vujanovic, M.; Priesching, P.; Wimmer, A. Development and Validation of 3D-CFD Injection and Combustion Models for Dual Fuel Combustion in Diesel Ignited Large Gas Engines. *Energies* **2018**, *11*, 643. [CrossRef]

183. Guimarães, C.; Maia, D.; Serra, E. Construction of Biodigesters to Optimize the Production of Biogas from Anaerobic Co-Digestion of Food Waste and Sewage. *Energies* **2018**, *11*, 870. [CrossRef]

184. Tic, W.; Guziałowska-Tic, J.; Pawlak-Kruczek, H.; Woźnikowski, E.; Zadorożny, A.; Niedźwiecki, Ł.; Wnukowski, M.; Krochmalny, K.; Czerep, M.; Ostrycharczyk, M.; et al. Novel Concept of an Installation for Sustainable Thermal Utilization of Sewage Sludge. *Energies* **2018**, *11*, 748. [CrossRef]

185. Klimeš, L.; Charvát, P.; Hejčík, J. Comparison of the Energy Conversion Efficiency of a Solar Chimney and a Solar PV-Powered Fan for Ventilation Applications. *Energies* **2018**, *11*, 912. [CrossRef]

186. Cottafava, D.; Sonetti, G.; Gambino, P.; Tartaglino, A. Explorative Multidimensional Analysis for Energy Efficiency: DataViz versus Clustering Algorithms. *Energies* **2018**, *11*, 1312. [CrossRef]

187. Oluleye, G.; Allison, J.; Kelly, N.; Hawkes, A. An Optimisation Study on Integrating and Incentivising Thermal Energy Storage (TES) in a Dwelling Energy System. *Energies* **2018**, *11*, 1095. [CrossRef]

188. Li, S.; Zhou, L.; Yang, J.; Wang, Q. Numerical Simulation of Flow and Heat Transfer in Structured Packed Beds with Smooth or Dimpled Spheres at Low Channel to Particle Diameter Ratio. *Energies* **2018**, *11*, 937. [CrossRef]

189. Sun, C.; Zhang, X.; Xi, L.; Tian, Y. Design of a Path-Tracking Steering Controller for Autonomous Vehicles. *Energies* **2018**, *11*, 1451. [CrossRef]

190. Borjigin, S.; Ma, T.; Zeng, M.; Wang, Q. A Numerical Study of Small-Scale Longitudinal Heat Conduction in Plate Heat Exchangers. *Energies* **2018**, *11*, 1727. [CrossRef]

*Article*

# A Numerical Study of Small-Scale Longitudinal Heat Conduction in Plate Heat Exchangers

**Saranmanduh Borjigin, Ting Ma, Min Zeng and Qiuwang Wang \***

Key Laboratory of Thermo-Fluid Science and Engineering, MOE, Xi'an Jiaotong University, Xi'an 710049, China; sarenmanduhu@163.com (S.B.); mating715@mail.xjtu.edu.cn (T.M.); zengmin@mail.xjtu.edu.cn (M.Z.)
* Correspondence: wangqw@mail.xjtu.edu.cn; Tel.: +86-29-8266-3502

Received: 22 May 2018; Accepted: 22 June 2018; Published: 2 July 2018

**Abstract:** Longitudinal heat conduction has a significant effect on the heat transfer performance of plate heat exchangers, but longitudinal heat conduction is usually neglected in numerical studies and the thermal design of a heat exchanger. In this paper, heat transfer models with and without longitudinal heat conduction are proposed to analyze the effect of small-scale longitudinal heat conduction in a plate heat exchanger. The performance of small-scale longitudinal heat conduction is illustrated by temperature and heat flux contours in the heat transfer models with and without longitudinal heat conduction. The results show that small-scale longitudinal heat conduction occurs in the plate and a more uniform temperature profile of the plate is obtained due to small-scale longitudinal heat conduction. In balanced flow, the contributions of longitudinal heat conduction for counter-flow, cross-flow and parallel-flow plate heat exchangers are −3.15%, −0.09% and 0, respectively, whereas, for the respective unbalanced flows they are evaluated to be −1.73%, 0.53% and 0.05%, respectively. Moreover, it is observed that small-scale longitudinal heat conduction in plates is influenced by the thermal conductivity of the plate. The higher the thermal conductivity, the larger is the reduction of thermal performance. The contribution of longitudinal heat conduction varies from −0.54% to −4.01%.

**Keywords:** plate heat exchanger; longitudinal heat conduction; heat transfer model; heat flux distribution

---

## 1. Introduction

In a surface heat exchanger, heat is transferred from the hot fluid to the cold fluid through a solid wall [1]. The log-mean temperature difference (LMTD) method and the number of heat transfer unit (NTU) method are always used for heat exchanger design. Longitudinal heat conduction (LHC), which is parallel to the solid walls, is usually neglected in these methods. The reason is that the effect of LHC is ignored in both original LMTD method and NTU method at the beginning. The computational fluid dynamics (CFD) simulation of heat exchanger is helpful for studying thermal-hydraulic performance [2]. LHC in the wall is usually neglected in the CFD simulations even though its effect is significant. LHC was neglected because the CFD simulation of heat exchanger usually just considers the fluid domain. However, the conjugate heat transfer model of the solid wall and fluid domain contains LHC in the solid wall.

Many researchers have studied the effect of LHC in the solid wall of different kinds of heat exchangers for special conditions. Ranganayakulu et al. [3] measured the heat transfer characteristics of plate-fin heat exchangers by the single-blow transient testing technique. The experimental results were 20% lower than their CFD results [4] due to deviation in the boundary conditions and LHC in the test core solid wall along the flow direction. In earlier studies [5–7], LHC, flow non-uniformity and temperature non-uniformity effects on the heat transfer characteristics of plate-fin heat exchangers were studied. They found that LHC in the wall had significant effects in cross-flow and counter-flow heat exchangers and the effect of LHC was negligible for the parallel-flow heat exchanger. They also found that heat transfer performance was deteriorated quite significantly by the combined effects of LHC, flow non-uniformity and temperature non-uniformity.

The effects of LHC on the inner and outer walls of the perforated plate matrix heat exchanger (MHE) were studied by Venkatarathnam and Narayanan [8,9]. There is only one inner wall in MHE. They found that the effectiveness of MHE was largely controlled by LHC through both the walls and the performance degradation due to LHC in the inner wall was more severe. Raju et al. [10] found that LHC through the outer wall could reduce effective NTU of a MHE significantly. The performance deterioration was more severe at low mass flow rates or at high NTU operations. Later [11] they presented an expression for effective NTU considering unbalanced flows and LHC through the both inner and outer walls.

Arici [12] examined the conjugate heat transfer of both parallel-flow and counter-flow concentric tube heat exchangers. The numerical results indicated that the effect of LHC was non-negligible only for counter-flow conditions especially for low heat capacity ratio.

For cross-flow indirect evaporative air coolers, Heidarinejad et al. [13] got more precise results by using a model with consideration of LHC through the wall than with the numerical results which ignore LHC. Madhawa et al. [14] found that the deterioration in thermal performance of the evaporative coolers due to LHC through the wall could be as high as 10%.

LHC has significant effect on the flow in microchannels. Lin and Kandlikar [15] developed a model to analyze the effect of LHC on heat transfer during single-phase flow in microchannels. The results showed for any tube material the effect of LHC in the wall was severe for gas flow and for water the effect of LHC in the wall was not negligible only for high conductivity channel material or thick channel wall. Numerical results of Rahimi and Mehryar [16] showed that LHC in the duct wall caused a reduction in the local Nusselt number at the entrance and also a deviation in the local Nusselt number at the ending regions of the microchannel. Maranzana et al. [17] proposed a non-dimensional axial conduction (longitudinal heat conduction) number to quantify the effect of LHC. They suggested that the effect of LHC could be neglected when the axial conduction number is less than 0.01. Lin et al. [18] studied the effect of LHC for laminar flow in a circular tube by three evaluation criterion parameters i.e., the axial conduction number, the modified axial conduction number and the temperature gradient number. They found that, for constant outside wall temperature boundary condition, the effect of LHC was non-negligible for both axial conduction numbers were below 0.01 and at this condition the temperature gradient number could be used to quantify the effect of LHC.

Plate heat exchangers are usually applied for liquid to liquid heat exchange [19]. Miao et al. [20] studied the heat transfer performance of plate heat exchanger by using the grey-box dynamic model. The simulation results predicted by grey-box method were more suitable and accurate than the results obtained by the white-box method. Ciofalo et al. [21] numerically simulated crossed-corrugated (CC) geometry with various computational approaches and the results were compared with the experimental measurements. The best agreements with the measured results were obtained by using both low Reynolds number *k-ε* model (LRKE) and large eddy simulation (LES). Rogiers and the co-workers [22,23] downsized both counter-flow and cross-flow plate heat exchanger without losing thermal-hydraulic performance. They found that when plate heat exchanger was downsized at a constant pressure drop the effectiveness exhibited a maximum due to LHC.

Plate heat exchangers developed for gas to gas heat exchange were reviewed by Wang et al. [1]. Ciofalo [24] separated the large-scale (end-to-end) LHC from the small-scale (local) LHC and studied the effect of small-scale LHC in the plate. The author developed a network of resistance to evaluate the effect of LHC in plate heat exchangers and compared the results with CFD predictions. However, most of the above studies focused on large-scale LHC and only the effect of LHC on thermal performance of the heat exchanger was studied. Doo et al. [25] investigated the effect of LHC in the plate of CC primary surface heat exchanger. The numerical model of CC unit cell was used to study the thermal performance including LHC. They evaluated the effect of LHC by theoretical method and the theoretically predicted results matched to that of the numerical simulations. Ma et al. [26,27] studied small-scale LHC in the plate of cross-wavy (CW) primary surface heat exchangers. They calculated the contribution of longitudinal heat conduction by using two different heat transfer models. They found that the contribution of longitudinal heat conduction changed with the flow conditions and parameters of the CW primary surface heat exchanger. They also found that the variation of some parameters made the contribution of longitudinal heat conduction below zero. It means that small-scale LHC degraded the thermal performance of the CW heat exchanger, so the effects of small-scale (local) LHC either enhanced or degraded the thermal performance caused by the parameters of CW primary surface heat exchanger.

A network of resistance-based method [24] calculated the effect of small-scale LHC from computational result of heat transfer model with LHC as general heat transfer model and only a simple two-dimensional model could be analyzed. In this paper, alternative three-dimensional heat transfer models with and without LHC of a plate heat exchanger are proposed to study the small-scale LHC. The small-scale LHC in the plate is prevented in heat transfer model without LHC and occurs in the plate in heat transfer model with LHC. Not only the thermal performance of the plate heat exchanger is analyzed, but also the local thermal field is obtained easily in this study. Moreover, it is clear that the thermal conductivity of the plate has a significant effect on LHC in the plate and it is conveniently to study different plate materials in heat transfer models with and without LHC. Effects of small-scale LHC in cross-flow, counter-flow and parallel-flow plate heat exchangers for balanced flow and unbalanced flow are studied. Plate thermal conductivity is also considered.

## 2. Physical and Numerical Model

The physical model of this numerical study is a gas-to-gas plate heat exchanger, which is used in a cabinet cooling system. The computational domain includes one hot channel and one cold channel. The channels are separated by two solid plates, as shown in Figure 1. The hot channel is detached from the center and a periodic boundary is used between the detached surfaces. The size of plate is 50 mm × 50 mm with a thickness of 0.2 mm and the distance between plates is 2.5 mm. The computational models are established for cross-flow, counter-flow and parallel-flow plate heat exchangers.

**Figure 1.** Computational domain of plate heat exchanger.

Both hot fluid and cold fluid are considered as incompressible, turbulent and steady. In the Cartesian tensor system, the continuity, momentum and energy equations are described as follow [26,28]:

$$\frac{\partial}{\partial x_i}(\rho u_i) = 0, \tag{1}$$

$$\frac{\partial}{\partial x_j}(\rho u_i u_j) = -\frac{\partial p}{\partial x_i} + \frac{\partial}{\partial x_j}\left[\mu\left(\frac{\partial u_i}{\partial x_j} + \frac{\partial u_j}{\partial x_i} - \frac{2}{3}\delta_{ij}\frac{\partial u_k}{\partial x_k}\right)\right] + \frac{\partial}{\partial x_j}(-\rho\overline{u'_i u'_j}), \tag{2}$$

$$\frac{\partial}{\partial x_i}[u_i(\rho E + p)] = \frac{\partial}{\partial x_j}\left[\left(\lambda + \frac{c_p \mu_t}{Pr_t}\right)\frac{\partial T}{\partial x_j}\right], \tag{3}$$

where $E$ is the total enthalpy, $\mu_t$ is the turbulent viscosity, $Pr_t$ is the turbulent Prandtl number, $\overline{\rho u'_i u'_j}$ is defined as:

$$-\overline{\rho u'_i u'_j} = \mu_t\left(\frac{\partial u_i}{\partial x_j} + \frac{\partial u_j}{\partial x_i}\right) - \frac{2}{3}\left(\rho k + \mu_t \frac{\partial u_k}{\partial x_k}\right)\delta_{ij}, \tag{4}$$

Plate heat exchangers are always studied by CFD simulations and a good agreement with experimental results were obtained by using low Reynolds number k-ε model (LRKE) for both liquid [21] and gas [29]. LRKE is used in this study. In LRKE the turbulence kinetic energy $k$ and turbulence dissipation rate $\varepsilon$ are solved using following equations [26]:

$$\frac{\partial}{\partial t}(\rho k) + \frac{\partial}{\partial x_j}\left[\rho u_j k - \left(\mu + \frac{\mu_t}{\sigma_k}\right)\frac{\partial k}{\partial x_j}\right] = \\ \mu_t(P + P_B) - \rho\varepsilon - \frac{2}{3}\left(\mu_t\frac{\partial u_i}{\partial x_j} + \rho k\right)\frac{\partial u_i}{\partial x_j} + \mu_t P_{NL} \tag{5}$$

$$\frac{\partial}{\partial t}(\rho\varepsilon) + \frac{\partial}{\partial x_j}\left[\rho u_j\varepsilon - \left(\mu + \frac{\mu_t}{\sigma_\varepsilon}\right)\frac{\partial\varepsilon}{\partial x_j}\right] = \\ C_{\varepsilon_1}\frac{\varepsilon}{k}\left\{\mu_t(P + P_{NL} + P') - \frac{2}{3}\left(\mu_t\frac{\partial u_i}{\partial x_i} + \rho k\right)\frac{\partial u_i}{\partial x_i}\right\} + C_{\varepsilon_3}\frac{\varepsilon}{k}\mu_t P_B - \\ C_{\varepsilon_2}\left(1 - 0.3e^{-R_t^2}\right)\rho\frac{\varepsilon^2}{k} + C_{\varepsilon_4}\rho\varepsilon\frac{\partial u_i}{\partial x_i} \tag{6}$$

The governing equations are solved by using the finite volume method. The second upwind differential scheme and SIMPLE algorithm are applied.

The Reynolds number $Re_H$ based on channel height, i.e., distance between the plates, is always applied for dimpled plate channel and baseline flat plate channel in both numerical [30,31] and experimental studies [32]. The Reynolds number $Re_H$ used in this study is given as:

$$Re_H = \frac{\rho U H}{\mu}, \tag{7}$$

where $U$ is the area-averaged velocity in the channel.

The Nusselt number $Nu$ is defined as:

$$Nu = \frac{\bar{q}H}{\lambda(T_w - T_f)}, \tag{8}$$

where $\bar{q}$ is the average heat flux through the solid plate and the flow average temperature $T_f$ is:

$$T_f = \frac{T_{in} + T_{out}}{2}, \tag{9}$$

The Fanning friction factor $f$ is defined as:

$$f = \frac{\Delta p H}{2\rho U^2 L} \tag{10}$$

where $\Delta p$ is the pressure loss in the channel and $L$ is the channel length in flow direction.

## 3. Heat Transfer Models with and without Longitudinal Heat Conduction

Actual heat transfer process in a plate heat exchanger includes transverse heat conduction (THC) and longitudinal heat conduction (LHC), as shown in Figure 2. The plate absorbs the heat from the hot fluid and releases it to the cold fluid by THC. The heat transfer model with LHC as general heat transfer model includes both THC and LHC. The heat flux between fluids is mainly transferred by the THC, whereas in plate the heat flux is transferred by LHC, the LHC heat flux $q_{LHC}$ for every infinitesimal element can be defined as:

$$q_{LHC} = q_c - q_h, \tag{11}$$

where $q_c$ and $q_h$ are cold side heat flux and hot side heat flux, respectively.

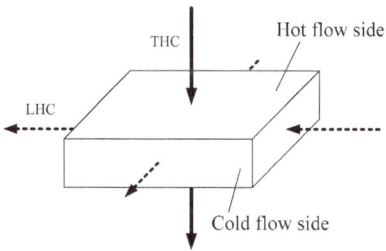

**Figure 2.** Actual heat transfer in plate.

The heat transfer model without LHC prevents small-scale LHC in the plate. The side surfaces of every infinitesimal element in the plate perpendicular to longitudinal directions are imposed adiabatic boundaries [26,27]. There is no LHC between infinitesimal elements, as shown in Figure 3.

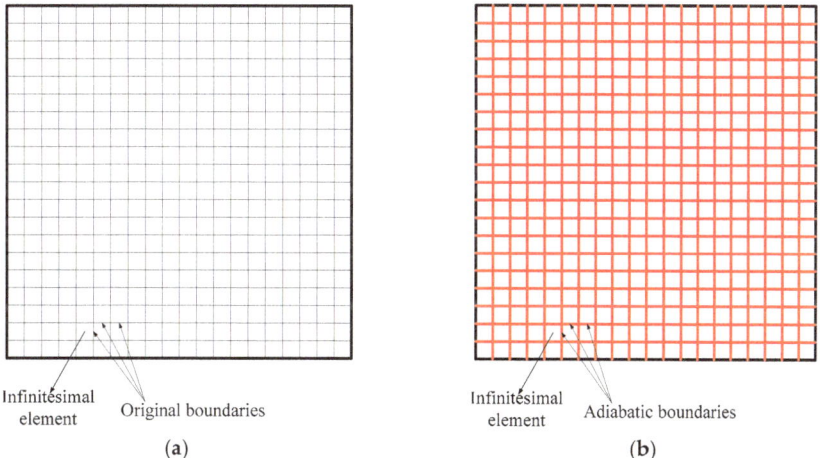

**Figure 3.** Boundaries of the plate (**a**) Heat transfer model with Longitudinal heat conduction (LHC); (**b**) Heat transfer model without LHC.

The heat transfer model without LHC is established to study the effect of small-scale LHC. The effect of small-scale LHC is clearly illustrated by temperature and heat flux contours of the heat transfer models.

In the present study, the contribution of longitudinal heat conduction $\eta$, which is obtained from the difference of average heat flux of the heat transfer models with LHC $\bar{q}_{wi}$ and without LHC $\bar{q}_{wo}$, is defined as:

$$\eta = \frac{\bar{q}_{wi} - \bar{q}_{wo}}{\bar{q}_{wi}}, \tag{12}$$

The contribution of longitudinal heat conduction $\eta$ can be used to describe the effect of small-scale LHC. If the value is above zero, the effect of small-scale LHC enhances the thermal performance of the heat exchanger. Else, the effect of small-scale LHC reduces the thermal performance of the heat exchanger.

## 4. Grid Independence and Code Validation

In this paper, the parameters in the heat transfer models with and without LHC for cross-flow, counter-flow and parallel-flow plate heat exchangers are the same. The heat transfer model with LHC of the cross-flow plate heat exchanger is used to test the grid independence. The average heat flux and the pressure loss of hot flow along with the cell number of the model are shown in Figure 4. The average heat flux and the pressure loss are stable when the cell number is greater than 0.4 million. Models with 0.52 million cells are chosen to get the stable results and save the computational time.

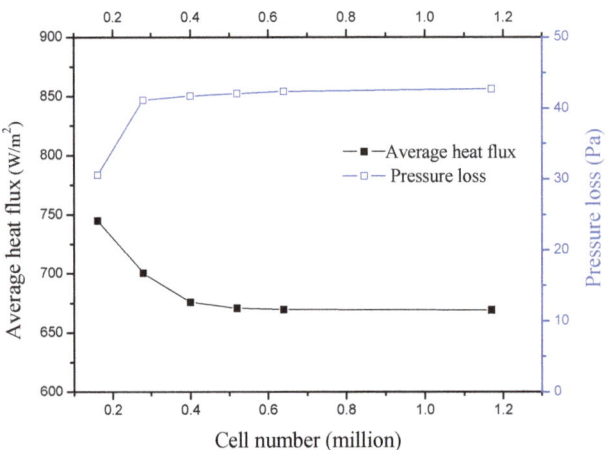

**Figure 4.** Grid independence.

The baseline Nusselt number $Nu_0$ and the Fanning friction factor $f_0$ in flat plate channel for transitional conditions are summarized by Zhang and Che [33], as follows:

$$\begin{cases} Nu_0 = 0.3726 \left( Re_H^{0.8} - 57.4 \right) Pr^{0.4} \left( \frac{T_f}{T_w} \right)^{0.45} \\ f_0 = 0.06651 Re_H^{-0.25} \end{cases} , 1150 \leq Re_H \leq 2500, \qquad (13)$$

They are used to validate the numerical code. The simulation results of this study for Reynolds number $Re_H$ from 1250 to 2430 show good agreements with both the baseline Nusselt number $Nu_0$ and the Fanning friction factor $f_0$. The maximum deviations of the Nusselt number and the Fanning friction factor are 4.9% and 9.7%, respectively. It validated that the present numerical method is reliable.

The distributions of heat flux on the plate surface of cross-flow plate heat exchanger are shown in Figures 5–7. It is shown that the absolute values of hot side heat flux $q_h$ distribution and cold side heat flux $q_c$ distribution of the heat transfer model with LHC are nearly symmetrical along the diagonal of the plate. The absolute values of hot side heat flux $q_h$ distribution and cold side heat flux $q_c$ distribution of the plate of the heat transfer model without LHC are the same, which means that only THC occurs in the plate. It is proved by LHC heat flux $q_{LHC}$ distribution in the plate of cross-flow plate heat exchanger of the heat transfer model without LHC as shown in Figure 7b. Compared with real LHC heat flux $q_{LHC}$ distribution of the heat transfer model with LHC, as shown in Figure 7a, the LHC heat flux $q_{LHC}$ in the heat transfer model without LHC is negligible. For counter-flow and parallel-flow plate heat exchangers, the heat transfer model without LHC acts the same. It is indicated that small-scale LHC is prevented in heat transfer model without LHC. So the effect of small-scale LHC is analyzed well by the heat transfer models with and without LHC.

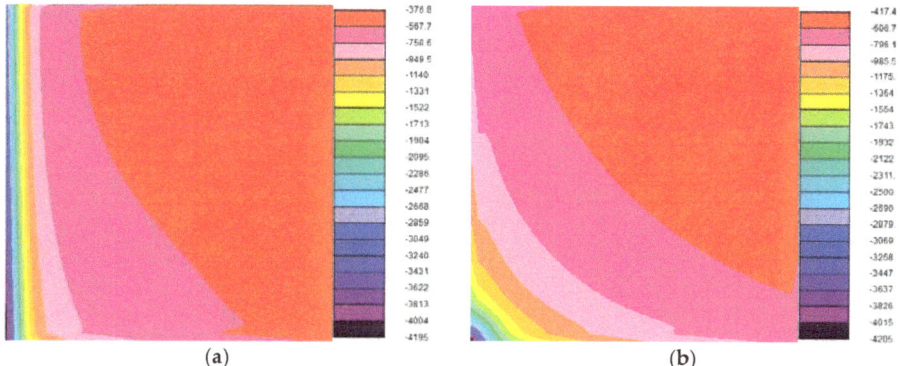

**Figure 5.** The plate hot side heat flux $q_h$ distributions of cross-flow plate heat exchanger (**a**) Heat transfer model with LHC; (**b**) Heat transfer model without LHC.

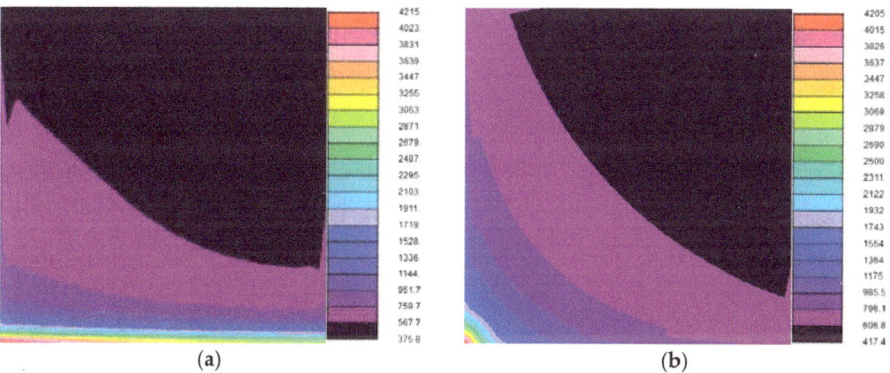

**Figure 6.** The plate cold side heat flux $q_c$ distributions of cross-flow plate heat exchanger (**a**) Heat transfer model with LHC; (**b**) Heat transfer model without LHC.

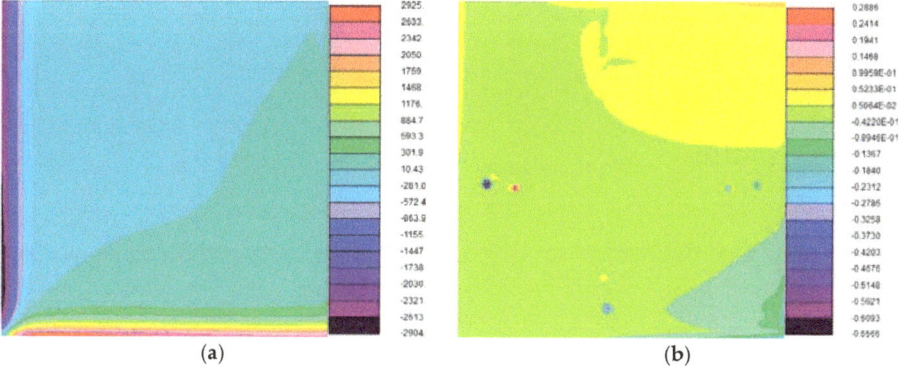

**Figure 7.** The plate LHC heat flux $q_{LHC}$ distributions of cross-flow plate heat exchanger (**a**) Heat transfer model with LHC; (**b**) Heat transfer model without LHC.

## 5. Results and Discussion

### 5.1. Effect of Small-Scale Longitudinal Heat Conduction on Balanced Flow

Balanced flow and unbalanced flow are always considered in the number of heat transfer unit (NTU) method for heat exchanger design [11]. Balanced flow means heat capacity rates of cold flow and hot flow are the same i.e., $C_c/C_h = 1$.

The small-scale LHC in the plate can be studied by the heat transfer models with and without LHC. The effect of LHC is clearly illustrated by the distributions of temperature and LHC heat flux. The temperature distributions of the plate in balanced flow for cross-flow and counter-flow plate heat exchangers are shown as Figures 8 and 9, respectively.

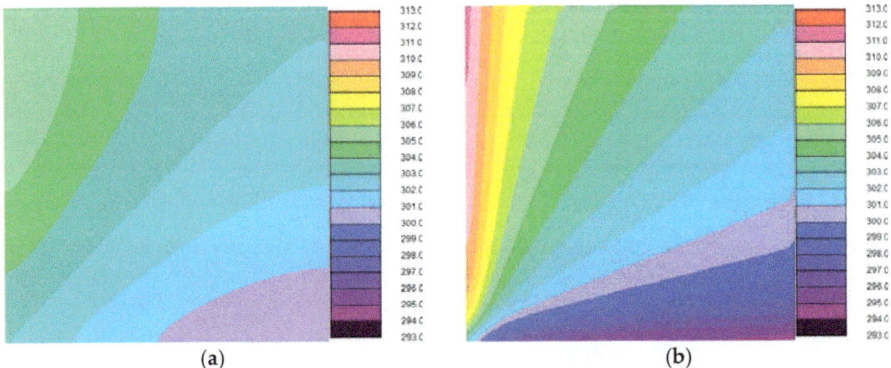

(a)                                                           (b)

**Figure 8.** The plate temperature distributions of cross-flow plate heat exchanger (**a**) Heat transfer model with LHC; (**b**) Heat transfer model without LHC.

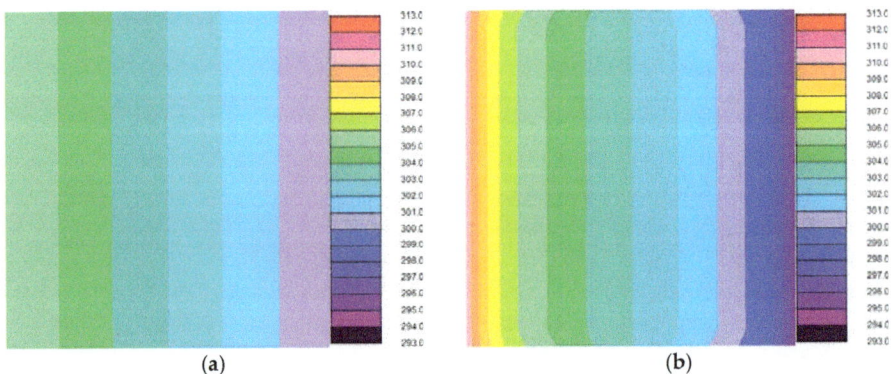

(a)                                                           (b)

**Figure 9.** The plate temperature distributions of counter-flow plate heat exchanger (**a**) Heat transfer model with LHC; (**b**) Heat transfer model without LHC.

For balanced flow, the average heat flux, contribution of longitudinal heat conduction, and plate temperature difference of different type of plate heat exchangers are summarized in Table 1. It is clear that a more uniform temperature profile of the plate is obtained due to small-scale LHC. The plate temperature differences in the cross-flow plate heat exchanger heat transfer models with and without LHC are 5.84 K and 16.44 K, respectively, and for counter-flow they are 5.37 K and 14.95 K. It means that in cross-flow and counter-flow plate heat exchangers non-negligible small-scale LHC occurs in the plate.

**Table 1.** Results of calculation for different heat exchangers in balanced flow: average heat flux $\bar{q}$; contribution of longitudinal heat conduction $\eta$; plate temperature difference.

| Type of Heat Exchanger | $\bar{q}$, W/m² | | $\eta$, % | Plate Temperature Difference, K | |
|---|---|---|---|---|---|
| | wi | wo | | wi | wo |
| Counter-flow | 680.67 | 702.13 | −3.15 | 5.37 | 14.95 |
| Cross-flow | 667.96 | 668.57 | −0.09 | 5.84 | 16.44 |
| Parallel-flow | 660.52 | 660.52 | 0 | 0.01 | 0.04 |

Actually, for balanced flow, small-scale LHC is very strong in both cross-flow and counter-flow plate heat exchangers. It is clearly shown from LHC heat flux distributions of the counter-flow and cross-flow plate heat exchangers heat transfer model with LHC, as indicated in Figures 10a and 11a, respectively. The contributions of longitudinal heat conduction $\eta$ of cross-flow and counter-flow plate heat exchangers are −0.09% and −3.15%, respectively. This means that, for balanced flow, the effect of small-scale LHC reduces the heat transfer performance of counter-flow plate heat exchanger. However, the small-scale LHC has little effect on the heat transfer performance of cross-flow plate heat exchanger in balanced flow.

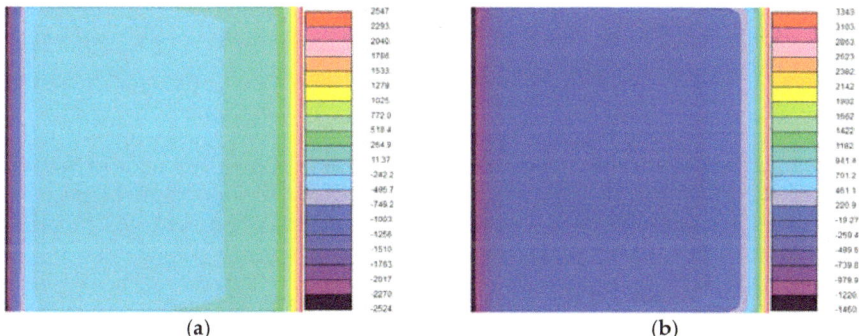

(a)  (b)

**Figure 10.** The plate LHC heat flux distributions of counter-flow plate heat exchanger heat transfer model with LHC (**a**) Balanced flow; (**b**) Unbalanced flow.

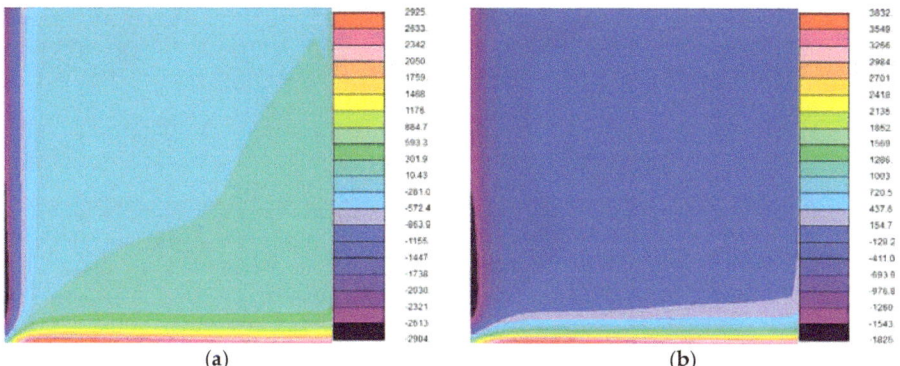

(a)  (b)

**Figure 11.** The plate LHC heat flux distributions of cross-flow plate heat exchanger heat transfer model with LHC (**a**) Balanced flow; (**b**) Unbalanced flow.

The plate temperature differences in parallel-flow plate heat exchanger heat transfer model with and without LHC are 0.01 K and 0.04 K, respectively. For balanced flow, LHC heat flux distribution of parallel-flow plate heat exchanger using heat transfer model with LHC is shown in Figure 12a. It observed that a very weak small-scale LHC occurs in the plate and small-scale LHC has no effect on the heat transfer performance of parallel-flow plate heat exchanger in balanced flow. Actually, the contribution of longitudinal heat conduction $\eta$ is zero in this condition.

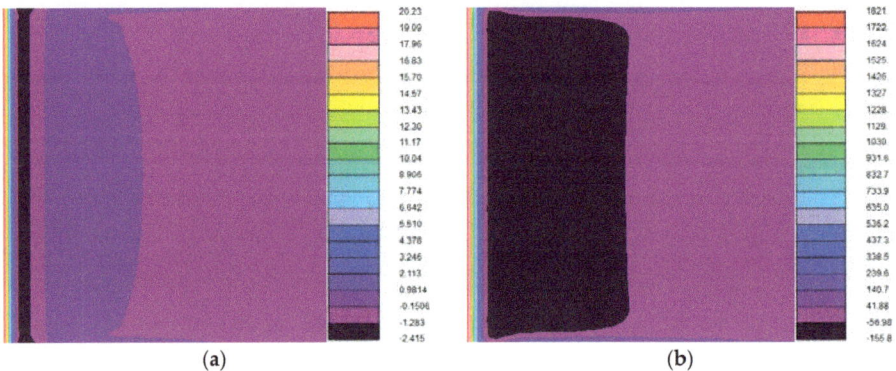

(a)                                          (b)

**Figure 12.** The plate LHC heat flux distributions of parallel-flow plate heat exchanger heat transfer model with LHC (a) Balanced flow; (b) Unbalanced flow.

### 5.2. Effect of Small-Scale Longitudinal Heat Conduction on Unbalanced Flow

In this section, two kinds of unbalanced flow are considered first. Their ratios of heat capacity rates are $C_c/C_h = 1.8$ and $C_h/C_c = 1.8$, respectively. It is found that small-scale LHC has similar effects in both conditions for counter-flow, cross-flow and parallel-flow plate heat exchangers, so, only one of them is considered in this study, for which the ratio of heat capacity rates is $C_h/C_c = 1.8$.

For unbalanced flow, the plate temperature difference in counter-flow plate heat exchanger heat transfer models with and without LHC is 5.53 K and 13.85 K, respectively. The former one is bigger than that in balanced flow and the latter one is smaller. It is indicated that small-scale LHC is weaker in unbalanced flow than that in balanced flow. It is clearly illustrated by the plate LHC heat flux distributions of counter-flow plate heat exchanger in Figure 10. For unbalanced flow, average heat flux; contribution of longitudinal heat conduction; plate temperature difference of different type of plate heat exchangers are summarized in Table 2. In this condition, the average heat flux of the plate for unbalanced flow and balanced flow are 869.495 W/m$^2$ and 680.67 W/m$^2$, respectively. Moreover, the variation of LHC heat flux for unbalanced flow and balanced flow are 4803 W/m$^2$ and 5071 W/m$^2$, respectively.

**Table 2.** Results of calculation for different heat exchangers in unbalanced flow: average heat flux $\bar{q}$; contribution of longitudinal heat conduction $\eta$; plate temperature difference.

| Type of Heat Exchanger | $\bar{q}$, W/m$^2$ | | $\eta$, % | Plate Temperature Difference, K | |
|---|---|---|---|---|---|
| | wi | wo | | wi | wo |
| Counter-flow | 869.495 | 884.515 | −1.73 | 5.53 | 13.85 |
| Cross-flow | 858.225 | 853.685 | 0.53 | 6 | 15.52 |
| Parallel-flow | 853.53 | 853.085 | 0.05 | 0.64 | 2.99 |

Similarly, for a cross-flow plate heat exchanger, the small-scale LHC is weaker in unbalanced flow than that in balanced flow. The plate LHC heat flux distributions are shown in Figure 11.

And, plate temperature difference in cross-flow plate heat exchanger heat transfer models with and without LHC are 6 K and 15.52 K, respectively.

Many researchers have pointed out that LHC degraded the heat transfer performance [6,9]. It is easily found that the heat transfer performance is greatly reduced by a strong LHC effect. In this study, a similar situation is obtained by using heat transfer model with LHC. It is indicated the weaker small-scale LHC achieves higher heat transfer performance, so less reduction of the heat transfer performance occurs for a counter-flow plate heat exchanger in the unbalanced flow as compared with the balanced flow. The contribution of longitudinal heat conduction $\eta$ for counter-flow heat exchanger in unbalanced flow and balanced flow are $-1.73\%$ and $-3.15\%$, respectively. For a cross-flow plate heat exchanger they are 0.53% and $-0.09\%$, respectively. For unbalanced flow, the heat transfer performance of cross-flow plate heat exchanger is enhanced slightly. For unbalanced flow, the LHC heat flux distribution of parallel-flow plate heat exchanger of heat transfer model with LHC is shown in Figure 12b. Small-scale LHC is very strong in this condition. However, the effect of small-scale LHC on the heat transfer performance is still very small. The contribution of longitudinal heat conduction $\eta$ is 0.05%.

## 5.3. Effect of Plate Thermal Conductivity on Small-Scale Longitudinal Heat Conduction

Both THC and LHC are strongly influenced by the thermal conductivity $\lambda_m$ of the plate material. Three different materials have been studied i.e., stainless-steel ($\lambda_m = 15.1$ W/(m·K)), brass ($\lambda_m = 109$ W/(m·K)) and silver ($\lambda_m = 427$ W/(m·K)). A counter-flow plate heat exchanger in balanced flow is used to study the effect of different materials.

For all the materials, the numerical results of the heat transfer model without LHC are the same. It is thus indicated that the thermal resistance in the transverse direction is negligible, even for stainless-steel due to the much lesser thickness of the plate, so the different materials have the same effect on the plate THC.

The silver plate heat exchanger achieves the strongest small-scale LHC due to its highest thermal conductivity $\lambda_m$. This is clearly shown in Figures 13a, 14a and 15a. The higher the thermal conductivity $\lambda_m$, the bigger is the plate LHC heat flux. As shown in Figures 13b, 14b and 15b, the stronger the small-scale LHC the more uniform a temperature profile of the plate is obtained. For different plate materials, the average heat flux, contribution of longitudinal heat conduction, and plate temperature difference of a counter-flow plate heat exchanger are summarized in Table 3.

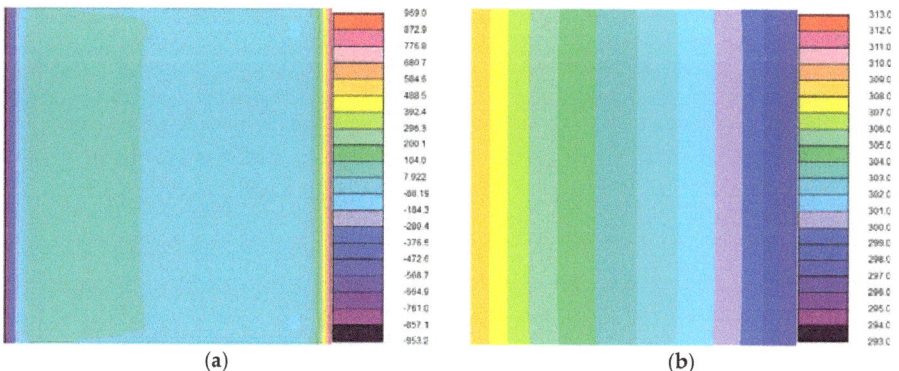

**Figure 13.** Counter-flow stainless-steel plate heat exchanger heat transfer model with LHC (a) The plate LHC heat flux distribution; (b) The plate temperature distribution.

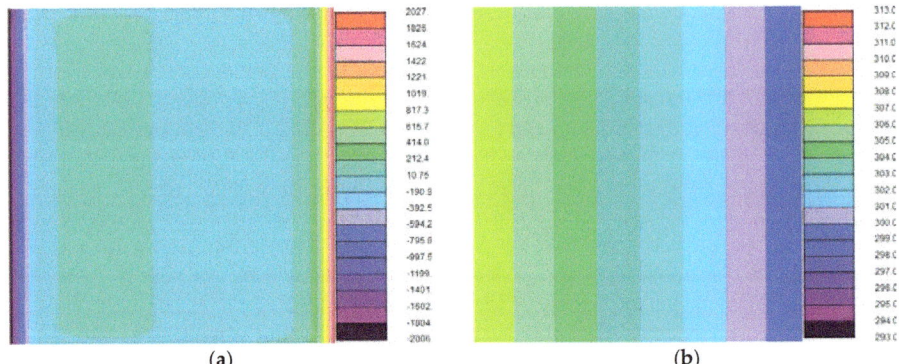

**Figure 14.** Counter-flow brass plate heat exchanger heat transfer model with LHC (**a**) The plate LHC heat flux distribution; (**b**) The plate temperature distribution.

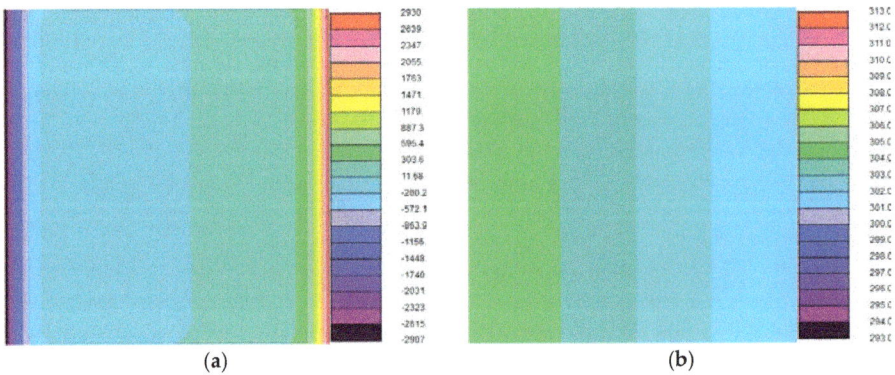

**Figure 15.** Counter-flow silver plate heat exchanger heat transfer model with LHC (**a**) The plate LHC heat flux distribution; (**b**) The plate temperature distribution.

**Table 3.** Results of calculation for different plate materials counter-flow plate heat exchanger in balanced flow: average heat flux $\bar{q}$; contribution of longitudinal heat conduction $\eta$; plate temperature difference.

| Plate Material/Thermal Conductivity, W/(m·K) | $\bar{q}$, W/m$^2$ | | $\eta$, % | Plate Temperature Difference, K | |
|---|---|---|---|---|---|
| | wi | wo | | wi | wo |
| Stainless-steel/15.1 | 698.04 | 701.81 | −0.54 | 11.55 | 14.94 |
| Brass/109 | 687.84 | 702.11 | −2.07 | 7.46 | 14.95 |
| Silver/427 | 675.05 | 702.145 | −4.01 | 3.82 | 14.95 |

The contributions of longitudinal heat conduction $\eta$ for stainless-steel, brass and silver under the same conditions are −0.54%, −2.07% and −4.01%, respectively. It is indicated that a plate with high thermal conductivity $\lambda_m$, such as silver plate, has a strong influence on LHC and reduces the thermal performance significantly.

## 6. Conclusions

The longitudinal heat conduction in plate heat exchangers is investigated in this paper. Heat transfer models with and without longitudinal heat conduction are established to study the

*Energies* **2018**, *11*, 1727

small-scale longitudinal heat conduction. The performance of small-scale longitudinal heat conduction is illustrated by temperature and heat flux contours in the heat transfer models with and without longitudinal heat conduction. The conclusions could be summarized as follows:

(1) Small-scale longitudinal heat conduction occurs in the plate and a more uniform temperature profile of the plate can be obtained due to small scale longitudinal heat conduction. In balanced flow, the contributions of longitudinal heat conduction of counter-flow and cross-flow plate heat exchangers are −3.15% and −0.09%, respectively. The effect of small-scale longitudinal heat conduction reduces the heat transfer performance significantly for counter-flow plate heat exchanger, whereas for cross-flow plate heat exchanger the heat transfer performance is reduced slightly. For the parallel-flow plate heat exchanger small-scale longitudinal heat conduction is very weak and the contribution of longitudinal heat conduction is zero. The small-scale longitudinal heat conduction has no effect on parallel-flow plate heat exchanger in balanced flow.

(2) In unbalanced flow, the small-scale longitudinal heat conduction is weakened for both counter-flow and cross-flow plate heat exchangers. The contributions of longitudinal heat conduction of these two heat exchangers are −1.73% and 0.53%, respectively. For counter-flow plate heat exchanger, the effect of small-scale longitudinal heat conduction reduces the heat transfer performance lesser than that in balanced flow, whereas the effect of small-scale longitudinal heat conduction enhance the heat transfer performance of cross-flow plate heat exchangers. For parallel-flow plate heat exchanger, the small-scale longitudinal heat conduction is strengthened significantly in unbalanced flow. However, the contribution of longitudinal heat conduction is only 0.05%.

(3) The small-scale longitudinal heat conduction is influenced by thermal conductivity of the plate. The contributions of longitudinal heat conduction for counter-flow stainless-steel, brass and silver plate heat exchangers in balanced flow are −0.54%, −2.07% and −4.01%, respectively. The higher the thermal conductivity of the plate, the stronger the small-scale longitudinal heat conduction and the larger the thermal performance reduction.

**Author Contributions:** S.B. proposed the model, performed simulation and wrote the paper; T.M. participated in analysis and revised the paper; M.Z. contributed to writing and revising the paper; Q.W. supervised the work and revised the paper. All authors contributed to this work.

**Funding:** This research was funded by State Key Program of National Natural Science of China (Grant No. 51536007) and the Foundation for Innovative Research Groups of the National Natural Science Foundation of China (Grant No. 51721004).

**Conflicts of Interest:** The authors declare no conflict of interest.

## Nomenclature

| | |
|---|---|
| $C$ | heat capacity rate, W/K |
| $c_p$ | specific heat, J/(kg·K) |
| $f$ | Fanning friction factor |
| $f_0$ | baseline Fanning friction factor |
| $H$ | channel height |
| $k$ | turbulence kinetic energy |
| $L$ | plate length, $m^2/s^2$ |
| $Nu$ | Nusselt number |
| $Nu_0$ | baseline Nusselt number |
| $p$ | static pressure |
| $Pr$ | Prandtl number |
| $Pr_t$ | turbulent Prandtl number |
| $q$ | heat flux, $W/m^2$ |
| $q_{LHC}$ | longitudinal heat conduction heat flux, $W/m^2$ |
| $\bar{q}$ | average heat flux, $W/m^2$ |

| $Re_H$ | Reynolds number based on channel height |
|---|---|
| $T$ | temperature, K |
| $U$ | area-averaged velocity in inlet section, m/s |
| $u$ | velocity, m/s |

*Greek*

| $\eta$ | contribution of longitudinal heat conduction, % |
|---|---|
| $\lambda$ | fluid thermal conductivity, W/(m·K) |
| $\lambda_m$ | solid thermal conductivity, W/(m·K) |
| $\mu$ | dynamic viscosity, Pa·s |
| $\mu_t$ | turbulent viscosity |
| $\rho$ | density, kg/m$^3$ |

*Subscripts*

| i, j | 1, 2, 3 |
|---|---|
| c | cold side |
| h | hot side |
| f | flow |
| w | solid wall |
| in | inlet |
| out | outlet |
| wi | heat transfer model with longitudinal heat conduction |
| wo | heat transfer model without longitudinal heat conduction |

## References

1. Wang, Q.W.; Zeng, M.; Ma, T.; Du, X.P.; Yang, J.F. Recent development and application of several high-efficiency surface heat exchangers for energy conversion and utilization. *Appl. Energy* **2014**, *135*, 748–777. [CrossRef]
2. Bhutta, M.M.A.; Hayat, N.; Bashir, M.H.; Khan, A.R.; Ahmad, K.N.; Khan, S. CFD applications in various heat exchangers design: A review. *Appl. Therm. Eng.* **2012**, *32*, 1–12. [CrossRef]
3. Ranganayakulu, C.; Luo, X.; Kabelac, S. The single-blow transient testing technique for offset and wavy fins of compact plate-fin heat exchangers. *Appl. Therm. Eng.* **2017**, *111*, 1588–1595. [CrossRef]
4. Ranganayakulu, C.; Pallavi, P. Development of heat transfer coefficient and friction factor correlations for offset fins using CFD. *Int. J. Numer. Methods Heat Fluid Flow* **2011**, *21*, 935–951.
5. Ranganayakulu, C.; Seetharamu, K.N.; Sreevatsan, K.V. The effects of longitudinal heat conduction in compact plate-fin and tube-fin heat exchangers using a finite element model. *Int. J. Heat Mass Transf.* **1997**, *40*, 1261–1277. [CrossRef]
6. Ranganayakulu, C.; Seetharamu, K.N. The combined effects of longitudinal heat conduction, flow nonuniformity and temperature nonuniformity in crossflow plate-fin heat exchangers. *Int. Commun. Heat Mass Transf.* **1999**, *26*, 669–678. [CrossRef]
7. Ranganayakulu, C.; Seetharamu, K.N. The combined effects of wall longitudinal heat conduction and inlet fluid flow maldistribution in crossflow plate-fin heat exchangers. *Heat Mass Transf.* **2000**, *36*, 247–256. [CrossRef]
8. Venkatarathnam, G.; Narayanan, S.P. Performance of a counter flow heat exchanger with longitudinal heat conduction through the wall separating the fluid streams from the environment. *Cryogenics* **1999**, *39*, 811–819. [CrossRef]
9. Narayanan, S.P.; Venkatarathnam, G. Performance degradation due to longitudinal heat conduction in very high NTU counterflow heat exchangers. *Cryogenics* **1998**, *38*, 927–930. [CrossRef]
10. Kumar, S.S.; Raju, L.R.; Nandi, T.K. Thermal performance of perforated plate matrix heat exchangers with effects from outer wall and flow channel geometry. *Cryogenics* **2015**, *72*, 153–160. [CrossRef]
11. Raju, L.R.; Nandi, T.K. Effective NTU of a counterflow heat exchanger with unbalanced flow and longitudinal heat conduction through fluid separating and outer walls. *Appl. Therm. Eng.* **2017**, *112*, 1172–1177. [CrossRef]
12. Arici, M.E. Heat transfer analysis for a concentric tube heat exchanger including the wall axial conduction. *Heat Transf. Eng.* **2010**, *31*, 1034–1041. [CrossRef]

13. Heidarinejad, G.; Moshari, S. Novel modeling of an indirect evaporative cooling system with cross-flow configuration. *Energy Build.* **2015**, *92*, 351–362. [CrossRef]
14. Hettiarachchi, H.D.M.; Golubovic, M.; Worek, W.M. The effect of longitudinal heat conduction in cross flow indirect evaporative air coolers. *Appl. Therm. Eng.* **2007**, *27*, 1841–1848. [CrossRef]
15. Lin, T.Y.; Kandlikar, S.G. A theoretical model for axial heat conduction effects during single-phase flow in microchannels. *J. Heat Transf.* **2012**, *134*, 020902. [CrossRef]
16. Rahimi, M.; Mehryar, R. Numerical study of axial heat conduction effects on the local Nusselt number at the entrance and ending regions of a circular microchannel. *Int. J. Therm. Sci.* **2012**, *59*, 87–94. [CrossRef]
17. Maranzana, G.; Perry, I.; Maillet, D. Mini- and micro-channels: Influence of axial conduction in the walls. *Int. J. Heat Mass Transf.* **2004**, *47*, 3993–4004. [CrossRef]
18. Lin, M.; Wang, Q.W.; Guo, Z.X. Investigation on evaluation criteria of axial wall heat conduction under two classical thermal boundary conditions. *Appl. Energy* **2016**, *162*, 1662–1669. [CrossRef]
19. Abu-Khader, M.M. Plate heat exchangers: Recent advances. *Renew. Sustain. Energy Rev.* **2012**, *16*, 1883–1891. [CrossRef]
20. Miao, Q.W.; You, S.J.; Zheng, W.D.; Zheng, X.J.; Zhang, H.; Wang, Y.R. A grey-box dynamic model of plate heat exchangers used in an urban heating system. *Energies* **2017**, *10*, 1398. [CrossRef]
21. Ciofalo, M.; Stasiek, J.; Collins, M.W. Investigation of flow and heat transfer in corrugated passages—II. Numerical simulations. *Int. J. Heat Mass Transf.* **1996**, *39*, 165–192. [CrossRef]
22. Rogiers, F.; Baelmans, M. Towards maximal heat transfer rate densities for small-scale high effectiveness parallel-plate heat exchangers. *Int. J. Heat Mass Transf.* **2010**, *53*, 605–614. [CrossRef]
23. Buckinx, G.; Rogiers, F.; Baelmans, M. Thermal design and optimization of small-scale high effectiveness cross-flow heat exchangers. *Int. J. Heat Mass Transf.* **2013**, *60*, 210–220. [CrossRef]
24. Ciofalo, M. Local effects of longitudinal heat conduction in plate heat exchangers. *Int. J. Heat Mass Transf.* **2007**, *50*, 3019–3025. [CrossRef]
25. Doo, J.H.; Ha, M.Y.; Min, J.K.; Stieger, R.; Rolt, A.; Son, C. Theoretical prediction of longitudinal heat conduction effect in cross-corrugated heat exchanger. *Int. J. Heat Mass Transf.* **2012**, *55*, 4129–4138. [CrossRef]
26. Ma, T.; Zhang, J.; Borjigin, S.; Chen, Y.T.; Wang, Q.W.; Zeng, M. Numerical study on small-scale longitudinal heat conduction in cross-wavy primary surface heat exchanger. *Appl. Therm. Eng.* **2015**, *76*, 272–282. [CrossRef]
27. Borjigin, S.; Peng, Y.Y.; Ma, T.; Chen, Y.T.; Zeng, M.; Wang, Q.W. Parameter study on longitudinal heat conduction in a cross-wavy primary surface heat exchanger. In Proceedings of the 1st Thermal and Fluid Engineering Summer Conference, New York, NY, USA, 9–12 August 2015.
28. CD-adapco; Methodology, STAR-CD. *Version 4.16 CD-adapco*; CD-adapco: Melville, NY, USA, 2011.
29. Zhang, L.; Che, D.F. Turbulence models for fluid flow and heat transfer between cross-corrugated plates. *Numer. Heat Transf. A* **2011**, *60*, 410–440. [CrossRef]
30. Elyyan, M.A.; Rozati, A.; Tafti, D.K. Investigation of dimpled fins for heat transfer enhancement in compact heat exchangers. *Int. J. Heat Mass Transf.* **2008**, *51*, 2950–2966. [CrossRef]
31. Won, S.Y.; Ligrani, P.M. Numerical predictions of flow structure and local Nusselt number ratios along and above dimpled surfaces with different dimple depths in a channel. *Numer. Heat Transf. A* **2004**, *46*, 549–570. [CrossRef]
32. Ligrani, P.M.; Burgess, N.K.; Won, S.Y. Nusselt numbers and flow structure on and above a shallow dimpled surface within a channel including effects of inlet turbulence intensity level. *J. Turbomach.* **2005**, *127*, 321–330. [CrossRef]
33. Zhang, L.; Che, D.F. Influence of corrugation profile on the thermalhydraulic performance of cross-corrugated plates. *Numer. Heat Transf. A* **2011**, *59*, 267–296. [CrossRef]

*Article*

# Design of a Path-Tracking Steering Controller for Autonomous Vehicles

Chuanyang Sun [1], Xin Zhang [1,*], Lihe Xi [1] and Ying Tian [2]

[1] Beijing Key Laboratory of Powertrain for New Energy Vehicle, School of Mechanical, Electronic and Control Engineering, Beijing Jiaotong University, Beijing 100044, China; sunchuanyang@bjtu.edu.cn (C.S.); xilihe@bjtu.edu.cn (L.X.)

[2] Beijing Jiaotong University Yangtze River Delta Research Institute, Zhenjiang 212009, China; ytian1@bjtu.edu.cn

* Correspondence: zhangxin@bjtu.edu.cn; Tel.: +86-010-5168-8404

Received: 25 April 2018; Accepted: 1 June 2018; Published: 4 June 2018

**Abstract:** This paper presents a linearization method for the vehicle and tire models under the model predictive control (MPC) scheme, and proposes a linear model-based MPC path-tracking steering controller for autonomous vehicles. The steering controller is designed to minimize lateral path-tracking deviation at high speeds. The vehicle model is linearized by a sequence of supposed steering angles, which are obtained by assuming the vehicle can reach the desired path at the end of the MPC prediction horizon and stay in a steady-state condition. The lateral force of the front tire is directly used as the control input of the model, and the rear tire's lateral force is linearized by an equivalent cornering stiffness. The course-direction deviation, which is the angle between the velocity vector and the path heading, is chosen as a control reference state. The linearization model is validated through the simulation, and the results show high prediction accuracy even in regions of large steering angle. This steering controller is tested through simulations on the CarSim-Simulink platform (R2013b, MathWorks, Natick, MA, USA), showing the improved performance of the present controller at high speeds.

**Keywords:** autonomous vehicles; model linearization; path tracking; steering controller; model predictive control

## 1. Introduction

Autonomous vehicle technology aims to increase driving safety, reduce traffic congestion and emissions, and improve energy efficiency [1,2]. The ability to track the desired path accurately and steadily plays a critical role in the control task of an autonomous vehicle, especially when operating at high speeds. Therefore, a great deal of research has been done on steering control of autonomous vehicles.

Full vehicle models and nonlinear tire models are usually used to simulate the vehicle response during high speeds and large-steering-angle driving [3]. However, the nonlinearity of vehicle and tire models leads to a high computational burden [4]. A bicycle model with a small-angle assumption and a proportional linear tire model are widely used in path-tracking research [5,6]. However, when the steering angle and lateral slip angle are larger than $5°$, the model becomes inaccurate, especially in the region of tire-force saturation. Erlien et al. [7] introduced an affine approximation linearization method to handle the nonlinearity of the tires in the model predictive control (MPC) scheme, however, this approach is inaccurate when the length of the prediction horizon is larger. Talvala et al. [8] proposed a tire slip angle-related parameter to capture the nonlinearity even when the tire is saturated. In general, tuning the parameters of a given model to realistic values can be challenging work.

Heading deviation and lateral deviation are usually chosen as the control reference states, and multiple approaches are presented to eliminate them [9–11]. Brown et al. [5] presented a path-tracking controller, based on an MPC with different prediction horizons, which achieved stabilization and obstacle avoidance simultaneously. Kritayakirana et al. [6] used the center of percussion as a reference point for calculating the steering command, in order to minimize both the heading and lateral deviations. Their algorithm has lower complexity and can maintain stability even when the rear tires are saturated. Katriniok et al. [12] proposed a combined longitudinal and lateral optimal control algorithm for the collision-avoidance system. The robustness of their algorithm is demonstrated through experiment.

While there has been some success in these studies, choosing the heading deviation as a reference state may not effectively minimize the lateral deviation, especially when operating at high speeds. Mammar et al. [13] utilized the yaw-rate error instead of the path-tracking error as the feedback input to minimize the deviation from the path, and good robustness was demonstrated through their tests. Tagne et al. [14] presented an adaptive controller and used the steady-state sideslip and yaw rate to help bring the operating point to the desired equilibrium quickly. However, their controller is not capable of accurate path tracking in the tire-friction limit. Kapania et al. [15] pointed out that lateral deviations would be minimized when vehicle sideslip was tangential to the path, and hence designed a feedback controller to keep the vehicle velocity vector at the desired heading. However, a fixed look-ahead distance is not always optimal over a range of vehicle speeds.

The capability to systematically include system constraints and future predictions in the design procedure makes model predictive control (MPC) an attractive method in the control of autonomous vehicles, where vehicle stability constraints, as well as changing vehicle and tire dynamics, exist in the system [16–19]. While MPC seems promising, when combined with the nonlinear plant model, it still faces convergence and high online computational complexity issues, making it unsuitable under high-speed conditions. On the contrary, the unique global minimum of the linear MPC can be efficiently calculated by various methods in a limited number of iterations [20,21]. Several methods that combine MPC with a linear model were proposed in References [5,7,17]. Raffo et al. [18] presented an MPC path-tracking controller with a linear kinematic model to achieve the desired performance during high-speed driving. Beal et al. [19] combined an MPC with linear vehicle and tire dynamic models to stabilize the vehicle at the limits of handling.

In this paper, a methodology is proposed to allow one to linearize the nonlinearities of the vehicle and tire models under the MPC scheme, with some additional assumptions. A linear-model MPC path-tracking steering controller, using the direction deviation between the vehicle's velocity vector and the path heading as the control reference state, is designed. Basic information about the MPC scheme is introduced in Section 2. The linearization method for the models and the simulation method for validating our approach are presented in Section 3. In Section 4, the use of different control reference states is discussed and the control objective is determined. Additionally, our proposed MPC controller is presented, including the constraints and the optimization problem. In Section 5, the simulation results verify the efficacy of the presented controller under high speeds and large lateral-acceleration conditions. Finally, Section 6 concludes the paper with a brief discussion of the results.

## 2. MPC Algorithms

Before constructing the linearization method of the vehicle model using the prediction information, we first briefly introduce the MPC control scheme. The system to be controlled is described with a difference equation:

$$x^+ = f(x, u), \tag{1}$$

where $x$, $x^+$ and $u$ are the state, successor state, and control input of the system, respectively, subject to the constraints:

$$x \in \mathbb{X}, \ u \in \mathbb{U}, \tag{2}$$

where $\mathbb{X} \subset \mathbb{R}^n$ is a closed, convex set and $\mathbb{U} \subset \mathbb{R}^m$ is a compact, convex set. We employ $x(k)$ and $u(k)$ to denote the state and the control action at sampling time $k$.

In the MPC scheme, the future behavior of the system can be predicted using the plant model $f(\cdot)$. The control objective is to steer the state trajectory, $x^u(\cdot)$, to the desired state, $x_r$, over a finite prediction horizon, $N_P$, by applying the control sequence, $u(\cdot)$, to the system. Figure 1 shows an example control process of the MPC.

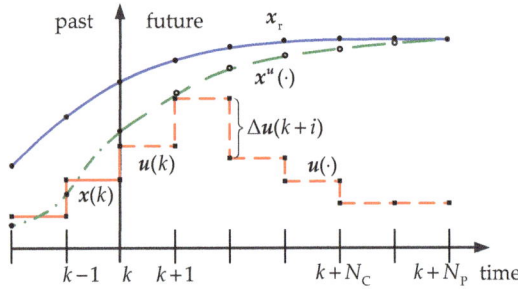

**Figure 1.** Control process of the model predictive control (MPC) algorithm.

We consider the typical cost function, $J_{N_C, N_P}(\cdot, \cdot)$, as defined by:

$$J_{N_C, N_P}(x^u(\cdot), \Delta u(\cdot)) = \sum_{i=0}^{N_P-1} (x(k+i) - x_r(k+i))^T Q(x(k+i) - x_r(k+i)) + \sum_{i=0}^{N_C-1} \Delta u(k+i)^T R \Delta u(k+i), \quad (3)$$

where $\Delta u(\cdot) = \{\Delta u(k), \Delta u(k+1), \dots, \Delta u(k+N_C-1)\}$ is the sequence of control input increments, $x^u(\cdot) = \{x(k+1), x(k+2), \dots, x(k+N_P)\}$, $N_C$ is the control horizon, with the constraint $N_C \leq N_P$, and $Q$ and $R$ are the weighting matrices.

At each sampling time $k$, MPC solves the following optimization problem:

$$\min_{\Delta u(\cdot), x^u(\cdot)} J_{N_C, N_P}(x(k), u(k-1), \Delta u(\cdot)), \quad (4)$$

$$\text{s.t. } x(k+i+1) = f(x(k+i), u(k+i)), \ i = 0, \dots, N_P - 1, \quad (5)$$

$$u(k+i) = u(k-1) + \sum_{j=0}^{i} \Delta u(k+j), \ i = 0, \dots, N_C - 1, \quad (6)$$

$$u(k+i) = u(k+N_C-1), \ i = N_C, \dots, N_P - 1, \quad (7)$$

$$\Delta u(k+i) \in \Delta \mathbb{U}, \ i = 0, \dots, N_C - 1, \quad (8)$$

$$u(k+i) \in \mathbb{U}, \ i = 0, \dots, N_C - 1, \quad (9)$$

$$x(k+i) \in \mathbb{X}, \ i = 1, \dots, N_P. \quad (10)$$

The optimal solution denoted by $\Delta u^*(\cdot)$ of Equations (4)–(10) is generated, and the control input is, therefore, defined by:

$$u(k) = u(k-1) + \Delta u^*(k). \quad (11)$$

Hence, the control $u(k)$ is applied to the system at time $k$. At the next sampling time, the optimization problem in Equations (4)–(10) is resolved over the shifted prediction horizon, and the process is thus repeated for every sampling time.

Solving the optimization problem of Equations (4)–(10) is a computationally demanding task. Furthermore, the system order, control horizon length $N_C$, and nonlinearities in the plant model $f(\cdot)$ are the main factors in determining the computational burden [22]. Usually, adequately lowering the system order and successively linearizing the process model to formulate a linear–quadratic optimization problem are efficient ways to improve computational efficiency.

## 3. Modelling

### 3.1. Nonlinear Model

#### 3.1.1. Vehicle Dynamic Model

The single-track 'bicycle' model, shown in Figure 2 with two speed states and three position states, can adequately capture the tracking performance and handling stability under various operating conditions. In this scheme, the front steering angle $\delta$ is the only actuation. To make the optimization problem of $\delta$ convex, longitudinal speed $U_x$ is not allowed to be variable. While an external speed controller could be used to track a desired speed profile, the speed over the prediction horizon is assumed to be fixed when building the process model.

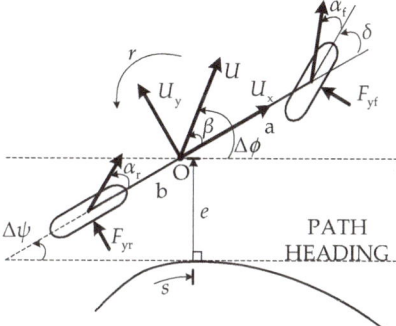

**Figure 2.** Bicycle model schematic. O is the center of mass; $U$ and $U_y$ are vehicle speed and the lateral speed at the center of gravity, respectively.

The sideslip $\beta$ and yaw rate $r$ are described by the equations of motion [23]:

$$\dot{\beta} = \frac{F_{yf}\cos(\delta) + F_{yr}}{mU_x} - r,$$
(12)

$$\dot{r} = \frac{aF_{yf}\cos(\delta) - bF_{yr}}{I_{zz}},$$
(13)

where $m$ and $I_{zz}$ are the vehicle mass and yaw inertia, respectively; $F_{y[f,r]}$ denotes the lateral tire force of the front and rear axle, respectively; and $a$ and $b$ are the distances from the center of mass O to the front and rear axles, respectively.

#### 3.1.2. Tire Model

The lateral tire force is modelled using the Fiala brush tire model [24]; $F_{y[f,r]} = f_{tire}(\alpha, F_z)$, where:

$$f_{tire}(\alpha, F_z) = \begin{cases} -C_\alpha \tan \alpha + \frac{C_\alpha^2}{3\mu F_z} |\tan \alpha| \tan \alpha - \frac{C_\alpha^3}{27\mu^2 F_z^2} \tan^3 \alpha, & |\alpha| < \arctan\left(\frac{3\mu F_z}{C_\alpha}\right) \\ -\mu F_z \text{sgn}\alpha, & \text{otherwise} \end{cases}.$$
(14)

In Equation (14), $\mu$ is the coefficient of friction, $F_z$ is the normal force, and $C_\alpha$ is the tire cornering stiffness. The tire slip angles $\alpha_f$ and $\alpha_r$, using small-angle approximations, can be expressed as:

$$\alpha_f = \beta + \frac{ar}{U_x} - \delta, \tag{15}$$

$$\alpha_r = \beta - \frac{br}{U_x}, \tag{16}$$

where $\delta$ is the front steer angle.

In order to simplify the nonlinear relationship between the actuation $\delta$ and the vehicle dynamic states while taking the saturation of the tire into account, the front lateral force $F_{yf}$ is considered as the control input of the model. The desired $F_{yf}$, generated by the MPC optimization, is then mapped to $\delta$ by:

$$\delta = \beta + \frac{ar}{U_x} - f_{tire}^{-1}(F_{yf}), \tag{17}$$

where $f_{tire}^{-1}(F_{yf})$ is the inverted tire model, which calculates the tire slip from the tire force via numerical methods.

### 3.1.3. Path-Tracking Model

The path-tracking model is shown in Figure 2, and the vehicle's relative position to the desired path can be determined by three state parameters: the lateral deviation $e$, the heading deviation $\Delta\psi$, and the distance $s$ along the path. The path tracking model can be written as [5]:

$$\Delta\dot{\psi} = r - U_x \kappa(s), \tag{18}$$

$$\dot{e} = U_x \sin(\Delta\psi) + U_y \cos(\Delta\psi), \tag{19}$$

$$\dot{s} = U_x \cos(\Delta\psi) - U_y \sin(\Delta\psi), \tag{20}$$

where $\kappa(s)$ is the curvature of the desired path at $s$.

### 3.2. Model Linearization

#### 3.2.1. Vehicle Dynamic Model

The most popular linearization method is the small-angle assumption (when $\delta < 5°$, $\cos(\delta) \approx 1$). The nonlinear model of Equations (12) and (13) can thus be expressed as:

$$\dot{\beta} \approx \frac{F_{yf} + F_{yr}}{mU_x} - r, \tag{21}$$

$$\dot{r} \approx \frac{aF_{yf} - bF_{yr}}{I_{zz}}. \tag{22}$$

However, when the vehicle tracks a path with high curvature, the steering angle can be very large, and the small-angle assumption is invalid. The model in Equations (21) and (22) will fail to simulate the vehicle response at these operational conditions.

Consider the following suppositions:

(1)   Suppose the steering angle increments are fixed at every step over the prediction horizon, and are independent of the control sequence $u(\cdot)$, that is:

$$\Delta\delta_a(k+i) = \Delta\delta_{N_P}, \ i = 0, \ldots, N_P - 1, \tag{23}$$

where $\Delta\delta_a(k+i)$ is the supposed steering angle increment at step $i + 1$ and $\Delta\delta_{N_P}$ is the corresponding fixed increment, which will be determined later.

(2) Suppose the vehicle will reach the desired path at step $N_P$ of the prediction horizon, and will subsequently track the path without deviation, as shown in Figure 3. Then, the vehicle at step $N_P$ is assumed to be in the steady state. Hence:

$$e_a(k + N_P) = 0, \tag{24}$$

$$\delta_a(k + N_P - 1) = \delta_{ss}(k + N_P - 1), \tag{25}$$

$$\alpha_{[f,r],a}(k + N_P - 1) = \alpha_{[f,r],ss}(k + N_P - 1), \tag{26}$$

where $e_a$, $\delta_a$ and $\alpha_{[f,r],a}$ are the supposed lateral deviation, steering angle, and tire slip angle, respectively, and $\delta_{ss}$, $\alpha_{[f,r],ss}$ are the steady-state steering angle and tire slip angle, respectively.

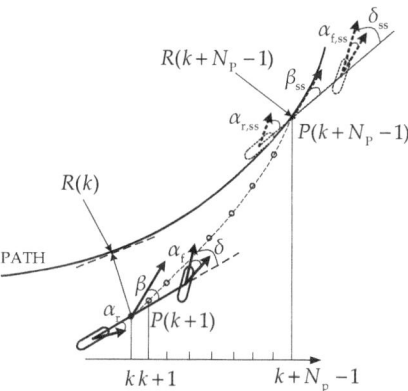

**Figure 3.** The supposed path over the prediction horizon for the linearized model. $R(k + i)$ and $P(k + i)$ denote the radius of the desired path and the supposed point at $s(k + i)$, respectively, $i = 0, \dots, N_P - 1$.

Under steady-state cornering conditions, setting $\dot{r} = 0$ in Equation (19), the front and rear tire forces are yielded:

$$F_{yf}^{ss} = \frac{mb}{2L} u_x^2 \kappa, \tag{27}$$

$$F_{yr}^{ss} = \frac{ma}{2L} u_x^2 \kappa. \tag{28}$$

Hence, the steady-state steering angle relates to the front and rear lateral tire slip by vehicle kinematics via:

$$\delta_{ss} = L\kappa - \alpha_{f,ss} + \alpha_{r,ss}, \tag{29}$$

where $L = a + b$ is the wheel base, and $\alpha_{f,ss}$ and $\alpha_{r,ss}$ can be calculated from Equations (14) and (27) and (28) by the inverted tire model $f_{tire}^{-1}(F_{y[f,r]})$.

Under the suppositions above, the supposed steering-angle increment can be written as:

$$\Delta\delta_a(k + i) = \frac{\delta(k) - \delta_a(k + N_P - 1)}{N_P}, \; i = 0, \dots, N_P - 1. \tag{30}$$

Finally, considering the slew-rate capabilities of the vehicle, the supposed steering-angle increment is determined via:

$$\Delta\delta_{N_P} = \begin{cases} \Delta\delta_a, \; |\Delta\delta_a| \le \Delta\delta_{max} \\ \text{sign}(\Delta\delta) \cdot \Delta\delta_{max}, \; \text{else} \end{cases}, \tag{31}$$

and the supposed steering angle for every step of the prediction horizon is thus:

$$\delta_a(k+i) = \delta(k) + i \cdot \Delta\delta_{N_P}, \ i = 0, \ldots, N_P - 1. \tag{32}$$

Combining Equation (32) with Equations (12) and (13) yields a linearized version of our nonlinear vehicle model:

$$\dot{\beta}(k+i) = \frac{F_{yf}(k+i) \cdot \cos(\delta_a(k+i)) + F_{yr}(k+i)}{mU_x} - r(k+i), \ i = 0, \ldots, N_P - 1, \tag{33}$$

$$\dot{r}(k+i) = \frac{aF_{yf}(k+i) \cdot \cos(\delta_a(k+i)) - bF_{yr}(k+i)}{I_{zz}}, \ i = 0, \ldots, N_P - 1. \tag{34}$$

### 3.2.2. Tire Model

The front lateral force is considered as the control input, and the nonlinear dynamics of the rear-tire force should be properly accounted for to accurately approximate the MPC with a linear optimization problem. Inspired by Reference [7], which assumes that the tire-cornering stiffness keeps constant in the prediction horizon, we propose a new online successive linearization method by combining both the information of time $k$ and step $N_P$ of the prediction horizon to accurately predict the propagation of lateral tire forces, as shown in Figure 4.

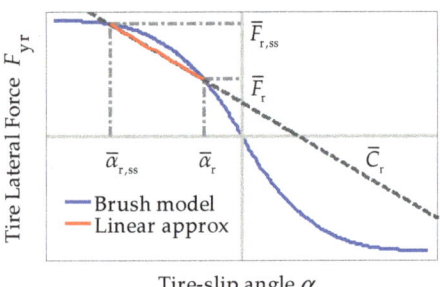

**Figure 4.** Linear approximation of the brush tire model. $\bar{\alpha}_r$ and $\bar{\alpha}_{r,ss}$ are the tire slip angle at time $k$ and steady-state tire slip angle at step $N_P$ of the prediction horizon, respectively. $\bar{F}_r$ and $\bar{F}_{r,ss}$ are the corresponding lateral forces.

In the linearized model, the equivalent cornering stiffness over the prediction horizon is:

$$\bar{C}_r = \frac{(\bar{F}_{r,ss} - \bar{F}_r)}{\bar{\alpha}_{r,ss} - \bar{\alpha}_r}. \tag{35}$$

Thus, the approximate expression for the predicted rear-tire lateral force is:

$$F_{yr}(k+i) = \bar{F}_r - \bar{C}_r(\alpha_r(k+i) - \bar{\alpha}_r), \ i = 0, \ldots, N_P - 1, \tag{36}$$

where $\alpha_r(k+i)$ is the predicted rear-tire slip angle which can be calculated by Equation (16).

The resulting linear expressions for the motion equations are described as follows:

$$\dot{\beta}(k+i) = \frac{F_{yf}(k+i) \cdot \cos(\delta_a(k+i)) + \left[\bar{F}_r - \bar{C}_r\left(\beta(k+i) - \dfrac{b \cdot r(k+i)}{U_x} - \bar{\alpha}_r\right)\right]}{mU_x} - r(k+i), \ i = 0, \ldots, N_P - 1, \tag{37}$$

$$\dot{r}(k+i) = \frac{aF_{yf}(k+i) \cdot \cos(\delta_a(k+i)) - b\left[\overline{F}_r - \overline{C}_r\left(\beta(k+i) - \frac{b\cdot r(k+i)}{U_x} - \overline{\alpha}_r\right)\right]}{I_{zz}}, \; i = 0,\ldots,N_P-1. \quad (38)$$

### 3.2.3. Tracking Model

Making the small-angle approximation for $\beta$ and $\Delta\psi$ yields:

$$\Delta\dot{\psi}(k+i) = r(k+i) - U_x\kappa(k+i), \; i = 0,\ldots,N_P-1, \quad (39)$$

$$\dot{e}(k+i) = U_x(\beta(k+i) + \Delta\psi(k+i)), \; i = 0,\ldots,N_P-1, \quad (40)$$

$$\dot{s}(k+i) = U_x, \; i = 0,\ldots,N_P-1. \quad (41)$$

Due to the assumption that the speed is fixed over the prediction horizon, the distance along the path can be given, a priori, as:

$$s(k+i) = s(k) + \sum_{j=0}^{i} U_x, \; i = 1,\ldots,N_P. \quad (42)$$

## 4. MPC Controller Design

### 4.1. Problem Statement

Unlike human driving, which cannot perceive the velocity direction of the vehicle, autonomous vehicles can obtain more accurate information from sensors and estimation technology. Taking the heading deviation $\Delta\psi$ as a control reference state does not maximize the capacity of an autonomous vehicle. The course deviation $\Delta\varphi$, which is the angle between the vehicle's velocity vector and the path heading, denotes the real deviation of the vehicle's moving direction and also indicates the trend of the lateral deviation, as shown in Figure 5. When the sideslip, $\beta$, is small and $\Delta\psi$ is close to $\Delta\varphi$, a controller based on $\Delta\psi$ can keep the tracking deviation in a small range. However, when the difference between $\Delta\psi$ and $\Delta\varphi$ becomes large, especially near the handling limits where a high rear-tire slip angle and a high yaw rate lead to high sideslip, $\beta$, as shown in Figure 5, a controller based on $\Delta\psi$ will fail to effectively minimize the tracking deviation. This is especially important for a vehicle traveling across a corner at the physical limits of tire friction, where vehicle sideslip, $\beta$, can reach $5°$, and cannot be ignored at this level.

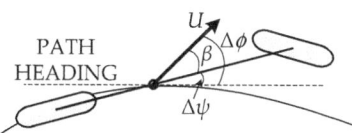

**Figure 5.** The relationship between the course-direction deviation $\Delta\varphi$, sideslip $\beta$ and the heading deviation $\Delta\psi$.

According to the vehicle kinematics, the course deviation is given by:

$$\Delta\varphi = \Delta\psi + \beta. \quad (43)$$

Zero steady-state lateral deviation requires the value of $\Delta\varphi$ to be zero, and the value of $\Delta\psi$ is, hence, nonzero. Due to this, the course-direction deviation $\Delta\varphi$ should be chosen as a reference state when designing a path-tracking steering controller.

*4.2. Control Model*

Using the zeroth-order hold discretization method, we can get the discrete vehicle model from Equations (37)–(42) as follows:

$$x(k+1) = A_c x(k) + B_{F_{yf}} F_{yf}(k) + B_\kappa \kappa(k) + d_{\bar{\alpha}_r}, \tag{44}$$

where $x = \begin{bmatrix} \beta & r & \Delta\psi & e \end{bmatrix}^T$ is the state vector, and

$$A_c = \begin{bmatrix} \frac{-2\bar{C}_r}{mU_x} & \frac{2\bar{C}_r b}{mU_x^2} - 1 & 0 & 0 \\ \frac{2\bar{C}_r b}{I_{zz}} & -\frac{2\bar{C}_r b^2}{I_{zz}U_x} & 0 & 0 \\ 0 & 1 & 0 & 0 \\ U_x & 0 & U_x & 0 \end{bmatrix}, \ B_{F_{yf}} = \begin{bmatrix} \frac{2\cos(\delta+i)}{mU_x} \\ \frac{2a\cos(\delta+i)}{I_{zz}} \\ 0 \\ 0 \end{bmatrix}, \ B_\kappa = \begin{bmatrix} 0 \\ 0 \\ -U_x \\ 0 \end{bmatrix}, \ d_{\bar{\alpha}_r} = \begin{bmatrix} \frac{2(\bar{F}_r+\bar{C}_r\bar{\alpha}_r)}{mU_x} \\ -\frac{2b(\bar{F}_r+\bar{C}_r\bar{\alpha}_r)}{I_{zz}} \\ 0 \\ 0 \end{bmatrix}.$$

In order to apply the integral action to eliminate the static offset caused by model uncertainties, the discrete model can be written in the incremental form:

$$\xi(k+1) = A\xi(k) + B_1 \Delta F_{yf}(k) + B_2\kappa + d, \tag{45}$$

$$\eta(k) = C\xi(k), \tag{46}$$

$$F_{yf}(k) = F_{yf}(k-1) + \Delta F_{yf}(k), \tag{47}$$

where $\xi(k) = \begin{bmatrix} x(k) & F_{yf}(k-1) \end{bmatrix}^T$ is the extended state vector; $\eta(k) = \begin{bmatrix} \Delta f(k) & e & (k) \end{bmatrix}^T$ is the output vector; and $A = \begin{bmatrix} A_C & B_{F_{yf}} \\ 0 & I \end{bmatrix}$, $B_1 = \begin{bmatrix} B_{F_{yf}} \\ I \end{bmatrix}$, $B_2 = \begin{bmatrix} B_\kappa \\ 0 \end{bmatrix}$, $d = \begin{bmatrix} d_{\bar{\alpha}_r} \\ 0 \end{bmatrix}$, and $C = \begin{bmatrix} 1 & 0 & 1 & 0 \\ 0 & 0 & 0 & 1 \end{bmatrix}^T$.

The control objective is to generate an optimal front force input $F_{yf}(k)$ by the steering controller, such that the lateral path-tracking deviation is minimized and that the vehicle maintains stability at the limits of handling.

*4.3. Constraints*

The design of the safety constraints is defined by the bounds of two vital indicators of vehicle stability. Under the assumptions of steady-state cornering and the given tire model, the bounds of $\beta$ and $r$ reflect the maximum capabilities of the vehicle's tires.

The maximum steady-state yaw rate can be expressed as follows:

$$r_{max} = \frac{g\mu}{U_x}, \tag{48}$$

where $g$ is the gravity. Given a yaw rate $r$, the vehicle sideslip, $\beta$, reaches a maximum when the rear tires approach saturation:

$$\beta_{ss,max} = \alpha_{r,sat} + \frac{br}{U_x}, \tag{49}$$

where $\alpha_{r,sat}$ is the saturated tire slip angle, which is expressed as:

$$\alpha_{r,sat} = \tan^{-1}\left( \frac{3mg\mu}{C_{\alpha_r}} \frac{a}{a+b} \right), \tag{50}$$

where $C_{\alpha_r}$ is the rear-tire cornering stiffness.

The constraints defined by Equations (49) and (50) can be concisely expressed via the inequality:

$$|H_v\xi(k)| \leq G_v, \tag{51}$$

where:

$$H_v = \begin{bmatrix} 0 & 1 & 0 & 0 & 0 \\ 1 & -\frac{b}{u_x} & 0 & 0 & 0 \end{bmatrix} \text{ and } G_v = \begin{bmatrix} r_{max} & \alpha_{r,sat} \end{bmatrix}^T.$$

## 4.4. MPC Formulation

The optimization problem for MPC, given in Equations (4)–(10), can be formulated as follows:

$$
\begin{aligned}
\min_{\Delta F_{yf}, \varepsilon_v} J_{N_P} &= \sum_{i=1}^{N_P} (\eta(k+i))^T Q\eta(k+i) + \sum_{i=1}^{N_C} R\left(\Delta F_{yf}(k+i)\right)^2 + W\varepsilon_v \\
&= \sum_{i=1}^{N_P} \xi(k+i)^T C^T Q C \xi(k+i) + \sum_{i=1}^{N_C} R\left(\Delta F_{yf}(k+i)\right)^2 + W\varepsilon_v
\end{aligned}
\tag{52}
$$

$$\text{s.t. } |H_v \xi(k+i)| \le G_v + \varepsilon_v, \ \forall i, \tag{53}$$

$$\left|\Delta F_{yf}(k+i)\right| \le \Delta F_{yf,max}, \ i = 0, \dots, N_C - 1, \tag{54}$$

$$\Delta F_{yf}(k+i) = 0, \ i = N_C, N_C + 1, \dots, N_P - 1, \tag{55}$$

$$\left|F_{yf}(k+i)\right| \le F_{yf,max}, \ i = 0, \dots, N_C - 1, \tag{56}$$

where $\Delta F_{yf} = [\Delta F_{yf}(k), \Delta F_{yf}(k+1), \dots, \Delta F_{yf}(k+N_C-1)]^T$ is the sequence of future input increments, and $Q$, $R$ and $W$ are weighting matrices of appropriate dimension. $\Delta F_{yf,max}$ and $F_{yf,max}$ are the slew rate capabilities and the maximum lateral force, respectively. As Equation (51) is based on steady-state assumptions, the vehicle state can exceed the bounds and still return back within the bounds after a short excursion. To ensure the optimization problem is always feasible, a nonnegative slack variable $\varepsilon_v$ is used.

The solution vector of the optimization problem in Equations (52)–(56) is expanded as follows:

$$\Delta U^* = [\Delta F_{yf}^*, \varepsilon_v^*]^T. \tag{57}$$

The optimal front lateral force input is obtained through the first element of the optimal solution sequence:

$$F_{yf}^*(k) = F_{yf}(k-1) + \Delta F_{yf}^*(k). \tag{58}$$

Additionally, the steering angle $\delta$ that will be applied to the vehicle is obtained by mapping $F_{yf}^*(k)$ through Equation (17).

In order to accurately capture the propagation of $\beta$ and $r$ at a high frequency, the sampling time $T_s = 0.02$ s is small enough. Considering the balance of the control performance and the computational complexity, $N_C$ and $N_P$ are chosen as 20 and 50, respectively. The following weighting matrices were obtained by iteratively tuning via simulation:

$$Q = \begin{bmatrix} 1000 & 0 \\ 0 & 5 \end{bmatrix}, \tag{59}$$

$$R = 1, \tag{60}$$

$$W = \begin{bmatrix} 10 & 10 & 10 & 10 \end{bmatrix}. \tag{61}$$

## 5. Simulations and Results

### 5.1. Model Validation

In order to demonstrate the improvements of the linearized method, two paths with different corner curvatures are designed. The general shape of both paths are the same, as shown in Figure 6, with three sections—corner entry, constant radius, and corner exit—and the distance of each section is the same. The center points of each section would be comparison points. The radii of the constant radius section (green in Figure 6) are 8 m and 30 m for the two paths. The nonlinear models of Equations (12)–(14) and (18)–(20) are used to compute the vehicle states. The Stanley method is used to track the desired path at two constant speeds, and the steering-angle control law is given by:

$$\delta = \Delta\psi + \tan^{-1}\left(\frac{k_P e}{U_x}\right), \tag{62}$$

where $k_P$ is the gain parameter.

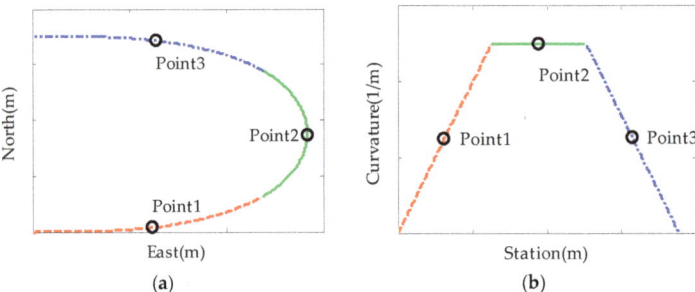

**Figure 6.** (**a**) The designed path and (**b**) the corresponding path curvature varies along the path.

The simulation is carried out with sampling time $T_s = 0.02$ s, and the simulated steering angle and the corresponding front lateral force are shown in Figure 7.

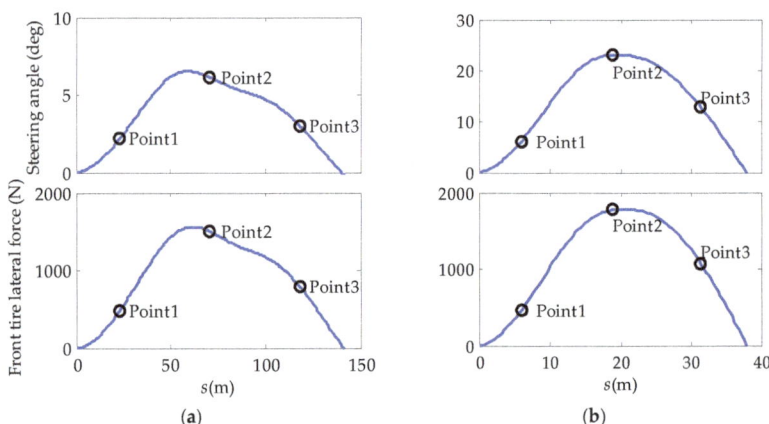

**Figure 7.** The steering angle and front tire lateral force of the model in (**a**) $R = 30$ m and $U_x = 45$ km/h and (**b**) $R = 8$ m and $U_x = 25$ km/h cases.

We use Equations (21) and (22) and the tire-linearization method proposed in Reference [7] as a baseline prediction model to compare with our model, which provides steady-state information and is a combination of Equations (36)–(38). The prediction horizon length is chosen as 0.4 s with 20 steps. Utilizing the simulated front lateral force as the input of the prediction model, the predicted vehicle state sequences can be generated, and the results are shown in Figure 8.

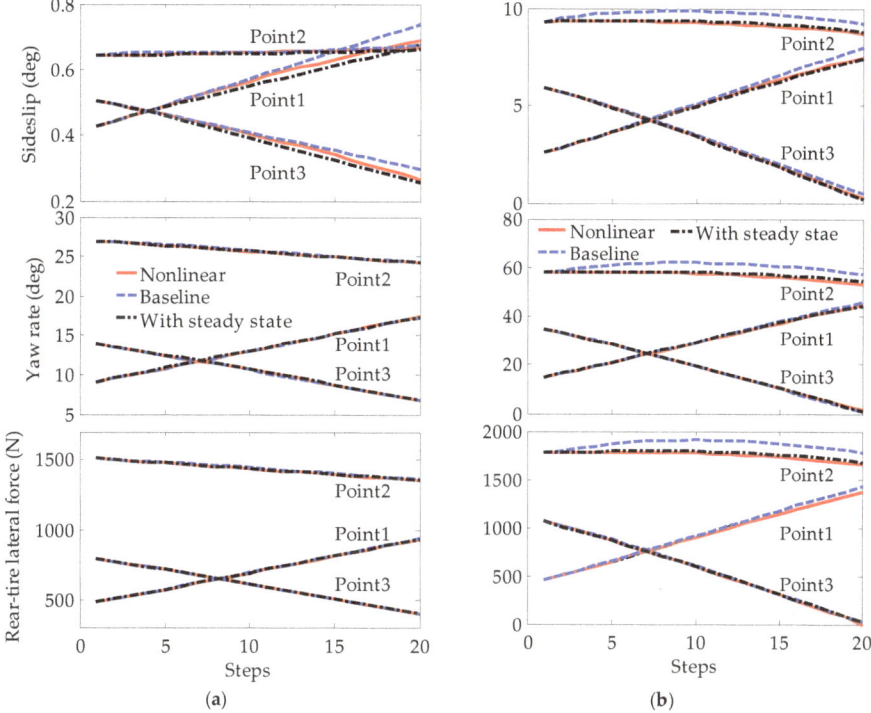

**Figure 8.** The prediction results of different linearization models in (**a**) $R = 30$ m, $U_x = 45$ km/h and (**b**) $R = 8$ m, $U_x = 25$ km/h cases.

For the first case, where the constant radius is 30 m and the speed is 45 km/h, the lateral acceleration can reach 6 m/s$^2$. The steering angle is less than 5° at points 1 and 3, and slightly higher at point 2. Therefore, the small-angle assumption is valid. As shown in Figure 8a, both linear models can predict the vehicle states accurately. However, the predicted sideslip $\beta$ at point 1 of the baseline linear model begins to deviate from the nonlinear value. For the second case, where the speed is 25 km/h and the maximum lateral acceleration is 7 m/s$^2$, the steering angle is much larger than 5° at all three points, especially at point 2. The small-angle assumption is no longer valid under these conditions. As shown in Figure 8b, the prediction errors of sideslip, yaw rate and rear-tire force for the baseline model increase with step number at point 2. At the same time, the linear model with steady-state information can still predict the vehicle state accurately.

*5.2. Controller Performance*

To validate the performance of the presented steering controller, a test was performed via simulation. The simulation is implemented based on the CarSim-Simulink platform with a validated high-fidelity full-vehicle dynamics model. The parameters of the vehicle and path are listed in Table 1.

**Table 1.** Parameter values of the vehicle.

| Parameter | Symbol | Value | Units |
|---|---|---|---|
| Vehicle mass | $m$ | 1230 | kg |
| Yaw inertia | $I_{zz}$ | 1343.1 | kg·m² |
| Front axle-O distance | $a$ | 1.04 | m |
| Rear axle-O distance | $b$ | 1.56 | m |
| Front cornering stiffness | $C_{\alpha_f}$ | 48,840 | N/rad |
| Rear cornering stiffness | $C_{\alpha_r}$ | 32,887 | N/rad |
| Friction coefficient | $\mu$ | 0.95 | n/a |

The path for the simulated steering controller to follow is a 584 m circuit generated by the path-generation method of Reference [25], as shown in Figure 9a. The desired path is parameterized as a curvature profile that varies with distance counterclockwise along the path, which is shown in Figure 9b. The desired longitudinal speed and lateral acceleration profile are generated by the speed controller proposed in Reference [26], as shown in Figure 10. The curvature varies between $-0.04$ m$^{-1}$ and $0.05$ m$^{-1}$, the longitudinal speed varies between 13.5 m/s to 28 m/s, and the lateral acceleration varies between $-9$ m/s² to 9 m/s².

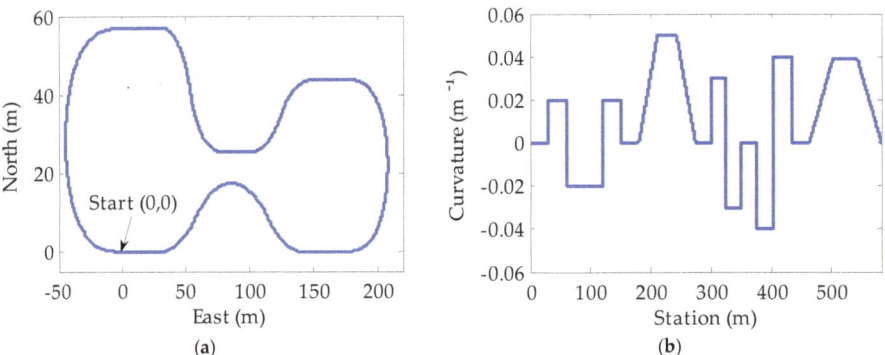

**Figure 9.** The designed path for steering controller to track. (**a**) Overhead view of path and (**b**) the corresponding curvature varies counterclockwise along the path.

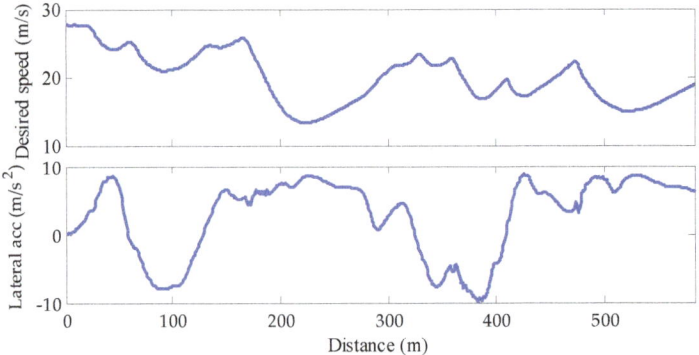

**Figure 10.** The desired speed and lateral acceleration.

Figure 11 shows the simulation results of three separate MPC steering controllers. The first (linear controller) utilizes the steering angle as the control input with the linear vehicle and tire models employed. The second (original controller) and the third (proposed controller) controllers both utilize the vehicle and tire models introduced in Section 3. The difference is that, as a baseline controller, the original controller uses the heading deviation $\Delta\psi$ and lateral deviation $e$ as a reference state. The longitudinal controller and stability bounds are used for all cases.

The simulation results for the lateral deviation and steering angle for three controllers are shown in Figure 11, and statistics of the results are shown in Table 2. As shown in Table 2, the average of the absolute lateral deviation $\overline{|e|}$ and the standard deviation of the absolute lateral deviation $\sigma(|e|)$, and the maximum absolute lateral deviation $\max(|e|)$ of the linear controller, are much higher than the other two MPC controllers. Additionally, as shown in Figure 11, the steering angle of the linear controller keeps increasing until it reaches a maximum, beyond which the lateral deviation continues to increase. This is because the linear model fails to predict the lateral tire force when the tire reaches the nonlinear region, especially near saturation. On the other hand, the controllers with the proposed dynamic model can maintain the lateral deviation in a small range, which indicates that the linearization method proposed in the Section 3 can properly retain the characteristics of the vehicle nonlinearity even under high-speed conditions.

The performances of the original and the proposed controllers are quite close, as shown in Figure 11. However, the $\overline{|e|}$ and $\sigma(|e|)$ of the proposed controller are lower as shown in Table 2, while the $\max(|e|)$ of the two controllers remains roughly the same. The reason for this can be concluded as follows: At some points, the front lateral force necessary to minimize deviation has exceeded the available friction, and there is nothing the steering controller can do to bring the vehicle back to the desired path with such a large lateral acceleration. To show more details of the control process, Figure 12 shows the simulation results of the two controllers over the range 0–200 m.

**Table 2.** Comparison of control results of different controllers.

| Model | $\overline{|e|}$ (m) | $\sigma(|e|)$ (m) | $\max(|e|)$ (m) |
|---|---|---|---|
| Linear Controller | 2.460 | 3.295 | 11.702 |
| Original Controller | 0.671 | 0.906 | 4.402 |
| Proposed Controller | 0.539 | 0.750 | 4.400 |

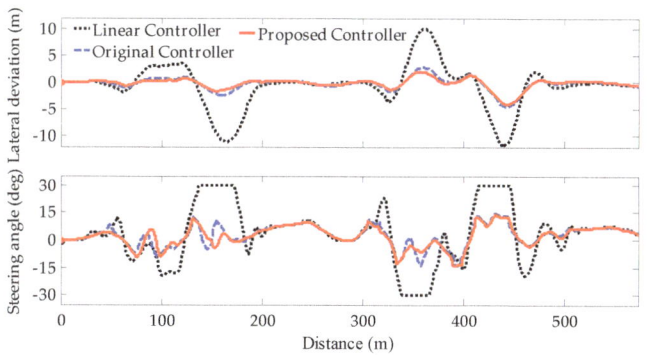

**Figure 11.** The simulation results of lateral deviation and steering angle of different controllers.

As shown in Figure 12, the lateral deviation $e$ is reduced when using the course-direction deviation $\Delta\varphi$ in the controlled state. Around $s = 50$, 80, 120 and 150 m along the track, vehicle sideslips increase with speed and curvature beyond $5°$, and even reach up to $10°$. This explains the noticeable difference between heading deviation $\Delta\psi$ and course-direction deviation $\Delta\varphi$ at the same points. Additionally, the lateral deviation $e$ of the original controller is much larger than the proposed controller, which matches the expected results in Section 4. Around $s = 140$ m, the heading deviation, $\Delta\psi$, of the original controller becomes positive (the left side of path-heading direction) and maintains about 10 m. However, the lateral deviation *e* keeps increasing on the right side of the desired path. This indicates that heading deviation cannot accurately reflect the real direction of the vehicle and the tendency of lateral deviation when the vehicle sideslip is large. From *s* = 130 to 150 m, the front lateral force is constant, while the steering angle of the front tires fluctuates. This confirms that using lateral force as the input to the model is a more direct approach to account for the nonlinearity of the tires.

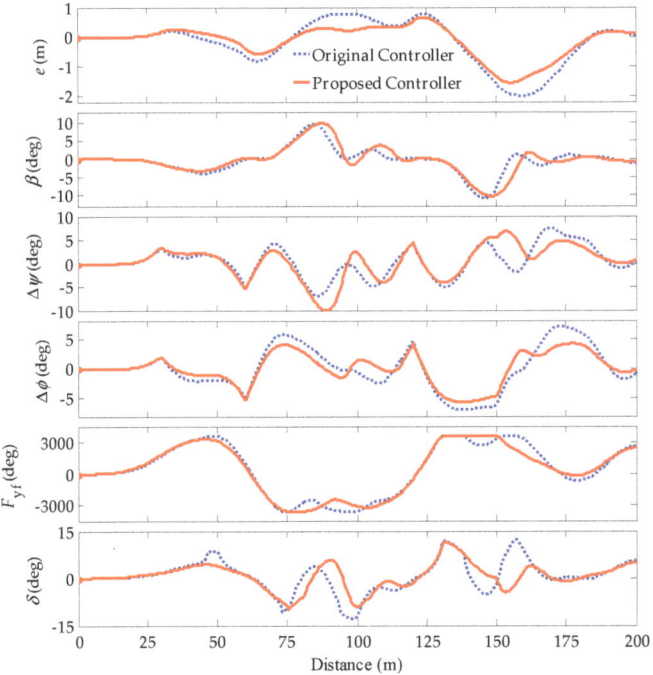

**Figure 12.** Simulation results over 0–200 m. The lateral deviation *e*, sideslip beta *β*, heading deviation $\Delta\psi$, course-direction deviation $\Delta\varphi$, front-tire lateral force $F_{yf}$ and steering angle *δ* of proposed controller are compared with that of original controller over 0–200 m.

## 6. Conclusions

The design of an MPC steering controller based on a linearized model for autonomous vehicle path tracking is described in this paper. The proposed steering controller can track the desired path accurately at high speeds and under large lateral-acceleration conditions. By including the predicted and steady-state information in the model, the proposed linearization method can properly retain the nonlinear characteristics of the vehicle and tire models. Additionally, a single-track 'bicycle' model and a brush tire model are linearized to accurately describe the motion of the vehicle at high speeds. A simulation has been conducted to validate the accuracy of the method. Problems with the effective control reference states were discussed. Based on the linearized model, the MPC controller utilizes course-direction deviation instead of heading deviation as the control reference state to eliminate the tracking deviation. Thus, an improved MPC controller with course-direction deviation and stability constraints was developed. Finally, by comparing with the linear controller, the simulation results demonstrate the controller with proposed linearized model can maintain the deviation in a small range even under large lateral acceleration conditions. Analysis of results indicates that the steering controller with course-direction deviation reduces the average of absolute lateral deviation, compared to the controller with heading deviation, by nearly 20%. This steering controller can ensure tracking accuracy and vehicle stability under high-speed conditions and can be applied to drive an autonomous vehicle.

**Author Contributions:** C.S. studied the linearization method and wrote the paper; X.Z. conceived the path tracking method; and L.X. and Y.T. analyzed the data and modified the paper.

**Acknowledgments:** This work supported by the National Key Research Development Program of China (2016YFB0101000) and the Fundamental Research Funds for the Central Universities (M17JB00170).

**Conflicts of Interest:** The authors declare no conflict of interest.

## References

1. Sun, C.; Zhang, X.; Xi, L. Design of a lateral Control Strategy for High-Speed Intelligent Vehicle. In Proceedings of the SDEWES2017, Dubrovnik, Croatia, 4–8 October 2017. No. 0131.
2. Li, A.; Zhao, W.; Wang, X. ACT-R Cognitive Model Based Trajectory Planning Method Study for Electric Vehicle's Active Obstacle Avoidance System. *Energies* **2018**, *11*, 75. [CrossRef]
3. Dixit, S.; Fallah, S.; Montanaro, U. Trajectory planning and tracking for autonomous overtaking: State-of-the-art and future prospects. *Annu. Rev. Control* **2018**, in press. [CrossRef]
4. Amer, N.H.; Zamzuri, H.; Hudha, K. Modelling and Control Strategies in Path Tracking Control for Autonomous Ground Vehicles: A Review of State of the Art and Challenges. *J. Intell. Robot. Syst.* **2017**, *86*, 225–254. [CrossRef]
5. Brown, M.; Funke, J.; Erlien, S.; Gerdes, J.C. Safe driving envelopes for path tracking in autonomous vehicles. *Control Eng. Pract.* **2017**, *61*, 307–316. [CrossRef]
6. Kritayakirana, K.; Gerdes, J.C. Using the centre of percussion to design a steering controller for an autonomous race car. *Veh. Syst. Dyn.* **2012**, *50*, 33–51. [CrossRef]
7. Erlien, S.M.; Funke, J.; Gerdes, J.C. Incorporating non-linear tire dynamics into a convex approach to shared steering control. In Proceedings of the American Control Conference, Portland, OR, USA, 4–6 June 2014; pp. 3468–3473.
8. Talvala, K.L.R.; Kritayakirana, K.; Gerdes, J.C. Pushing the limits: From lanekeeping to autonomous racing. *Annu. Rev. Control* **2011**, *35*, 137–148. [CrossRef]
9. Hu, C.; Jing, H.; Wang, R.; Yan, F. Robust H∞ output-feedback control for path following of autonomous ground vehicles. *Mech. Syst. Signal Process.* **2015**, *70*, 414–427.
10. Kapania, N.R.; Gerdes, J.C. Path tracking of highly dynamic autonomous vehicle trajectories via iterative learning control. In Proceedings of the 2015 IEEE American Control Conference, Chicago, IL, USA, 1–3 July 2015; pp. 2753–2758.
11. Gray, A.; Gao, Y.; Hedrick, J.K.; Borrelli, F. Robust Predictive Control for semi-autonomous vehicles with an uncertain driver model. In Proceedings of the 2013 IEEE Intelligent Vehicles Symposium, Gold Coast, QLD, Australia, 23–26 June 2013; pp. 208–213.
12. Katriniok, A.; Maschuw, J.P.; Christen, F.; Eckstein, L. Optimal vehicle dynamics control for combined longitudinal and lateral autonomous vehicle guidance. In Proceedings of the 2013 IEEE Control Conference, Zürich, Switzerland, 17–19 July 2013; pp. 974–979.
13. Mammar, S.; Koenig, D. Vehicle Handling Improvement by Active Steering. *Veh. Syst. Dyn.* **2002**, *38*, 211–242. [CrossRef]
14. Tagne, G.; Talj, R.; Charara, A. Design and Comparison of Robust Nonlinear Controllers for the Lateral Dynamics of Intelligent Vehicles. *IEEE Trans. Intell. Transp. Syst.* **2016**, *17*, 796–809. [CrossRef]
15. Kapania, N.R.; Gerdes, J.C. Design of a feedback-feedforward steering controller for accurate path tracking and stability at the limits of handling. *Veh. Syst. Dyn.* **2015**, *53*, 1687–1704. [CrossRef]
16. Rawlings, J.B.; Mayne, D.Q. *Model Predictive Control: Theory and Design*; Nob Hill Pub: San Francisc, CA, USA, 2009.
17. Waschl, H. *Optimization and Optimal Control in Automotive Systems*; Springer: Cham, Switzerland, 2014.
18. Raffo, G.V.; Gomes, G.K.; Kelber, C.R. A predictive controller for autonomous vehicle path tracking. *IEEE Trans. Intell. Transp. Syst.* **2009**, *10*, 92–102. [CrossRef]
19. Beal, C.E.; Gerdes, J.C. Model Predictive Control for Vehicle Stabilization at the Limits of Handling. *IEEE Trans. Control Syst. Technol.* **2013**, *21*, 1258–1269. [CrossRef]

20. Ławryńczuk, M. Computationally efficient model predictive control algorithms. In *A Neural Network Approach, Studies in Systems, Decision and Control*; Janusz, K., Ed.; Springer: Cham, Switzerland, 2014.

21. Borrelli, F. MPC-based approach to active steering for autonomous vehicle systems. *Int. J. Veh. Auton. Syst.* **2005**, *3*, 265–291. [CrossRef]

22. Falcone, P.; Borrelli, F.; Tseng, H.E. Linear time-varying model predictive control and its application to active steering systems: Stability analysis and experimental validation. *Int. J. Robust Nonlinear Control* **2011**, *21*, 862–875. [CrossRef]

23. Rajamani, R. *Vehicle Dynamics and Control*; Springer: Boston, MA, USA, 2011.

24. Pacejka, H. *Tire and Vehicle Dynamics*; Elsevier: Amsterdam, The Netherlands, 2005.

25. Li, X.; Sun, Z.; Cao, D. Development of a new integrated local trajectory planning and tracking control framework for autonomous ground vehicles. *Mech. Syst. Signal Process.* **2017**, *87*, 118–137. [CrossRef]

26. Kritayakirana, K.; Gerdes, J.C. Autonomous vehicle control at the limits of handling. *Int. J. Veh. Auton. Syst.* **2012**, *10*, 271–296. [CrossRef]

*Article*

# Explorative Multidimensional Analysis for Energy Efficiency: DataViz versus Clustering Algorithms

**Dario Cottafava [1],* , Giulia Sonetti [2] , Paolo Gambino [3] and Andrea Tartaglino [4]**

[1]   Department of Culture, Politics and Society, University of Turin, Turin 10100, Italy
[2]   Interuniversity Department of Regional & Urban Studies and Planning, Politechnic of Turin,
       Turin 10100, Italy; giulia.sonetti@polito.it
[3]   Department of Physics, University of Turin, Turin 10100, Italy; paolo.gambino@unito.it
[4]   Energy Management, University of Turin, Turin 10100, Italy; andrea.tartaglino@unito.it
*   Correspondence: dario.cottafava@unito.it; Tel.: +39-(0)110912079

Received: 10 April 2018; Accepted: 10 May 2018; Published: 21 May 2018

**Abstract:** We propose a simple tool to help the energy management of a large building stock defining clusters of buildings with the same function, setting alert thresholds for each cluster, and easily recognizing outliers. The objective is to enable a building management system to be used for detection of abnormal energy use. We start reviewing energy performance indicators, and how they feed into data visualization (DataViz) tools for a large building stock, especially for university campuses. After a brief presentation of the University of Turin's building stock which represents our case study, we perform an explorative analysis based on the Multidimensional Detective approach by Inselberg, using the Scatter Plot Matrix and the Parallel Coordinates methods. The k-means clustering algorithm is then applied on the same dataset to test the hypotheses made during the explorative analysis. Our results show that DataViz techniques provide quick and user-friendly solutions for the energy management of a large stock of buildings. In particular, they help identifying clusters of buildings and outliers and setting alert thresholds for various Energy Efficiency Indices.

**Keywords:** Energy Efficiency Indices; data visualization; clustering algorithms; university campus; energy management

---

## 1. Introduction

Energy efficiency programs as well as policies for the reduction of greenhouse gas (GHG) emissions have been adopted worldwide by national governments, international organizations, and public administrations [1]. The reduction of energy consumption and the shift toward a more sustainable use of resources are increasingly becoming a challenge for any sector and activity related to the built environment [2].

The buildings sector is indeed a high energy-consumer, accounting for over one-third of the global final energy consumption [3]. Energy demand is expected to rise by 50% by 2050 if no action is urgently taken [4]. This means that major efforts are required to go beyond existing technical and economic barriers for improving the efficiency of our energy use in buildings. The power to characterize the energy consumption of a complex building stock, for instance, can reduce cost barriers for energy efficient solutions. The improvement of reliable indicators to measure building energy performance at a neighbourhood/city scale is therefore an important contribution for achieving urban sustainability targets [5,6].

For instance, Ascione et al. [7] proposed a methodology for energy analysis in large building stocks of building categories, named Simulation-based Large-scale uncertainty/sensitivity Analysis of Building Energy performance (SLABE) exploiting software as EnergyPlus and Matlab. Ciulla et al. [8] focused on the energy performance of historical building envelopes and thanks to TRNSYS software

(17, Thermal Energy System Specialists, LLC, Madison, WI, USA) (http://www.trnsys.com/) energy performance analyses were conducted for the residential sector. Abu Bakar et al. [9] measured buildings' energy performance based on heating, ventilating and conditioning (HVAC) system consumption. Moghimi et al. [10] studied commercial buildings, analyzing indicators related to the occupied air conditioning area. González et al. [11] suggested adopting a reference building to compare the energy consumption within a building stock. The work of Ballarini et al. [12] pointed towards the same direction, explaining the experience of the Typology Approach for Building Stock Energy Assessment (TABULA) project (www.episcope.eu), applied to residential sector. In this case, researchers defined a reference building for each category (i.e., single family, terraced, multi-family houses, etc.) according to the Directive 2010/31/EU by considering the optimal trade-off costs and energy savings. At EU level, the Directive 2002/91/EC (i.e., Energy Performance Building Directive) introduced the compulsory energy certification from 2006. Andaloro et al. [13] reviewed the adopted of these directives within the 27 EU Member States. In particular, related to the residential sector, in the literature, various works have been done on energy retrofit analysis at national level in past years [14], focused on identifying clusters of buildings depending on architectural features and historical period. Ciulla et al. [15] extrapolated data to allow a fast analysis of heating energy consumption based on TRNSYS software and identified clusters depending on energy demand, climate and office building features at a European Level.

Although Energy Efficiency Indices (EEIs) are widely studied, there is still a lack of research in the literature related to energy decision-making tools relying on these indices [16]. As energy analysis and energy retrofit studies improve results and accuracy, the learning curve for energy managers becomes steeper and harder. In fact, the majority of energy analyses focus on Energy+ (8.9.0, the U.S. Department of Energy's Building Technologies Office, Washington, DC, USA) (https://energyplus.net/), TRNSYS and other highly specialized engineering software. For this reason, current research challenges are envisaged in developing links between EEIs and more general energy assessment frameworks, to enable sounding comparisons among buildings with different architectural features, functions and/or occupations schedules [17].

In this respect, university campuses may represent a valuable test bed, being often a joint resemble of buildings with very different characteristics, yet having the same purpose. For their physical scale in the city, university campuses have a significant role to play with respect to local energetic and socioeconomic impacts, going far beyond the university scale itself [18]. Universities are increasingly conceived as hubs for innovation, serving as test bed for new energy reduction strategies [19–21].

However, a major focus among all the initiatives is generally devoted to energy performance improvement, and its monitoring [22], justified by the increased investments in energy efficient technologies [23]. Living labs monitoring infrastructure provide an appropriate way for answering energy data queries while displaying all the necessary information for performance self-assessment and external reporting purposes [24]. There is, however, a gap between these energy performances oriented experiences and the international ranking systems for green labeling of campuses which are not based on performance indicators but relying on ranges of total energy consumption [25].

Towards the same direction, a work of the National Bureau of Statistics of China [26] highlights that universities or megaversities with different building functions have energy consumption per square meter that cannot be compared and classified with the same criteria. Those challenges are also linked to the diversity of material utilization, $CO_2$ emissions, energy source and regulatory compliance, which is different from country to country, and from city to city [27].

### 1.1. Motivation and Problem Identification

At both city scale and campus scale, as already noted by Haas [28], the most difficult task when dealing with EEIs is to provide the corresponding data by end use to obtain suitable numbers for cross-country evaluations. Many of the parameters needed for time series and cross-country analyses depend on the obtainability of disaggregated data from wide-ranging surveys and cross-section analyses, and there are several critical methodological problems that still impede the creation of such operational indicators of energy efficiency [29]. Regarding specifically the university campus realm, Sonetti et al. [30] already argued the lack of a precise analysis based on building types or functions, in one of the most popular green ranking for universities, the UI GreenMetric—World Universities Rankings. The need of three clusters based on urban morphology, climate zones and university functions has been highlighted for sound comparisons among campuses.

In the field of building engineering, softwares such as SPARK, EnergyPlus, MODELICA, TRNSYS, HVACSIM+ are the most common tools for analyzing complex non linear thermodynamic interactions between building shells, the surrounding environment, HVAC systems and building control strategies. All these software packages numerically simulate the behaviour of the real systems with various degrees of accuracy, but require specialised staff and a large computing time.

Several papers have recently investigated new techniques to develop precise predictive algorithms for energy consumption analysis or to design clustering algorithms. For example, some work focused on the assignment of a fixed (or predictable) amount of energy resources to areas of a city or to different buildings of the same district [31], the identification of energy outliers [32] or the possibility of demand side management and local balancing [33]. Furthermore, Hong et al. [34] developed a methodology to support decision-makers during the retrofit process estimating future energy savings in order to identify the priorities for selecting building energy retrofit intervention areas, while Yalcintas [35,36] exploited an artificial neural network for large benchmarking analysis allowing inter-comparison, Fan et al. [37] presented a data mining approach to predict sources of peak power demand, and so forth.

While building simulation softwares play a fundamental role in the energy management, they require a large investment of resources (computing time, staff, etc) and are unsuitable for online or near real-time applications. Unsupervised approaches combined with a multidimensional data visualization approach in the area of the data analysis, on the other hand, have the advantage of deliver a quick, user-friendly and easy online visualization of the status quo of a whole stock. They cannot replace simulation tools, but play a complementary role.

*1.2. Current Paper Aim and Structure*

The aim of this paper are i) to propose a simple, efficient and precise analysis tool able to compare buildings within a large stock, inputting only energy efficiency indices; ii) to explore how to use this tool to cluster buildings within a stock according to their specific function. The proposed tool tries to fill the gap between very detailed energy audits analysis and the lack of precise user-friendly and immediate tools for energy efficiency comparisons among buildings. The proposed approach needs basic energy data input for each building—i.e., monthly energy bills—and, starting from those, it adopts interactive data visualization tools to analyze the dataset. The Multidimensional Detective approach, as described by Inselberg [38], has been adopted to define the cluster alert thresholds.

The paper is structured as follows. In Section 2, current data visualization techniques and clustering algorithms are explained. In Section 3, the adopted approach for developing a simple energy monitoring tool exploiting the University of Turin's building stock, defining clusters of buildings with the same function, setting alert thresholds for each cluster, and easily recognizing outliers is described. For both data visualization and clustering algorithm processes, we discuss two possible approaches to choose the right number of clusters and the identification of alert thresholds and outliers, after a brief presentation of the University of Turin's building stock case study. Finally, Sections 4 and 5 report a comparison between the two approaches with considerations on the obtained clusters and their accuracy.

## 2. Large Scale Buildings Energy Monitoring Methods

### 2.1. Data Visualization

In the Big Data decade, data visualization becomes fundamental to extract useful and valuable information from the enormous amount of data available today. Each specific dataset, in fact, potentially has a huge amount of hidden information and could reveal important tips for managers and policy makers, as well as for data miners and data scientists. According to Card et al. [39], Information Visualization, the most general definition of data visualization (DataViz), is defined as visual representations that are computer-supported and able to amplify human cognition. Keim et al. [40], in fact, defined DataViz as the process to "translate" complex dataset into visual tips and immediate qualitative information and they identified three main aims: presentation, confirmative and explorative. For these three aims, one of the fundamental aspects of DataViz is based on the interactive process allowed by modern DataViz coding libraries, such as D3.js [41], Julia [42], GoogleCharts and others tools, which permit users to manipulate datasets to better understand hidden information in datasets. Within this framework, interactive data visualizations are crucial for explorative analysis where data miners have no quantitative insights to model a particular datasets. This is particularly important for data driven research, such as for energy efficiency studies, or more in general for analysis aimed at policy makers and managers, where the main aim of an analysis should be to identify alert thresholds, outliers or anomalies [43].

Generally speaking, each multidimensional dataset $X$ is composed by $n$ arrays—i.e., the number of observations/the size of the dataset, $x_i = (x_{i1}, x_{i2}, ..., x_{im}), i = 1, ..., n$ with $m$ attributes/dimensions—and it may be represented by a matrix $nxm$. With this representation, $x_{ij}$ is the datum of the real observation $i$ with attribute $j$. Data visualization techniques may be grouped into four main approaches: (1) axis reconfiguration [44]; (2) dimensional embedding [45]; (3) dimensional sub-setting [46]; and (4) dimensional reduction [47]. Two of the four approaches, axis reconfiguration and dimensional sub-setting, are discussed within this paper, exploiting, respectively, the Scatter Plot Matrix (dimensional sub-setting) and Parallel Coordinates (axis reconfiguration), two of the most popular techniques.

#### 2.1.1. The Scatter Plot Matrix

It highlights, as described by Keller [48], relationships among variables as in a correlation matrix, where single scatter plots between two attributes of the datasets are plotted within the same graph. The Scatter Plot Matrix can be understood as a generalization of a single Scatter Plot. With respect to the energy field, for instance, Corgnati et al. [49] proposed the use of a single Scatter Plot based on two attributes, i.e., the annual building consumption and the annual electrical building consumption per square meter, to identify the top intervention priorities within a large building stock, while Cottafava et al. [50] proposed two other attributes to identify buildings with the most inefficient lighting and heating schedules: electrical building consumption per square meter and the day/night energy efficiency index (a ratio between energy consumption during the weekday working hours and during the night/weekend). Thus, the Scatter Plot Matrix could be exploited as a preliminary analysis method useful to identify the top/bottom priorities with respect to three, or more, attributes of a datasets.

#### 2.1.2. The Parallel Coordinates

This method, introduced by Inselberg [44], allows visualizing a multidimensional dataset thanks to $m$ equidistant copies of the $y$-axis, perpendicular to the $x$-axis. Thanks to this method, the observation $x_i = (x_{i1}, x_{i2}, ..., x_{im})$ is represented as a polygonal line which intersects each vertical axis. It is noteworthy to highlight that, in this visualization, each vertical axis represents a different attribute/dimension of a multidimensional dataset, and each polyline represents a different observation. To exploit the Parallel Coordinates method, it is crucial to cite one fundamental property, named *Bumping the Boundaries*, which ensures that a polygonal line lying in-between two other polygonal lines represents an interior point of the corresponding hypersurface in *m* dimensions [38].

## 2.2. Data Clustering Algorithms

Data Clustering is a process of detection of different groups within a specific dataset to identify patterns or subsets, i.e., clusters, as well as outliers. Clustering process aims to identify clusters where "Instances, in the same clusters, must be similar as much as possible", while "Instances, in different clusters, must be different as much as possible" [51]. Clustering, in particular, is an unsupervised process where instances (objects) have no initial label (i.e., assigned cluster) given by data scientists and researchers but the cluster configuration depends on the chosen algorithm and on the adopted similarity measures and distance metrics.

### 2.2.1. Distance Metrics

Metrics depend on, as reviewed by Xu et al. [52], the adopted definition of distance. The most commonly used definition, for quantitative measures, is the Minkowski distance of order *p*:

$$D_{il} = \left( \sum_{j=1}^{m} |x_{ij} - x_{lj}|^p \right)^{1/p}$$

where *m* = n. of dimensions, $x_{ij}$ = value of the attribute *j* of the object/point *i* and $D_{il}$ is the distance between the point *i* and the point *l*. For specific values of *p* the Minkowski distance corresponds to the Euclidean distance (Minkowski order 2), the Manhattan distance (order 1) or the Cebysev distance (order ∞). Other common distance metrics are based on the Mahalanobis distance, $D_{il} = (x_i - x_l)^T S^{-1} (x_i - x_l)$ and the Jaccard distance $J_\delta (A, B) = 1 - |A∩B|/|A∪B| = |A∪B|-|A∩B|/|A∪B|$, where *S* is the Covariance Matrix of the cluster where $x_i$ and $x_l$ belong to the same group and $|X|$ is the number of element in subset *X* [53].

### 2.2.2. Evaluation

Evaluation consists in the process of testing of the validity of the chosen algorithm. Evaluation indicators may be subdivided into two categories: internal evaluation and external evaluation. The first one refers to data within the same cluster, while the second one refers to similarity evaluation among data lying in different clusters [53]. Some of the most widely adopted internal evaluation methods are:

(i)   The *within-cluster sum of square* [54]

$$Q_T = \frac{1}{k} \sum_{j=1}^{k} \sigma_j = \frac{1}{k} \sum_{j=1}^{k} \sum_{i=1}^{|Z_j|} \frac{d(x_i^j, c_j)}{|Z_j|} \tag{1}$$

(ii)  The *Davies–Bouldin index* [55]

$$DB = \frac{1}{k} \sum_{i=1}^{k} \max_{i \neq j} \left( \frac{\sigma_i + \sigma_j}{d(c_i, c_j)} \right) \tag{2}$$

(iii) The *silhouette index* [56]

$$S = \frac{1}{k} \sum_{j=1}^{k} S_j = \frac{1}{k} \sum_{j=1}^{k} \frac{1}{|Z_j|} \sum_{i=1}^{|Z_j|} \frac{b_i^j - a_i^j}{\max \left[ a_i^j, b_i^j \right]} \tag{3}$$

where

$$a_i^j = \frac{1}{|Z_j|} \sum_{l=1, l \neq i}^{|Z_j|} d(x_i, x_l) \ and \ b_i^j = \min_{p=1,\dots,k; k \neq j} \left[ \frac{1}{|Z_p|} \sum_{l=1}^{|Z_p|} d(x_i^j, x_l^p) \right]$$

where $n$ is the total number of points, $x_i^j$ = point $i$ lying in cluster $j$, $k = n$. of clusters, $c_x$ = the centroid of the cluster $x$, $\sigma_x$ = the mean distance between any data in cluster $x$ and the centroid of the cluster, $|Z_x| = n$. of point in cluster $Z_x$, $d\left(x_i, x_j\right)$ = the distance between points $x_i$ and $x_j$ (both centroids or observations). Finally, there are various external evaluation indices, as reported by Dongkuan et al. [53] (i.e., Rand index [57], Jaccard index [58], Fowlkes–Mallows index [59], etc.), useful to evaluate the efficiency of clustering algorithms in terms of finding true (false) positives and negatives with respect to a reference cluster configuration.

### 2.2.3. Clustering Algorithms

In the literature, clustering algorithms are mainly split into two categories—*Hierarchical* and *Partition* clustering methods—but various sub-classifications have been proposed to categorize the dozens of clustering algorithms. Dongkuan et al. [53] subdivided algorithms into traditional ones and modern algorithms. Traditional algorithms have been aggregated into nine categories—partition-, hierarchy-, Fuzzy Theory-, distribution-, density-, graph theory-, grid-, fractal- and model-based—while modern algorithms count more than 40, divided into 10 categories. Nagpal et al. [60], instead, proposed a classification where algorithms are partition-, hierarchy-, density-, grid-, model- and category-based. Partition clustering algorithms arrange the $n$ data into $k$ different clusters [61]. The number $k$ of cluster is an input parameter of the algorithm. The partitioning is obtained by minimizing an objective function, and it depends on the distance from the centroid to any point within a single cluster or on some similarity functions. Basically, the initialization of a partition algorithm consists in: (a) assigning randomly $k$ seed points, the initial centroids and (b) every point in the dataset must be labeled to the nearest cluster centroid. Then, in each step, (c) a new centroid for each cluster must be computed by averaging over all points lying in the same cluster and (d) the nearest centroid for every point in the dataset must be checked again. Steps (c) and (d) continue until a local optimum is found. The two most famous partition clustering algorithms are the $k$-means [62] and the $k$-medoids ($k$-means for discrete data) [63] directly developed from the core concept of partition algorithms. A typical way to choose seed points, for instance, as reviewed by Nagpal et al. [60], is to choose randomly from the existing points, to avoid empty clusters. Other partition algorithms, instead, such as Clustering for Large Applications (CLARA) [64], Clustering Large Applications based on RANdomized Search (CLARANS) [65] and Partition Around Medoids (PAM) [66], choose seed points randomly in a grid based way. Generally, the advantage of these algorithms is a high efficiency and low time complexity while the disadvantage consists in the necessity of defining the number of clusters $k$ as an algorithm input, taking into account that the choice of $k$ affects results and the identification of outliers. Hierarchical algorithms find clusters in an iterative way starting from the whole dataset in a unique cluster, divisive mode (top-down approach), or from a single point, agglomerative mode (bottom-up approach). The basic idea of hierarchical algorithms is to find nested clusters starting from one group to $n$ groups or vice versa in an iterative way merging (or splitting) the nearest clusters (or the furthest ones). Typical algorithms are CURE [67], BIRCH [68], CHAMELEON [69] and many others. For instance, Balanced Iterative Reducing and Clustering using Hierarchies (BIRCH) is based on saving only the Cluster Features triple $n, LS, SS$ where $n$ = total number of points within a cluster, $LS$ is the sum of attributes of all points within a cluster and $SS$ is the sum of square. CURE (Clustering Using REpresentatives) is considered, for large database, insensitive to outliers, while CHAMELEON merges two clusters only if they are close "enough". Many algorithms, such as $k$-means, need the number of cluster $k$ as an input, while many others determine the right number in a dynamic way. The problem of the identification of the number of clusters can be solved thanks to various methods. For instance, Ketchen et al. [70] analyzed the elbow method based on the within-cluster sum of square method introduced by Robert L. Thorndike [71] in 1953. The elbow method consists in plotting the within-cluster sum of square, i.e., the average distance of any point within a cluster with respect to its centroid, in a scatter plot with the number of cluster $k$, looking for the "elbow", the point where the within-cluster sum of square (WSS) stops to rapidly decrease. The elbow point shows the best number of cluster $k$. Pollard et al. [72] used the Mean Split Silhouette (MSS), a measure of cluster heterogeneity, and minimized it to choose the best $k$. Tibshirani et al. [73], instead, proposed the gap statistic, a methodology based on the comparison of the change in within-cluster sum of square dispersion with respect to a proper reference null distribution. Other methods, widely adopted in the literature, are based on MonteCarlo simulations cross validation [74,75]. Consensus Clustering [76] and Resampling [77] try to find $k$ looking for the most "stable" configuration through different MonteCarlo simulations but with the same number of clusters. On the contrary, Wang [78] proposed selecting the number of clusters by minimizing the algorithm's instability, a simple measure of the robustness of any algorithm against the initial random seeds.

### 3. Methodology

To design a simple, user-friendly approach for energy efficiency analysis for large building stock, we compared different data visualization tools applying a specific clustering algorithm, $k$-means. An explorative analysis based on the general Multidimensional Detective approach [44] was performed as first step. We exploited two multidimensional analysis tools, the Scatter Plot Matrix and the Parallel coordinates method. Secondly, the $k$-means clustering algorithm was applied on the same dataset to test the hypothesis made during the explorative analysis. The first step, the Multidimensional Detective approach proposed by Inselberg [44], identified the most meaningful clusters. As described in Cottafava et al. [50], the process consists of few steps, and it is able to identify outliers and "junk attributes" as well as to define boundaries and alert thresholds, a minimum and a maximum value, such as $x_{min,j} \leq x_{ij} \leq x_{max,j}, \forall x_i \epsilon Z_k$ where $Z_k$ is the $k$-th subset of $X$ for every cluster. The three steps—(i) define building types; (ii) test the assumptions; and (iii) identify thresholds and outliers —consists in choosing the building types (e.g., libraries, hospitals, research centres, etc.) and labeling each data relying on the knowledge background of the data source organization. When each datum has been labeled, alert thresholds can be identified and outliers can be recognized. The three steps were accomplished via the Scatter Plot Matrix and the Parallel Coordinates methods. After defining clusters and thresholds, the $k$-means algorithm tests the validity of the clusters hypothesis. Finally, we proposed a tool to monitor historical trends based on an interactive application of the Parallel Coordinates method.

### 3.1. Dataset and Indices Description

As briefly mentioned in the Introduction, the selected case study for testing the simple tool for large scale building stocks energy analysis was the University of Turin (Unito) in Italy. The advantage of choosing the Unito campus relies in the availability of a wide historical data set and the precise match of energy-related information and the locus of its consumption, thanks to a wide net of smart meters, periodical human-based control on data trends and an open access website prompting all data. The University of Turin is a little city within a city: Unito's building stock is very heterogeneous with respect to functions of the buildings, their construction year (ranging from the XVI century to 2014) and architectural features. It sums more than 800,000 m$^2$, with about 120 buildings sprout all over the city and in Piedmont region, for a total of 2.08 TOE of methane gas and 23.5 GWh of electrical energy consumption per year. The building stock comprises museums, administrative offices, libraries, and hospitals, as well as research centres, a botanical garden and departments of humanities and sciences [79]. The Unito energy data, obtained from monthly energy bills, are related to a whole year and have been adopted as the training dataset for this study. Analyzed data refers to 46 buildings, with 59 electricity meters and 77 methane gas meters. Four attributes for each point have been chosen: the absolute annual energy consumption (kWh), the annual energy consumption per meter square (kWh/m$^2$), the annual energy consumption per user (kWh/user) and the "night/day energy efficiency index" $EEI_{year,kWh,night/day} = 1/12 \sum_{i=1}^{12} E_{i,kWh,night}/E_{i,kWh,day}$ where $E_{i,kWh,day} =$ kWh during working hours and $E_{i,kWh,night} =$ kWh during night/holiday for month $i$.

*3.2. k-Means Algorithm*

The *k*-means algorithm has been used for the same dataset to compare results obtained by the algorithms with the results obtained by the Multidimensional Detective approach. Each real observation $x_{ij}$, for each dimension $j$ has been normalized so that $x_{ij} = (x_{ij} - \min x_j)/(\max x_j - \min x_j) \in (0,1)$ to allow computing a meaningful Euclidean distance metric among points. The initial centroids for each cluster were picked at random among the existing points of the dataset to avoid empty clusters. Three internal evaluation indices were used to validate results and to choose the right number of clusters *k*: the within-cluster sum of square, the Davies–Bouldin index and the silhouette index. The final result, for each *k* (from *k* = 2 up to *k* = 15), was chosen as the best configuration—the one with the minimum WSS index over 1000 independent MonteCarlo simulations. The right number of cluster *k*, as described by the Elbow method, was obtained by identifying the elbow in the scatter graph WSS vs. *k*. Finally, once the right *k* was defined, the best cluster configuration was selected by choosing the highest external evaluation indices, the Rand index and the Fowlkes–Mallows index, over 1000 MonteCarlo simulations, with respect to the algorithm result and the target cluster configuration. The target cluster configuration is the one chosen during the Multidimensional Detective process.

**4. Results**

*4.1. Cluster Identification*

4.1.1. Cluster Hypothesis

A general hypothesis has been made due to the heterogeneity of the Unito's building stock. The whole stock has been categorized into nine clusters with respect to the functions of the buildings: Scientific Departments (with laboratories), Scientific Departments (without laboratories), Medical, Agrarian and Humanities Departments, libraries and administrative offices, and, finally, sport infrastructures and large complexes.

4.1.2. Data Visualization Techniques

The proposed clusters were tested with two types of visualization: the Scatter Plot Matrix, a dimensional sub-setting method (Figure 1), and the Parallel Coordinates method, an axis reconfiguration technique (Figure 2). First, our approach separates the chosen cluster from all the other ones to define, in a qualitative way, cluster thresholds and to look for anomalies and outliers. Second, hypothesis have to be tested to identify alert thresholds and outliers. The first step can be achieved thanks to the brush functions of the two proposed visualizations. As shown in Figure 1a,b, for the Scatter Plot Matrix and in Figure 2a,b, for the Parallel Coordinates method, the identification of the pre-defined clusters is straightforward and outliers emerge in a very clear way.

The Scatter Plot Matrix is the generalization of the Scatter Plot, as described by Cottafava et al. [79], and is publicly available at https://goo.gl/o4nn4f. Figure 1 shows the whole building stock of the University of Turin and reports 16 different single Scatter Plots. Respectively, the *x*-axis and *y*-axis, starting from the bottom-left graph, report the following attributes: Type of building, the day/night energy efficiency index, the annual energy consumption per user and the annual energy consumption per meter square. The four graphs on the diagonal, as for a correlation matrix, has the same attribute on both *x*-axis and *y*-axis. Each cluster is identified with a different color and it can be highlighted simply selecting the type of the building in the bottom-left graph. The nine labeled colors are: red (Agrarian Depts.), green (Medical Depts.), blue (Humanities Depts.), black (Scientific Depts.—with lab.), grey (Scientific Depts.—without lab.), sky-blue (Large complexes), yellow (Libraries) and pink (Sport infrastructure). In particular, Figure 1a reports, as an example, the Humanities Departments and Figure 1b shows the Administrative Offices of the University of Turin. This visualization configuration allows checking if buildings with the same label lie on the same 1-D cluster, simply observing points distribution on the left and bottom plots. The tool here described is publicly available at https://goo.gl/ZJem9h. The Parallel Coordinates method also allows displaying various attributes for hundred points with a different visualization configuration. This approach permits a data miner to analyze dependent, or independent, attributes and to detect anomalies or precise trends and correlation among different attributes as in a pattern recognition problem. Figure 2 shows the entire Unito building stock with respect to four different attributes: the type of the building, the annual energy consumption per square meter, the absolute annual energy consumption and the day/night energy efficiency index. In this case, the nine clusters are labeled with number 1–9 and represented by the first vertical axis. Respectively, 1–9, the clusters correspond to the following: Agrarian Depts., Medical Depts., Humanities Depts., Scientific Depts.—with lab, Scientific Depts.—without lab, Large Complexes, Libraries and Sport Infrastructure. As for the Scatter Plot Matrix, in this case, the brush function allows a data miner, or a policy maker/energy manager, to highlight precise subset of the whole dataset. This feature permits to exploit the property Bumping the boundaries = to bound the clusters. Figure 2a,b, respectively, shows Humanities Depts. and Agrarian Depts. It is possible to notice quite precise fluxes/patterns of polygonal lines with a high density. The tool we used is publicly available at: https://goo.gl/4aHYuj.

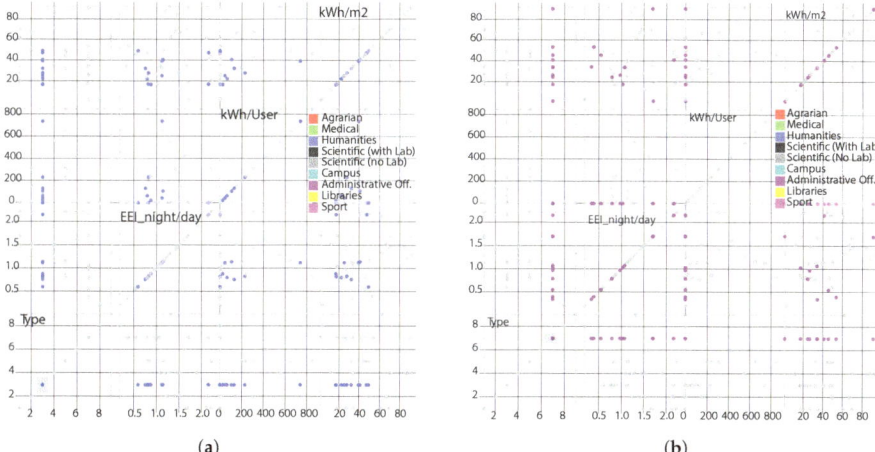

**Figure 1.** Scatter plot matrix for the Unito's buildings stock with respect to four attributes: type of building (1–9), the night/day energy efficiency index, the energy consumption per user and the energy consumption per square meter.

### 4.1.3. Clustering Algorithm

The $k$-means algorithm has been used to identify and recognize clusters depending on three main attributes, annual absolute energy consumption, annual energy consumption per square meter and the day/night energy efficiency index, avoiding the energy consumption per user due to lack of data for administrative offices and other buildings. In this section, first, we report some considerations on the right number of clusters found thanks to the elbow method. We select the best configuration for each $k$—i.e., the lowest WSS—running 1000 MonteCarlo simulations. The elbow method suggests, as previously defined in data visualization analysis, that the right number of $k$ is between 9 and 10, where the WSS slightly stops decreasing. Figure 3 shows the elbow plot with the WSS index on the $y$-axis and $k$, the number of clusters, on the $x$-axis. In Table 1, we report data obtained related to WSS, to the Davies–Bouldin index and to the silhouette index. Silhouette index is constant for different $k$ while WSS and DB index decrease as $k$ increases. Since silhouette index lies in $-1 \leq Sil \leq 1$, where a silhouette index of $-1$ means a bad cluster correlation and 1 a good one, the obtained clusters represent a quite good configuration.

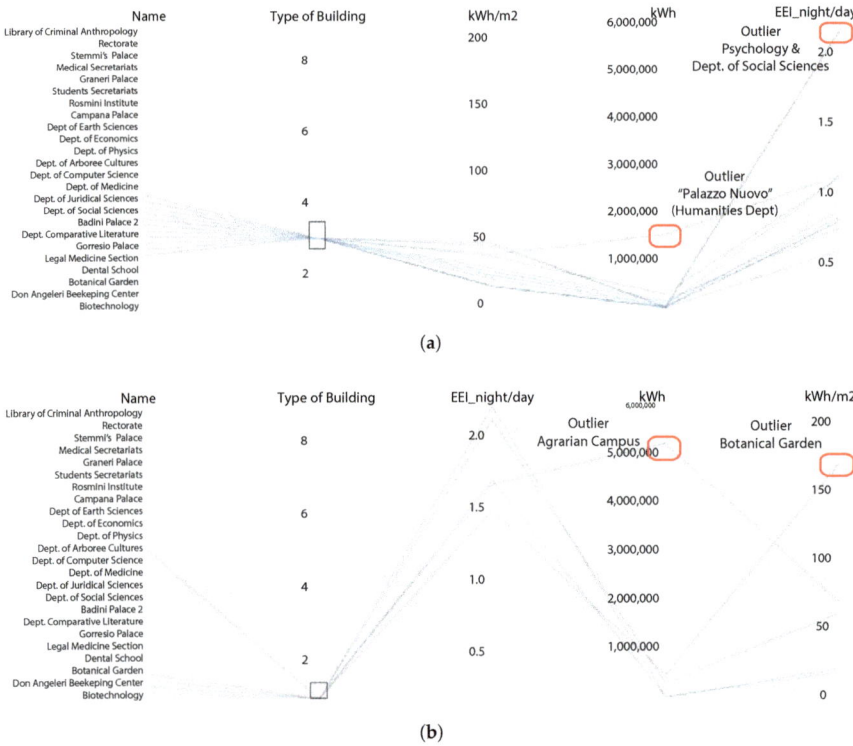

(a)

(b)

**Figure 2.** Parallel coordinates method for the Unito's buildings stock for two building functions—(**a**) Humanities Depts. and (**b**) Agrarian Depts.—with respect to four attributes: type of building (1–9), the night/day energy efficiency index, absolute annual energy consumption and the energy consumption per square meter.

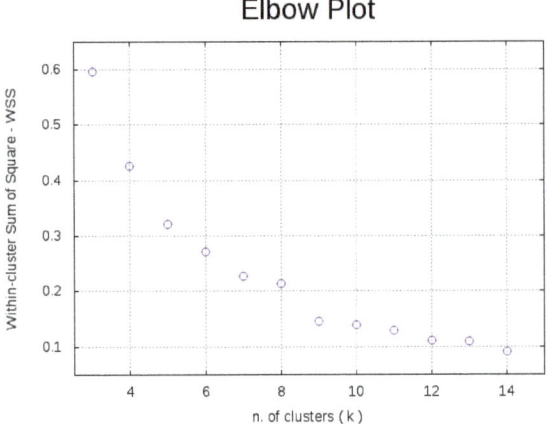

**Figure 3.** Elbow method. The plot shows within-cluster sum of square vs. $k$ ($n$. of clusters). The right $k$ number is between 9 and 10.

**Table 1.** Best configuration evaluation index.

| k | WSS | DB Index | Sil Index |
|---|---|---|---|
| 3 | 0.597 | 2.165 | 0.407 |
| 4 | 0.426 | 1.985 | 0.505 |
| 5 | 0.321 | 1.966 | 0.479 |
| 6 | 0.271 | 1.726 | 0.466 |
| 7 | 0.228 | 1.701 | 0.490 |
| 8 | 0.213 | 1.694 | 0.411 |
| 9 | 0.146 | 1.501 | 0.680 |
| 10 | 0.140 | 1.539 | 0.531 |

### 4.1.4. Comparison between DataViz and k-Means Clusters

Once the best number of clusters ($k = 9$) was chosen, two external evaluation indices—the Rand index and the Fowlkes–Mallows index—were computed comparing clusters obtained by the k-means and the previously defined clusters within the Data Visualization Section. To obtain the best configuration, further 10,000 MonteCarlo simulations were run with the chosen $k = 9$ maximizing the Rand index and choosing the respective cluster configuration. Table 2 reports the best cluster configuration result with respect to the Rand index.

**Table 2.** Best external evaluation index.

| Rand Index | Fowlkes Index |
|---|---|
| 0.769 | 0.645 |

### 4.2. Setting Thresholds

Starting from the Parallel Coordinates graph, we defined alert thresholds for the main six clusters, i.e., Scientific Depts. (without lab.), Scientific Depts. (with lab.), Humanities, Agrarian and Medical Depts. and Administrative Offices. Results and alert thresholds are reported in Table 3 with respect two main attributes $EEI_{year,kWh,night/day}$ and kWh/year*m$^2$. We do not report absolute energy consumption per year because it is not interesting as a general index for energy efficiency. Table 3 shows that clusters corresponding to Scientific Depts. (with lab.), Agrarian and Medical Depts. have an high day/night energy efficiency index, as expected. Scientific Depts. (with lab.) shows a higher energy consumption per meter square with respect to Agrarian and Medical Depts. and in general with respect to all other clusters. Administrative Offices, Scientific Depts. (without lab.) and Humanities Depts., instead, have a common behavior with low kWh/year*m$^2$ and $EEI_{year,kWh,night/day}$. Scientific Depts. (without lab), generally, present a slightly higher energy consumption at night.

**Table 3.** Thresholds for consumption per square meter and for day/night energy efficiency index.

| Building | kWh/year*m$^2$ | $EEI_{night/day}$ |
|---|---|---|
| Scientific Depts without lab | 30–50 | 0.8–1.1 |
| Scientific Depts with lab | 70–110 | 1.1–1.9 |
| Humanities Depts | <50 | 0.6–1.1 |
| Agrarian Depts | 20–70 | 1.5–2.5 |
| Medical Depts | 50–70 | 1.2–1.5 |
| Administrative Offices | <50 | 0.4–1 |

## 4.3. Monitoring Trends

The final step of the presented process is based on an application of the parallel coordinates method. In this case, we plot different annual energy consumptions on a different axis (each axis represents a different year) where only one attribute may be plotted. This tool, shown in Figure 4, allows visualizing the historical trend of a chosen energy efficiency index. A useful feature is the possibility to highlight simultaneously various buildings to observe their historical trends. By simply hoovering the mouse on each polyline, the building energy consumption for the chosen year is shown. By clicking on it, the polyline is highlighted, as shown in Figure 4, where "Library Dept. Philological Sciences" (violet) and the "Rectorate" (orange) stand out. The tool here described is publicly available at https://goo.gl/YuPTRB.

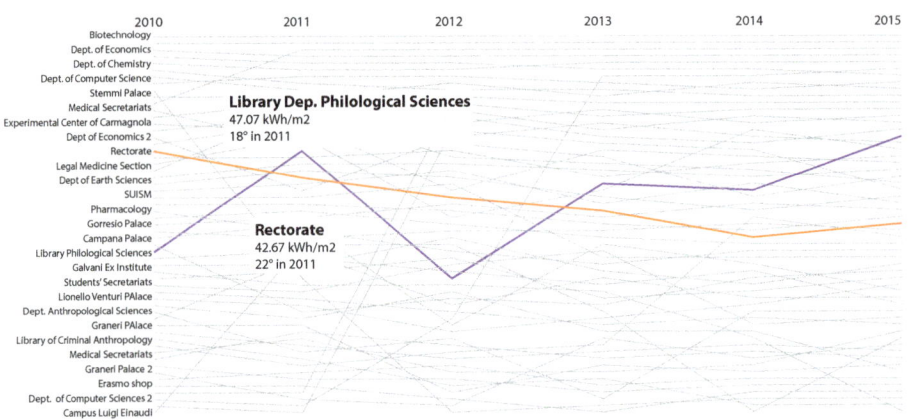

**Figure 4.** Interactive data visualization tool to monitor historical trends based on the Parallel Coordinates method.

## 5. Discussion

The first aim of this paper is to determine a process to set general hypotheses on building clusters with respect to energy efficiency indices. A clusters hypothesis has been previously stated relying on the background knowledge of the energy management staff at the University of Turin. We envisage this step as a limit of this study, since it requires a preliminary effort by a human task force that is not always reliable, available, competent or even present. However, the time required in this phase is widely compensated by the easiness of the subsequent steps and the replicability of the monitoring phase in each institution able to offer at least the energy bill data source.

The clusters hypothesis has been made based on main building function and then it has been verified via two methodologies for the identification of buildings clusters: a data visualization approach and a clustering algorithm.

The data visualization approach allowed recognizing the validity of the clustering hypothesis. In fact, after labeling each building with a precise function, it is possible to match each building within a precise cluster straightforwardly (via the brush function). In this way, it is possible to immediately identify outliers and set rough alert thresholds, as described in Section 4.2 and in Table 3.

This method made us identify some outliers in the Unito case study. For instance, the Physics Dept. and the Biotechnology Dept. are two outliers within the cluster "Scientific Depts.—with laboratory". High consumption per square meter and high day/night energy efficiency index are due both to large IT centres and to electric chillers running 24/24 h. Within the "Agrarian Depts." cluster, the botanical garden is another outlier, with its very high consumption per square meter. The Agrarian Campus has been identified as an outlier, too, with respect to its annual energy consumption. Looking into that, one can infer that, since it hosts many thousands of students and very specific function related to field experiments and greenhouses maintenance, its energy behavior must be different and must be treated differently. Within the "Medical Depts." cluster, the Dental School and the Legal Medicine Section are outliers compared to an average energy consumption or a day/night energy efficiency index. Again, a more detailed data source analysis reveals that the Dental School has a large electrical energy consumption due to the large amount of electrical technical machineries. As for the Legal Medicine Section, the reason of the high night consumption is the presencee and the mortuary rooms, asking for a constant air conditioning system, which is very costly especially during spring and summer seasons. Within the "Humanities Depts." cluster, there are two outliers, the Social Science Dept. and the Psychology Dept., with respect to the day/night index: the reasons for this anomalous consumption is still under study at the Unito's facility management office after a signaling coming from this work. "Palazzo Nuovo" (Humanities Dept) has one of the highest number of students and classrooms within the same building, thus explaining its higher energy request. Within the "Scientific Depts—without laboratory", two outliers emerge: the Management Dept and Torino Exhibitions. The first one has a high consumption per square meter and a high annual consumption because the energy meter counts also the consumption of the Regional IT center, while "Torino Exhibitions" has a very high day/night index because of the secondary function of the building (art exhibitions, fairs and other types of events). Finally, the "Administrative offices" cluster has three outliers: Stemmi's Palace, Tobacco Factory and Students' secretariat. These three buildings have a high day/night index due to different reasons. The first one is the main building for the technical directions of the university and it hosts a lot of IT servers. Other reasons are under investigation. The other two buildings, instead, are two multifunctional buildings hosting public events for the City of Turin.

As a second step, a clustering algorithm has been used to test the initial hypothesis. The test was made exploiting two external indices—i.e., the Rand index and the Fowlkes–Mallows index—comparing the clusters configuration hypothesis (hp0) and the obtained clusters thanks to the $k$-means algorithm. The obtained clusters configuration with $k = 9$ may be compared with the clusters hypothesis (Rand index = 0.769). $k$-means, due to its algorithm's basic principles, similar to many other clustering algorithm, is strongly affected by local optimum and outliers. In fact, with a deeper analysis on clusters details, $k$-means algorithm is able to well-identify outliers—e.g., Management Dept., Biotechnology Dept. or Agrarian Campus—but it recognizes some clusters without physical explanation due to local optimum. For instance, the Department of Arboree Cultures (hp0: Agrarian Depts. cluster) and the Tobaccoes Factory (hp0: administrative offices cluster) or the Don Angeleri Beekeeping Center (Agrarian Depts. cluster) and the Psychology Dept. (hp0: Humanities Depts. cluster) always lie within the same cluster without any other point because they have a very common energy consumption behavior. The three clusters hp0: Administrative Offices, Humanities Depts. and Scientific Depts. (without laboratories) are mixed together in only two clusters. Scientific Depts. (with laboratories) cluster is well-recognized losing one of the outliers described in data visualization approach, the Biotechnology Dept., and gaining two outliers from other clusters, the Botanical Garden and the Dental School. Many outliers, identified in the DataViz approach, are aggregated into the same cluster—e.g., Tobaccoes Factory, Torino Exhibitions, Legal Medicine Section, Social Science Dept and Students' secretariat. This behavior reveals that a possible new cluster hypothesis should include a multifunctional building cluster. Finally, the two main campuses Campus Luigi Einaudi and the Agrarian Campus are always grouped together, representing a reasonable choice. The Management Dept., outlier within the Scientific Depts. (without lab) cluster, and the Biotechnology Dept. are clustered alone. In conclusion, $k$-means clustering algorithm recognizes very accurately the main clusters—identified as campuses, service industry buildings and Scientific Depts.—confirming our initial hypothesis but is not able, as expected, to recognize slight differences between Humanities Depts., Scientific Buildings (without lab) and Administrative Offices.

## 6. Conclusions

To conclude, this data visualization approach offers a simple way to identify outliers and to set alerts energy consumption thresholds for each buildings function, but the reasons for the inefficiency have to be explained with deeper analyses. For instance, with the help of specific mapping tools (e.g., GIS) to localise buildings and visualize their location within the city or with the support of the facility management, further features and informations, that did not emerge during the preliminary labelling phase, may be revealed. As a methodological caveat, this approach reveals outliers within clusters defined ex-ante; therefore, every multifunctional cluster is shown as an outlier of its own cluster, and that can be a limit if a cluster is the result of a preliminary wrong human inference. However, DataViz techniques revealed to be very useful to explore quickly and simply a large building stock, identifying the least efficient buildings and clustering buildings according to their functions. Moreover, the results are presented in a user-friendly graphical interface, understandable and accessible to facility managers, even if they are not expert in energy monitoring. This improves the decision process, supporting the diagnosis of energy faults and therefore prioritizing alternatives and energy-efficient strategies, achieving faster energy saving. The implementation of easy energy diagnostics tools and underlying analysis algorithms can lead to a systematic identification of more efficient operations. Last but not least, such clusterization can be used to group relatively similar buildings for comparison inside the university realm, thus providing precious and still lacking national/international thresholds for setting limits for energy consumptions. Similarly, since benchmarking policies require an accurate measurement of buildings relative to one another, this easy way of energy monitoring can act as a kick-off for a quick assessment of universities as an independent building/district class, and then tailor appropriate energy policies for it.

Our results revealed that clustering algorithms—$k$-means in our case—cannot be exploited to design useful clusters based on the function of the building, except for some macro—cluster like tertiary service buildings or campuses and scientific buildings. Moreover, the results show how the most interesting part of information in energy efficiency analysis is lost. In fact, data analysts or energy managers are usually interested in inefficient buildings, thus in outliers with respect to their cluster, even when clustering algorithms tend to aggregate outliers in the wrong cluster. This makes a humanized process always necessary and irreplaceable. At the city level, such data driven tools require a large penetration of metering systems and the possibility to use the private data of the entire building stock; these conditions are still not commonly met, but combined techniques need to be taken into account for future researches to achieve the desired level of granularity in the data source. Of course, identifying and removing causes of abnormal energy use ensures a more efficient environment and not just in terms of the building energy costs.

The algorithms we have applied appear computationally efficient and robust, and can be easily integrated into existing university campus building energy management and warning systems. Of course, further work is needed: (i) to build on this clustering technique; (ii) to provide additional datasets for training the algorithm; (iii) to adopt language processing tools for the automated reading of metered building/energy bills data.

**Author Contributions:** D.C. was the principal investigator. G.S. contributed in finding proper energy efficiency indices and A.T., Energy Manager of the University of Turin, helped to verify the results and to identify proper explanations for the outliers. P.G. supervised the work.

**Acknowledgments:** Support by the Fondazione Goria, program "Talenti per la Società Civile", is gratefully acknowledged.

**Conflicts of Interest:** The authors declare no conflict of interest.

## Nomenclature

*List of Abbreviations*

| | |
|---|---|
| GHG | Greenhouse Gas |
| HVAC | Heating, ventilating and conditiong |
| TRNSYS | Transient System Simulation Tool |
| EEI | Energy Efficiency Index |
| DataViz | Data Visualization |
| TOE | Tonne of oil equivalent |
| kWh | Kilowatt hour |
| WSS | Within-Cluster Sum of Square |
| $X, A, B$ | Multidimensional Dataset |
| $n$ | Number of elements in dataset $X$ |
| $m$ | Number of attributes/dimensions of dataset $X$ |
| $x_i$ | Observation $i$ of dataset $X$ |
| $x_{ij}$ | Real value of attribute $j$ of observation $i$ |
| $i, l$ | Observation subscript |
| $D_{il}$ | Minkowski (or Mahalanobis) distance between observation $i$ and $l$ |
| $p$ | Minkowski order |
| $J_\delta (A, B)$ | Jaccard Distance |
| $Q_T$ | Within-Cluster Sum of Square |
| $DB$ | Davies-Bouldin Index |
| $S$ | Silhouette Index |
| $k$ | Number of Clusters |
| $x_i^j$ | Observation $i$ lying in cluster $j$ |
| $c_x$ | The centroid of the cluster $x$ |
| $\sigma_x$ | Mean distance between any data in cluster $x$ and the centroid of the cluster |
| $|Z_x|$ | Number of points in cluster |
| $d \left( x_i, x_j \right)$ | distance between points $x_i$ and $x_j$ |
| $x_{min,j}, x_{max,j}$ | Alert thresholds for attribute $j$ |
| $EEI_{year,kWh,night/day}$ | Annual night/day Electrical Energy Efficiency Index |
| $E_{i,kWh,day}$ | Electrical energy consumption during working day for month $i$ |
| $E_{i,kWh,night}$ | Electrical energy consumption during night and weekend for month $i$ |

## References

1.  Powell, J.B. Green Building Services. *J. Int. Commer. Econ.* **2015**.
2.  Wilkinson, P.; Smith, K.; Beevers, S.; Tonne, C.; Oreszczyn, T. Energy, energy efficiency, and the built environment. *Lancet* **2007**, *370*, 1175–1187. [CrossRef]
3.  Newman, P. The environmental impact of cities. *Environ. Urban.* **2006**, *18*, 275–295. [CrossRef]
4.  Staff, I.E.A. *Transition to Sustainable Buildings: Strategies and Opportunities To 2050*; Organization for Economic Cooperation and Development: Paris, France, 2013.
5.  Lombardi, P.; Trossero, E. Beyond energy efficiency in evaluating sustainable development in planning and the built environment. *Int. J. Sustain. Build. Technol. Urban Dev.* **2013**, *4*, 274–282. [CrossRef]
6.  Brandon, P.S.; Lombardi, P.; Shen, G.Q. *Future Challenges in Evaluating and Managing Sustainable Development in the Built Environment*; John Wiley & Sons: Southern Gate, Chichester, UK, 2017.
7.  Ascione, F.; Bianco, N.; Stasio, C.D.; Mauro, G.M.; Vanoli, G.P. Addressing Large-Scale Energy Retrofit of a Building Stock via Representative Building Samples: Public and Private Perspectives. *Sustainability* **2017**, *9*, 940. [CrossRef]
8.  Giuseppina, C.; Galatioto, A.; Ricciu, R. Energy and economic analysis and feasibility of retrofit actions in Italian residential historical buildings. *Energy Build.* **2016**, *128*, 649–659.

9.  Bakar, N.N.A.; Hassan, M.Y.; Abdullah, H.; Rahman, H.A.; Abdullah, M.P.; Hussin, F.; Bandi, M. Sustainable energy management practices and its effect on EEI: A study on university buildings. In Proceedings of the Global Engineering, Science and Technology Conference, Dubai, UAE, 1–2 April 2013.

10. Moghimi, S.F.A.; Mat, S.; Lim, C.; Salleh, E.; Sopian, K. Building energy index and end-use energy analysis in large-scale hospitals case study in Malaysia. *Energy Effic.* **2014**, *7*, 243–256. [CrossRef]

11. González, A.B.R.; Díaz, J.J.V.; Caamano, A.J.; Wilby, M.R. Towards a universal energy efficiency index for buildings. *Energy Build.* **2011**, *43*, 980–987. [CrossRef]

12. Ballarini, I.; Corgnati, S.P.; Corrado, V. Use of reference buildings to assess the energy saving potentials of the residential building stock: The experience of TABULA project. *Energy Policy* **2014**, *68*, 273–284. [CrossRef]

13. Andaloro, A.P.; Salomone, R.; Ioppolo, G.; Andaloro, L. Energy certification of buildings: A comparative analysis of progress towards implementation in European countries. *Energy Policy* **2010**, *38*, 5840–5866. [CrossRef]

14. Galatioto, A.; Ciulla, G.; Ricciu, R. An overview of energy retrofit actions feasibility on Italian historical buildings. *Energy* **2017**, *137*, 991–1000. [CrossRef]

15. Ciulla, G.; Lo Brano, V.; D'Amico, A. Modelling relationship among energy demand, climate and office building features: A cluster analysis at European level. *Appl. Energy* **2016**, *183*, 1021–1034. [CrossRef]

16. Yun, G.; Steemers, K. Behavioural, physical and socio economic factors in household cooling energy consumption. *Appl. Energy* **2011**, *88*, 2191–2200. [CrossRef]

17. Wu, L.-M.; Chen, B.-S. Modeling of energy efficiency indicator for semi-conductor industry. In Proceedings of the IEEE International Conference on Industrial Engineering and Engineering Management, Singapore, 2–4 December 2007; IEEE: Piscataway, NJ, USA, 2007.

18. Ferrer-Balas, D.; Lozano, R.; Huisingh, D.; Buckland, H.; Ysern, P.; Zilahy, G. Going beyond the rhetoric: System-wide changes in universities for sustainable societies. *J. Clean. Prod.* **2010**, *18*, 607–610. [CrossRef]

19. Agdas, D.; Srinivasan, R.; Frost, K.; Masters, F. Energy Use Assessment of Educational Buildings: Toward a Campus-wide Sustainable Energy Policy. *Sustain. Cities Soc.* **2015**, *17*, 15–21. [CrossRef]

20. Chung, M.; Rhee, E. Potential opportunities for energy conservation in existing buildings on university campus: A field survey in Korea. *Energy Build.* **2014**, *78*, 176–182. [CrossRef]

21. Escobedo, A.; Briceño, S.; Juárez, H.; Castillo, D.; Imaz, M.; Sheinbaum, C. Energy consumption and GHG emission scenarios of a university campus in Mexico. *Energy Sustain. Dev.* **2014**, *18*, 49–57. [CrossRef]

22. Evans, J.; Jones, R.; Karvonen, A.; Millard, L.; Wendler, J. Living labs and co-production: University campuses as platforms for sustainability science. *Curr. Opin. Environ. Sustain.* **2015**, *16*, 1–6. [CrossRef]

23. Robinson, O.; Kemp, S.; Williams, I. Carbon management at universities: A reality check. *J. Clean. Prod.* **2014**, *106*, 109–118. [CrossRef]

24. Del Mar Alonso-Almeida, M.; Marimon, F.; Casani, F.; Rodriguez-Pomeda, J. Diffusion of sustainability reporting in universities: Current situation and future perspectives. *J. Clean. Prod.* **2015**, *106*, 144–154. [CrossRef]

25. Lauder, A.; Sari, R.F.; Suwartha, N.; Tjahjono, G. Critical review of a global campus sustainability ranking: GreenMetric. *J. Clean. Prod.* **2015**, *108*, 852–863. [CrossRef]

26. NBS. *China Statistical Yearbook*; Technical Report; China Statistics Press: Beijing, China, 2012.

27. Shriberg, M. Institutional assessment tools for sustainability in higher education: Strengths, weaknesses, and implications for practice and theory. *Int. J. Sustain. High. Educ.* **2002**, *3*, 254–270. [CrossRef]

28. Haas, R. Energy efficiency indicators in the residential sector: What do we know and what has to be ensured? *Energy Policy* **1997**, *25*, 789–802. [CrossRef]

29. Jollands, N.; Patterson, M. Four theoretical issues and a funeral: Improving the policy-guiding value of eco-efficiency indicators. *Int. J. Environ. Sustain. Dev.* **2004**, *3*, 235–261. [CrossRef]

30. Sonetti, G.; Lombardi, P.; Chelleri, L. True Green and Sustainable University Campuses? Toward a Clusters Approach. *Sustainability* **2016**, *8*, doi:10.3390/su8010083. [CrossRef]

31. Yik, F.; Burnett, J.; Prescott, I. Predicting air-conditioning energy consumption of a group of buildings using different heat rejection methods. *Energy Build.* **2001**, *33*, 151–166. [CrossRef]

32. Howard, B.; Parshall, L.; Thompson, J.; Hammer, S.; Dickinson, J.; Modi, V. Spatial distribution of urban building energy consumption by end use. *Energy Build.* **2012**, *45*, 141–151. [CrossRef]

33. Yang, C.; Létourneau, S.; Guo, H. Developing Data-driven Models to Predict BEMS Energy Consumption for Demand Response Systems. In *Modern Advances in Applied Intelligence*; Ali, M., Pan, J.S., Chen, S.M., Horng, M.F., Eds.; Springer International Publishing: Cham, Switzerland, 2014; pp. 188–197.

34. Hong, T.; Yang, L.; Hill, D.; Feng, W. Data and analytics to inform energy retrofit of high performance buildings. *Appl. Energy* **2014**, *126*, 90–106. [CrossRef]

35. Yalcintas, M. An energy benchmarking model based on artificial neural network method with a case example for tropical climates. *Int. J. Energy Res.* **2006**, *30*, 1158–1174. [CrossRef]

36. Yalcintas, M.; Ozturk, U.A. An energy benchmarking model based on artificial neural network method utilizing US Commercial Buildings Energy Consumption Survey (CBECS) database. *Int. J. Energy Res.* **2007**, *31*, 412–421. [CrossRef]

37. Fan, C.; Xiao, F.; Wang, S. Development of prediction models for next-day building energy consumption and peak power demand using data mining techniques. *Appl. Energy* **2014**, *127*, 1–10. [CrossRef]

38. Inselberg, A. Multidimensional Detective. In Proceedings of the IEEE Symposium on Information Visualization, Phoenix, AZ, USA, 20–21 October 1997.

39. Card, S.K.; Mackinlay, J.; Shneiderman, B. *Readings in Information Visualization: Using Vision to Think*; Morgan Kaufman: San Francisco, CA, USA, 1999.

40. *NIST-SEMATECH. E-Handbook of Statistical Methods*; NIST: Gaithersburg, MD, USA, 1997.

41. Bostock, M.; Ogievetsky, V.; Heer, J. D3: Data-Driven Documents. *IEEE Trans. Vis. Comput. Graph.* **2011**, *12*, 2301–2309. [CrossRef] [PubMed]

42. Bezanson, J.; Edelman, A.; Karpinski, S.; Shah, V.B. Julia: A fresh approach to numerical computing. *arXiv* **2014**, arXiv:1411.1607. [CrossRef]

43. Keim, D. *Visual Techniques for Exploring Databases*; Technical Report; NIST: Gaithersburg, MD, USA, 2003.

44. Inselberg, A. The plane with parallel coordinates. *Vis. Comput.* **1985**, *1*, 69–97. [CrossRef]

45. Feiner, S.; Beshers, C. Worlds within worlds: Metaphors for exploring n-dimensional virtual worlds. In Proceedings of the 3rd Annual ACM SIGGRAPH Symposium on User Interface Software and Technology, Snowbird, UT, USA, 3–5 October 1990; pp. 76–83.

46. Cleveland, W. *Visualizing Data*; Hobart Press: Summit, NJ, USA, 1993.

47. Borg, I.; Groenen, P.J.F. Modern Multidimensional scaling: Theory and Applications. *Vis. Comput.* **2005**, *2*, 276–278. [CrossRef]

48. Keller, P.R.; Keller, M.M. Visual Cues-Practical Data Visualization. *IBM Syst. J.* **1993**, *33*. [CrossRef]

49. Ariaudo, F.; Balsamelli, L.; Corgnati, S.P. Il Catasto Energetico dei Consumi come strumento di analisi e programmazione degli interventi per il miglioramento dell'efficienza energetica di ampi patrimoni edilizi. In Proceedings of the 48th International Conference AICARR, Baveno, VCO, Italy, 22–23 September 2011; pp. 547–559.

50. Cottafava, D.; Gambino, P.; Baricco, M.; Tartaglino, A. Multidimensional analysis tools for energy efficiency in large building stocks. In Proceedings of the 12th Conference on Sustainable Development of Energy, Water and Environment Systems, Dubrovnik, Croatia, 4–8 October 2017.

51. Jain, A.; Dubes, R. *Algorithms for Clustering Data*; Prentice-Hall: Upper Saddle River, NJ, USA, 1988.

52. Xu, R.; Wunsch, D. Survey of clustering algorithms. *IEEE Trans. Neural Netw.* **2005**, *16*, 645–678. [CrossRef] [PubMed]

53. Xu, D.; Tian, Y. A Comprehensive Survey of Clustering Algorithms. *Ann. Data Sci.* **2015**, *2*, 165–193. [CrossRef]

54. Kassambara, A. *Practical Guide To Cluster Analysis in R*; CreateSpace: North Charleston, SC, USA, 2017.

55. Maulik, U.; Bandyopadhyay, S. Performance evaluation of some clustering algorithms and validity indices. *IEEE Trans. Pattern Anal. Mach. Intell.* **2002**, *24*, 1650–1654. [CrossRef]

56. Starczewski, A.; Krzyżak, A. Performance Evaluation of the Silhouette Index. In *Artificial Intelligence and Soft Computing*; Rutkowski, L., Korytkowski, M., Scherer, R., Tadeusiewicz, R., Zadeh, L.A., Zurada, J.M., Eds.; Springer International Publishing: Cham, Switzerland, 2015; pp. 49–58.

57. Rand, W.M. Objective Criteria for the Evaluation of Clustering Methods. *J. Am. Stat. Assoc.* **1971**, *66*, 846–850. [CrossRef]

58. Kosub, S. A note on the triangle inequality for the Jaccard distance. *arXiv* **2016**, arXiv:1612.02696. [CrossRef]

59. Fowlkes, E.B.; Mallows, C.L. A Method for Comparing Two Hierarchical Clusterings. *J. Am. Stat. Assoc.* **1983**, *78*, 553–569. [CrossRef]

60. Nagpal, A.; Jatain, A.; Gaur, D. Review based on data clustering algorithms. In Proceedings of the 2013 IEEE Conference on Information Communication Technologies, Thuckalay, Tamil Nadu, India, 11–12 April 2013; pp. 298–303.

61. Ahmad, A.; Dey, L. A K-mean Clustering Algorithm for Mixed Numeric and Categorical Data. *Data Knowl. Eng.* **2007**, *63*, 503–527. [CrossRef]

62. Macqueen, J. Some methods for classification and analysis of multivariate observations. In Proceedings of the 5-th Berkeley Symposium on Mathematical Statistics and Probability, Berkeley, CA, USA, 21 June–18 July 1967; pp. 281–297.

63. Park, H.; Jun, C. A simple and fast algorithm for K-medoids clustering. *Expert Syst. Appl.* **2009**, *36*, 3336–3341. [CrossRef]

64. Kaufman, L.; Rousseeuw, P. *Partitioning around Medoids (Program Pam)*; Wiley: Hoboken, NJ, USA, 1990; pp. 126–160.

65. Ng, R.T.; Jiawei, H. CLARANS: A method for clustering objects for spatial data mining. *IEEE Trans. Knowl. Data Eng.* **2002**, *14*, 1003–1016. [CrossRef]

66. Kaufman, L.; Rousseeuw, P. *Partitioning around Medoids (Program Pam)*; Wiley: Hoboken, NJ, USA, 1990; pp. 68–120.

67. Guha, S.; Rastogi, R.; Shim, K. CURE: An Efficient Clustering Algorithm for Large Data sets. In Proceedings of the ACM SIGMOD Conference, Seattle, WA, USA, 2–4 June 1998.

68. Zhang, T.; Ramakrishnan, R.; Livny, M. BIRCH: An efficient data clustering method for very large databases. In Proceedings of the ACM SIGMOD International Conference on Management of Data (SIGMOD), Montreal, QC, Canada, 4–6 June 1996; pp. 103–114.

69. Karypis, G.; Han, E.H.; Kumar, V. Chameleon: Hierarchical clustering using dynamic modeling. *Computer* **1999**, *32*, 68–75. [CrossRef]

70. Ketchen, J.D.; Shook, C.L. The application of cluster analysis in strategic management reasearch: An analysis and critique. *Strateg. Manag. J.* **1996**, *17*, 441–458. [CrossRef]

71. Thorndike, R.L. Who belongs in the family? *Psychometrika* **1953**, *18*, 267–276. [CrossRef]

72. Pollard, K.S.; Van Der Laan, M.J. A method to identify significant clusters in gene expression data. In Proceedings of the SCI (World Multiconference on Systemics, Cybernetics and Informatics), Orlando, FL, USA, 14–18 July 2002; Volume 2, pp. 318–325.

73. Tibshirani, R.; Walther, G.; Hastie, T. Estimating the number of clusters in a data set via the gap statistic. *J. R. Stat. Soc. Ser. B (Stat. Methodol.)* **2001**, *63*, 411–423. [CrossRef]

74. Sheikholeslami, G.; Chatterjee, S.; Zhang, A. Wavecluster: A Multi-Resolution Clustering Approach for Very Large Spatial Databases; *VLDB* **1998**, *98*, 428–439.

75. Smyth, P. Clustering Using Monte Carlo Cross-Validation. In Proceedings of the Second International Conference on Knowledge Discovery and Data Mining, KDD'96, Portland, Oregon, 2–4 August 1996; Volume 1, pp. 26–133.

76. Monti, S.; Tamayo, P.; Mesirov, J.; Golub, T. Consensus clustering: A resampling-based method for class discovery and visualization of gene expression microarray data. *Mach. Learn.* **2003**, *52*, 91–118. [CrossRef]

77. Roth, V.; Lange, T.; Braun, M.; Buhmann, J. A resampling approach to cluster validation. In *Compstat*; Springer: Heidelberg, Germany, 2002; pp. 123–128.

78. Wang, J. Consistent selection of the number of clusters via crossvalidation. *Biometrika* **2010**, *97*, 893–904. [CrossRef]

79. Cottafava, D.; Gambino, P.; Baricco, M.; Tartaglino, A. Energy efficiency in a large university: The UniTo experience. In Proceedings of the Sustainable Built Environment. Towards Post Carbon Cities, Turin, Italy, 18–19 February 2016; pp. 92–101.

*Article*

# Hydrogen Economy Model for Nearly Net-Zero Cities with Exergy Rationale and Energy-Water Nexus

**Birol Kılkış [1],\* and Şiir Kılkış [2]**

[1]  Energy Engineering Graduate Program, Başkent University, Ankara 06790, Turkey
[2]  The Scientific and Technological Research Council of Turkey, Ankara 06100, Turkey; siir.kilkis@tubitak.gov.tr
\*  Correspondence: birolkilkis@hotmail.com

Received: 15 March 2018; Accepted: 2 May 2018; Published: 10 May 2018

**Abstract:** The energy base of urban settlements requires greater integration of renewable energy sources. This study presents a "hydrogen city" model with two cycles at the district and building levels. The main cycle comprises of hydrogen gas production, hydrogen storage, and a hydrogen distribution network. The electrolysis of water is based on surplus power from wind turbines and third-generation solar photovoltaic thermal panels. Hydrogen is then used in central fuel cells to meet the power demand of urban infrastructure. Hydrogen-enriched biogas that is generated from city wastes supplements this approach. The second cycle is the hydrogen flow in each low-exergy building that is connected to the hydrogen distribution network to supply domestic fuel cells. Make-up water for fuel cells includes treated wastewater to complete an energy-water nexus. The analyses are supported by exergy-based evaluation metrics. The Rational Exergy Management Efficiency of the hydrogen city model can reach 0.80, which is above the value of conventional district energy systems, and represents related advantages for $CO_2$ emission reductions. The option of incorporating low-enthalpy geothermal energy resources at about 80 °C to support the model is evaluated. The hydrogen city model is applied to a new settlement area with an expected 200,000 inhabitants to find that the proposed model can enable a nearly net-zero exergy district status. The results have implications for settlements using hydrogen energy towards meeting net-zero targets.

**Keywords:** hydrogen; hydrogen economy; renewable energy; photovoltaic thermal; wind turbine; biogas; geothermal energy; exergy; low-exergy buildings; net-zero targets

---

## 1. Introduction

Hydrogen production from renewable energy sources based on options for power-to-gas or power-to-liquid is one of the essential components of smart energy systems, which require the integration of smart electricity, thermal, and gas grids [1]. Smart energy systems are deemed as the most feasible approach towards 100% renewable energy solutions [2]. In this context, electrolysers and fuel cells are options to allow energy systems to gain flexibility [3]. A hydrogen economy that encompasses an entire supply chain based on hydrogen energy from production to usage [4] is also a valid option for supporting progress towards cleaner, smarter, and integrated energy systems.

Among related studies, an outlook for hydrogen as an energy storage medium and energy carrier in renewable energy systems for islands, including water, waste treatment, and wastewater treatment, was put forth for Porto Santo Island [5,6]. Future scenarios for the energy system of Denmark [7] were undertaken with the aim of enabling a hydrogen economy. Those for Italy [8] involved the use of hydrogen energy to increase energy system flexibility. In contrast, studies that undertake the integration of hydrogen-based options at the urban level as a whole for districts and cities are still limited. One of the examples may be given from the analyses of Sveinbjörnsson et al. [9] who evaluated a smart energy system for Sønderborg in Denmark. As a contribution to these and other studies, the present research work provides a hydrogen economy based model for districts, including original metrics and an extended outlook to energy-water relations in the urban context.

At the building level, Singh et al. [10] had presented the selection and analysis of a hybrid energy system for an academic building, including a system configuration that involved a system of solar photovoltaic (PV) arrays, an electrolyser, a hydrogen fuel cell, and a hydrogen storage tank. Cao et al. [11] analyzed a zero-energy building with a ground-source heat pump (GSHP), solar PV panels and/or a wind turbine according to the geographical context, and a hydrogen vehicle in a vehicle-to-building (V2B) scheme. As other related developments in the field of hydrogen systems, Reuß et al. [12] analyzed hydrogen production from electrolysis and its seasonal storage, transport, and fuelling means, including liquid organic hydrogen carrier tanks, trailers, and stations. Nabgana et al. [13] overviewed developments in hydrogen production from biomass using steam reforming. In addition, Qolipour et al. [14] compared options to produce hydrogen from wind power plants, PV, and hybrid PV-wind power plants, of which the latter was found to be more feasible. Tebibel et al. [15] proposed an off-grid system with a PV array, an aqueous methanol ($CH_3OH$) tank, an electrolyser that produces hydrogen from $CH_3OH$, and a hydrogen tank to supply hydrogen on demand. The proposed system was found to be more suitable than the selection of an option for hydrogen production based on water electrolysis at the location of Algiers. In contrast, these studies did not provide a district energy model with hydrogen, solar, and wind energy utilization.

In the urban transport context, Xu et al. [16] calculated the quantity of fuel cell vehicles on the road and the daily hydrogen demand in Shenzhen, China to the year 2025. The quantities were estimated based on cautious, moderate, and optimistic scenarios. Mohareb and Kennedy [17] used the Pathways to Urban Reductions in Greenhouse Gas Emissions modeling tool to analyze possible scenarios for Toronto, including hydrogen fuel cell vehicles. Miranda et al. [18] analyzed the energy management system of a prototype city bus using a hybrid electric-hydrogen fuel cell powertrain that was demonstrated during the Rio Olympics. In addition, Franzitta et al. [19] evaluated the use of electricity from wind and wave farms as well as solar energy to produce hydrogen for fuel cells to substitute diesel fuel in the public transport fleets of the city of Trapani and island of Pantelleria in Italy. Briguglio et al. [20] further analyzed possible uses of hydrogen energy for urban mobility in another Italian city. At the country level, Moreno-Benito et al. [21] modeled the required quantity of hydrogen production to satisfy transport demands in the next 50 years for the United Kingdom. In contrast, additional recommendations to shift modes of transport from the use of private vehicles to public mass transit were not given, which could further reduce carbon dioxide ($CO_2$) emissions.

It is possible to evaluate multiple sectors with relevance for urban areas from an urban systems perspective. Oldenbroek et al. [22] analyzed the possibility of a 100% local renewable energy system to provide for the energy needs of power, heat, and transport in an urban area. The options were based on solar, wind, and fuel cell options with hydrogen as an energy carrier. The proposed energy system was applied to a hypothetical smart city area as an average city based on European statistics. The possibility of eliminating high and medium voltage electricity grids was assessed. This study, however, did not involve energy self-sufficiency or near-zero targets and exergy-based analyses.

Other studies focused on hydrogen production from available sources at the city or industrial complex vicinity with a technological focus. For example, Kumar et al. [23] evaluated the prospects of valorizing industrial wastewater for biological hydrogen production and techniques to increase the hydrogen yield. Nahar et al. [24] reviewed the technological options for producing hydrogen from biogas in India, including industrial wastewater and landfill gas. Khan et al. [25] concluded on the applicability of the use of microbial electrolysis cells in replacing conventional technologies for municipal wastewater treatment technologies. In contrast, none of these studies addressed the need to plan for a more closed urban water cycle or compare possibilities to progress in net-zero targets.

Among other necessities, the need to address an energy-water nexus in the water treatment sector is crucial [26]. This need also extends to processes of water desalination when this option may be valid or required in a given local context. Rather than the use of fossil fuels, solar thermal, solar PV [27], hybrid solar PV-wind, geothermal, and wave energy [28] as well as hybrid wave-solar [29] systems can be used to satisfy the energy intense demands of water desalination. In this respect, Viola et al. [26] used an island as a laboratory to experiment with the use of wave energy to support cleaner energy options for water desalination. At the same time, studies that span across hydrogen energy, an urban systems perspective that extends to the water sector, and net-zero targets remain to be addressed. For example, Sanseverino et al. [30] conceptualized a "net zero energy island" based on the use of solar, wind and geothermal energy while hydrogen energy as an energy carrier was not involved. In contrast, Da Silva et al. [31] analysed prospects for a hydrogen production plant in Brazil based on electricity from solar, wind and hydropower for export to neighbouring countries. Despite the combined use of renewable energy sources for hydrogen production, the study focused on a centralized approach at the country level without considerations of an energy-water nexus.

Most recently, Alanne and Cao [32] reviewed small-scale options for hydrogen economy in buildings and communities and proposed that future research work may be directed to "zero-energy hydrogen economy" (ZEH$_2$E) concepts where hydrogen is the main energy carrier. Based on the most recent literature, it is therefore evident that there is a knowledge gap for integrating hydrogen economy models for urban renewable energy systems, especially those that involve net-zero targets.

Moreover, hydrogen economy models for urban systems may be supported with guidance based on metrics that involve the quality of energy, namely, exergy. Exergy is a measure of the useful work potential of energy. Unlike energy, exergy is irreversibly destroyed according to the Second Law of Thermodynamics while temperatures converge to thermal equilibrium with a given reference environment [33]. In this way, this research work seeks to put forth hydrogen economy models in the urban context based on renewable energy using exergy metrics and net-zero targets. The framework and the analytical results are expected to be instrumental for engineers and city planners in integrating a multitude of renewable and waste energy resources at the urban level.

*Aims of the Research Work*

The main objective of this research is to develop a hydrogen economy model for nearly net-zero cities with a holistic approach. The metrics involve those from the Rational Exergy Management Model (REMM), which provides an analytical framework based on exergy in planning for $CO_2$ mitigation measures, including those for districts that may seek to reach net-zero targets [34]. This necessitates that energy resources, including renewable energy, are allocated with the priority of ensuring better compatibility in exergy levels to streamline primary energy spending [34]. Among others, REMM has been applied to districts [35,36], university campuses [37], airports [38] and dairy farms [39] while applications that involve hydrogen production based on renewable energy and its utilization within the urban context remain to be analyzed as a further basis for the present study.

The paper proceeds to the method of the research work and the metrics that are utilized. As an additional novelty of the research work, net-zero targets for a hydrogen community are combined with an energy/exergy and water nexus perspective. To achieve the main aim, multiple hydrogen cycles for the urban context are envisioned and analyzed, including comparisons to conventional district energy systems. The analyses are extended to an application that involves a new settlement.

## 2. Method of the Research Work

The large-scale mobilization of renewable and waste energy resources is required for a net-zero or net-positive concept based on exergy at large, such as at the district and city levels. In addition, the hybridization of systems with energy conversion and distribution systems that are connected to respective demand points is necessary. This must be planned at an optimum mix based on local conditions, constraints as well as options for effective and efficient distribution, energy storage, and cogeneration. The concept of a hydrogen economy can provide a valid response in several aspects:

- Hydrogen may be produced by renewable energy resources to provide a suitable energy storage and distribution system.

- Hydrogen may be distributed even with existing natural gas pipelines [40] given upgrades involving hydrogen meters and sensors [41].
- Hydrogen is a suitable fuel for fuel cells, which are in essence a cogeneration system.
- With optimum design and operation, exergy destruction in a hydrogen economy may be minimal.
- Hydrogen production may be realized in a closed-cycle energy-water nexus in a district energy system.

In addition, a hydrogen distribution network based on existing natural gas pipelines can consume less pumping energy than the district hot and cold-water piping in conventional systems. Western Europe already has a hydrogen gas pipeline network with a total length of 1500 km [40].

These and other aspects indicate that hydrogen economy can have multiple attributes for a more efficient energy supply base in districts. This research work acknowledges that the existing unresolved issues of future net or near net-zero cities and districts based on hydrogen economy is an important knowledge gap in the literature. To fill such a gap, an exergy-based hydrogen economy model is put forth with proper evaluation metrics and compared to a baseline district energy system.

The proposed hydrogen economy for nearly net-zero districts based on exergy is coupled with the hybridization of several systems, such as solar photovoltaic thermal (PVT), wind turbines, fuel cells, poly-generation systems, organic Rankine cycle (ORC) and heat pumps with biogas and/or geothermal energy. According to REMM, the level of exergy matches in a district must be improved to minimize related $CO_2$ emission responsibilities. This includes comparisons based on the avoidable $CO_2$ emissions impact due to exergy destruction that takes place within the boundaries of the district. Improvements in the level of exergy match are compared based on respective Exergy Flow Bars [42].

In the proposed energy system, two cycles of a hydrogen economy at district and building levels are analyzed in an exergy-based framework. Comparisons with a geothermal energy option are further put forth to evaluate integration possibilities. The model is applied to the planning of a new settlement with 200,000 inhabitants that is conceived as a case study of the research work.

In the first cycle, hydrogen gas is produced by electrolysis of water in the district power plant based on wind turbines with double-blade arrangement [43] and third-generation PVT panels. PVT panels were designed such that coolant fluid has minimum pumping requirements by extensively using heat pipes in the PVT modules with internal thermal energy storing capability. The embedded layer contains phase change material (PCM) to obtain efficiency improvements. Experimental data on the PVT modules are conducted and integrated into the analyses and the case study. Accordingly, low-pressure hydrogen is supplied to the district through a network of hydrogen pipelines.

The second cycle is the hydrogen utilization in each low-exergy building based on building scale fuel cells to satisfy virtually all types of domestic energy demands. Power that is produced by all energy systems, including the fuel cell unit, is in direct current (DC) electricity form. Buildings are equipped with low-exergy heat distribution/absorption equipment, such as radiant wall, ceiling and floor panels, chilled beams, desiccant type of humidity controls, and high-efficiency appliances, faucets, and drainage systems. In the buildings, fuel cells also produce water and heat. The heat is used in low-exergy space heating systems and for domestic hot water (DHW) subject to temperature peaking. Absorption chillers produce cold and their waste heat is collected. Separate large-scale thermal energy storage systems (TES) with different exergy levels are utilized in the buildings.

Moreover, rainwater is collected and utilized in the water supply system. In an energy-water nexus, water is cycled between the plant where it is first electrolyzed to produce hydrogen and then recovered mostly in the fuel cells at the power plant and the buildings. Make-up water is supplied by the building fuel cells, treated wastewater from the district grid, and sea (lake) water, if nearby or feasible to transport. In the latter case, seawater is converted to fresh water by light-assisted catalysis oxidation where power is received from the plant fuel cells. This is an important aspect of the system to close the energy-water nexus. The possibilities of directly connecting biogas generation based on city waste and low-enthalpy geothermal energy resources are also evaluated for further utilization.

Prior to the application of the method to realize the analyses in this research work, a justification of an exergy-based framework is put forth based on two examples, particularly those that involve net-zero buildings and Coefficient of Performance (*COP*) based on exergy principles. These examples are used to emphasize the crucial role of the Second Law of Thermodynamics in addressing major urban challenges, most importantly $CO_2$ mitigation. Needs for the exergy-based metrics that are used to evaluate the hydrogen city model are further put forth with discussions.

## 2.1. Near-Zero Targets for Buildings and Districts

Net-zero energy buildings (NZEB), near zero energy buildings (nZEB) and net positive-energy buildings (NPEB) [44] are gaining importance in the quest of reducing $CO_2$ emissions towards reaching the goals of the Paris Agreement. At the same time, there are still issues to be resolved [45]. A major issue that is not addressed in the building and energy sector is the fact that renewable energy resources and systems in the built environment have or require different energy quality or exergy levels. With an increasing share of renewable energy resources, differences in exergy levels need to be identified to ensure an exergy balance between the supply (resource) and the demand points in the built environment. In addition, the importance of renewable energy resources in optimum and net-positive solutions has to be acknowledged [46]. The First Law of Thermodynamics is necessary but not sufficient to address these problems as demonstrated in the following contexts.

### 2.1.1. Necessity for Net-Zero Exergy Targets

In addition to the exchange of electricity, the exchange of heat through NZEBs can support district networks [47]. At the same time, thermal energy at different temperatures means variation in quality. Several shortcomings of the NZEB definition may be inferred from references [34,48]:

- Thermal energy exchange definitions must distinguish between different forms of heat with different exergy levels, such as steam, hot water, service water, and cold water.
- The quality of energy exchange needs to be embedded into the nZEB definition.
- The impact of the exchanged energy quality must be considered when calculating emissions.

Hence, differences in the energy received from and supplied to a district energy system must be considered. For example, a NZEB may exchange electrical and thermal power with a district energy system. The building may receive 10,000 kWh of alternating current (AC) electrical energy with an average power rms of 5% and provide 10,000 kWh AC electrical energy with an average power rms of 10% annually. The building may also receive 15,000 kWh of heat in the form of hot water from the district at an average supply temperature of 353 K (80 °C) and provide 15,000 kWh of thermal energy to the district at an average temperature of 343 K (70 °C). From the ideal Carnot cycle with reference environment temperature of 283 K, the thermal exergy exchange between the building and district, namely $E_{xsup}$ as the supplied exergy (Equation (1)) and $E_{xret}$ as the returned exergy (Equation (2)) is:

$$E_{xsup} = \left(1 - \frac{283 \text{ K}}{353 \text{ K}}\right) \times 15,000 \text{ kWh} = 2974.5 \text{ kWh} \tag{1}$$

$$E_{xret} = \left(1 - \frac{283 \text{ K}}{343 \text{ K}}\right) \times 15,000 \text{ kWh} = 2623.9 \text{ kWh} \tag{2}$$

By definition, this building is a net-zero energy building with an exact annual exchange of 15,000 kWh with the district but has a deficit based on the exergy levels of the energy amount that is exchanged. The qualities of the exchanged electrical energy are also different in terms of power quality characteristics, possibly due to the electronics involved in the DC to AC power conversion.

Evidently, the building in Equations (1) and (2) is not a building that satisfies the NZEXB target. In order to account for an exergy balance, a Net Zero Exergy Building (NZEXB) was defined, which generates energy at the same grade and quality as consumed on an annual basis while involving exchanges with the grid [48]. Such a definition is important especially when renewable energy systems become more diversified and coupled to the district at different exergy levels [49–51].

### 2.1.2. Exergy-Based Coefficient of Performance

Figure 1 represents the energy and exergy flow of a GSHP driven by grid electricity [52,53]. The electrical power input to a GSHP is utilized with a given *COP* value at given operating conditions to supply thermal energy. From an exergy perspective, the GSHP needs to have such a *COP* value that the exergy of the electrical power supply ($\varepsilon_{in}$) is at least equal to the exergy of the thermal output ($\varepsilon_{out}$). Equation (3) defines a minimum *COP* as $COP_{min}$ that reaches this threshold for a temperature output ($T_{out}$) of 55 °C (328 K) and an environment reference temperature ($T_{ref}$) that is equal to 283 K.

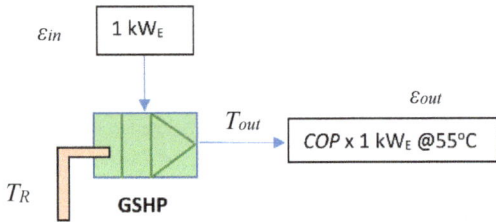

**Figure 1.** Exergy input and output for GSHP.

$$COP_{min} = \frac{1}{\left(1 - \frac{283 \text{ K}}{328 \text{ K}}\right)} = 7.28 \tag{3}$$

The example shown in Figure 1 indicates that most conventional heat pumps will have an exergy-based *COP* value ($COP_{EX}$) that is less than one according to Equation (4) even if an optimum $T_{out}$ is found. In Equation (5), an optimum $T_{out}$ is based on maximum $COP_{EX}$ for a given reservoir temperature and $T_R$ considering function constants $a$ and $b$ that are linearized for a given heat pump. Combining Equations (4) and (5), taking a derivative of the product, and equating it to zero gives the optimum $T_{out}$ value in Equation (6) as put forth within the method of this research work:

$$COP_{EX} = COP \times \frac{\varepsilon_{out}}{\varepsilon_{in}} = COP \times \frac{\left(1 - \frac{T_{ref}}{T_{out}}\right)}{\varepsilon_{in}} \tag{4}$$

$$COP = a - b(T_{out} - T_R) \tag{5}$$

$$T_{out} = \sqrt{T_{ref}\left(T_R + \frac{a}{b}\right)} \tag{6}$$

New developments are promising in making heat pumps exergetically feasible above the threshold value in Equation (3). These include water-source heat pumps with heat recovery that has a heating *COP* of 8.15 and a cooling Energy Efficiency Ratio (EER) of 5.02 [54]. With technological advances, heat pumps may perform better in hybridized applications that involve hydrogen energy (see Section 3). The Primary Energy Ratio (PER) definition can also be advanced with a Primary Exergy Ratio (PEXR) definition as put forth in Equation (7) that considers a power plant with a First Law efficiency $\eta_I$ and a heat pump with $COP_{EX}$. If $\eta_I$ is 0.3 for a conventional power plant running on fossil fuels and $COP_{EX}$ is 0.49 as in Figure 2 (the blue circled point), PEXR is 0.147. This means that a heat pump uses only 14.7% of the exergy available in the fossil fuel consumed at the power plant. In contrast, the *PER* definition would give a result of 0.3 times 2.85, which is 0.86:

$$PEXR = \eta_I \times COP_{EX} \quad \{\text{Quality flow of energy from the primary resource}\} \tag{7}$$

Evidently, the utilization of exergy-based analyses is necessary to effectively show the quality flow of energy rather than the quantity flow [55] in related analyses, design, and operation steps.

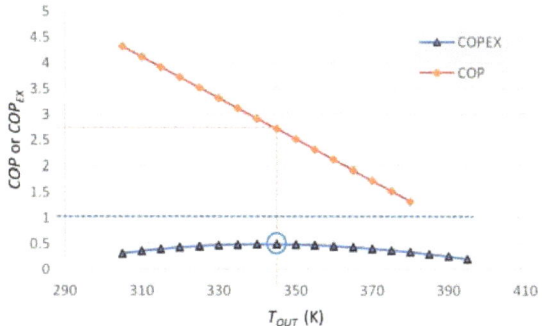

**Figure 2.** Sample variation of *COP* and $COP_{EX}$ with $a = 5$, $b = 0.04 \text{ K}^{-1}$, $T_R = 288 \text{ K}$, $T_{ref} = 283\text{K}$.

### 2.1.3. Exergy-Based Formulations for a Nexus Approach

Equations (8)–(13) put forth additional formulations that are used in the evaluation of the hydrogen city model. The unit exergy of each 1 kWh of the supply heat ($\varepsilon_{sup}$) according to the ideal Carnot cycle is given in Equation (8). Here, $T_{sup}$ is the supply temperature. Similarly, Equation (8) is adapted for unit destroyed exergy ($\varepsilon_{des}$), unit demand exergy ($\varepsilon_{dem}$), and unit returned exergy ($\varepsilon_{ret}$):

$$\varepsilon_{sup} = \left(1 - \frac{T_{ref}}{T_{sup}}\right) \times (1 \text{ kWh}) \quad \{\text{Unit Exergy}\} \tag{8}$$

$$E_x = \varepsilon_{sup} \times Q_{sup} \quad \{\text{Energy and Exergy}\} \tag{9}$$

The basis for establishing the energy, exergy, and environment nexus is provided by the exergy magnitude $E_x$, which is based on $\varepsilon_{sup}$ and magnitude of thermal energy $Q_{sup}$ (Equation (9)), REMM efficiency (see Equations (10) and (11)) and $CO_2$ emissions (see Equations (12) and (13)), respectively. The latter formulations are based on REMM in which a mismatch in the supply and demand of exergy is linked to additional primary energy spending in the energy system and related $CO_2$ emissions [34].

In Equation (10), $\psi_R$ is the metric for the exergy utilization rationale, namely the Rational Exergy Management Efficiency [34]. The formulation is for cases that involve power generation. If in any process, major exergy destruction takes place upstream of the useful application at the absence of power generation, then Equation (10) is replaced based on a re-arrangement of terms as in Equation (11) [34]. A weighted mean value is used when multiple energy outputs are involved:

$$\psi_R = 1 - \frac{\sum \varepsilon_{des}}{\varepsilon_{sup}} \quad \{\text{Rationality of Exergy Use}\} \tag{10}$$

$$\psi_R = \frac{\varepsilon_{dem}}{\varepsilon_{sup}} \tag{11}$$

By definition, the annual average of $\psi_R$ must be at least equal to 0.80 for any connected building in a hydrogen economy district with the aim of obtaining a better exergy match. This is instrumental for reducing available $CO_2$ emission impacts in the energy supply due to any need to re-supply primary energy resources. Equation (12) defines the compound $CO_2$ emissions, which includes avoidable emissions due to exergy destruction in a process as represented by the term $(1 - \psi_R)$ [34]:

$$\sum CO_2 = \left[\frac{c_l}{\eta_l} + \frac{c_m}{\eta_m \eta_T}(1 - \psi_R)\right]Q_H + \frac{c_m}{\eta_m \eta_T}E \quad \{\text{Environment}\} \tag{12}$$

Equation (12) as formulated in REMM [34] establishes the metric to evaluate the environmental dimension of the nexus. The first term within the square brackets is the direct $CO_2$ emissions from an on-site (local) energy conversion unit, such as a boiler with a thermal efficiency, $\eta_l$, which satisfies a thermal load $Q_H$. Here, $c_l$ is the $CO_2$ intensity of the energy resource that is used locally on-site. In conventional thermal systems, exergy is usually destroyed upstream of the thermal load. Hence, the second term within the square brackets derives from the forgone power generation opportunity as a function of the destroyed exergy $(1 - \psi_R)$ while satisfying a thermal load $Q_H$. This second term is the avoidable $CO_2$ emissions impact, which is associated with a power plant at the energy system level that in effect has to compensate for the forgone opportunity of generating power on-site. The variable $c_m$ is the $CO_2$ intensity of the energy resource that is used at the power plant and $\eta_m$ is the power generation efficiency of the power plant. According to an energy system boundary, the variable $\eta_T$ is the overall efficiency of power transmission and power feeding. The last term in Equation (12) is the $CO_2$ emissions that take place to satisfy the on-site electrical power demand, $E$.

For a net-zero $CO_2$ building (NZCB) or district, Equation (12) implies that renewable energy resources must be used ($c_l$ and $c_m$ approach zero) and exergy mismatches must be reduced for $\psi_R$ to approach one. In addition, the Ratio of Emissions Difference (*EDR*) as given in Equation (13) must be close to one. Here, the $CO_{2base}$ term is the standardized emission rate with unit defaults for 0.5 kWh thermal ($Q_H$) and 0.5 kWh electrical power demand ($E$) with a power to heat ratio ($C$) of one. Other default values include 0.2 for $\psi_R$ for an energy system that does not involve any combined heat and power (CHP) with renewables. In Equation (14), $CO_{2base}$ is 0.63 kg $CO_2$ per 1 kWh total energy load based on Equation (12). The $CO_{2base}$ is compared within *EDR* for a given hydrogen economy option:

$$EDR = 1 - \frac{[CO_2/(Q_H + E)]}{CO_{2base}} \tag{13}$$

$$\Sigma CO_{2base} = \left[ \frac{0.2 \text{ kg } CO_2/\text{kWh}}{0.85} + \frac{0.2 \text{ kg } CO_2/\text{kWh}}{0.35}(1 - 0.2) \right] \times 0.5 \text{ kWh} + \frac{0.2 \text{ kg } CO_2/\text{kWh}}{0.35} \times 0.5 \text{ kWh} = 0.63 \text{ kg } CO_2 \tag{14}$$

### 2.1.4. Definition of a Composite Rationality Indicator

The efficiency of energy activities can be improved based on at least six major parameters:

1. Type of fuel or renewable energy source
2. Equipment and plant energy efficiency
3. Rational Exergy Management Efficiency ($\psi_R$)
4. Thermal loads
5. Plant and grid power transmission efficiency, transformer losses, etc.
6. Power loads

The trend of transitioning to renewable energy is already improving the first parameter. The second parameter, namely the equipment efficiency, is also improving as CHP, condensing boilers, and other energy technologies are approaching theoretical limits so that there is limited room for improvement. Parameters 4, 5, and 6 are also on the right track with smart grids, DC underground lines, and energy saving measures for thermal and electrical loads. In contrast, the third parameter $\psi_R$ remains unresolved although it has large room for improvement. This parameter is important since the current average value for most cities is less than 0.3 [51]. This value will substantially improve by addressing more structural issues in the energy system, namely imbalances between the supply and demand of exergy. Re-thinking exergy aspects can support innovative combinations of technology in a circular economy approach, improve urban quality, and reduce $CO_2$ emissions.

Given both quantity and quality oriented efficiency aspects, a new indicator that combines the First and Second Law efficiencies is defined as a Composite Rationality Indicator, $C_R$. Equation (15) is valid for the use of energy efficiency values that may also be *COP* in Equation (16). The defined $C_R$ is used to compare proposed options, including possible uses of geothermal energy.

$$C_R = \eta_I \times \psi_R \quad \text{or,} \tag{15}$$

$$C_R = COP \times \psi_R \tag{16}$$

### 2.1.5. Exergy-Based Net and Near-Zero Definitions

Net-zero targets based on exergy are valid for buildings and districts as developed in previous phases of the research work and summarized in Table 1. Prior to these definitions, various applied definitions for a Low-Exergy Building (LowExB) were present [56], which may be considered as a building that satisfies its heating loads with low-exergy sources at about 40 °C and sensible cooling loads at about 15 °C to 18 °C [57]. All such definitions have been put forth for approval in ASHRAE Technical Committees, namely Exergy Analysis for Sustainable Buildings and Terminology based on [48–51]. In Table 1, related definitions are also harmonized based on above Equation (9) or Equation (13). Based on Table 1, for example, a nearly Zero Exergy Building (nZEXB) is a building or building cluster that is connected to the district returning at least 80% of the total exergy of heat and power to the district as the total exergy of heat and power supplied from the district annually.

**Table 1.** Building and District Level Net and Near Zero Definitions Based on Exergy.

| Building or District Target | Acronym | Ref. | Definition | Equation |
|---|---|---|---|---|
| Net-Zero Exergy Building | NZEXB | [34,48] | $E_{xsup} = E_{xret}$ | (9) |
| Nearly Zero Exergy Building | nZEXB | [38] [a] | $E_{xret} \geq E_{xsup} \times 0.8$ | (9) |
| Net Positive Exergy Building | NPEXB | [38] [a] | $E_{xret} \geq E_{xsup}$ | (9) |
| Net-Zero Exergy District | NZEXD | [35,36] | $E_{xsup} = E_{xret}$ | (9) |
| Near Net-Zero Exergy District | nZEXD | [35,36] | $E_{xret} \geq E_{xsup} \times 0.8$ | (9) |
| Net-Zero $CO_2$ Building | NZCB | [51] [a] | $EDR = 1.0$ | (13) |
| Near Zero $CO_2$ Building | nZCB | [51] [a] | $0.8 \leq EDR < 1.0$ | (13) |
| Net-Zero $CO_2$ (Emissions) District | NZCD/NZCED | [36] [a] | $EDR = 1.0$ | (13) |
| Near Zero $CO_2$ District | nZCD | [36] [a] | $0.8 \leq EDR < 1.0$ | (13) |

[a] Extended in the present manuscript based on the defined $E_{xret}$ or $EDR$ conditions.

A Net Positive Exergy Building (NPEXB) supplies a surplus of total exergy of heat and power to the local district energy system when compared to the total exergy of heat and power received from the district energy system on an annual basis.

At a district level, a Net-Zero Exergy District (NZEXD) [35,36] is a district that has its own local centralized and/or distributed energy system with any sub-stations in the same district so that the same total exergy of heat and power is supplied by the local district energy system as the total exergy of heat and power used in the district on an annual basis. In this context, lower temperature supply networks [1,2] that take place in Fourth Generation District Energy Systems (4GDE) can support the NZEXD target. Figure 3 shows the relation between NZEXD and NZEXB targets. By definition, the parameter $\psi_R$ must be equal to or greater than 0.80.

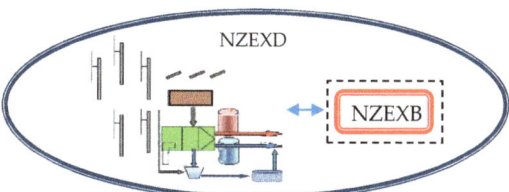

**Figure 3.** NZEXD and NZEXB Targets.

### 3. Characterization of the Hydrogen City Model

Based on the method, a hydrogen city model is characterized based on two cycles at the district and building levels as described subsequently. Both of the cycles support the hybridization of energy options in the district energy system for the effective use of renewable energy sources.

*3.1. Main Cycle of the Hydrogen City at the District Level*

The first cycle consists of the central CHP plant (Figure 4). The CHP system runs on locally produced biogas from city wastes. Wind turbines and solar PVT systems are further combined to generate on-site electricity. Surplus renewable electricity is utilized in an on-site hydrogen production facility by the electrolysis of water. The produced hydrogen is stored in high-pressure tanks and upon demand, de-pressurized below 100 bar and then served to the city-wide grid. The central fuel cell system generates DC electricity that is supplemented by the DC electricity, which is generated by wind and solar energy systems. A smart low voltage DC (LVDC) micro-grid serves the district along with all information and data services. Hydrogen is about 1.5 times more energy dense compared to natural gas. The higher heating value (HHV) of hydrogen is 142 MJ/kg that favorably compares with natural gas that has a HHV of 52 MJ/kg [58]. This allows hydrogen to be better suited for being distributed in the district. In addition, the stored hydrogen is partly used to enrich the biogas that is used in the central CHP plant to generate AC electricity for the city infrastructure, mass transport systems, and industry. Biogas enriched with hydrogen increases the net reaction rate with higher addition ratios of hydrogen, thereby improving combustion [59]. The reaction of the $CO_2$ in biogas with hydrogen in a Sabatier process substitutes conventional upgrading units [60].

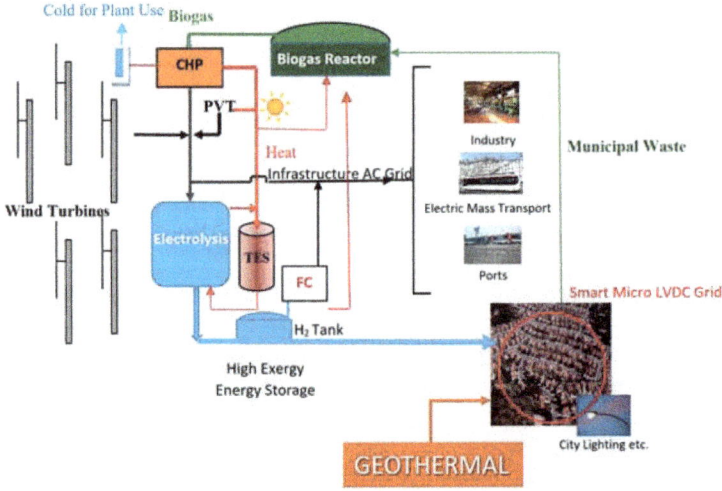

**Figure 4.** Hydrogen-Solar-Wind District Plant in the Energy-Water-Environment Nexus.

The production of hydrogen from numerous renewable energy sources as given in Figure 4 can provide the basis for a more stable and sustainable energy supply profile for the district. Among the renewable energy options, double-blade wind turbines are considered to expand the feasible operational wind speed range by starting at low speeds and sustaining generating power [43]. These turbines are located only in and around the district plant due to the relatively high turbine noise.

Figure 5 shows the water cycle in the main cycle of the hydrogen city where water recycling takes place between the plant where it is first electrolyzed to produce hydrogen and the fuel cells at the central plant. Additional water input as make-up water includes treated wastewater and any light-assisted catalysis oxidation from seawater with the partial use of the power that is generated by the fuel cells. The integration of treated wastewater into the hydrogen production plant alongside any additional fresh water sources provides an opportunity to attain a more closed water cycle.

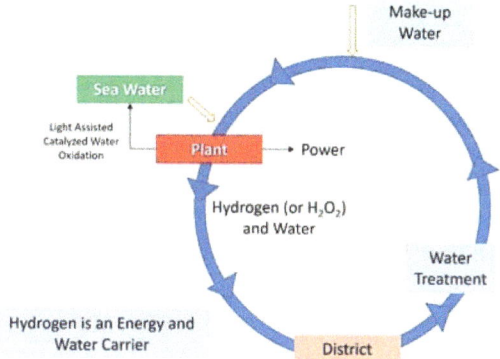

**Figure 5.** Closing the Water Cycle and Generating Fresh Water from the Sea or a Large Lake.

Integration with Solar PVT-3 System

In solar PV systems, the electrical efficiency is a function of temperature with higher panel temperatures resulting in lower efficiency. PVT systems can stabilize the PV efficiency despite hotter panel surfaces. In many cases, water is circulated through heat exchanging pipes on the backside of PV panels. However, proper control of the circulation pump flow rate is essential to minimize motor power consumption. The water output temperature needs to be low if the PV is to be cooled effectively and vice versa. It is also important to maximize the total exergetic efficiency by controlling the coolant flow rate by recognizing that power and heat have different exergy levels.

PVT systems become more feasible in warmer and hot climates in which PV systems need to be cooled frequently and the temperature of the heated water can satisfy useful applications on-site. Figure 6 provides the feasibility contours of PVT systems based on average solar radiation on a flat surface in Europe. The plant size also makes a difference since unit costs reduce with total surface area of solar radiation, including costs for automation software, hardware, and equipment (e.g., pyranometers).

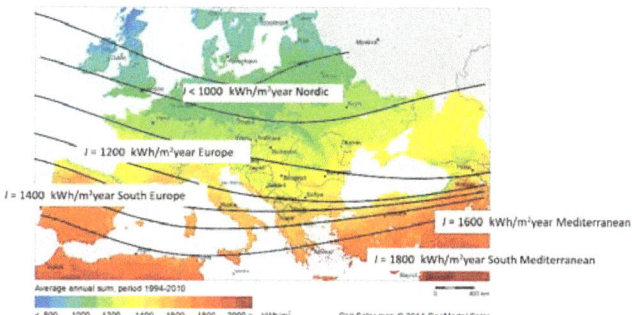

**Figure 6.** Solar PVT Feasibility Map for Different Levels of Solar Irradiation in Europe.

The simple payback periods are evaluated based on PVT area in Figure 7. Based on Figure 7, even smaller systems in residential applications become more feasible and can payback the initial financial investment in a shorter time if the annual solar insolation level, $I$ is high as in Southern Europe and the Mediterranean. The payback period is three years if $I$ is 1800 kWh/m$^2$-year and the PVT area is 200 m$^2$ as denoted in the marking in Figure 7. In contrast, the same-sized PVT plant will have a payback period of 5.2 years in a climatic region with $I$ equal to 1200 kWh/m$^2$-year.

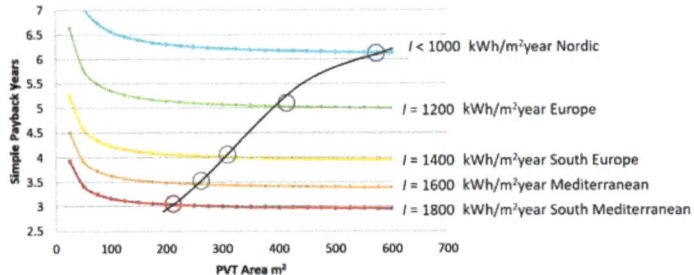

**Figure 7.** Solar PVT Feasibility Diagram for Europe with Different PVT Plant Size.

The proposed hydrogen city includes a novel third-generation solar PVT system, namely PVT-3 that involves multiple layers as shown in Figure 8, including a thermoelectric generator (TEG) layer [61,62]. Thermal energy storage is achieved with an embedded layer of PCM. The circulation pump is eliminated by using heat pipes (HP), which transfer the heat when there is thermal demand and according to the level of solar insolation at the site. The glass cover (GC) and the air gap (AG) over the PV surfaces form a flat plate collector, which is optimized to maximize PVT performance.

After sunrise, solar irradiation enables the generation of power while the undesired heating of the PV panel surfaces takes place. Cooling is effectively achieved by transferring the additional solar heat to the backside of the TEG modules with a heat-conducting nano-sheet (NS). While the packed-bed type PCM layer is thermally charging at a relatively cool temperature, a temperature difference across the TEG units takes place. This temperature difference generates additional DC power. Depending on the thermal demand, heat may be transferred to the external manifold via the heat pipes. After sunset, the PVT-3 module starts to back radiate to the cooler atmosphere from the top surface. This generates a reverse heat flow starting from the bottom of the TEG units via the heat conducting sheet. In turn, additional electrical power with a reverse polarity is generated. A polarity switch corrects the DC output sign. Power generation can be extended after sunset depending on the total PCM mass, temperature distribution, thermal mass, and the material of the module.

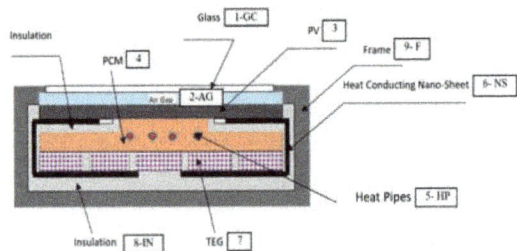

**Figure 8.** Photo-Heat-Voltaic-Thermal (PVT-3) Module (not to scale) [61,62].

Figure 9 shows the PVT-3 test set-up in a horizontal position with packets of PCM material that eliminates the gravity effect of molten PCM in operation. The PVT-3 unit may also be positioned vertically for integration to building façades. In practice, it is difficult to control the flow in a heat pipe. For this reason, a device to control the heat pipes was developed, which eliminates this problem mechanically that is depicted in Figure 10. Figure 11 shows a power output performance curve of the PVT-3 prototype on a typical summer day on a flat surface with 1 m$^2$ area and $I_n$ at 750 W/m$^2$ where $I_n$ is the net solar insolation intensity reaching perpendicular to the solar PV surface. Here, $E_1$ and $E_2$ are the power generated by the PV layer and TEG elements of the PVT-3 module.

**Figure 9.** Experimental Set-up. (Photo courtesy of Varışlı and Aydoğan).

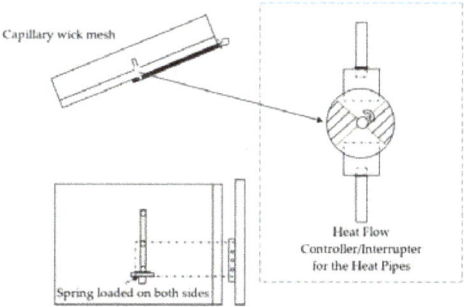

**Figure 10.** Heat Pipe Controls of the PVT-3.

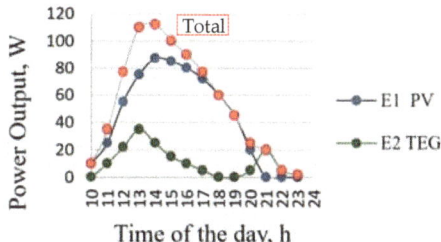

**Figure 11.** Combined power performance of PVT-3 on a typical summer day [61,62].

### 3.2. Second Cycle of the Hydrogen City at the Building Level

A hydrogen city may be considered either for retrofit cities or new green cities in brownfield area developments. One of the first essential steps, however, is to retrofit buildings accordingly or to construct new buildings of a plug-in type that are ready for innovative hydrogen energy systems.

Figure 12 represents a transition to a hydrogen city through the phased introduction of net-positive exergy buildings that are connected to the hydrogen pipeline. A domestic fuel cell is the centerpiece of the building system with close to or even higher than 60% energy efficiency for power generation. The distributed power and heat system in Figure 12 enables a downsizing of the central fuel cell system and eliminates a thermal grid previously servicing buildings. Rather, the central fuel cell system is dedicated to other city infrastructure, mass transit, and industrial applications.

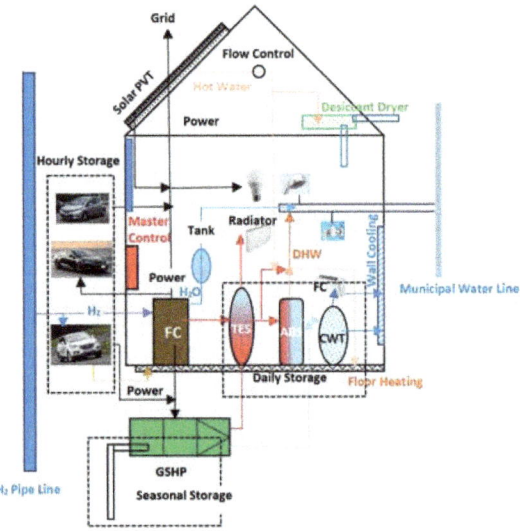

**Figure 12.** All DC-Solar-Central Hydrogen Hybrid Net-Zero/Positive Exergy Building.

The building fuel cell system primarily satisfies the base loads. Other renewables are assisted by domestic daily or weekly TES and are used mainly for satisfying the peak loads, such as cooling in the summer months. Separate TES units at different exergy levels serve the building heating and cooling system. During the summer season, part of the heat is used to temperature peak the reject heat from the absorption cooling system (ABS) for DHW supply to avoid the risk of Legionella bacteria. Cold energy is used in fan-coils for peak loads and in-wall cooling panels are used for the base loads. In the winter season, if there is a cooling load present in the building, the reject heat of the ABS is used for low-temperature space heating through radiant floor systems. If thermal loads are too high, then a GSHP is installed that also serves for seasonal thermal storage in the ground.

The energy supply is complemented by roof-top and façade integrated PVT-3 that generates both power and warm water. The warm water charges the desiccant dehumidification system. In the net-positive exergy building, rainwater, fuel cell water condensate, and wastewater are domestically treated and returned to the plant in a separate water line to close an energy-water nexus (Figure 12).

Hydrogen Building to Hydrogen Car Interaction

There can be four power inputs to the net-positive exergy building of Figure 12, namely the domestic fuel cell, the solar PVT, the grid electricity provided by the central fuel cell system in DC current as well as power inputs from private vehicles, including those from any hydrogen cars.

Private vehicles spend almost 95% of their time in a parked position in or around the buildings [63]. Hourly electrical energy storage is possible by connecting the hydrogen and electric cars to the building power system. In addition, any gasoline-engine car may be a part of the hourly/nightly electrical energy storage system based on car batteries. In total, three types of cars may be docked to the building, namely those with a conventional gasoline engine, an electric car, or a hydrogen car.

The source of supply to electric cars depends on the context of the energy system in which they operate. If an electric car is parked in the building of Figure 12, then car batteries may be charged by the fuel cell system at a much higher efficiency of power conversion using hydrogen gas and with almost zero emissions due to the fact that hydrogen is produced by renewable energy. Even in the case of the conventional gasoline engine car, the car may be connected to the electrical system of the building to provide electricity from its battery that is charged during the daytime while driving.

In the presence of a parked hydrogen car, hydrogen fuel may be received at the building site after pressurization. In turn, the hydrogen car may provide power, water, and heat to the building. Figure 13 depicts a "hydrogen building to hydrogen car" interaction. In this case, a micro hydrogen generating system may be added to the hydrogen building, driven by the excess power generated by the PVT-3 system, possibly during the daytime when the hydrogen car is away off-site and the power load of the building may be at a minimal level according to lower occupancy. This additional generation supplements the main hydrogen grid supply. When the car docks back to the building, the hydrogen stored in the dedicated storage tank is pressurized more and supplied to the hydrogen car. In the meantime, electrical energy derived from the building may charge the backup battery of the car or vice versa to enable nightly electrical energy exchange between the car and the building.

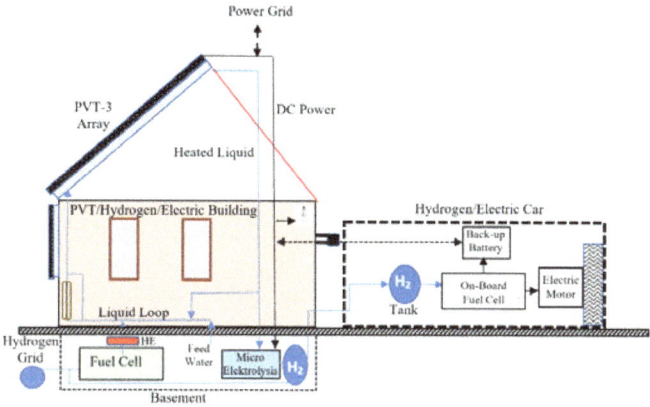

**Figure 13.** Schematic for Hydrogen Building to Hydrogen Car Interaction.

## 4. Results for the Exergy Rationale of the Hydrogen City Model

Figure 14 shows the Exergy Flow Bars [42] for a baseline district energy (DE) system to which the hydrogen city model is compared. The baseline DE system uses natural gas with a combustion temperature of 2000 K while the reference environment temperature is 283 K. The available unit exergy (blue bars) is initially equal to $\varepsilon_{sup}$ that reduces at each proceeding application and point of exergy destruction. The first application is electricity production after which an alternating order of exergy destruction and applications at lower temperature levels take place for heat and cold production. The REMM Efficiency $\psi_R$ of the baseline DE system is calculated from Equation (10) as 0.25 based on two temperature intervals that represent points of exergy destruction. The values of the temperature intervals that determine $\varepsilon_{des(1)}$ and $\varepsilon_{des(2)}$ are given in Table 2 that represent un-used temperature intervals between demanded applications. If the same system involves an additional steam generation process starting from 600 K and ending at 450 K, then the temperature interval for $\varepsilon_{des(1)}$ will be split and reduce to $(1 - 600\text{ K}/700\text{ K})$ plus $(1 - 365\text{ K}/450\text{ K})$, which would increase the REMM efficiency only to 0.42. Exergy losses due to the pumping of any steam if generated and the hot and cold-water circulation in separate circuits in the district are not included in this value.

**Table 2.** Un-Utilized Temperature Intervals between Demanded Applications.

| Case | $\varepsilon_{des}$ | Temperature Intervals (K) | | | |
|---|---|---|---|---|---|
| DE Baseline | $\varepsilon_{des(1)}$ | $T_{dem(1)out}$ | 700 | $T_{dem(2)in}$ | 365 |
| | $\varepsilon_{des(2)}$ | $T_{dem(2)out}$ | 345 | $T_{dem(3)in}$ | 288 |

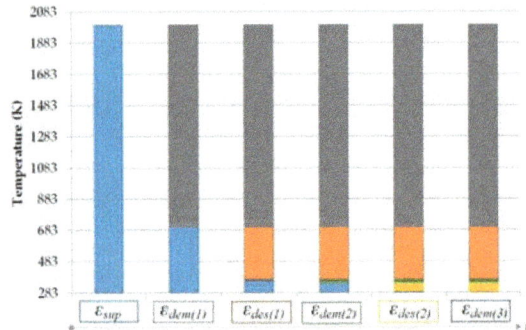

**Figure 14.** Exergy Flow Bars for a Conventional DE System with Power, Heat, and Cold Supply.

The baseline DE system that provides a basis of comparison is to be upgraded to a hydrogen city. Figure 15 puts forth the linkages between the various components in providing exergy supply and useful applications. In the hydrogen city model, the circulation of hot and cold water and any steam is eliminated through the circulation of hydrogen gas, which is less energy intensive. In the upper left component of Figure 15, solar energy is utilized in the PVT-3 system in the plant. First, DC electric power is generated in the PV modules. The heat that is absorbed by the PV coolant is utilized in the thermal charging of the biogas reactor. The exergy flow bar for a typical PVT-3 application at a solar insolation level of 600 W/m² is depicted in the form of the upper left bar in Figure 15. Here, the Carnot cycle equivalent temperature for solar energy $T_{fs}$ is by definition the mapped equivalent source temperature for solar energy at a given insolation level $I_n$ as given by Equation (17) [64]. This enables an exergy accounting with a more consistent boundary other than a Sun-Earth boundary.

$$T_{fs} = \frac{T_{ref}}{1 - 6.96 \times 10^{-4} \times I_n} \tag{17}$$

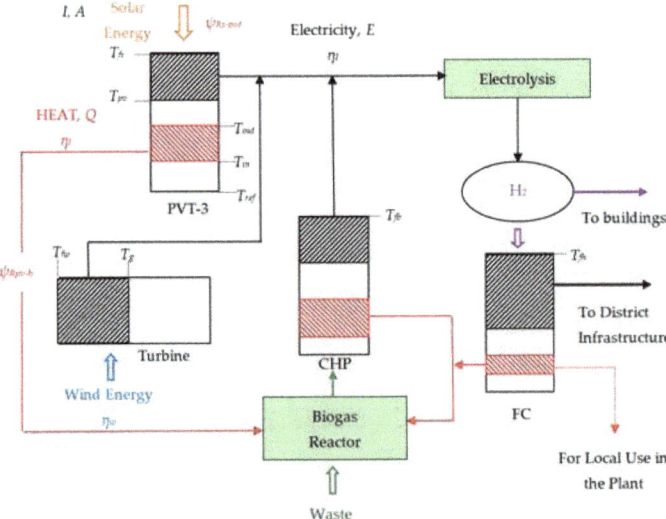

**Figure 15.** Exergy Flow Bars for the District Energy Plant of the Main Cycle Generating Hydrogen from Renewables for Power, Heat, and Cold Supply.

According to Equation (17), if $I_n$ is 600 W/m$^2$, then $T_{fs}$ is 486 K. From Equation (10), the value of $\psi_R$ for the PVT-3 will then be established as given below. This involves typical operating temperature values of $T_E$ as the PV layer average temperature with cooling at 313 K, $T_{out}$ as the average supply water temperature from PVT-3 to the biogas reactor at 303 K, and return temperature $T_{in}$ at 298 K.

$$\psi_{Rs-PV} = 1 - \frac{\left(1 - \frac{303\ K}{313\ K}\right) + \left(1 - \frac{283\ K}{298\ K}\right)}{\left(1 - \frac{283\ K}{486\ K}\right)} = 0.803 \tag{18}$$

In addition to the PVT-3 system, the Carnot cycle equivalent temperature $T_f$ for wind energy at a given mechanical energy of $\eta_W$, namely $T_{fw}$, is given by Equation (19) [42]. Accordingly, if the wind turbine efficiency $\eta_W$ is 0.4, then $T_{fw}$ will be 471.6 K.

$$T_{fw} = \frac{T_{ref}}{1 - \eta_W} \tag{19}$$

The DC power generated by the wind turbine system is combined with the PVT-3 DC output. This power combination is further supported by the CHP system, which is driven by biogas. Cumulative DC power is utilized mainly in the electrolysis process. The generated hydrogen drives the central fuel cell system to generate electricity as needed. As a means of utilizing residual heat to a maximum extent, the residual heat of the CHP system mainly goes to the biogas reactor (see arrows from the red component in Figure 15). Part of the residual heat may be also utilized within the plant such as in the form of DHW. In contrast, the CHP power output may be partially inverted to AC for district infrastructure that is not shown. Based on an analysis of the district energy plant, Figure 15 provides the Exergy Flow Bars in each step and marks the respective exergy components.

### 4.1. Partial REMM Efficiencies and $CO_2$ Emissions Avoidance

Partial REMM Efficiencies, $\psi_{Ri-j}$ between any node $i$ and $j$ is calculated by using Equation (10) or Equation (11) as provided in the method of Section 2. These nodes depend on the locations of the points of exergy destruction either upstream or downstream of the major application. As an overall summation, the averaged REMM Efficiency is found as given in Equation (20) [64]:

$$\overline{\psi}_R = \frac{\sum\limits_{i=1}^{u} \sum\limits_{j=1}^{v} \psi_{Ri-j} E_{xi-j} / \eta_{i-j}}{\sum\limits_{i=1}^{u} \sum\limits_{j=1}^{v} E_{xi-j} / \eta_{i-j}} \tag{20}$$

In Equation (20), $E_{xi-j}$ is the exergy flow between nodes $i$ and $j$. The term $E_{xi-j}$ depends on the energy flow between the same nodes $i$ and $j$ multiplied by the temperature factor $(1 - T_{ref}/T_{out})$ and divided by the First Law energy efficiency. After applying typical operational temperatures and efficiencies for the hydrogen economy cycle, the value of $\psi_R$ bar varies between 0.75 and 0.85. Such a range represents typical daily variation of the efficiency as well as the overall performance. The value of $\psi_R$ bar is higher than district energy systems using fossil fuels. Comparing a value of $\psi_R$ bar at 0.80 for the hydrogen economy cycle and 0.42 for the baseline DE system, the hydrogen city model has a potential of 32% more reduction in $CO_2$ emissions from the built environment stock ($\{[(2 - 0.42)/(2 - 0.80)] - 1\} \times 100 = 31.67\%$).

### 4.2. Additional Exergy Benefit of the Hydrogen City

In conventional district energy systems, thermal energy in the form of heating, cooling, DHW, and sometimes steam is distributed in the district in separate circuits, each having different flow rated pumping requirements. In particular, water circulation that requires pumping power and a piping network is energy/exergy intensive both in embedded and operational terms. These parasitic losses may be up to 10% of the load and even 15% during cooling. The parasitic pumping energy demand for district energy systems and the parasitic losses will be much less in a hydrogen economy as another important advantage. In hydrogen piping, the circuit length practically has no limit. In contrary, depending on the amount of thermal power of different forms to be distributed, there are economical and technical limits on the maximum piping length as given in Equation (21) for heating.

$$L_{max} = a_0 + \left(\frac{Q_H}{1000}\right)^n \times \left(\frac{\Delta T}{20}\right)^{1.3} \qquad \{Q_H > 1000 \text{ kW}_H, \ \Delta T \leq {}^\circ C\} \tag{21}$$

In Equation (21), $Q_H$ is the useful thermal power to be transmitted and $L_{max}$ is the farthest point that a closed thermal circuit may feasibly reach. Here, $a_0$ is an empirical constant, which is generally taken as 0.6 km. The power $n$ depends on the temperature, thus the exergy of the heat supplied, as provided in Equation (22). $T_{ref}$ is 283.15 K while 333.15 K is the traditional supply temperature. For cooling circuits, a similar formulation is applicable.

$$n = 0.6 \times \left(\frac{\left(1 - \frac{T_{ref}}{T_f}\right)}{\left(1 - \frac{T_{ref}}{333.15 \text{ K}}\right)}\right)^{0.33} \qquad \{\text{For heating}\} \tag{22}$$

*4:3. Comparison with a Circular Geothermal Option*

Low-enthalpy geothermal energy sources provide another option for the hydrogen city model if such resources exist in the vicinity (see previous Figure 4). This option emerges from the fact that low-enthalpy geothermal energy sources have about a 30% share among different heat sources that drive ORC systems for electricity generation [65]. The ORC market is rapidly increasing but their expansion is dependent on economic incentives, subsidies, and special tariffs [65,66]. For this reason, the ORC industry is reliant on the economic benefits of producing and selling electrical energy based on favorable conditions without considering the existing possibilities of improving exergy efficiency and acting upon the additional benefits of utilizing the available waste heat [67].

Exergy analysis mainly focuses on the ORC operation and design without a holistic approach based on its connection between the energy source and demand points in the built environment. For example, Rowshanzadeh [68] underlined the wide-ranging applications of ORC technology while pointing out the need for exergy analysis. Sun et al. [69] investigated the suitable application conditions of ORC-Absorption Refrigeration Cycle (ARC) and ORC-Ejector Refrigeration Cycle (ERC) and compared results based on exergy analyses. Marini et al. [70] analyzed an ORC system driven by solar energy with vacuum-tube collectors that provided electrical power for a building. The performance of different working fluids was simulated based on the objective of minimizing exergy destruction to conclude that ORC can be exergetically feasible given careful optimization.

Other studies that evaluated the benefits, risks, and potential disadvantages of ORC systems from a sustainability perspective indicate that ORC units may not be ecologically sound if used in a stand-alone format to generate only electric power [67]. ORC systems need to be bundled with other renewable energy resources, systems, and energy storage units to be acceptable from an exergy point of view [67]. From this perspective, exergy analyses can be used to quantify the advantages and disadvantages of using stand-alone ORC units versus different bundling alternatives with other renewable energy systems. Kılkış et al. [53] indicated that the First Law of Thermodynamics is not sufficient to evaluate ORC systems for maximum performance and environmental sustainability. Different renewable energy systems and energy storage need to be bundled to form a hybrid system.

In this context, ground heat and geothermal energy is combined in a circular exergy flow to support the hydrogen city model. The option in heating mode is shown in Figure 16 in which each unit power of geothermal energy at 80 °C is utilized in an ORC unit, which produces 0.08 kW$_E$ that is used in a GSHP. The GSHP can generate 0.32 kW$_H$ given that the average *COP* is 4.0 at an output temperature of 55 °C in heating mode. This is coupled with the waste heat of the ORC at the same output temperature and directed to the district buildings in a local sub-district heating network.

If needed, the saved natural gas from the buildings' previous on-site thermal systems is utilized in a poly-generation unit based on fuel cells. TES that are suited to two different levels of exergy are used to match the loads and shave-off peak loads. Electricity and additional high-exergy heat is generated at 90 °C for high-temperature applications in the district. In the cooling season, this heat may be used in absorption chillers for cold generation. From the geothermal production well to the re-injection well, the overall performance results are obtained as the total output. The three thermal power terms that include the later term at 35 °C for the preheating of DHW provides 1 $kW_H$.

$$\text{Total Output} = (0.62 \text{ kW}_H \text{ @ } 55 \text{ °C} + 0.34 \text{ kW}_H \text{ @ } 90 \text{ °C} + 0.04 \text{ kW}_H \text{ @ } 35 \text{ °C}) + 0.348 \text{ kW}_E \quad (23)$$

In the case that the displaced natural gas, which was originally used in the district, is consumed internally in the fuel cell unit, then the gross *COP* of the Circular Geothermal option becomes 1.348 based on 1 $kW_H$ of geothermal thermal power input (Equation (24) and Figure 16). *COP* is greater than a value of one, since ground heat is utilized in the GSHP in addition to the geothermal energy.

$$COP = (1 \text{ kW}_H + 0.348 \text{ kW}_E)/1 \text{ kW}_H = 1.348 \quad \{\text{First Law}\} \quad (24)$$

Starting from a unit geothermal power at 80 °C (353 K), the Circular Geothermal option provides 0.348 $kW_E$ and 1 $kW_H$ at different supply temperatures. This output compares favorably with the 0.08 $kW_E$ supplied by the ORC unit without reject heat recovery (see Table 3) and 1 $kW_H$ at 80 °C supply if the geothermal power is utilized in the district in the form of heating only. In contrast, the above Equation (24) algebraically combines heat and power although their exergy values are quite different. While *COP* is greater than one, this definition is misleading and requires the use of the $COP_{Ex}$ definition that considers the quality of the outputs as put forth in Equation (4).

$$COP_{Ex} = \frac{0.62 \times \left(1 - \frac{283 \text{ K}}{328 \text{ K}}\right) + 0.34 \times \left(1 - \frac{283 \text{ K}}{363 \text{ K}}\right) + 0.04 \times \left(1 - \frac{283 \text{ K}}{308 \text{ K}}\right) + 0.348 \times (1)}{1 \times \left(1 - \frac{283 \text{ K}}{353 \text{ K}}\right) + \left(\frac{0.62}{0.80}\right) \times \left(1 - \frac{283 \text{ K}}{2000 \text{ K}}\right)} = 0.59 \quad (25)$$

**Table 3.** Comparison of the Circular Geothermal Option with Conventional Options.

| System | Output | | | |
|---|---|---|---|---|
| | Electricity | Heat at 90 °C | Heat at 55 °C | Heat at 35 °C |
| Circular Geothermal | 0.348 $kW_E$ | 0.34 $kW_H$ | 0.62 $kW_H$ | 0.04 $kW_H$ |
| DH with NG | - | - | 0.775 $kW_H$ | - |
| ORC | 0.08 $kW_E$ | - | - | - |

In the above application of Equation (4), $COP_{Ex}$ is the exergy-based *COP* for the entire cycle. All systems operate at constant base load. Optional solar and wind energy systems in the district contribute to peak loads with thermal storage. The grid is also acting for electrical energy storage at large. The entire collection of systems operates in a cascaded form, similar to a single, large-scale heat pump. If only an ORC unit would be used, then $COP_{Ex}$ would be 0.092 and only 0.08 $kW_E$ would be generated. For this reason, the bundling of renewable energy systems can be warranted.

In some countries, including Italy, New Zealand, and Turkey, geothermal reservoirs are located in carbonate-rich rock grabens that contain calcium carbonate ($CaCO_3$). This means that geothermal wells extract $CO_2$ that needs to be recaptured, which is a rather expensive process. Consequently, most of the applications release $CO_2$ emissions into the atmosphere, nearly at a rate of 0.5 kg $CO_2$/kWh. In extreme cases, such as in the Menderes and Gediz grabens with high-enthalpy geothermal energy sources, the $CO_2$ emissions per kWh as *c* is 0.9 to 1.3 kg/kWh that can be much higher than coal-based thermal plants [71,72]. If, however, a pumped binary power plant is used, then the emission factor is zero [72]. An effective $CO_2$ capture and selection of the right technology are necessary if such reservoirs are to contribute favorably towards the aims of the Paris Agreement.

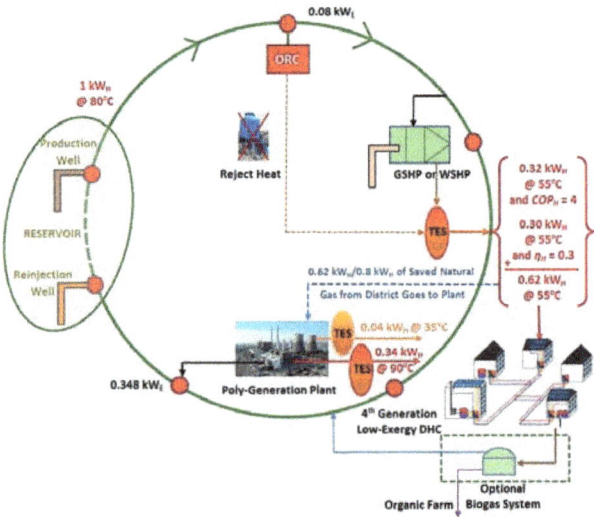

**Figure 16.** Combined Heat and Power in Circular Geothermal Model in Heating Mode.

The Circular Geothermal option as shown in Figure 16 comprises of geothermal production and re-injection wells, an ORC system, GSHP or water-source heat pumps (WSHP), a district energy distribution and collection system, a poly-generation system, and TES at different exergy levels. Such an option couples and mobilizes ground thermal energy and geothermal energy, including with the aim of displacing the use of natural gas in boilers for space heating in buildings. Instead, the displaced natural gas is used in the poly-generation plant with the generated electric power fed into the local grid. In distributed applications, the evaporator side of the heat pumps may be coupled to PV systems if this option is used in buildings to absorb the heat collected by the panels to improve the *COP* of the GSHP. The flow rate needs to be dynamically optimized according to instantaneous solar insolation, heat demand, and other operating conditions to maximize the total exergy output (both power and heat) of the PVT system [61,62]. A biogas system that could be mixed with natural gas that is saved from the boilers and also provide organic fertilizer for local farming is optional.

The multiple outputs of the Circular Geothermal option surpass the outputs of conventional options. In this respect, Figure 16 can be contrasted to options that have singular outputs, namely a DE system based on natural gas and the use of geothermal energy in ORC to produce only electricity. The integration that enables the multiplicity of outputs in Figure 16 also provides for improvements in the value of $\psi_R$. This is further valid if the Circular Geothermal option is compared to the direct use of geothermal energy for separate heat and power production. The direct use of geothermal energy for district heating (Figure 17) and the direct use of geothermal energy for ORC power generation (Figure 18) are compared based on Exergy Flow Bars as provided below.

In Figure 17 for the direct use of geothermal energy for district heating, exergy destruction takes place both upstream ($\varepsilon_{des(1)}$) and downstream ($\varepsilon_{des(2)}$) of energy usage. Since exergy is also destroyed upstream, Equation (11) is used for the REMM Efficiency $\psi_R$ based on an ideal Carnot cycle [42]:

$$\psi_R = \frac{\varepsilon_{dem}}{\varepsilon_{sup}} = \frac{\left(1 - \frac{323\ \text{K}}{343\ \text{K}}\right)}{\left(1 - \frac{283\ \text{K}}{353\ \text{K}}\right)} = 0.294 \tag{26}$$

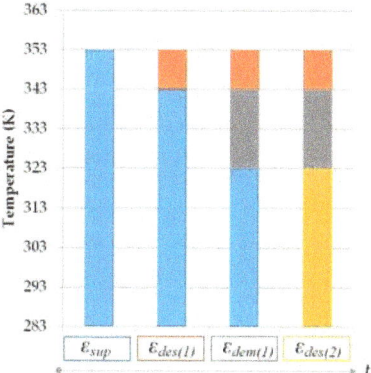

**Figure 17.** Exergy Flow Bars for Geothermal District Heating.

Here, $\varepsilon_{dem}$ represents the demand exergy of the district heating system application between 70 °C (343 K) and 50 °C (323 K) for buildings that are connected to the system (see also Equation (11)). Another feature of the analysis is the ability to identify the exergetic match of the exergy supply with the final application. The final application is comfort heating at 20 °C indoor air temperature in buildings so that the $\varepsilon_{dem}$ term is replaced by (1 − 283 K/293 K). In this case, $\psi_R$ reduces to 0.172.

Figure 18 provides the Exergy Flow Bars for the ORC power generation case. The un-utilized thermal output of the ORC is taken at temperatures about 60 °C (333 K) onwards. Since practically no exergy destruction takes place upstream, Equation (10) is applied as given in Equation (27):

$$\psi_R = 1 - \frac{\varepsilon_{des}}{\varepsilon_{sup}} = 1 - \frac{\left(1 - \frac{283\,K}{333\,K}\right)}{\left(1 - \frac{283\,K}{353\,K}\right)} = 0.243 \tag{27}$$

The resource temperature $T_f$ is the geothermal fluid temperature at the wellhead that is taken as 353 K also in Figures 17–19. If any fuel like biogas or natural gas is used, then this temperature is equal to the Adiabatic Flame Temperature (AFT). As previously defined, an equivalent temperature is put forth for solar, wind, and any other renewable energy resource without a direct $T_f$ value.

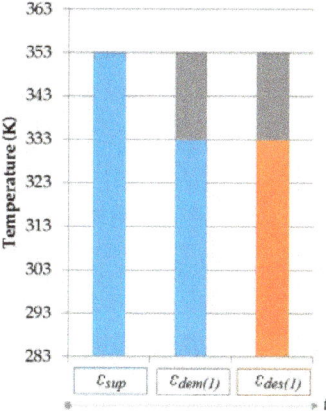

**Figure 18.** Exergy Flow Bars for Direct Geothermal Power with ORC.

The primary characteristic of the Circular Geothermal option is that it represents an integrated, compound power and heat system at large. The option may be also applied to single buildings and scaled up to large district energy systems with possible integration with a hydrogen economy cycle. While most suitable for 4GDE, the Circular Geothermal option can be further applicable to district cooling applications. In this case, cold storage and absorption/adsorption units may be used.

The two options in Figures 17 and 18 are further compared based on the $C_R$ indicator as defined in Equations (15) and (16). From Equation (15), the values of $C_R$ for geothermal district heating and power-only ORC options are 0.19 and 0.019, respectively. Here, the value of the net energy efficiency after parasitic losses $\eta_l$ is taken as 0.65 for district heating while it is taken as 0.08 for ORC.

The approach of $C_R$ further reveals advantages when applied to the Circular Geothermal option for which the Exergy Flow Bars are provided below in Figure 19. Other minor exergy destructions in heating are neglected. The respective values based on the application of Equation (10) are:

$$\psi_R = 1 - \frac{\varepsilon_{des}}{\varepsilon_{sup}} = 1 - \frac{\left(1 - \frac{283\ \text{K}}{293\ \text{K}}\right)}{\left(1 - \frac{283\ \text{K}}{353\ \text{K}}\right)} = 0.827 \tag{28}$$

and from Equation (16):

$$C_R = COP \times \psi_R = 1.348 \times 0.827 = 1.114 \tag{29}$$

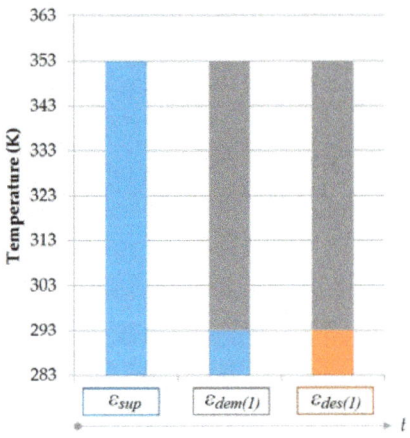

**Figure 19.** Exergy Flow Bars for the Circular Geothermal Option.

After comparing the above results, the $C_R$ values are further used to evaluate a $CO_2$ reduction potential ratio $R$ according to REMM [38]. Equation (30) compares the geothermal district heating only case with $C_R = 0.19$ and the Circular Geothermal option with $C_R = 1.114$. A similar comparison with the power-only ORC case with $C_R = 0.09$ indicates 2.15 times higher $CO_2$ reduction potential. The degree of improvement increases with the geothermal reservoir temperature and applications.

$$R = \frac{(2 - C_R)_{district\ heating}}{(2 - C_R)_{geotherm}} = \frac{2 - 0.19}{2 - 1.114} = 2.04 \tag{30}$$

*4.4. Comparison of Options for the Hydrogen City Model*

Table 4 compares each option based on $\psi_R$ values and the $CO_2$ avoidance capacity based on unit $CO_2$ emissions, including an avoidable $CO_2$ emissions impact due to exergy mismatches in the energy system according to REMM [34]. The reference values represent separate heat and power production.

The comparisons in Table 4 indicate that the hydrogen city model with renewables has important advantages over the reference and conventional DE systems. The inclusion of a Circular Geothermal option can further improve the values of the proposed model. For example, the hydrogen city model with all renewable energy and a contribution from the geothermal energy option has a $\psi_R$ value of 0.83 as given in Table 4, which takes place above the targeted value of 0.80.

Such improvements are further represented in the *EDR* values based on Equation (13) so that the exergetic advantages of these options are compared to reductions in $CO_2$ emissions over the reference case. In particular, while the reference case has no reduction (*EDR* = 0), the evaluated options represent incremental or significant improvements in *EDR* values when compared with $CO_{2base}$. The greatest improvements take place for the hydrogen city model with all renewable energy sources involving the inclusion of a Circular Geothermal option. The *EDR* values of these options are calculated as 0.91 and 0.92, respectively, which indicates greater $CO_2$ savings over $CO_{2base}$. These *EDR* values can qualify the district as a nZCD based on the net-zero definitions that were provided in Table 1. In contrast, none of the options have *EDR* values equal to 1 for a strictly NZCD status although these values have closely approached to 1 already with the respective values in Table 4.

**Table 4.** Overall Comparison of Conventional and Proposed District Energy System Options.

| Compared Options | $\psi_R$ | $CO_2$ per kWh (kg $CO_2$/kWh) | EDR (Equation (13)) |
|---|---|---|---|
| Reference Values | 0.20 | 0.63 ($CO_{2base}$) | 0.00 |
| Basic DE System without Steam Generation [a] | 0.25 | 0.62 | 0.02 |
| DE System with Steam Generation [b] | 0.42 | 0.27 | 0.58 |
| Hydrogen City Model (All Renewables) | 0.80 | 0.06 | 0.91 |
| With Circular Geothermal Option (All Renewables) | 0.83 | 0.05 | 0.92 |

[a] The energy source is natural gas; [b] Supported with renewable energy.

## 5. Discussions on an Application of the Hydrogen City Model to a New Settlement

About half of the carbon budget that remains to have a chance of limiting global warming to at most 1.5 °C by the end of this century could be consumed with the emissions impact from new urban development alone unless prompt action is taken to avoid lock-in to incumbent technologies [73]. One implication of the present research work requires that new settlements, most preferably at brownfield sites to reduce land use changes, are equipped with an energy system that maximizes the rational use of renewable energy sources based on exergy matches between the supply and demand.

For this reason, an application of the hydrogen city model was considered for a new district development in the province of Ankara, Turkey (40.13° N, 33.00° E) that has a population target of 200,000 inhabitants. Degree days for seasonal heating (less than or equal to 15 °C) and cooling (greater than 22 °C) are 2493 and 289 degree-days, respectively [74]. Although Ankara appears to have dominance of the heating season, practices in other recent urban development projects that are in operation indicate considerable increases in comfort cooling loads. Figure 20 shows the annual variation of monthly average outdoor dry-bulb temperatures [75]. The local climate is dry and latent cooling loads may be negligible except in gathering places. Design outdoor dry-bulb temperatures are $-12$ °C for winter and 35 °C for summer with a cooling season that is less than three months.

The new district development is expected to consist of residential areas, social service buildings, offices, shopping plazas, and mixed mode (office, residential, commercial) buildings. Consequently, low-rise residential buildings, single story homes, and higher-rise office buildings will be common. The total required floor area of all buildings, $A_F$ in the new settlement is obtained by Equation (31):

$$A_F = q \cdot P^y \tag{31}$$

Here, *P* is the population and *q* is the floor area per person averaged for the settlement based on different building functions, typology, and their relative mix in the settlement plan.

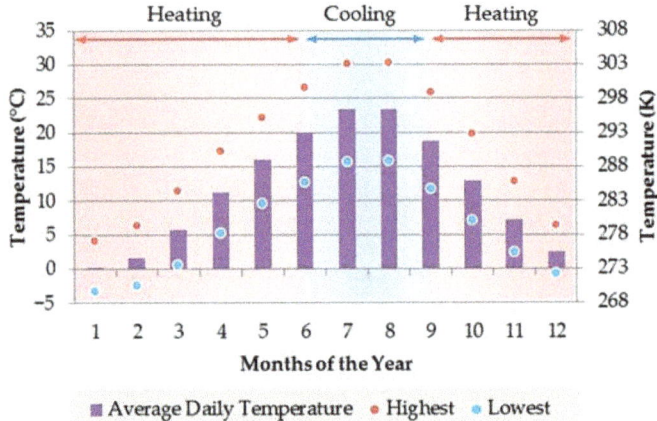

**Figure 20.** Annual Variation of Monthly Average Outdoor Dry-Bulb Temperatures in Ankara.

In Equation (31), the multiplier $q$ generally decreases with $P$. Conversely, density per given floor area increases with higher population [76]. This relation is represented by the power of $P$, namely $y$ that depends on climate, population, culture, average affluence, geographic location, population of neighboring built environments, and the degree of daily or weekly commuting, among other factors. Therefore, $y$ is a number smaller than 1 (0.8 in this case) while $q$ is taken to be 10 $m^2$ per person for the purposes of the case study. The multiplier $q$ also varies based on the development of the region or country and building function(s). About 65% of the world population has less than 20 $m^2$ per person [77]. Moreover, the average number of floors in the buildings for minimum $CO_2$ emissions responsibility is optimized. The optimum number of floors in the buildings with conventional walls (no glass façade) that are partially integrated with PVT-3 panels on south facing walls was determined to be 25 floors for the present case study based on the optimization approach as developed in related research [76]. The evaluations are used to find the number of buildings and to determine energy loads using the tool available for the Building Energy Performance of Turkey, namely BEP-TR [78].

Table 5 provides typical data for the hydrogen city model at the building level for a building with total area of 250 $m^2$ based on loads and equipment differentiated for the heating and cooling seasons. Peak thermal and power loads of the city were predicted for nominally well-insulated hydrogen homes. The two TES options and cold storage are further integrated into the building. Electrical loads of the typical building increase in the summer months since the *COP* values of heat pumps decrease in the cooling season. More power is required even if the cooling capacity is the same as the heating capacity. In addition, cooling appliances require more power in the summer months. However, this increase in individual power demands can level-off at the district scale since the diversity factor is greater in the summer months with inhabitants locating to coastal areas. Such a load balancing factor that is typical for the context of Ankara makes it possible to size the system with respect to the winter season without any redundancy for the cooling season. The same holds true for the fuel cells, which are also sized for the heating season. The only seasonal redundancy will be the absorption chillers, which are not used in the heating season. However, this capacity may be directed to the industry for such applications as cold warehouses. Moreover, the PVT-3 system produces more power and heat in the summer months due to the increase in solar insolation. This helps to overcome the increase in the power demand in the summer months and at the same time, charges the desiccant system. Absorption cooling is activated by hydrogen heat upon demand. The heat output of the PVT-3 in the summer goes to DHW demands. Therefore, load amounts and type variations in the summer months are largely compensated at the building scale. This enables the central power system to operate with the same overall demand loads year-round.

**Table 5.** Typical Building Level Data for the Hydrogen City Model.

| Loads and Equipment | | Demand or Capacity | |
|---|---|---|---|
| | | Heating Season | Cooling Season |
| Loads | Electrical Load | 7 kW$_E$ | 10 kW$_E$ |
| | Heating Load | 16 kW$_H$ | 2 kW$_H$ (DHW only) |
| | Cooling Load (Sensible, Latent) | - | 5 kW$_C$, 3 kW$_C$ |
| Equipment | Fuel Cell (Building Level) | 6 kW$_E$, 8 kW$_H$ at 40 °C | 6 kW$_E$, 8 kW$_H$ at 40 °C |
| | TES 1, TES 2, Cold Storage [a] | 10 kWh, 7 kWh | 4 kWh |
| | Radiant Floor Panels | 2 kW$_H$ (Heating) | 1.5 kW$_C$ (Cooling) |
| | Radiant Wall [b] | 2 kW$_H$ | 5.5 kW$_C$ |
| | GSHP [c] | 2 kW$_H$ | 2 kW$_C$ |
| | PVT-3 (Power, Thermal at 40 °C) | 1 kW$_E$ 2 kW$_H$ | 4 kW$_E$, 4kW$_H$ |
| | Absorption Cooling System | - | 5 kW$_C$ (Hydrogen activated) |
| | Desiccant System [c] | - | 1 kW$_C$ (Latent) |

[a] Enables the storage of energy that is not concurrent with the exergy demands. [b] Optimized for low-exergy circuit temperature for maximum exergetic efficiency [79]. [c] The cooling component of the equipment satisfies the latent cooling loads.

At the district level, Table 6 tabulates the power and thermal loads for the main cycle of the hydrogen city model. The power and thermal loads include those for mass transport, city lighting, and miscellaneous service loads. The base loads are taken to be about 50% of the peak loads, including diversity factors, while peak loads are to be satisfied by various energy storage. Overall, the district is evaluated to have electrical loads of 10,000 kW$_E$, heating loads of 20,000 kW$_H$ and cooling loads for sensible and latent cooling needs at 14,000 kW$_C$ and 10,000 kW$_C$, respectively. As implied by the loads that can be provided by equipment and energy systems in the hydrogen city, including hydrogen storage, the district produces sufficient energy at the required exergy levels to qualify in becoming a nearly self-sufficient district towards reaching a possible NZEXD status. The region is nominally rich in low-exergy geothermal energy resources with wellhead temperatures in the range of 80 °C to 70 °C, which can also permit integrating a Circular Geothermal option.

**Table 6.** Power and Thermal Loads for the District Infrastructure.

| Loads and Equipment | | Demand or Capacity |
|---|---|---|
| Loads | Electrical Loads | 10,000 kW$_E$ |
| | Heating Loads | 20,000 kW$_H$ |
| | Cooling Loads (Sensible, Latent) | 14,000 kW$_C$, 10,000 kW$_C$ |
| Equipment and Energy Systems | Fuel Cells (Central Plant) [a,b] | 8000 kW$_E$, 12,000 kW$_H$ at 40 °C |
| | Hydrogen Storage [c] | 50,000 kWh |
| | PVT-3 (Power, Thermal at 40 °C) | 400 kW$_E$, 700 kW$_H$ |
| | Cogeneration (Hydrogen Enriched Biogas) [d] | 1500 kW$_E$, 2500 kW$_H$ or 1000 kW$_C$ |
| | Biogas Production Plant | 20,000 kW-equivalent fuel |
| | Wind Turbines [e] | 20,000 kW$_E$ |

[a] Thermal energy is not distributed from the central plant to the district. Hydrogen distribution has lower pumping needs. [b] The thermal energy is used for local demands, e.g., greenhouses and agricultural drying. [c] Hydrogen is mainly used for hydrogen-enriched biogas for the district gas supply. [d] The 1000 kW$_c$ includes an absorption cycle. [e] Surplus from the electrical load is utilized for hydrogen production.

## 6. Conclusions

The rapidly emerging penetration of renewable energy systems makes it more feasible to establish a hydrogen economy, which needs to be optimized by both the First and Second Laws of Thermodynamics. In this research work, third-generation PVT modules, double-blade wind turbines, high-efficiency fuel cells at the central plant and building levels, as well as cogeneration based on hydrogen-enriched biogas, are gathered around a district level hydrogen economy approach. The renewable energy-based hydrogen city model was found to provide an end result that is particularly rewarding for DE systems. The analysis of the renewable energy-based hydrogen city model with PVT-3 and high-efficiency wind energy indicated the possibility of obtaining values of the REMM efficiency reaching 0.80 and savings in $CO_2$ emission impacts when compared to a conventional DE system based on lower primary energy spending. Values of the parameter $\psi_R$ at or beyond 0.80 were targeted as a criterion for a net or nearly net-zero exergy status for the district. In addition to exergy benefits, a more closed-loop water cycle is proposed to provide a way in which hydrogen economy can support a more self-sustaining urban water cycle in an energy-water nexus.

The two cycles of a hydrogen city model and the Circular Geothermal option are found to provide a useful approach in contributing to the success of future hydrogen economy cities as well as a more rational utilization and storage of renewable energy sources. Dedicated metrics to evaluate related improvements, including an assessment of *COP* values for heat pumps based on exergy, namely $COP_{EX}$ were also integrated into the analyses. The dedicated metrics have enabled an effective comparison with conventional DE systems, including natural gas based DE systems as well as those in which there may be only district heating or only electricity generation. These metrics underlined the importance of utilizing renewable energy resources in combined energy systems.

The results for the hydrogen city model and Circular Geothermal option have addressed a gap in the literature for analyzing renewable energy oriented hydrogen economy solutions for urban energy systems. The two hydrogen cycles at the district and building levels that are analyzed from an exergy framework can provide districts with new opportunities to reach near-zero exergy targets. The comparative analyses have shown that this target is achievable and the level of match between the supply and demand of exergy can be effectively increased in comparison to the basic DE system. The effective utilization of renewable energy sources also increased the emissions difference with the reference values, including avoidable $CO_2$ emissions. Indeed, utilization of renewables in hydrogen economy can have a key role in the success of urban energy systems. For example, the Athlete's Village of the Tokyo 2020 Olympics will be the first hydrogen city example relying on renewables [80]. The possible case study of a 200,000 inhabitant hydrogen city in a new settlement in Ankara, Turkey as analyzed in this research work is important to diversify the options that may be adapted to reach net-zero targets in the urban context with an outlook for the more rational use of exergy. The integration of hydrogen economy principles, considerations of an energy-water nexus and net-zero targets can provide an effective option for the future of urban settlements and districts.

**Author Contributions:** B.K. conceived the Hydrogen City model that provided the basis for the research work and also designed the experiments for the PVT-3 system. Ş.K. contributed to analyses in formulations and comparisons of the energy system options based on REMM metrics and net-zero targets.

**Acknowledgments:** The manuscript is a revised and expanded version of an original scientific contribution that was presented at the 12th Conference on Sustainable Development of Energy, Water and Environment Systems (SDEWES) held during 4–8 October 2017 in Dubrovnik, Croatia entitled "Hydrogen Economy-Based Net-Zero Exergy Cities of the Future with Water-Energy Nexus." A case study of a new settlement is added to the present version among other original elaborations. Funding has not been received to undertake the research work.

**Conflicts of Interest:** The authors declare no conflict of interest.

## Nomenclature

| | |
|---|---|
| *AF* | Total floor area of buildings to be occupied by population *P*, m² |
| *a, b* | Heat pump *COP* versus heat output temperature function constants |

| | |
|---|---|
| $a_o$ | Constant of $L_{max}$ |
| $C$ | Power to heat ratio, dimensionless |
| $c$ | Emissions ratio (Factor), kg $CO_2$/kWh |
| $CO_2$ | Compound carbon dioxide emissions, kg $CO_2$ |
| $COP$ | Coefficient of Performance (First Law) |
| $COP_{EX}$ | Exergy-Based Coefficient of Performance (Second Law) |
| $C_R$ | Composite Rationality Indicator |
| $E$ | Electrical energy (load), kWh |
| $EDR$ | Ratio of Emissions Difference to the base case $CO_2$ emissions, dimensionless |
| $E_x$ | Exergy, kW or kWh |
| $I$ | Annual solar insolation on horizontal surface, kWh/m$^2$/year |
| $I_n$ | Net solar insulation intensity reaching perpendicular to the solar PV surface |
| $L_{max}$ | Maximum length of the district circuit (one way) |
| $PER$ | Primary energy ratio |
| $PEXR$ | Exergy-based primary energy ratio |
| $Q, Q_H$ | Thermal energy (load), kWh |
| $Q$ | Floor area per person, m$^2$/person |
| $R$ | $CO_2$ reduction potential ratio |
| $T$ | Temperature, K |

## Greek Symbols

| | |
|---|---|
| $\eta_{EX}$ | Second Law Efficiency, dimensionless |
| $\eta_T$ | Power transmission and distribution efficiency |
| $\psi_R$ | Rational exergy management efficiency, rationality ratio |
| $\varepsilon$ | Unit exergy, kW/kW |
| $\eta_I$ | First Law Efficiency |
| $\Delta$ | Difference |

## Subscripts

| | |
|---|---|
| $b$ | Biogas |
| $base$ | Base |
| $c$ | Cold |
| $dem$ | Demand |
| $des$ | Destroyed |
| $E$ | Electric |
| $f$ | Resource temperature, or Adiabatic Flame Temperature (Real or virtual), K |
| $g$ | Generator |
| $H$ | Thermal (Heat) |
| $h$ | Hydrogen |
| $i, j$ | Node indexes for partial REMM efficiencies between two nodal connections |
| $in, out$ | Inlet and outlet connections of a hydronic circuit |
| $l, m$ | Local power plant, distant power plant, respectively |
| $min, max$ | Minimum, maximum |
| $opt$ | Optimum |
| $R$ | Rational or Reservoir |
| $ref$ | Reference |
| $ret$ | Return |
| $s$ | Solar |
| $sup$ | Supply |
| $ref$ | Reference |
| $T$ | Power transmission |
| $w$ | Wind |
| $X, EX$ | Exergy, exergetic |

## Superscripts

| | |
|---|---|
| $n$ | Maximum district circuit length coefficient |
| $y$ | Population coefficient |

## Chemical Symbols

| | |
|---|---|
| $CaCO_3$ | Calcium carbonate |
| $CH_3OH$ | Methanol |
| $CO_2$ | Carbon dioxide |
| $H_2$ | Hydrogen |
| $H_2O_2$ | Hydrogen peroxide |

## Abbreviations

| | |
|---|---|
| 4GDE | Fourth generation district energy system |
| ABS | Absorption chiller |
| AC | Alternating current |
| AG | Air gap |
| AFT | Adiabatic flame temperature, K |
| ARC | Absorption Refrigeration Cycle |
| CR | Composite Rationality Index |
| CHP | Combined heat and power |
| CWT | Cold water tank |
| DC | Direct current |
| DE | District energy |
| DHC | District heating and cooling |
| DHW | Domestic hot water |
| $E_1$ | Power generated by PV layer of the PVT-3 module |
| $E_2$ | Power generated by the TEG elements in the PVT-3 module |
| EER | Energy Efficiency Ratio |
| ERC | Ejector Refrigeration Cycle |
| F | Frame of the PVT-3 module |
| FC | Fuel cell, Fan-coil |
| GC | Glass cover |
| GSHP | Ground-source heat pump |
| HHV | Higher heating value |
| HP | Heat pipe |
| HWT | Hot water tank |
| IN | Insulator |
| LowEx | Low-exergy |
| LVDC | Low-voltage DC power |
| NPEB | Net Positive Energy Building |
| NPEXB | Net Positive Exergy Building |
| NS | Heat-conducting nano sheet |
| NZCB | Net-Zero $CO_2$ Building |
| nZCB | Near Zero $CO_2$ Building |
| NZCD | Net-Zero $CO_2$ (Emissions) District |
| nZCD | Near Zero $CO_2$ District |
| NZEXB | Net-Zero Exergy Building |
| nZEXB | Nearly Zero Exergy Building |
| NZEXD | Net-Zero Exergy District |
| nZEXD | Near Net-Zero Exergy District |
| ORC | Organic Rankine cycle |

| PCM | Phase change material |
|---|---|
| PV | Photovoltaic |
| PVT | Photovoltaic-thermal |
| PVT-3 | Third-generation photovoltaic-thermal |
| REMM | Rational Exergy Management Model |
| rms | Root-mean square |
| TEG | Thermo-electric generator |
| TES | Thermal energy storage |
| V2B | Vehicle to Building or Vehicle to Grid (V2G) |
| WSHP | Water-source heat pump |
| $ZEH_2E$ | Zero-energy hydrogen economy |

## References

1.  Lund, H.; Østergaard, P.; Connolly, D.; Vad Mathiesen, B. Smart energy and smart energy systems. *Energy* **2017**, *137*, 556–565. [CrossRef]
2.  Lund, H. Renewable Heating Strategies and their Consequences for Storage and Grid Infrastructures Comparing a Smart Grid to a Smart Energy Systems Approach. *Energy* **2018**, *151*, 94–102. [CrossRef]
3.  Mathiesen, B.; Skov, I.; Hansen, K.; Connolly, D.; Wunsch, J. *Applications of SOECs in Different Types of Energy Systems: German and Danish Case Studies*; Department of Development and Planning, Aalborg University: Aalborg, Denmark, 2015.
4.  Veziroğlu, N. Conversion to Hydrogen Economy. *Energy Procedia* **2012**, *29*, 654–656. [CrossRef]
5.  Duić, N.; Krajačić, G.; Graça Carvalho, M. RenewIslands methodology for sustainable energy and resource planning for islands. *Renew. Sustain. Energy Rev.* **2008**, *12*, 1032–1062. [CrossRef]
6.  Krajačić, G.; Martins, R.; Busuttil, A.; Duić, N.; Graça Carvalho, M. Hydrogen as an energy vector in the islands' energy supply. *Int. J. Hydrog. Energy* **2008**, *33*, 1091–1103. [CrossRef]
7.  Sørensen, B.; Petersen, A.; Juhj, C.; Ravn, H.; Søndergren, C.; Simonsen, P.; Jørgensen, K.; Nielsen, L.H.; Larsen, H.V.; Morthorst, P.E.; et al. Hydrogen as an energy carrier: Scenarios for future use of hydrogen in the Danish energy system. *Int. J. Hydrog. Energy* **2004**, *29*, 23–32. [CrossRef]
8.  Calise, F.; D'Accadia, M.; Barletta, C.; Battaglia, V.; Pfeifer, A.; Duic, N. Detailed Modelling of the Deep Decarbonisation Scenarios with Demand Response Technologies in the Heating and Cooling Sector: A Case Study for Italy. *Energies* **2017**, *10*, 1535. [CrossRef]
9.  Sveinbjörnsson, D.; Amer-Allam, S.; Hansen, A.; Algren, L.; Pedersen, A. Energy supply modelling of a low-$CO_2$ emitting energy system: Case study of a Danish municipality. *Appl. Energy* **2017**, *195*, 922–941. [CrossRef]
10. Singh, A.; Baredar, P.; Gupta, B. Techno-economic feasibility analysis of hydrogen fuel cell and solar photovoltaic hybrid renewable energy system for academic research building. *Energy Convers. Manag.* **2017**, *145*, 398–414. [CrossRef]
11. Cao, S.; Klein, K.; Herkel, S.; Sirén, K. Approaches to enhance the energy performance of a zero-energy building integrated with a commercial-scale hydrogen fueled zero-energy vehicle under Finnish and German conditions. *Energy Convers. Manag.* **2017**, *142*, 153–175. [CrossRef]
12. Reuß, M.; Grube, T.; Robinius, M.; Preuster, P.; Wasserscheid, P.; Stolten, D. Seasonal storage and alternative carriers: A flexible hydrogen supply chain model. *Appl. Energy* **2017**, *200*, 290–302. [CrossRef]
13. Nabgana, W.; Abdullah, T.; Nabgan, R.; Gambo, Y.; Ibrahim, M.; Ahmad, A.; Jalil, A.; Triwahyono, S.; Saeh, I. Renewable hydrogen production from bio-oil derivative via catalytic steam reforming: An overview. *Renew. Sustain. Energy Rev.* **2017**, *79*, 347–357. [CrossRef]
14. Qolipour, M.; Mostafaeipour, A.; Tousi, O. Techno-economic feasibility of a photovoltaic-wind power plant construction for electric and hydrogen production: A case study. *Renew. Sustain. Energy Rev.* **2017**, *78*, 113–123. [CrossRef]
15. Tebibel, H.; Khellaf, A.; Menia, S.; Nouicer, I. Design, modelling and optimal power and hydrogen management strategy of an off-grid PV system for hydrogen production using methanol electrolysis. *Int. J. Hydrog. Energy* **2017**, *42*, 14950–14967. [CrossRef]

16. Xu, X.; Xu, B.; Dong, J.; Liu, X. Near-term analysis of a roll-out strategy to introduce fuel cell vehicles and hydrogen stations in Shenzhen China. *Appl. Energy* **2017**, *196*, 229–237. [CrossRef]

17. Mohareb, E.; Kennedy, C. Scenarios of technology adoption towards low-carbon cities. *Energy Policy* **2017**, *66*, 685–693. [CrossRef]

18. Miranda, P.; Carreira, E.; Icardi, U.; Nunes, G. Brazilian hybrid electric-hydrogen fuel cell bus: Improved on-board energy management system. *Int. J. Hydrog. Energy* **2017**, *42*, 13949–13959. [CrossRef]

19. Franzitta, V.; Curto, D.; Rao, D.; Viola, A. Hydrogen production from sea wave for alternative energy vehicles for public transport in Trapani (Italy). *Energies* **2016**, *9*, 850. [CrossRef]

20. Briguglio, N.; Andaloro, L.; Ferraro, M.; Di Blasi, A.; Dispenza, G.; Matteucci, F.; Breedveld, L.; Antonucci, V. Renewable energy for hydrogen production and sustainable urban mobility. *Int. J. Hydrog. Energy* **2010**, *35*, 9996–10003. [CrossRef]

21. Moreno-Benito, M.; Agnolucci, P.; Papageorgiou, L. Towards a sustainable hydrogen economy: Optimisation-based framework for hydrogen infrastructure development. *Comput. Chem. Eng.* **2017**, *102*, 110–127. [CrossRef]

22. Oldenbroek, V.; Verhoef, L.; van Wijk, A. Fuel cell electric vehicle as a power plant: Fully renewable integrated transport and energy system design and analysis for smart city areas. *Int. J. Hydrog. Energy* **2017**, *42*, 8166–8196. [CrossRef]

23. Kumar, G.; Sivagurunathan, P.; Pugazhendhi, A.; Thi, N.; Zhen, G.; Chandrasekhar, K.; Kadier, A. A comprehensive overview on light independent fermentative hydrogen production from wastewater feedstock and possible integrative options. *Energy Convers. Manag.* **2017**, *141*, 390–402. [CrossRef]

24. Nahar, G.; Mote, D.; Dupont, V. Hydrogen production from reforming of biogas: Review of technological advances and an Indian perspective. *Renew. Sustain. Energy Rev.* **2017**, *76*, 1032–1052. [CrossRef]

25. Khan, M.; Nizami, A.; Rehan, M.; Ouda, O.; Sultana, S.; Ismail, I.; Shahzad, K. Microbial electrolysis cells for hydrogen production and urban wastewater treatment: A case study of Saudi Arabia. *Appl. Energy* **2017**, *185*, 410–420. [CrossRef]

26. Viola, A.; Franzitta, V.; Trapanese, M.; Curto, D. Nexus water & energy: A case study of wave energy converters (WECs) to desalination applications in Sicily. *Int. J. Heat Technol.* **2016**, *34*, S379–S386. [CrossRef]

27. Rizzuti, L.; Ettouney, H.M.; Cipollina, A. (Eds.) *Solar Desalination for the 21st Century: A Review of Modern Technologies and Researches on Desalination Coupled to Renewable Energies*, 1st ed.; Springer Netherlands Science+Business Media B.V.: Dordrecht, The Netherlands, 2007.

28. García-Rodríguez, L. Renewable energy applications in desalination: State of the art. *Sol. Energy* **2003**, *75*, 381–393. [CrossRef]

29. Franzitta, V.; Curto, D.; Milone, D.; Viola, A. The Desalination Process Driven by Wave Energy: A Challenge for the Future. *Energies* **2016**, *9*, 1032. [CrossRef]

30. Sanseverino, E.; Sanseverino, R.; Favuzza, S.; Vaccaro, V. Near zero energy islands in the Mediterranean: Supporting policies and local obstacles. *Energy Policy* **2014**, *66*, 592–602. [CrossRef]

31. Da Silva, E.; Marin Neto, A.; Ferreira, P.; Camargo, J.; Apolinário, F.; Pinto, C. Analysis of hydrogen production from combined photovoltaics, wind energy and secondary hydroelectricity supply in Brazil. *Sol. Energy* **2005**, *78*, 670–677. [CrossRef]

32. Alanne, K.; Cao, S. Zero-energy hydrogen economy (ZEH2E) for buildings and communities including personal mobility. *Renew. Sustain. Energy Rev.* **2017**, *71*, 697–711. [CrossRef]

33. Schmidt, D.; Torío, H. (Eds.) Low Exergy Systems for High Performance Buildings and Communities; Annex 49, Summary Report. Available online: https://www.annex49.info/download/Annex49_guidebook.pdf (accessed on 3 January 2018).

34. Kılkış, Ş. A Rational Exergy Management Model to Curb $CO_2$ Emissions in the Exergy-Aware Built Environments of the Future. Ph.D. Thesis, KTH Royal Institute of Technology School of Architecture and the Built Environment, Stockholm, Sweden, 2011.

35. Kılkış, Ş. Energy system analysis of a pilot net-zero exergy district. *Energy Convers. Manag.* **2014**, *87*, 1077–1092. [CrossRef]

36. Kılkış, Ş. Exergy transition planning for net-zero districts. *Energy* **2015**, *92*, 515–531. [CrossRef]

37. Kilkiş, Ş.; Wang, C.; Björk, F.; Martinac, I. Cleaner energy scenarios for building clusters in campus areas based on the Rational Exergy Management Model. *J. Clean. Prod.* **2017**, *155*, 72–82. [CrossRef]

38. Kılkış, B.; Kılkış, Ş. New exergy metrics for energy, environment, and economy nexus and optimum design model for nearly-zero exergy airport (nZEXAP) systems. *Energy* **2017**, *140*, 1329–1349. [CrossRef]

39. Kılkış, Ş.; Kılkış, B. Integrated circular economy and education model to address aspects of an energy-water-food nexus in a dairy facility and local contexts. *J. Clean. Prod.* **2017**, *167*, 1084–1098. [CrossRef]

40. Hydrogen Transport by Pipeline, Roads2Hycom, Institut für Kraft-Fahr-Zeuge. Available online: https://www.ika.rwth-aachen.de/r2h/index.php/Hydrogen_Transport_by_Pipeline.html (accessed on 18 August 2017).

41. Dodds, P.; Demoullin, S. Conversion of the UK gas system to transport hydrogen. *Int. J. Hydrog. Energy* **2013**, *38*, 7189–7200. [CrossRef]

42. Kılkış, B.; Kılkış, Ş. *Yenilenebilir Enerji Kaynakları ile Birleşik Isı ve Güç Üretimi (Combined Heat and Power Generation with Renewable Energy Resources)*, 1st ed.; TTMD Pub. No: 32; Doğa Publishers: İstanbul, Turkey, 2015; ISBN 978-975-6263-25-9.

43. Yavuz, T.; Kılkış, B.; Koç, E.; Erol, Ö. Flow and performance characteristics of a double-blade hydrofoil. *Adv. Mater. Res.* **2012**, *433–440*, 7218–7222. [CrossRef]

44. Net Zero Energy Buildings, Whole Building Design Guide (WBDG). National Institute of Building Sciences. Available online: https://www.wbdg.org/resources/net-zero-energy-buildings (accessed on 11 June 2017).

45. A Common Definition for Zero Energy Buildings. *Prepared for the U.S. Department of Energy by The National Institute of Building Sciences*. Available online: https://energy.gov/sites/prod/files/2015/09/f26/A%20Common%20Definition%20for%20Zero%20Energy%20Buildings.pdf (accessed on 11 June 2017).

46. Kılkış, B. Analysis of Fourth-Generation District Energy Systems with Renewable Energy Cogeneration by Using Rational Exergy Management Model. In Proceedings of the SBE 16 Smart Metropoles Conference, İstanbul, Turkey, 13–15 October 2016; pp. 512–522.

47. Nielsen, S.; Möller, B. Excess heat production of future net zero energy buildings within district heating areas in Denmark. *Energy* **2012**, *48*, 23–31. [CrossRef]

48. Kılkış, Ş. A net-zero building application and its role in exergy-aware local energy strategies for sustainability. *Energy Convers. Manag.* **2012**, *63*, 208–217. [CrossRef]

49. Kılkış, B. Energy and Exergy Analysis of Water and Air-Cooled PVT Systems with Heat Pipe Technology. In Proceedings of the Seminar 72 Low Energy Building Design Using Exergy Modeling, ASHRAE Winter Meeting, Las Vegas, NV, USA, 27 January–1 February 2017.

50. Kılkış, B. An Economic Analysis Tool for Tri-generation Systems in Net-Zero Exergy Buildings (NZEXB). In Proceedings of the XIIth International TTMD HVAC+R Symposium, Istanbul, Turkey, 31 March–2 April 2016.

51. Kılkış, B. Energy and Exergy Metrics in Zero $CO_2$ Emission Buildings. Which Comes First to Raise Cost Effectiveness of Deep Energy Refurbishment? In Proceedings of the 12th REHVA World Congress, Aalborg, Denmark, 22–25 May 2016.

52. Kılkış, B.; Kılkış, Ş. Rational Exergy Management Model for Effective Utilization of Low-Enthalpy Geothermal Energy Resources. In Proceedings of the 21th National Heat Science and Technique Conference, Çorum, Turkey, 13–16 September 2017.

53. Kılkış, B.; Kılkış, S.; Kılkış, Ş. Optimum Hybridization of Wind Turbines, Heat Pumps, and Thermal Energy Storage Systems for Near Zero-Exergy Buildings Using Rational Exergy Management Model. In Proceedings of the 12th IEA Heat Pump Conference, Rotterdam, The Netherlands, 15–18 May 2017.

54. Aldağ. Hundred Percent Fresh Air Comfort with Water-Source Heat Pumps with Heat Recovery. Available online: https://www.termodinamik.info/arsiv (accessed on 25 June 2017).

55. Kılkış, B. *Sustainability and Decarbonization Efforts of the EU: Potential Benefits of Joining Energy Quality (Exergy) and Energy Quantity (Energy) in EU Directives, A State of the Art Survey and Recommendations*; Exclusive EU Position Report; TTMD: Ankara, Turkey, 2017.

56. Meggers, F.; Ritter, V.; Goffin, P.; Baetschmann, M.; Leibundgut, H. Low exergy building systems implementation. *Energy* **2012**, *41*, 48–55. [CrossRef]

57. EBC. Annex 37 Low Exergy Systems for Heating and Cooling. Markku Virtanen, VTT Building and Transport. 2003. Available online: http://www.iea-ebc.org/projects/completed-projects/ebc-annex-37/ (accessed on 12 June 2017).

58. Khan, B. *Non-Conventional Energy Resources*, 2nd ed.; Tata McGraw-Hill: New Delhi, India, 2009.

59. Li, J.; Huang, H.; Osaka, Y.; Bai, Y.; Kobayashi, N.; Chen, Y. Combustion and Heat Release Characteristics of Biogas under Hydrogen and Oxygen-Enriched Condition. *Energies* **2017**, *10*, 1200. [CrossRef]

60. Wall, D.; McDonagh, S.; Murphy, J. Cascading biomethane energy systems for sustainable green gas production in a circular economy. *Bioresour. Technol.* **2017**, *243*, 1207–1215. [CrossRef] [PubMed]

61. Kılkış, B.; Kılkış, Ş.; Kılkış, Ş. Next-Generation PVT System with PCM Layer and Heat Distributing Sheet. In Proceedings of the Solar TR2016 Conference and Exhibition, İstanbul, Turkey, 6–9 December 2016; Paper No: 0006, pp. 20–28.

62. Kılkış, B. Optimum Operation of Solar PVT Systems: An Exergetic Approach. In Proceedings of the Solar TR2016 Conference and Exhibition, İstanbul, Turkey, 6–9 December 2016; Paper No: 0025, pp. 72–79.

63. Meis, K. Designing Cities for a Car-Light Future. Available online: https://www.greenbiz.com/article/designing-cities-car-light-future (accessed on 5 March 2017).

64. Kılkış, B. Exergetic comparison of wind energy storage with ice making cycle versus mini-hydrogen economy cycle in off-grid district cooling. *Int. J. Hydrog. Energy* **2017**, *42*, 17571–17582. [CrossRef]

65. Quoilin, S.; Lemort, V. Technological and Economical Survey of Organic Rankine Cycle Systems. In Proceedings of the 5th European Conference Economics and Management of Energy in Industry, Vilamoura, Portugal, 14–17 April 2009.

66. Shoshan, G. Application of ORC systems in Geothermal Energy. In Proceedings of the Plant Technologies Session, Turkish Geothermal Workshop and Congress, Ankara, Turkey, 8–9 February 2017.

67. Kılkış, B.; Kılkış, Ş. Energy and Exergy Based Comparison of Utilizing Waste Heat of a Cogeneration System for Comfort Cooling Using ORC Driven Chillers or Heat Pumps versus Absorption/Adsorption Cycles. In Proceedings of the ASME ORC 2013, Rotterdam, The Netherlands, 7–8 October 2013.

68. Rowshanzadeh, R. Performance and Cost Evaluation of Organic Rankine Cycle at Different Technologies. Master's Thesis, Department of Energy Technology, KTH, Stockholm, Sweden, 2010.

69. Sun, W.; Yue, X.; Wang, Y. Exergy efficiency analysis of ORC (Organic Rankine Cycle) and ORC-based combined cycles driven by low-temperature waste heat. *Energy Convers. Manag.* **2017**, *135*, 63–73. [CrossRef]

70. Marini, A.; Alexandru, D.; Grosu, L.; Gheorghian, A. Energy and Exergy Analysis of an Organic Rankine Cycle. *U.P.B. Sci. Bull. Ser. D* **2014**, *76*, 127–136. Available online: https://www.scientificbulletin.upb.ro/ (accessed on 30 August 2017).

71. Baba, A. Hydro-Geochemical Properties of Geothermal Systems and its Effect on the System in Turkey. In Proceedings of the IGC 2017 Conference, İzmir, Turkey, 22–24 May 2017.

72. *Greenhouse Gases from Geothermal Power Production*; ESMAP Technical Report 009/16; The World Bank Group: Washington, DC, USA, 2016.

73. Bai, X.; Dawson, R.; Ürge-Vorsatz, D.; Delgado, G.; Barau, A.; Dhakal, S.; Dodman, D.; Leonardsen, L.; Masson-Delmotte, V.; Roberts, D. Six research priorities for cities and climate change. *Nature* **2018**, *555*, 23–25. [CrossRef] [PubMed]

74. General Directorate of Meteorology. Heating and Cooling Degree Days. 2017. Available online: https://www.mgm.gov.tr/veridegerlendirme/gun-derece.aspx (accessed on 7 January 2018).

75. General Directorate of Meteorology. Monthly Average Temperatures by Province. Available online: https://www.mgm.gov.tr/veridegerlendirme/il-ve-ilceler-istatistik.aspx (accessed on 4 April 2018).

76. Kılkış, B. Most Effective Means of Utilizing Solar Energy in Urban Development for Different Building Typologies and Climates. In Proceedings of the 7th Solar Energy Symposium and Exhibition, Turkish Chamber of Mechanical Engineers, Mersin, Turkey, 22–23 September 2017.

77. United Nations Population Division. *Charting the Progress of Populations Section XII Floor Area per Person*; Department of Economic and Social Affairs: New York, NY, USA, 2000; pp. 79–83. Available online: http://www.un.org/en/development/desa/population/publications/trends/progress-of-populations.shtml (accessed on 15 December 2017).

78. BEP-TR. Ministry of Environment and Urbanization. Available online: http://www.bep.gov.tr/ (accessed on 25 December 2017).

79.  Kılkış, B. COOLP: a computer program for the design and analysis of ceiling cooling panels. *ASHRAE Transactions* **1995**, *101*, 703–710. Available online: https://www.ashrae.org/technical-resources/ashrae-transactions (accessed on 2 March 2018).

80.  Tokyo Aims to Realize "Hydrogen Society" by 2020—Metro Government Undertakes Pioneering Initiative. Available online: https://www.japan.go.jp/tomodachi/2016/spring2016/tokyo_realize_hydrogen_by_2020.html (accessed on 18 August 2017).

*Article*

# Portuguese Plan for Promoting Efficiency of Electricity End-Use: Policy, Methodology and Consumer Participation

**José L. Sousa [1,2,\*] and António G. Martins [2,3]**

1   School of Technology of Setúbal, Polytechnic Institute of Setúbal, 2910-761 Setúbal, Portugal
2   INESC Coimbra, Pólo II, R. Silvio Lima, 3030-290 Coimbra, Portugal; agmartins@uc.pt
3   Department of Electrical and Computer Engineering, Pólo II University of Coimbra, 3030-290 Coimbra, Portugal
\*   Correspondence: jose.luis.sousa@estsetubal.ips.pt; Tel.: +351-265-790-000

Received: 26 February 2018; Accepted: 25 April 2018; Published: 3 May 2018

**Abstract:** The Portuguese Electricity Demand-Side Efficiency Promotion Plan (PPEC) is a voluntary financial mechanism, under which several entities, among them electric utilities, may submit proposals of measures aiming at the reduction of electricity consumption or load management. It is one of the alternative options followed by the Portuguese government to the Energy Efficiency Obligations (EEO) stated in Article 7 of the EU Energy Efficiency Directive. A brief review is presented of the state of the implementation of Article 7 in EU. PPEC is one of the schemes that provide financial support to the implementation of measures whose results contribute to the commitments made under the Portuguese National Energy Efficiency Action Plan (NEEAP), the framework under which the alternatives to the EEO were designed. In the first edition of the PPEC, only three energy services were addressed, while, in the most recent PPEC edition, the sixth, measures addressed nine energy services. In addition, the co-funding by participating consumers and other agents has increased, raising the investment in energy efficiency from actors other than the program administrator. PPEC, although a voluntary mechanism, has proven to be a very competitive one, involving an increasing number of economic agents, measures and addressed energy services.

**Keywords:** Energy Efficiency Promotion; electric utilities; voluntary schemes; Portuguese experience; consumer participation; costs of avoided kWh

---

## 1. Introduction

On the grounds of Article 7 of the Energy Efficiency Directive (EED), several European Union Member States (MS) have adopted Energy Efficiency Obligations (EEO) [1]. EEO require that energy companies meet a savings target of 1.5% of annual sales to final consumers. However, an alternative was available for MS to fulfill their energy savings obligations. This was the road Portugal decided to follow. Among the policy measures taken to achieve the savings target is the Electricity Demand-Side Efficiency Promotion Plan (PPEC), a voluntary mechanism. Portugal has been implementing PPEC, with a track record of six calls for proposals for energy efficiency measures, since 2007. Under this mechanism, several entities, among them electric utilities, may submit proposals of measures that contribute to the reduction of electricity consumption or load management measures. Load management measures are those that allow a reduction of the costs of supply, without necessarily involving the reduction of energy consumption, namely the transfer of consumption from peak to off-peak hours. The proposed measures are evaluated according to a set of criteria, and the best performing ones are selected to be financed with funds raised from all electric energy ratepayers. Several changes to the regulations have been done over the years. Some of the changes were setting mandatory contribution of beneficiaries to the cost of the measures, setting maximum funding limits, and also setting criteria to assess the contribution of the measures to the national energy policy, among others. The number of participating actors has been increasing, as has also the total projected cost of the candidate measures. In the last PPEC edition, the total cost of the proposed measures accounted to 63 million euros, almost three times the available budget. This trajectory of the program implementation represents a source of data and relevant information on the role played by voluntary mechanisms in the involvement of electric utilities in the promotion of energy efficiency at the end-use.

One of the purposes of this paper is to highlight the role of the PPEC mechanism in the national energy policy and in the Portuguese commitments with the EU. Furthermore, light will be shed on how the participation in the program has been evolving, highlighting the energy services addressed, the sharing of the costs of the measures among the different agents, the evolution of the program administrator and societal costs of each saved kWh and the expected investments. Following a previous study regarding the Portuguese decision on the Article 7 of the EED that was presented at the 2015 ECEEE Summer Study [2], the authors intention was to assess how influential PPEC, as an example of a voluntary program, shows to be on several market agents' behaviors.

## 2. Energy Efficiency Mechanisms

In 2017, the International Energy Agency (IEA) published the report "Market-based Instruments for Energy Efficiency: Policy Choice and Design [3], which aimed at the characterization of energy efficiency instruments, such as energy efficiency obligations on utilities and white certificate programs. The study was requested by G7 countries at the Kitakyushu Energy Ministerial Meeting in 2016. MBI were defined as "MBIs for energy efficiency set a policy framework specifying the outcome (e.g., energy savings, cost-effectiveness) to be delivered by market actors, without prescribing the delivery mechanisms and the measures to be used" [3]. Fifty-two MBIs were studied with focus on energy savings and on investment costs. Some of the MBI are mandatory (energy efficiency obligations, white certificates, and energy efficiency resource standards) and others are voluntary (auctions or tendering programs). Regarding the mandatory schemes, 46 where found: 24 in the United States, 12 in Europe (Austria, Bulgaria, Denmark, France, Ireland, Italy, Luxemburg, Malta Poland, Slovenia, Spain, and United Kingdom), four in Australia, and one each in Canada, China, Brazil, Uruguay, Korea, and South Africa. By the time the study was done, three European countries were about to adopt EEO: Croatia, Greece and Latvia. Regarding non-mandatory involvement, auctions, six mechanisms were found, four of them in Europe: United Kingdom, Germany, Switzerland, and Portugal. In the United Kingdom, the voluntary scheme is a capacity market but, due to its design, does not support energy efficiency projects.

A 2017 update of the EEO in Europe [4] found out that Greece and Latvia had already started their EEO; that Malta was preparing an important revision of its obligations mechanism; Lithuania decided to adopt voluntary agreements; and Estonia was not imposing obligations, under the Article 7 of the EED.

The Swiss tender mechanism is managed by the Swiss Federal Office of energy and has been in place since 2010 [5]. The competition is launched annually and both projects and programs can compete for funds. For projects, and since 2015, there are two rounds each year. Projects are energy efficiency measures submitted by the beneficiary. On the other hand, projects are submitted and implemented by intermediaries, entities not directly beneficiaries of the measures. Projects and programs do not compete for the same budget. The budget available has been increasing from 5.5 million Euros, in 2010, to 41 million Euros in 2016. Over the six editions, about 93 million euros were awarded by this scheme. Since the funds come from a levy on the electricity grid, only measures addressing electricity uses can compete for funds. From 2010 to 2015, pumps and circulators received 21.4% of the budget, and lighting, electric motors and water heat pumps have shares of from 9.4% to 8.6%. In fact, an important part of the funds goes to a small number of technologies. In Switzerland, a project can only receive support of up to 40% of the investment cost [3]. The size of the projects must be between

18,000 Euros and 1.8 million Euros (from USD 20,000 to USD 2,060,000) and the programs between 130,000 Euros and 2.8 million Euros (USD 150,000 to USD 3,090,000).

The German mechanism that started in 2016 has two tenders: an "open" tender and a "close" tender [3]. "Specific" measures are those addressing a specific sector, beneficiary or technology, and compete in the close tender. In the open tender compete "neutral" measures, those measures that can be technology- or sector-neutral. The technologies must have a life time of at least ten years to be eligible. The mechanism is financed through the German Energy Efficiency Fund, partly funded by revenues from the EU Emissions Trading system. The available budget to fund electricity related measures was of EUR 50 million in 2016, EUR 100 million in 2017 and 150 million in 2018. A measure can only receive funds of up to 30% of the additional costs of the measure. Additional costs are the difference between the cost of the proposed technology and the cost of the market standard one. A project must have the best cost–benefit ratio and the savings must cost less than 0.10 Euro/kWh (USD 0.11/kWh) to be selected. The size of the projects must be between 27,000 Euros and 1.5 million Euros (from USD 30,000 to USD 1,680,000) and the programs (aggregated projects) between 200,000 Euros and 950,000 (USD 270,000 and 1,070,000). The measures may address all sectors although focusing only on electricity.

In the case of Portugal, where no EEO were adopted, two types of situations can be identified. On the one hand, the Distribution System Operator (DSO) involvement in the PPEC can be easily understood since it is a regulated entity, under a revenue cap kind of regulatory scheme—meaning that revenues are decoupled from sales. On the other hand, other companies in the electricity sector, namely free market retailers, do not necessarily experience decoupling between sales and revenues. In fact, even if they make a big effort to diversify their products, by trading services beyond selling electricity, their revenues depend heavily on sales volume, as profits thereof. In this case, the promotion of energy efficiency can originate revenue losses. The voluntary involvement of these agents in the promotion of energy efficiency may only be justified by the expected market advantage of a "green" image, due to the growing public awareness of the importance of energy efficiency, to preserve or increase market share. This becomes even more relevant to the definition of public policy since, in the PPEC case, cost sharing by promoters and beneficiaries has been increasing. The apparent paradox of the involvement of electric utilities in the promotion of energy efficiency was previously discussed by the authors [6].

In next section, the Portuguese energy efficiency policy is presented, and the role of PPEC is highlighted.

*Portuguese Energy Efficiency Policy*

In the 2016 Portuguese National Energy Efficiency Action Plan, NEEAP 2016—the Energy Efficiency Strategy, a new goal regarding a maximum limit of primary energy consumption was set. The actions and targets set for the 2013–2016 period included the concerns on energy consumption reduction set to 2020, part of the Directive on Energy Efficiency, Directive No. 2012/27/EU, from the European Parliament and the Council, from 25 October. The Government set a goal of 25% reduction in energy consumption by 2020, based on PRIMES projections made in 2007, setting the maximum consumption limit at approximately 22.5 Mtoe [7]. This commitment goes a little bit further than the 20% reduction set by Directive on Energy Efficiency. Besides the 25% reduction goal, a specific reduction goal for the State was set to 30% of the primary energy consumption by 2020. The NEEAP 2016 was developed in articulation with the 2020 National Renewable Energy Action Plan, NREAP 2020—the Renewable Energy Strategy [8].

The measures that were set in the previous NEEAP, the NEEAP 2008, were analyzed and some changes were made to cope with the new commitment. Some measures were discarded and others were introduced. The NEEAP 2016 includes ten programs covering six distinct areas (Table 1).

**Table 1.** Areas and programs considered in the NEEAP 2016 [8].

| Areas | | | | | | |
|---|---|---|---|---|---|---|
| | Transport | Residential and services | Industry | State | Behavior | Agriculture |
| **Programs** | Eco Car | House and Office Renovation | - | - | - | - |
| | Urban Mobility | Building Energy Efficiency System | Intensive Energy Consumption Management System | State Energy Efficiency | Communicate Energy Efficiency | Efficiency in Agricultural sector |
| | Transport Energy Efficiency System | Solar Thermal | - | - | - | - |

These programs can be briefly specified as follows:

- Eco car—measures towards improving energy efficiency of vehicles;
- Urban Mobility—measures to promote the use of the public transportation system and soft modes of transportation;
- Energy Efficiency System for Transport—measures addressing the promotion of railway systems and the energy management of transport fleets;
- House and Office Renovation—measures aiming the improvement of energy efficiency in lighting, home appliances and building retrofits;
- Buildings Energy Efficiency Systems—measures resulting from the energy certification system;
- Solar Thermal—measures addressing the adoption of renewable energy sources in buildings;
- Intensive Energy Consumption Management System—transversal measures in the industrial sector, and the revision of the Portuguese Intensive Energy Consumption Management System (SGCIE—Portuguese acronym for Sistema de Gestão dos Consumos Intensivos de Energia);
- State Energy Efficiency—measures aiming the energy certification of public buildings, as well as the Public Administration Energy Efficiency Program (ECO.AP), State transport fleets and Public lighting;
- Communicate Energy Efficiency—measures promoting communication and awareness campaigns to disseminate more energy efficient habits and attitudes; and
- Efficiency in the Agriculture sector—transversal measures addressing energy efficiency regarding the specificities of the sector.

In 2017, the Portuguese Government presented the third NEEAP that describes the measures to be adopted until 2020, and also the savings expected and obtained for the period from 2008 to 2015 [9]. The impacts of the measures implemented in each area, between 2008 and 2014, as well as the cumulative impact value, are presented in Figure 1. According to the third NEEAP, the slowdown in the economy during the "economic adjustment" period may justify the reduced adoption of energy efficiency measures in the years between 2011 and 2014.

The impact of the measures adopted in the buildings sector (state, residential and services) was still under evaluation at the date of the report. According to the authors of the report, it is expected that the impact of the measures for the 2015 and 2016 will be translated into higher energy savings due to some measures, such as electric mobility and the upgrade of the building stock.

The degrees of implementation of the measures within the NEEAP for 2016 and 2020 are presented in Table 2.

The implementation of the NEEAP is supported by regulatory measures, fiscal differentiation measures and financial support to the implementation of the energy efficiency measures. PPEC is one of the schemes that may provide financial support to the implementation of measures to be considered to account for the NEEAP targets. Other financial resources are the Energy Efficiency Fund, Fund to Support Innovation, the Portuguese Carbon Fund, and the National Strategic Reference Framework, among others. According to the 2017 NEEAP report, the Energy Efficiency Fund supported tangible measures leading to 12,220 toe of reduction in final energy consumption and the estimated savings due to PPEC financed measures is of 13,720 toe, after the implementation of the 2013–2014 edition [9].

**Table 2.** Degrees of implementation of the measures within the NEEAP for 2016 and 2020 [9].

| Type of Measures | Primary Energy Saved * | Primary Energy Savings Target 2016 | Primary Energy Savings Target 2016 (Implementation) | Primary Energy Savings Target 2020 | Primary Energy Savings Target 2020 (Implementation) |
|---|---|---|---|---|---|
| | (Toe) | (Toe) | (%) | (toe) | (%) |
| Transports | 297,923 | 343,683 | 87% | 406,815 | 73% |
| Residential and Services | 511,738 | 836,277 | 61% | 1,098,072 | 47% |
| Industry and agriculture | 260,167 | 407,221 | 64% | 561,309 | 46% |
| State | 36,245 | 153,634 | 24% | 295,452 | 12% |
| Behavior | 24,058 | 32,417 | 74% | 32,416 | 74% |
| Total | 1,130,131 | 1,773,232 | 64% | 2,394,064 | 47% |

\* Primary energy saved for the period 2008–2014.

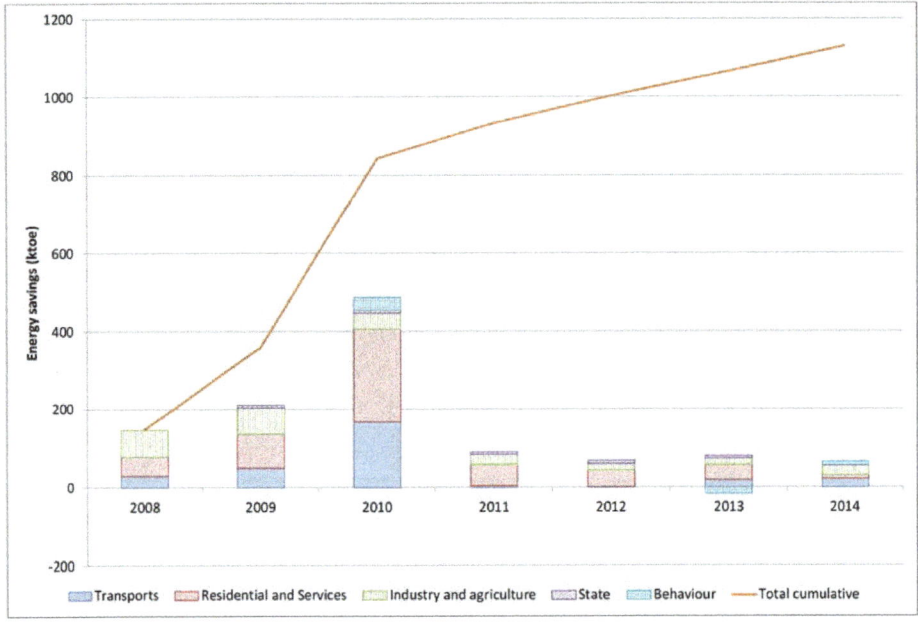

**Figure 1.** Impacts of the measures implemented in the different areas as well as the cumulative impact value [9].

In addition to verifying the fulfillment of the commitments assumed, it would be interesting to evaluate the energy and environmental performance of Portugal and to compare it with other countries [10].

PPEC is a mechanism, developed by the Portuguese Energy Services Regulatory Authority (ERSE—Entidade Reguladora dos Serviços Energéticos, in Portuguese), for the promotion of electric energy efficiency at the demand-side, whose first call for proposals occurred in 2007. However, the existence of stimuli to the involvement of electric utilities in the promotion of energy efficiency at the demand-side dates back to 1998. By then, the costs associated to demand-side projects were included in the revenues from the tariffs applied to all electricity consumers [11]. This was in force between 1999 and 2001. Changes were imposed by the Tariff regulation of 2001 [11], which defined a benefit-sharing scheme of 50% for each part, the utility and the consumers. The participation of the public electricity distributors was mandatory and Demand-Side Management Plans (PGP—Planos de

Gestão da Procura, in Portuguese) should be presented every year, between 2002 and 2005. Due to uncertainties regarding the regulatory evolution following the reform of the electricity sector, the PGP was suspended and then replaced by PPEC. Unlike PGP, PPEC is a voluntary scheme where, besides electric utilities, other entities can compete for funds to finance energy efficiency improving measures. The measures proposed are subjected to a competition, leading to the selection of the "best" energy efficiency measures, according to a set of criteria defined in advance.

PPEC rules have evolved over the years, motivated either by the experience gained by the regulator with previous editions or by energy policy requirements. With the publication of Ordinance No. 26/2013, on 24 January [12], which established new rules for the evaluation criteria and procedures, to be observed in the ranking and selection of the measures submitted to the competitions, it was determined that the assessment of these measures, in addition to being carried out by ERSE, should also be subjected to the appreciation of the Directorate-General for Energy and Geology (DGEG), a government department, in order to reflect energy policy criteria. A paper presenting the experience of the first two PPEC editions was presented at the 20th International Conference on Electricity Distribution [13]. Some regulatory characteristics of PPEC were presented, as well as the avoided consumption and environmental impacts of the 2007 and 2008 PPEC editions. Another paper was presented at the 2012 9th International Conference on the European Energy Market (EEM) [14]. In this paper, the results from the ex-post evaluation of the first PPEC edition were presented. In addition, in 2012, a paper was published addressing the methodology used for the ranking procedure [15]. In this study, an alternative was proposed that used avoided consumption and cost of saved kWh as decision variables in the definition of the weights of the criteria, for the selection process.

The evaluation of the measures submitted to PPEC in each call for proposals, which happens every two years, is carried out considering two main sets of evaluation criteria, both equally valued: (i) evaluation criteria regarding the efficiency of electricity consumption, from the perspective of economic regulation; and (ii) evaluation criteria related to energy policy objectives and instruments defined by decision of the member of the Government responsible for energy [12].

The evaluation criteria set by ERSE depends on the type of the measures: tangible or intangible measures. The intangible measures are those aimed at providing consumers with relevant information on the efficiency of electricity consumption and its benefits to adopt more efficient consumption habits, namely, training actions, information dissemination campaigns and energy audits. On the other hand, tangible measures are measures that address the installation of equipment with energy efficiency performance superior to the market standard or the replacement of inefficient equipment with more efficient ones. Tangible measures must address a consumption segment, Industry and Agriculture, Commerce and Services, and Households. The evaluation of the tangible measures is done taking into account the following criteria (a) cost–benefit analysis; (b) scale risk; and (c) weight of equipment investment in the total cost of the measure. Regarding intangible measures, their evaluation is made taking into account the following evaluation criteria: (a) quality of the presentation of the measure; (b) ability to overcome market barriers and spillover effect; (c) equity; (d) innovation; and (e) promoters' experience with similar programs. Each criterion has its own weight in the final score on the measure's performance.

The promoters, the entities that submit proposals of measures and that are responsible for their implementation, may be electric energy traders, operators of electricity transmission and distribution networks, associations and entities that defend consumers' interests, municipal associations, business associations, energy agencies, and higher education institutions and research centers.

The evaluation criteria set by the member of the Government responsible for energy, for the last PPEC call for proposals were: (a) alignment with the national energy policy and legislation in force; (b) alignment with the national energy efficiency policy and legislation in force; (c) support for the development and implementation of measures to promote energy efficiency; (d) diversification of promoters; and (e) coordination with other instruments to encourage energy efficiency.

Each PPEC call for proposals has six different competitions: four tenders for tangible measures and two tenders for intangible measures. The tenders for tangible measures go as follows: (a) three calls for tender for all promoters, one for each consumption segment (Industry and Agriculture, Commerce and Services, and Households); and (b) a tender for tangible measures for promoters which are not companies in the electricity sector.

The competitions of intangible measures are: (a) a tender for all promoters; and (b) a tender for promoters who are not companies of the electric sector.

Each candidate measure must include in its application a measurement and verification (M&V) plan, defining the methodology for the verification of savings. This M&V plan should be done by entities independent from the promoter. In addition, ERSE will carry out audits of various measures, by subset, subject to a budget that will not exceed 1% of the annual PPEC budget.

PPEC participation in the costs of tangible measures must be 80% or less of the total cost of the measure, fostering the participation and responsibility of the promoters and beneficiaries. In addition, if the budget for the first year of implementation of a tangible measure is less than 25% of the total PPEC candidate cost, the measure is excluded. Some other budget limits are imposed, such as measures submitted to the competitions for all promoters, with program administrator costs higher than 1/3 of the budget set for their contest and segment; and measures of the competitions for promoters who are not companies of the electric sector with candidate costs to the PPEC greater than 1/6 of the budget defined for the respective competition. The Net Present Value (*NPV*) from the societal perspective is an indicator of the societal value of the measure. A positive *NPV* is a screening criterion for a tangible measure to be competing in a tender. The *NPV* is computed according with the following expression:

$$NPV = \sum_{t=1}^{n} \frac{B_{S_t} - C_{S_t}}{(1+i)^t},$$ (1)

where $B_{S_t}$ is the total benefits from the societal perspective associated to the measure in year t; $C_{S_t}$ is the total costs from the societal perspective associated to the measure in year *t*; *i* is the discount rate; and *n* is the lifetime, in years.

The benefits, from the societal perspective, are the sum of the environmental benefits with the avoided supply costs. The societal costs include the financial costs incurred by the participant consumers, by all electric energy consumers (financed through PPEC), and by the promoters or any other entities. The costs and benefits are calculated in an incremental perspective against the market standard technology. Whenever it is considered that the proposed tangible measure does not contribute to the breakdown of market barriers, a free-ridership factor that penalizes the savings announced by the promoter may be applied.

In the next section, some results from the six editions of PPEC are presented, with emphasis on the costs sharing and the evolution of the program administrator costs and the societal costs of the measures. The analysis will be presented both for eligible and selected measures. Eligible measures are measures that are submitted and accepted into the competition, since they passed the eligibility criteria, and selected measures are the ones that were selected to be financed. PPEC costs are assumed as the costs to be financed by PPEC budget, similar to the perspective of the costs of the Program Administrator (PA) cost test of the California Standard Practice Manual [16].

A previous analysis of the first four editions of PPEC was published before, although with the focus on the involvement of electric utilities in the promotion of energy efficiency on the demand side [17].

## 3. PPEC Results

### 3.1. Costs and Budgets

Generally speaking, the participation in PPEC editions has been increasing from one edition to the following one. The number of promoters has been steadily increasing from eight in the first PPEC edition to 87 in the last one (Figure 2). In addition, the number of measures that have been submitted and considered eligible has been increasing. The number of eligible measures rose, from 62 in 2007 to 223 in the last edition, representing an increase of more than three times. The increase was sharper from 2007 to 2008, and there was a slight decrease from 2008 to 2009. This decrease was mostly due to a decrease in the number of tangible measures and a reduction in the pace of increase of intangible measures. The 2008 PPEC rules [18] came into force in the 2009–2010 edition. This change in the rules could also have influenced the reduction of the number of eligible measures.

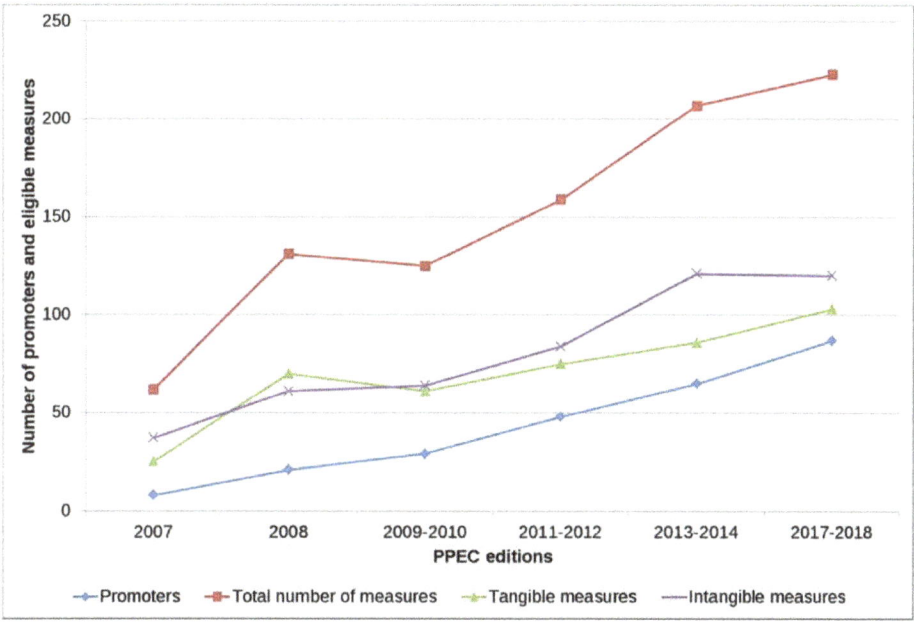

**Figure 2.** Evolution of the number of promoters and eligible measures in each PPEC edition [19–25].

In Table 3 the allocation of the PPEC funds to each type of measures, consumer segment, and competition, is shown. As mentioned before, the tangible measures in the all promoters' competition have distinct budgets, depending on the consumer segment addressed.

**Table 3.** Allocation of funds (in millions of Euros) by type of measure, consumer segment and competition, to each PPEC edition [19–25].

| Type of Measures | 2007 | 2008 | 2009–2010 | 2011–2012 | 2013–2014 | 2017–2018 |
|---|---|---|---|---|---|---|
| *Tangible measures* | - | - | - | - | - | - |
| Industry and Agriculture | 3 | 3 | 5.8 | 5.8 | 5.1 | 7 |
| Commerce and Services | 2.5 | 2.5 | 4.9 | 4.9 | 4.3 | 4 |
| Households | 2.4 | 2.4 | 5.3 | 5.3 | 4.6 | 3 |
| Non-Electric Sector Promoters | - | - | 2 | 2 | 3 | 4 |
| Total | 8 | 8 | 18 | 18 | 17 | 18 |
| *Intangible measures* | - | - | - | - | - | - |
| All promoters competition | - | - | 1.5 | 1.5 | 3 | 3 |
| Non-Electric Sector Promoters | 2 | 2 | 3.5 | 3.5 | 3 | 2 |
| Total | 2 | 2 | 5 | 5 | 6 | 5 |
| Total | 10 | 10 | 23 | 23 | 23 | 23 |

The PA costs of the eligible measures increased sharply in the first three editions and remained without significant variations since then (Figure 3). In the last four editions, the difference between the highest (61.6 M€ in 2017–2018) and the lowest (57 M€, in 2011–2012) is around of 4.5 M€, 8% of the lowest value. The main variations were an increase in the cost of intangible measures and a decrease in tangible ones, in the same PPEC edition (the 2013–2014 edition). These variations were probably due to the transfer of one million Euros from the tangible measures budget to intangible measures (Table 3).

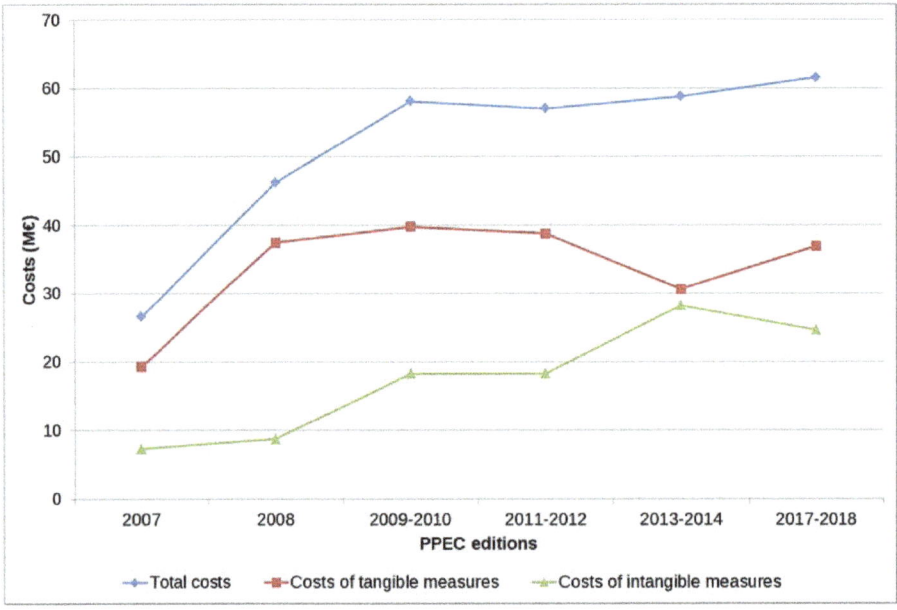

**Figure 3.** PA costs of the eligible measures throughout PPEC editions [19–25].

It is important to have in mind that the first two editions were annual, with a budget of 10 million Euros, and the other four were biennial. The biennial editions have a budget of 23 million Euros (equivalent to 11.5 million Euros/year). In the first two PPEC editions, tangible measures could have a multiannual implementation period of up to three years. However, only the costs to be spent in the first year were considered to be financed by that edition's budget. For example, for a measure selected in the 2007 edition that had an implementation cost plan for 2007 and 2008, only the costs pertaining to 2007 were financed by the 2007 budget. The costs to be spent in 2008 were to be financed by 2008 budget. Nowadays, only tangible measures with an implementation plan of two years are eligible.

As can be seen in Figure 4, although Commerce and Services has not been the consumer segment with the highest budgets, it has been the one with highest PA costs of the eligible tangible measures. Interesting is the fact that the PA cost of those measures has been very similar for all biennial editions. The same cannot be said to the PA costs of the measures addressing the other two segments.

Although the PA costs of the measures addressing Commerce and Services have been almost constant over the last four editions, the number of eligible measures in this segment more than doubled for the same editions (Figure 5), corresponding to the highest number of measures when compared to the number of those addressing the other two consumer segments. In fact, in the last PPEC edition, nearly 2/3 of the total number of eligible tangible measures addressed the Commerce and Service segment. Regarding the number of selected measures, the Commerce and Services consumer segment is also the one with the highest number of selected measures actually funded by PPEC.

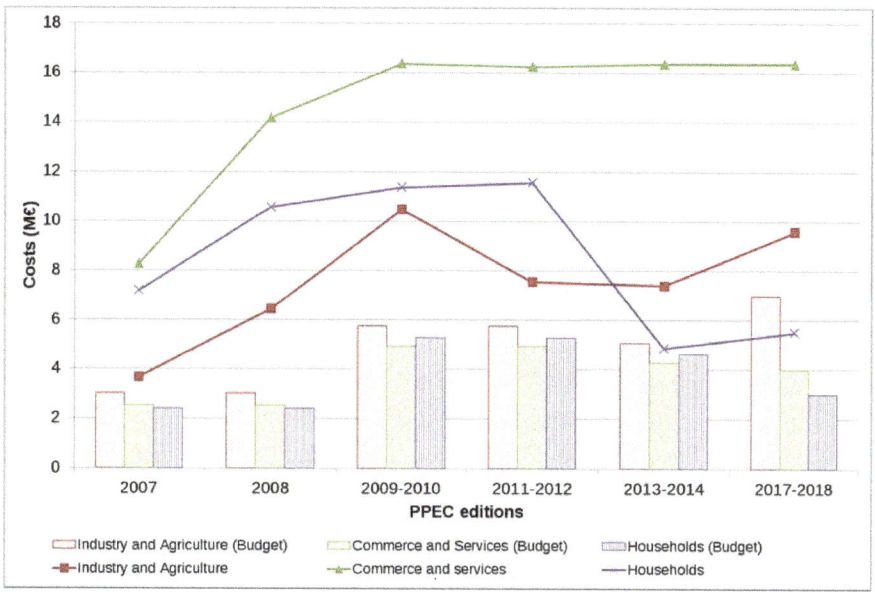

**Figure 4.** PPEC costs of the eligible tangible measures and corresponding available budget, for each PPEC edition [19–25].

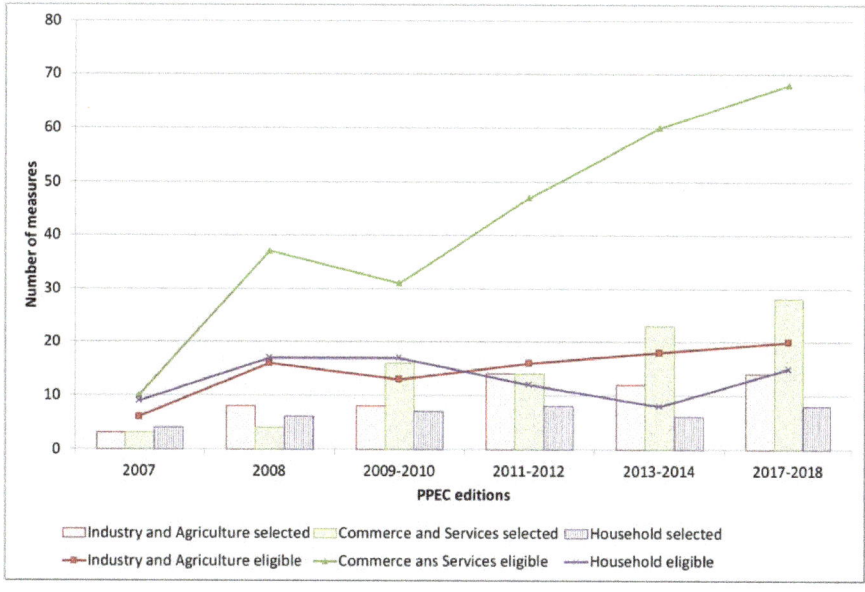

**Figure 5.** Number of eligible and selected measures by consumer segment in each PPEC edition [19–26].

## 3.2. Costs Sharing among Actors

The total cost of a measure, also addressed as the societal cost, can be financed by several agents. Besides the part of the cost of the measures that is financed by PPEC budgets, tangible measures are financed by the promoter, the consumer and/or other agents, such as promoter's partners. Over time, the societal costs sharing among agents has been changing. In Figure 6, it is possible to see the average costs, by agent, of the eligible tangible measures in each PPEC edition.

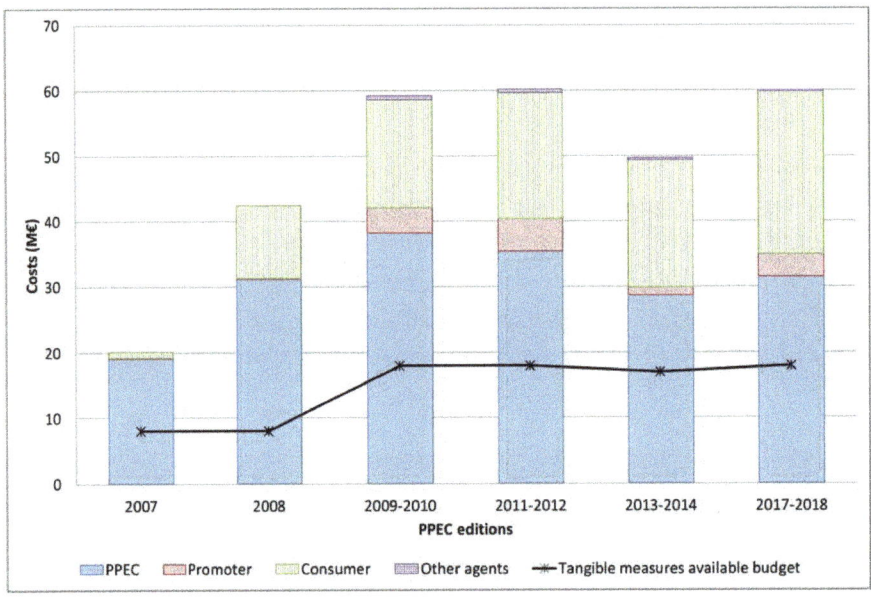

**Figure 6.** Societal costs of the eligible tangible measures and related available budget [19–25].

Each agent's average share of the societal costs of eligible measures is presented in Figure 7.

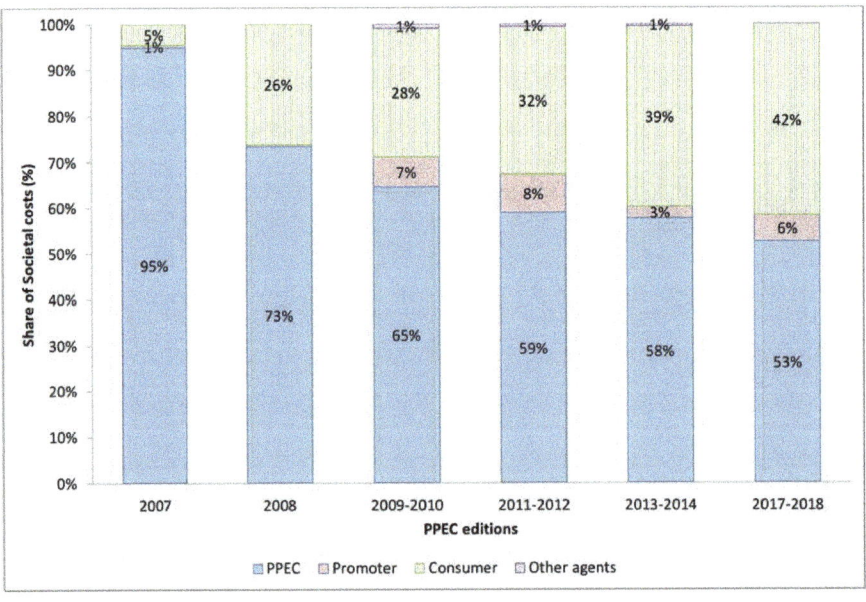

**Figure 7.** Share of societal costs of the eligible tangible measures in each PPEC edition [19–25].

In Figure 7 the PPEC share in total implementation costs has been decreasing. From the 2009–2010 PPEC edition, PPEC costs cannot exceed 80% of societal costs. Nevertheless, the decreasing tendency is probably because only PPEC costs are considered in the measures evaluation procedures. Then, the lower the participation of PPEC funds in each measure costs, the more interesting tends to be the benefit–cost ratio and the probability for the measure to be selected. Promoters are trying to involve beneficiaries and partners in societal costs. The promoter's shares tend to be quite small, which is understandable, since they share the costs but do not share the benefits. If the promoter is an electric utility, the incentive to participate in costs sharing is even smaller, since energy efficiency measures will reduce sales. In addition, the participation of partners in costs sharing has been residual or inexistent.

The average share of societal costs of the selected tangible measures (Figure 8) is quite similar to the costs sharing of eligible measures (Figure 6). Since the cost–benefit analysis is made under the perspective of the PA costs, it could be expected that the share of PPEC funds should be less than the share of costs of eligible measures. However, since the Cost–Benefit ratio is not the only criterion, there is no evidence of lower share of PPEC costs in selected measures, when compared to eligible measures.

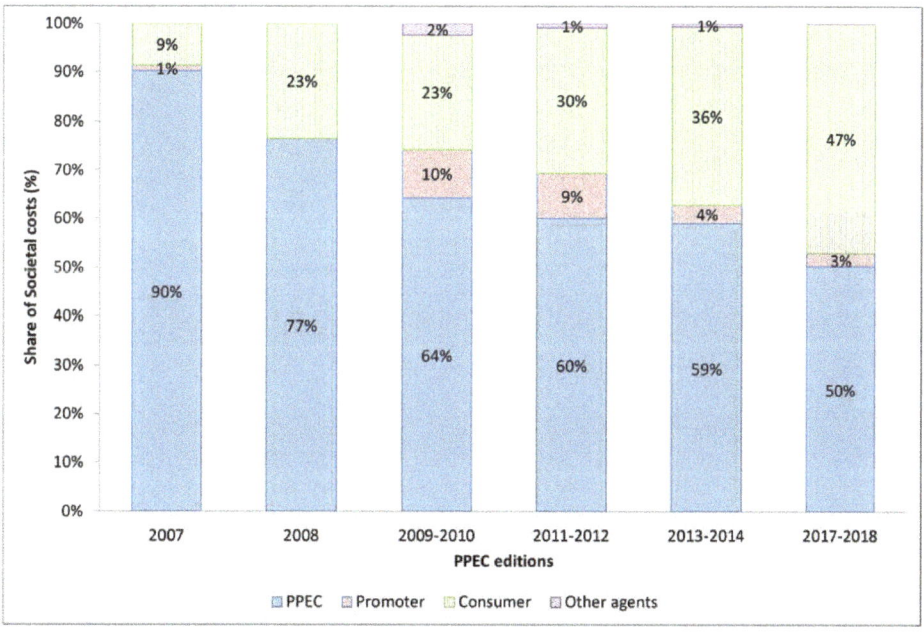

**Figure 8.** Share of societal costs of the selected tangible measures in each PPEC edition [19–26].

In the next three figures, societal cost sharing among agents is presented for the selected measures and the three consumption segments: Industry and Agriculture (Figure 9), Commerce and Services (Figure 10) and Households (Figure 11). Regarding PA costs, the measures for Industry and Agriculture follow closely the trend verified for eligible measures (Figure 6). It can be seen that the participation of other agents in costs, besides the PA and the beneficiary, is quite small or inexistent.

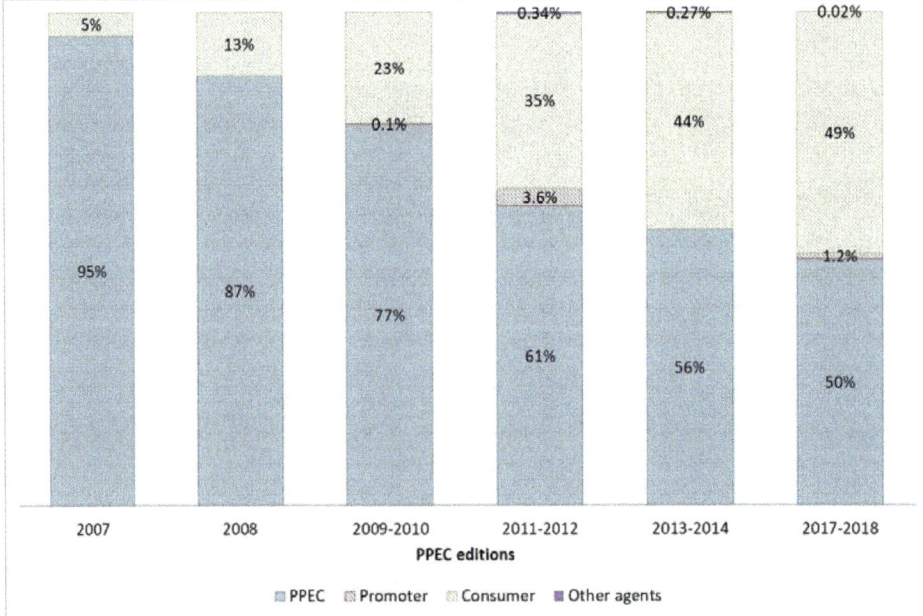

**Figure 9.** Share of societal costs of the selected tangible measures for the Industry and Agriculture consumption segment, in each PPEC edition [19–26].

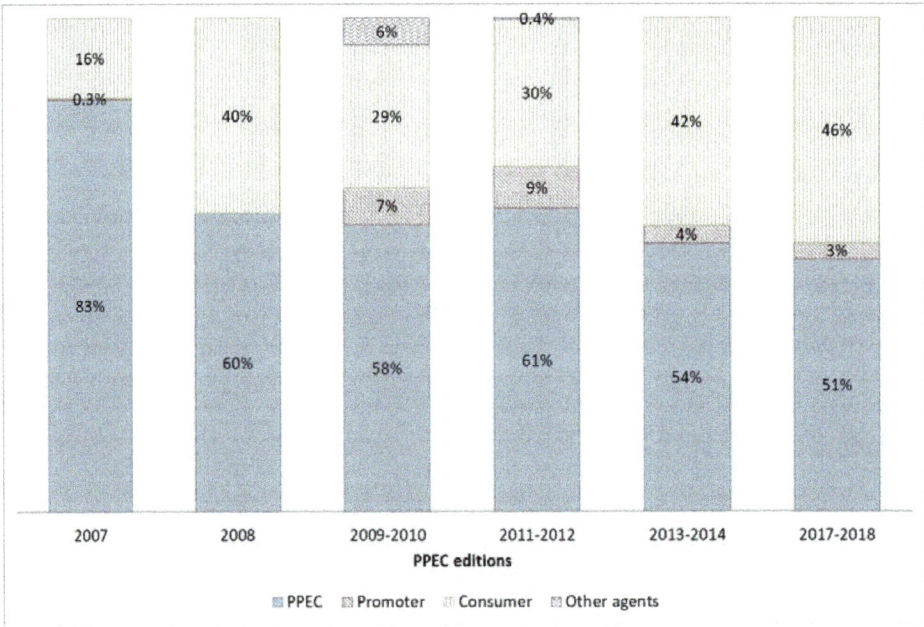

**Figure 10.** Share of societal costs of the selected tangible measures for the Commerce and Services consumption segment, in each PPEC edition [19–26].

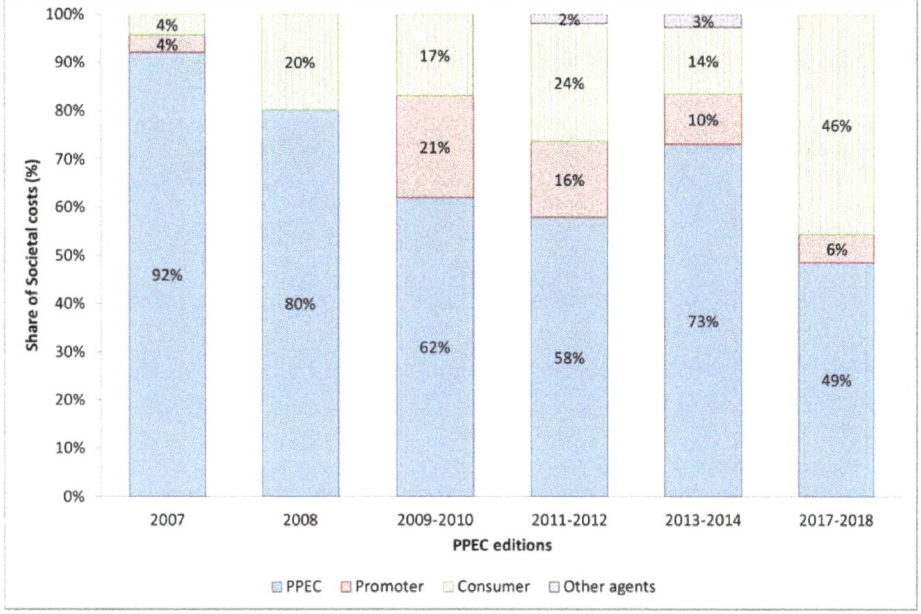

**Figure 11.** Share of societal costs of the selected tangible measures for the Households consumption segment, in each PPEC edition [19–26].

In the case of the measures addressing the Commerce and Services segment, PPEC relative participation in the costs of the selected measures has been under 61% since the 2008 edition, even before the 80% limit was defined. The average participation of the promoters in costs, although very small, is higher than the one verified in the case of the Industry and Agriculture measures.

For the Households sector, except for the 2013–2014 edition, the PA share in the societal costs of the measures has also been showing a decreasing tendency. The 73% of PPEC participation in costs, for the 2013–2014 edition, is mostly due to one measure whose PA costs represents 33% of the available budget and 78% of the societal costs. It can also be seen that the promoter's share in costs is above the one verified in the measures addressing the other two consumer segments.

In the first PPEC edition, total investment in tangible measures was 9.2 M€, being 8.3 M€ from the PPEC budget. In the last PPEC edition, the total expected investment is 35.9 M€, the participation of the PA reaching 18 M€. The expected investment increased 3.9 times, from 9.2 M€ in 2007 to 35.9 M€, in the last PPEC edition, while the PA costs increased only 2.2 times. The existence of a leverage factor [27] is clear. A leverage factor of 1.0 means that all the investment is made from program funds. A leverage factor of 2.0 means that only half the investment is made from program funds. Figure 12 shows the leverage factors for all selected tangible measures (all segments) and for each segment individually, in each PPEC edition. As can be seen, the leverage factors were, in the last PPEC edition, near 2.0 (1.99), meaning that, for each euro invested from the program funds, an additional ninety-nine cents were invested. In the first PPEC edition, the leverage factor was 1.11. In the Households sector, it is expected that 1.06 € will be invested for each euro from PPEC funds, since its leverage factor is 2.06. The lowest leverage factor, of 1.95, occurred in the Commerce and Services sector. Nevertheless, it is very close to the leverage factors found in the other consumer segments. The lowest value for the leverage factor found for the Households segment, in the 2013–2014 edition, is mostly due to that same measure, identified earlier, with the PA costs being 78% of the societal costs.

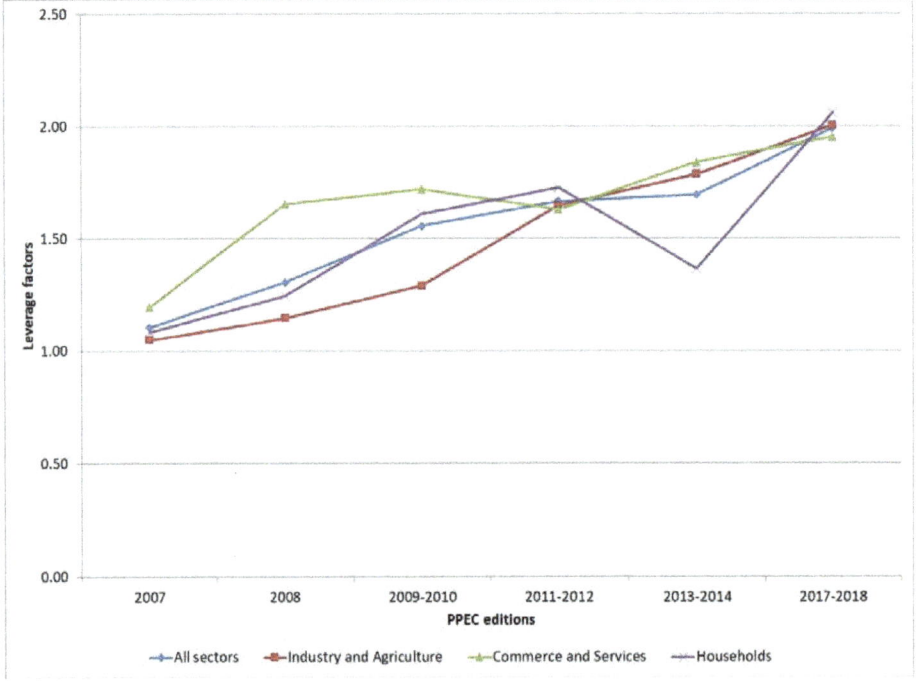

**Figure 12.** Leverage factors, for each PPEC edition, total and for each consumption segments [19–26].

Looking from another perspective, the extra investment added to each euro invested through PPEC evolved from eleven cents, in the first PPEC edition, to ninety-nine cents in the last edition.

## 4. Energy Services Addressed

In Figure 13 the PPEC costs of the proposed measures according to the energy services they address (for tangible measures) and the energy related services (for intangible measures) are presented. As can be seen, the most popular energy service addressed is, by far, lighting, mostly involving technology replacement actions. There are, in fact, three energy services related to lighting: interior lighting (mostly in buildings), public and traffic lighting. In the first three editions, most of the measures proposed a change in technology from incandescent to compact fluorescent lamps (CFL). In more recent editions, the standard of the market was considered the CFL, leading promoters to propose the replacement of existing lighting devices with LED technology based devices. The experience obtained by the promoters in previous editions made it possible for them to address more energy services, promoting nowadays a more diversified set of measures.

Regarding tangible measures, the diversity of energy services addressed and the number of selected measures in each PPEC edition is presented in Figure 14. As can be seen, in the first PPEC edition, only three energy services were addressed and the diversity has been increasing since then. Altogether, the number of measures addressing lighting (in buildings, public and traffic) represents almost 51% of the selected measures.

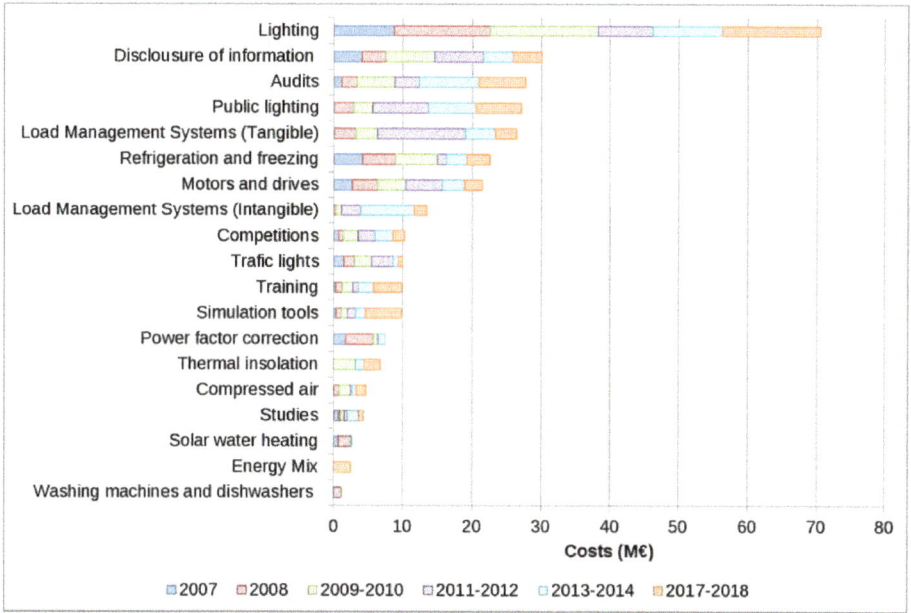

**Figure 13.** Costs of the measures by targeted energy services, for each PPEC edition [19–25].

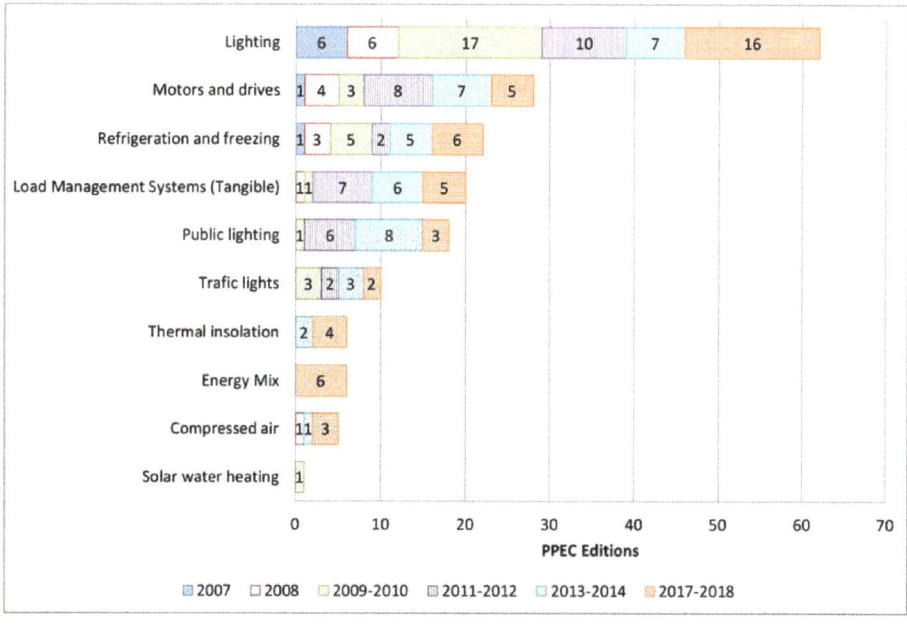

**Figure 14.** Number of selected measures, according to each energy services addressed, in each PPEC edition [19–26].

In Table 4, with the clear exception of the first PPEC edition, the PPEC cost of each saved MWh are very competitive. Comparing the values in Table 4 with the values in Table 5, it is possible to see that, due to the very high participation of consumers in costs, generally speaking, the energy services mostly addressed are also the ones with more competitive PPEC costs per each saved MWh.

**Table 4.** Average PA costs of each avoided MWh of the selected measures in each PPEC edition [19–26].

| Type of Measures | 2007 | 2008 | 2009–2010 | 2011–2012 | 2013–2014 | 2017–2018 |
|---|---|---|---|---|---|---|
| Compressed air | - | 20.36 | - | - | 29.82 | 17.14 |
| Energetic Mix | - | - | - | - | - | 12.81 |
| Lighting | 20.86 | 6.08 | 7.93 | 8.93 | 13.58 | 16.68 |
| Load Management Systems (Tangible) | - | 10.31 | 5.50 | 12.29 | 18.45 | 14.11 |
| Motors and drives | 10.25 | 8.29 | 4.44 | 6.94 | 6.57 | 8.70 |
| Public lighting | - | - | 8.95 | 6.73 | 6.71 | 29.06 |
| Refrigeration and freezing | 24.25 | 8.78 | 2.83 | 4.84 | 7.15 | 16.41 |
| Solar water heating | - | - | 4.95 | - | - | - |
| Thermal insolation | - | - | - | - | 13.34 | 13.05 |
| Traffic lights | - | - | 39.82 | 12.51 | 17.05 | 9.91 |

**Table 5.** Average Societal costs of each avoided MWh of the selected measures in each PPEC edition (in €/MWh) [19–26].

| Type of Measures | 2007 | 2008 | 2009–2010 | 2011–2012 | 2013–2014 | 2017–2018 |
|---|---|---|---|---|---|---|
| Compressed air | - | 28.83 | - | - | 51.38 | 39.84 |
| Energetic Mix | - | - | - | - | - | 23.78 |
| Lighting | 23.99 | 8.34 | 12.72 | 15.15 | 31.02 | 32.60 |
| Load Management Systems (Tangible) | - | 10.31 | 10.16 | 21.33 | 26.06 | 22.35 |
| Motors and drives | 10.25 | 9.33 | 6.17 | 10.65 | 10.29 | 16.41 |
| Public lighting | - | - | 11.19 | 12.10 | 10.29 | 56.27 |
| Refrigeration and freezing | 24.25 | 15.45 | 3.67 | 6.24 | 11.30 | 39.98 |
| Solar water heating | - | - | 64.84 | - | - | - |
| Thermal insolation | - | - | - | - | 27.84 | 35.64 |
| Traffic lights | - | - | 54.92 | 20.02 | 25.04 | 17.83 |

In more recent PPEC editions, with the investment in LED lamps, the costs of each saved MWh increased. The PA costs of each saved MWh among the selected measures ranged from 8 € to 39 €, in the last PPEC edition.

## 5. Conclusions

The promotion of energy efficiency at the demand side by electric utilities in Portugal has been in place since 1999: at first with the PGP, where the participation of the regulated utility was mandatory, and then, after 2007, with the PPEC, a voluntary scheme that allows the participation of other promoters not belonging to the electricity sector. The number of promoters over the six PPEC editions increased more than 10 times. In addition, the number of measures increased 3.6 times, with tangible measures increasing at a slightly higher rate (4.1 times) than intangible measures, although the number of intangible measures is higher than the number of eligible measures in almost every PPEC edition. The program costs of the eligible measures have also been 2.5 to 2.7 times the available budget. Thus, if one looks at the number of proponents, the number of measures, or the program costs of the eligible measures, it is safe to say that the PPEC scheme is an effective instrument to foster energy efficiency in electricity consumption.

The consumer participation in costs has been increasing, representing 42% in the case of eligible measures and 47% in the case of the selected measures, in the last PPEC edition. The share of the programs costs, although limited to 80% of the total cost of the measure, represented 53% and 50% of the total cost of eligible and selected measures, respectively, in the last PPEC edition.

Looking at the tangible measures, only three energy services were addressed in the first PPEC edition. In the last edition, nine energy services were addressed. The main energy services addressed are related to lighting, be it in buildings, public lighting or traffic lighting, representing more than 30% of the available budget, for eligible measures, and more than 40% for the selected measures.

The expected investment caused by PPEC increased nearly four times over the six editions. In addition, for each euro invested by PPEC in improving the efficiency in electricity consumption, another euro is invested by consumers, promoters and other agents.

After the 2009–2010 PPEC edition, CFL were considered the market standard and underwent a depreciation in the valuation of savings, resulting in a reduced number of eligible measures supporting this technology and an increase in the measures addressing LED technology, a more expensive but more efficient one.

As can be inferred from the above, the PPEC scheme, although voluntary, is a very effective one, with an increasing number of promoters, from different sectors of the economy. The promoters that have been answering PPEC calls have now experienced teams in the design of new proposals, addressing different energy services and technologies. In addition, participation in costs by actors other than the program administrator leads to a greater accountability of these agents in the effectiveness of the measure.

The PPEC is a very informative case study for assessing the motivations of different market actors, namely electricity sector companies, to be actively involved in the promotion of energy efficiency.

The voluntary nature of this program also allows the participation of a larger number of actors, reaching a more diversified number of consumers that become increasingly aware of the importance of energy efficiency. On the other hand, since it is a voluntary mechanism, projections of energy savings are prone to uncertainty.

**Author Contributions:** J.L.S. performed the literature research, the data gathering, mining and processing; both authors analyzed the data, have put up the reference framework and derived the conclusions.

**Funding:** This work has been supported by the Portuguese Foundation for Science and Technology (FCT) under project grant UID/MULTI/00308/2013.

**Conflicts of Interest:** The authors declare no conflicts of interest.

## References

1. European Parliament and Council. Directive 2012/27/EU of the European Parliament and of the Council of 25 October 2012 on Energy Efficiency, Amending Directives 2009/125/EC and 2010/30/EU and repealing Directives 2004/8/EC and 2006/32/EC. Available online: https://eur-lex.europa.eu/legal-content/EN/TXT/?uri=celex:32012L0027 (accessed on 14 April 2017).
2. Sousa, J.L.; Martins, A.G.; Jorge, H.M. Are energy efficiency obligations an alternative? The case-study of Portugal. In Proceedings of the ECEEE Summer Study Proceedings, Hyères, France, 1–6 June 2015; pp. 371–381.
3. Rosenow, J.; Cowart, R.; Thomas, S.; Kreuzer, F. *Market-Based Instruments for Energy Efficiency—Policy Choice and Design*; International Energy Agency: Paris, France, 2017.
4. Association for Teacher Education in Europe. Snapshot of Energy Efficiency Obligations schemes in Europe: 2017 update. In Proceedings of the Fourth European Workshop of the White Certificates Club, Paris, France, 30 June 2017.
5. Radgen, P.; Bisang, K.; Koenig, I. Competitive tenders for energy efficiency—Lessons learnt in Switzerland. In Proceedings of the ECEEE Industrial Summer Study Proceedings, Berlin, Germany, 12–14 September 2016; pp. 81–89.
6. Sousa, J.L.; Martins, A.G.; Jorge, H.M. Dealing with the paradox of energy efficiency promotion by electric utilities. *Energy* **2013**, *52*, 251–258. [CrossRef]

7.  Capros, P.; Mantzos, L.; Papandreou, V.; Tasios, N. *Energy and Transport Outlook to 2030—Update 2007*; European Communities: Brussels, Belgium, 2008.
8.  Council of Ministers. Council of Ministers Resolution No. 20/2013. Available online: https://ec.europa.eu/energy/sites/ener/files/documents/Portugal_NEEAP_en.pdf (accessed on 14 April 2017).
9.  PNAEE—Plano Nacional de Acção Para a Eficiência Energética—Third NEEAP 2017–2020. Available online: https://ec.europa.eu/energy/en/topics/energy-efficiency/energy-efficiency-directive/national-energy-efficiency-action-plansaccessedon (accessed on 30 March 2018).
10. Diário da República. *Despacho No. 16 288-A/98*; Diário da República: Lisbon, Portugal, 1998; pp. 13286–13292.
11. Diário da República. *Despacho No. 18 412-A/2001*; Diário da República: Lisbon, Portugal, 2001; pp. 149441–149491.
12. Diário da República. *Portaria No. 26/2013*; Diário da República: Lisbon, Portugal, 2013; pp. 489–491.
13. Apolinário, I.; Correia de Barros, I.; Coutinho, H.; Ferreira, L.; Madeira, B.; Oliveira, P.; Trindade, A.; Verdelho, P. Promoting demand side management and energy efficiency in Portugal. In Proceedings of the 20th International Conference on Electricity Distribution, Prague, Czech Republic, 8–11 June 2009.
14. Apolinário, I.; Correia de Barros, I.; Espírito Santo, C.; Ferreira, A.; Ferreira, L.; Madeira, B.; Verdelho, P. Results from a competitive tender mechanism to promote energy efficiency in Portugal. In Proceedings of the 2012 9th International Conference on the European Energy Market, Florence, Italy, 10–12 May 2012.
15. Sousa, J.L.; Martins, A.G.; Jorge, H.M. Societal objectives as drivers in the search for criteria weights when ranking energy efficiency measures. *Energy Policy* **2012**, *48*, 562–575. [CrossRef]
16. California Public Utilities Commission. *California Standard Practice Manual—Economic Analysis of Demand-Side Programs and Projects*; California Public Utilities Commission: San Francisco, CA, USA, 2001.
17. Sousa, J.L.; Martins, A.G.; Jorge, H.M. World-wide non-mandatory involvement of electricity utilities in the promotion of energy efficiency and the Portuguese experience. *Renew. Sustain. Energy Rev.* **2013**, *22*, 319–331. [CrossRef]
18. Diário da República. *Despacho No. 15546/2008*; Diário da República: Lisbon, Portugal, 2008; pp. 24886–24895.
19. Entidade Reguladora dos Serviços Energéticos. *Plano de Promoção de Eficiência no Consumo de Energia Eléctrica Para 2007*; Entidade Reguladora dos Serviços Energéticos: Lisboa, Portugal, 2007.
20. Entidade Reguladora dos Serviços Energéticos. *Plano de Promoção da Eficiência no Consumo de Energia Eléctrica para 2008*; Entidade Reguladora dos Serviços Energéticos: Lisboa, Portugal, 2007.
21. Entidade Reguladora dos Serviços Energéticos. *Plano de Promoção da Eficiência no Consumo de Energia Elétrica PPEC—Documento de Discussão*; Entidade Reguladora dos Serviços Energéticos: Lisboa, Portugal, 2008.
22. Entidade Reguladora dos Serviços Energéticos. *Plano de Promoção da Eficiência no Consumo de Energia Eléctrica para 2009–2010*; Entidade Reguladora dos Serviços Energéticos: Lisboa, Portugal, 2009.
23. Entidade Reguladora dos Serviços Energéticos. *Plano de Promoção da Eficiência no Consumo de Energia Eléctrica para 2011–2012*; Entidade Reguladora dos Serviços Energéticos: Lisboa, Portugal, 2010.
24. Entidade Reguladora dos Serviços Energéticos. *Plano de Promoção da Eficiência no Consumo de Energia Eléctrica para 2013–2014*; Entidade Reguladora dos Serviços Energéticos: Lisboa, Portugal, 2013.
25. Entidade Reguladora dos Serviços Energéticos. *Plano de Promoção da Eficiência no Consumo de Energia Eléctrica para 2017–2018—Avaliação na Perspetiva da Regulação Económica*; Entidade Reguladora dos Serviços Energéticos: Lisboa, Portugal, 2016.
26. Direção Geral de Energia e Geologia e Entidade Reguladora dos Serviços Energéticos. *Plano de Promoção da Eficiência no Consumo de Energia Eléctrica Para 2017–2018*; Direção Geral de Energia e Geologia e Entidade Reguladora dos Serviços Energéticos: Lisboa, Portugal, 2016.
27. Rohde, C.; Rosenow, J.; Eyre, N.; Giraudet, L.-G. Energy Saving Obligations—Cutting the Gordian Knot of leverage? *Energy Effic.* **2015**, *8*, 129–140. [CrossRef]

*Article*

# An Optimisation Study on Integrating and Incentivising Thermal Energy Storage (TES) in a Dwelling Energy System

**Gbemi Oluleye [1,*], John Allison [2] , Nicolas Kelly [2] and Adam D. Hawkes [1]**

[1]  Centre for Process Systems Engineering, Department of Chemical Engineering, Imperial College London, London SW7 2AZ, UK; a.hawkes@imperial.ac.uk

[2]  Department of Mechanical and Aerospace Engineering, University of Strathclyde, Glasgow G1 1XQ, UK; j.allison@strath.ac.uk (J.A.); nick@esru.strath.ac.uk (N.K.)

*  Correspondence: o.oluleye@imperial.ac.uk

Received: 1 April 2018; Accepted: 25 April 2018; Published: 29 April 2018

**Abstract:** In spite of the benefits from thermal energy storage (TES) integration in dwellings, the penetration rate in Europe is 5%. Effective fiscal policies are necessary to accelerate deployment. However, there is currently no direct support for TES in buildings compared to support for electricity storage. This could be due to lack of evidence to support incentivisation. In this study, a novel systematic framework is developed to provide a case in support of TES incentivisation. The model determines the costs, $CO_2$ emissions, dispatch strategy and sizes of technologies, and TES for a domestic user under policy neutral and policy intensive scenarios. The model is applied to different building types in the UK. The model is applied to a case study for a detached dwelling in the UK (floor area of 122 m$^2$), where heat demand is satisfied by a boiler and electricity imported from the grid. Results show that under a policy neutral scenario, integrating a micro-Combined Heat and Power (CHP) reduces the primary energy demand by 11%, CO2 emissions by 21%, but with a 16 year payback. Additional benefits from TES integration can pay for the investment within the first 9 years, reducing to 3.5–6 years when the CO2 levy is accounted for. Under a policy intensive scenario (for example considering the Feed in Tariff (FIT)), primary energy demand and CO2 emissions reduce by 17 and 33% respectively with a 5 year payback. In this case, the additional benefits for TES integration can pay for the investment in TES within the first 2 years. The framework developed is a useful tool is determining the role TES in decarbonising domestic energy systems.

**Keywords:** TES; multi-period mixed integer linear program; incentives; techno-economic analysis

---

## 1. Introduction

### 1.1. Background

Integration of Low-to-Zero Carbon (LZC) technologies in domestic buildings might go a long way in achieving the UK target for 80% reduction in greenhouse emissions by 2050 (from 1990 levels). Major barriers to implementation are: (1) large seasonal variations in space heating and electricity requirements; (2) sporadic nature of renewable energy; and (3) high capital costs. Thermal Energy Storage (TES) allows better integration of heat and electricity systems, and storage of energy for use during peak times, thereby addressing the first two barriers. However, despite these benefits, the penetration rate of TES in buildings in Europe is low [1].

Micro-generation can protect against future end-use energy cost and encourage household self-sufficiency [2]. Due to low penetration rates, a number of fiscal instruments have been introduced for micro-generation in the UK. There is a strong argument that fiscal incentives designed to increase uptake of technologies can in the long run decrease the cost of technologies [3]. An example is the Feed in Tariff (FIT). FITs are fixed electricity prices paid to micro-generation producers per unit of energy produced and injected into the electricity grid. The ability of FIT to stimulate development of less mature LZC technologies in many countries has been established [3,4]. However, the potential of the FIT to increase uptake of TES has not been analysed. In the UK, there is no incentive for homeowners to accommodate TES in their dwellings and communities. Policy makers in the UK have yet to consider storage a critical component of the UK energy future [5]. This could be due to uncertainty about the benefits of TES integration and lack of a reliable cost recovery mechanism for TES. This work seeks to address these challenges through the development of a systematic framework to evaluate the benefits of TES integration in a dwelling energy system, and determine if the additional investment in TES is economically viable. The framework is first mentioned in Oluleye et al. [6], and is extended in this paper.

Modelling and optimisation studies exist to determine the design and dispatch of LZC technologies and TES. However, there are still challenges in: (1) balancing complexity and robustness through the selection of appropriate temporal precision and time bands to represent energy demand and technology characteristics; and (2) representing storage and allowing the TES size to be determined optimally. These challenges are addressed in this work, through the development of a more accurate integrated framework for optimal design and operation of a low carbon dwelling energy system. The framework forms the basis of a case to support incentivising TES in dwellings.

### 1.2. Literature Review

A high share of micro-generation technology for satisfying the energy demand of the building sector could result in reduction in $CO_2$ emissions, if an optimal operation strategy is pursued [7]. Reduction in primary energy consumption is also possible with a Micro-Combined Heat and Power (CHP) [8], and further reductions when TES is integrated [9]. Fubara et al. [10] records a 6–10% reduction in primary energy from integrating a micro-CHP. Murugan and Horak [11] identify primary energy savings and emissions reduction from micro-CHP, and their large market potential. Further reductions are possible from integrating heat storage with micro-CHP [8]. Improved capacity factor, and reducing low utilisation plant are benefits of generating and storing heat during periods of low demand and regenerating at periods of high demand [12]. Diaz and Moreno [9] records cost improvements when a micro-CHP is supplemented by TES. Storing heat improves the synchronisation of the demand for heat and electricity in buildings [13]. In spite of these benefits, LZC technologies and TES still have a small percentage of the market [14], implying low adoption by homeowners.

Furthermore, it is unclear if the benefits from TES integration make the additional investment worthwhile. The economic viability could be determined by estimating how long it takes to payback the additional investment in TES and if the investment has a positive return. Such analysis has not been done before.

Economic benefits in terms of operational savings are made possible by charging an accumulator when electricity is high, and releasing heat during cheap night hours [15,16]. However, analysis was done on a district scale. An advantage of district heating is that bigger size technologies are required thereby benefitting from economies of scale. This implies, the additional cost of the TES may be less than in an individual home. Even though district heating has been identified as a way of decarbonising domestic heating in the UK, the estimated deployment of district heating in 2030 is 6% of the total heat demand according to the 4th carbon budget [17]. Furthermore, district heating is not economically viable for lower population density areas. There is a need to evaluate the benefits of TES integration in an individual home (considering different dwelling types), as such an analysis has not been done before.

The development of a framework for system design is required before a techno-assessment of the design, and impacts of TES can be determined. These models account for technical and economic parameters, and boundary conditions like heat and electricity demands. Frameworks for system design are based on detailed buildings and technology physics, heuristics and optimisation. A detailed building simulation model is applied by Kelly et al. [18] to determine the amount of thermal storage needed to shift heat pump operation to off-peak periods. Salata et al. [19] also applies a detailed simulation model to evaluate the impact of micro-CHP. Detailed building models determine the potential effect on the end user in terms of comfort; however, technology selection and dispatch determination is not addressed. Furthermore, capturing design trade-off is not done systematically. Therefore, designs obtained are suboptimal in relation to delivery of energy and costs, making them insufficient as a framework to quantify the benefits of TES integration, and determine if the additional investment in TES is worthwhile.

Heuristics based on operational logic have been applied to size thermal stores [20]. A heuristic approach for economic optimisation for a single CHP plant and TES capacity is also applied in Vogelin et al. [21]. Heuristics are often iterative in nature, therefore, solving larger models becomes challenging; making them insufficient to form the basis to provide a case to support incentivizing TES.

Optimisation frameworks are able to capture capital-energy trade-offs systematically, and produce optimal designs. LZC technologies have been modelled to determine capacity and dispatch based on linear models [22,23], non-linear models [24] and mixed integer linear models [25,26]. Linear models underestimate the optimal unit capacity by 15% [27], whilst non-linear models suffer from long solution times and local optima [25]. Furthermore, selection of technologies and their capacities is not addressed in linear and non-linear models. Mixed integer linear models are suitable for solving scheduling problems with binary variables and obtaining a global optimal solution rapidly. Thus providing a useful tool to provide evidence to support TES incentivisation.

Such models are often complex due to the varying nature of energy demand and technology characteristics. To reduce model size and complexity, the size of the thermal store is sometimes predetermined [8,13], resulting in non-optimal designs. However, within the Mixed Integer Linear Program (MILP) modelling framework appropriate use of integer variables can obtain good compromise with model complexity whilst ensuring the designs are optimal. Furthermore, integers have been used in a crude way, that is, only associated with selection of technologies. A compromise with model complexity is possible using integers for operation of technologies, charging/discharging TES, import/export of electricity per temporal precision. The later use of integers ensures the operations do not occur simultaneously. In previous research, charging/discharging the store was only based on excess heat available, neglecting price signals. In this present work, in addition to the heat balance, TES is actively regulated to account for gas and electricity price signals.

A model's environmental and economic outcome is influenced by the temporal precision [28] and the banding structure used to represent energy demand [16]. Different temporal precisions have been applied by various authors: 1 h [9,13,21,29–32] and 15 min [25]. Too large temporal precisions could introduce errors in technology sizing and ignore peak demands. Fine temporal precision (5–10 min) is required to adequately capture the characteristics of demand from an economic and environmental viewpoint [28]. Banding structures are used to simplify demand representation; 39 time bands (7 for weekdays, 6 for weekends and 3 seasons in a year) [16], bands of 3 weeks taken from 3 seasons with a time zoom factor for result scale up [25], 3 representative days [9], monthly time bands [33], 3 non-consecutive weeks for 3 seasons and multiplied by a weighing factor [26] and transition and winter days [8]. Choice of technology capacity becomes challenging with large time bands, and they may be errors in the techno-economic assessment [16]. A reduced time slice and band can guarantee a more reliable analysis of the system, as real peaks will be accounted for. The challenge in time slice and band selection is the trade-off between model complexity and accuracy. In the novel framework developed, a temporal precision of 5 min for all the days in the year is used to accurately determine the economic and environmental outcomes. The multi-period MILP model is selected to guarantee a global optimum and prevent the need to iterate when non-linear models are used. In addition to the heat balance, TES is actively regulated to account for gas and electricity price signals.

*1.3. Contributions of This Work*

A novel systematic framework based on a multi-period MILP model is proposed for the integration and assessment of TES in an existing dwelling. To improve the model accuracy finer temporal precisions are used considering 365 days in the year. The novel model is also able to determine the optimal LZC technology size. The assessment of the benefits of TES considers how long it will take to payback the additional investment compared to the system payback, and if the additional investment in TES has a positive return. Such an analysis of the benefits of TES integration has not been done before. This could provide evidence to support TES incentivisation. The analysis also considers different house types representative of the UK building sector, that is, detached, semi-detached, terrace and flat.

## 2. Methodology

A techno-economic assessment of the system is required to provide a case to support TES incentivisation. First, the energy system is designed systematically and the results forms the basis for the techno-economic study. A multi-period mixed integer linear program is proposed to design the energy system. Additional binary variables are included to address electricity import/export and TES charging/discharging. The model summary is provided in Figure 1. The scope considers integrating a micro-CHP, an auxiliary boiler and TES into an existing building.

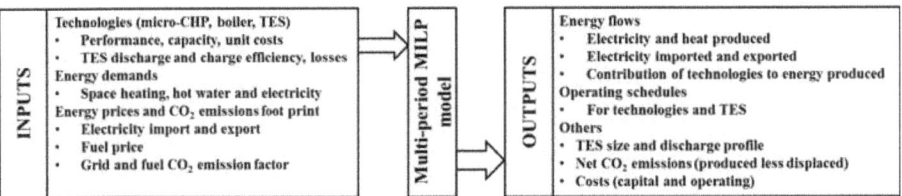

**Figure 1.** Model summary. CHP: Combined Heat and Power; TES: Thermal Energy Storage.

*2.1. Design Statement*

A more specific definition of the design problem is:

- Given:

    ○ Set of LZC technologies (different sizes of micro-CHP).

    ○ The energy demands of a dwelling: space heating, hot water and electricity provided in 5 min temporal precision for all the days in a year. The electricity demand is for combined appliances, electronics and lighting.

    ○ Energy prices (electricity and fuel), electricity tariff structure, technology costs.

- Determine:

    ○ Optimal energy system design

        ✧ Technologies selected, capacities and dispatch strategy

        ✧ Contribution of technologies to heat and power provision

    ○ Energy flows, economics and CO2 emissions

- Subject to:

    ○ Energy (both heat and electricity) balances

    ○     Technology capacity constraints

- In order to:

    ○     Minimise the EAC of meeting dwelling energy demand to a residential user

A detailed building model was developed in Environmental Systems Performance - Research (ESP-r) to determine the space heating and hot water demands [34]. The demands are evaluated based on building geometry, thermal characteristics and weather forecasts. The European Electrical Standard profiles, by Annex 42 of the IEA's ECBCS [35], were used for the electricity load. The micro-CHP applied is the gas engine due to maturity, fuel adaptability, fast responses to load changes and presence in the market [9]. The alternative to these is a gas boiler for meeting the heating demand, and electricity imported from the grid.

*2.2. Mathematical Formulation*

The optimal design is obtained by minimising the systems Equivalent Annual Cost (EAC), defined as the sum of the annualised capital, fuel and maintenance cost less the net electricity cost (Equation (1)). Breakdown of each component of the objective is provided in Equations (2)–(5), and the annualisation factor in Equation (6).

The annualized capital cost in Equation (2) takes into account the installed capital (IC) of the micro-CHP and TES. The TES size ($Size^{TES}$) is a degree of freedom in the model. Several sizes of the micro-CHP are included in the analysis, that is, 1, 2 and 4 kWe, and the binary variable $Z^{CHP}$ selects from the available sizes. The fuel costs (Equation (3)) is a sum of the fuel consumed in the micro-CHP and the boiler. Assumptions of the CHP performance ($Perf^{CHP}$) and boiler performance ($Perf^{BOI}$) are in Table A1. t is the time resolution used (5 min) and r is the number of time periods used to represent the annual energy demand and technology characteristics. The net electricity cost (Equation (5)) is the cost of electricity import less the revenue from electricity exported from the micro-CHP. Assumptions of electricity prices are in the Appendix A.

Equality and inequality constraints to describe the feasible region are also defined. Energy balances for heat and electricity is provided in Equations (7) and (8). In Equation (7), the heat produced from the micro-CHP, the boiler, and heat diverted out of storage must satisfy the thermal energy demand. No dumping of heat is allowed. Also in Equation (8), electricity produced from the micro-CHP and any electricity imported less electricity exported must satisfy the electricity demand. The technology operation is constrained in Equations (9) and (10). When $Y^{CHP}$ is 1, the unit operates between the minimum and maximum allowed. In Equation (11) if a unit is not selected (determined by the binary variable $Z^{CHP}$), it does not operate.

$$\text{Min}: ACC + FC + MC + W_{\text{GRID,NET}} \tag{1}$$

$$ACC = AF \times \left( \left( Size^{CHP} \times Z^{CHP} \times IC^{CHP} \right) + \left( Size^{TES} \times IC^{TES} \right) \right) \tag{2}$$

$$FC = \left[ \sum_r \left( \frac{\left( \left( Q^{CHP}_r + W^{CHP}_r \right) \times ts_r \right)}{Perf^{CHP}} \right) + \sum_r \left( \frac{\left( \left( Q^{BOI}_r \right) \times ts_r \right)}{Perf^{BOI}} \right) \right] \times NGP \tag{3}$$

$$MC = \left( \sum_r \left( W^{CHP}_r \times ts_r \right) \times MC^{CHP} \right) + \left( \sum_r \left( Q^{TES,OUT}_r \times ts_r \right) \times MC^{TES} \right) \tag{4}$$

$$W_{\text{GRID,NET}} = \sum_r \left( \left( W^{IMP}_r \times ts_r \times GIMP_r \right) - \left( W^{EXP}_r \times ts_r \times GEXP_r \right) \right) \tag{5}$$

$$AF = \frac{IR \times (1 + IR)^n}{(1 + IR)^{n-1}} \tag{6}$$

$$Q^{CHP}_r + Q^{BOI}_r - Qdemand_r + Q^{TES,OUT}_r - Q^{TES,IN}_r = 0, \forall r \in R \tag{7}$$

$$W^{CHP}_r + W^{IMP}_r - Wdemand_r - W^{EXP}_r = 0, \forall r \in R \tag{8}$$

$$W^{CHP}_r - Lo \times Y^{CHP}_r \geq 0; \forall r \in R \tag{9}$$

$$W^{CHP}_r - Uo \times Y^{CHP}_r \leq 0; \forall r \in R \tag{10}$$

$$Y^{CHP}_r - Z^{CHP} \leq 0; \forall r \in R \tag{11}$$

In order to define whether electricity is imported or exported, a binary variable, $Y^{EI}$, and a series of constraints are formulated in Equations (12) and (13). Electricity export and import cannot occur together in any time slice. $U_{EI}$ is a number some orders of magnitude larger than electricity import/export. The same formulation is done for charging/ discharging the store in Equations (14) and (15). The heat diverted out of the store, $Q^{TES,OUT}$ and heat diverted into the store $Q^{TES,IN}$ are degrees of freedom in the model. $Y^{TES}$ represents the binary variable for operating the store. Equation (14) and (15) are formulated to ensure charging and discharging the store does not occur simultaneously in any time period r.

$$W^{IMP}_r - U_{EI} \times Y^{EI}_r \leq 0, \forall r \in R \tag{12}$$

$$W^{EXP}_r - U_{EI} \times \left(1 - Y^{EI}_r\right) \leq 0, \forall r \in R \tag{13}$$

$$Q^{TES,OUT}_r - U_{TES} \times Y^{TES}_r \leq 0, \forall r \in R \tag{14}$$

$$Q^{TES,IN}_r - U_{TES} \times \left(1 - Y^{TES}_r\right) \leq 0, \forall r \in R \tag{15}$$

In each time slice, the energy content of the store is subject to the below constraint in Equation (16). Where θ is the daily storage loss, ISH is the Initial Store Heat.

$$0 \leq \sum_{r=r}^{r=1} \left( \left( \left(Q^{TES,IN}_r \times \eta_{charge}\right) - \left(\frac{Q^{TES,OUT}_r}{\eta_{discharge}}\right) - \left(Size^{TES} \times \theta\right)\right) \times ts_r \right) + \left(Size^{TES} \times ISH\right) \leq Size^{TES} \tag{16}$$

Equation (17) states that the heat transferred into the store is equal to the heat recovered from it during 24 h. Therefore, at the end of each day:

$$\left(Size^{TES} \times ISH\right) - 0.1 \leq \left[ \sum_{r=r}^{r=1} \left( \left( \begin{array}{c} \left(Q^{TES,IN}_r \times \eta_{charge}\right) \\ -\left(\frac{Q^{TES,OUT}_r}{>\eta_{discharge}}\right) \\ -\left(Size^{TES} \times >\theta\right) \end{array} \right) \times ts_r \right) + \left(Size^{TES} \times ISH\right) \right] \leq \left(Size^{TES} \times ISH\right) + 0.1 \tag{17}$$

After obtaining the optimal design based on Equations (1)–(17), a techno-economic analysis of the design follows. The criteria for assessment are the Equivalent Annual Income (EAI), the designed system payback and Net Present Value (NPV), payback and NPV associated with the additional investment in TES, the total delivered energy (TDE) and the net $CO_2$ emissions (NTCO$_2$).

The EAI is the difference between the cost of energy for a Business as Usual (BAU) system and the EAC in Equation (1). Traditionally, a dwelling space heating and hot water demand is satisfied by a boiler and electricity imported from the grid. The energy cost of a BAU system is calculated using Equation (18). The difference in the energy cost for a BAU system and the EAC for the optimized design with TES, gives the EAI associated with integrating TES. A positive EAI implies the homeowner has some savings from integrating the new system. The payback and the NPV for the additional investment in the TES is calculated using the difference in investment, and savings for design without TES and design with TES. The TDE is the fuel value of energy flows calculated in Equation (19), and the net $CO_2$ emissions is estimated using Equation (20) taking into account $CO_2$ from fuel, electricity import and $CO_2$ displaced when electricity is exported. The model was implemented using the GAMS 24.7.3 software (24.7.3, GAMS Development Corporation, Fairfax, VA, USA) and solved with the CPLEX solver on a 64 bit 3.40 GHz Intel® Core™ i7-6700 CPU with 32 GB RAM machine.

$$C^{BAU} = \left( \sum_r \left( \frac{Qdemand_r \times ts_r}{Perf^{BOI}} \right) \times NGP \right) + \sum_r (Wdemand_r \times ts_r \times GIMP_r) \tag{18}$$

$$TDE = \left( \sum_r \left( \frac{((Q^{CHP}_r + W^{CHP}_r) \times ts_r)}{Perf^{CHP}} \right) + \left( \sum_r \left( \frac{((Q^{BOI}_r) \times ts_r)}{Perf^{BOI}} \right) \right) \right) + \frac{\left( \sum_r \left( W^{IMP}_r \times ts_r \right) \right) - \left( \sum_r \left( W^{EXP}_r \times ts_r \right) \right)}{GEE} \tag{19}$$

$$NTCO_2 = \left( \sum_r \left( \frac{((Q^{CHP}_r + W^{CHP}_r) \times ts_r)}{Perf^{CHP}} \right) + \left( \sum_r \left( \frac{((Q^{BOI}_r) \times ts_r)}{Perf^{BOI}} \right) \right) \right) \times FEF + \left( \begin{array}{c} \left( \sum_r \left( W^{IMP}_r \times ts_r \right) \right) \\ - \left( \sum_r \left( W^{EXP}_r \times ts_r \right) \right) \end{array} \right) \times GEE \tag{20}$$

## 3. Case Study

The case study is on integrating micro-CHP and TES into an existing dwelling with an objective to analyse the benefits from its integration, and determine if the additional investment is economically viable, and how it can be incentivised. Four building types representative of UK housing stock are investigated.

### 3.1. Design Problem

Given the energy demands for different dwelling types, the space heating demand for a detached house, semi-detached house, terrace and flat are presented in Figure 2, associated hot water demand in Figure 3, summary provided in Table 1. Simulation models of four key UK house types is presented in Allison et al. [34]. The simulation models enable the dynamic thermal demand (heating and hot water) of each house to be quantified over a range of different operating conditions. Occupancy, insulation levels, and construction thermal characteristics are defined in the earlier work done in Allison et al. [34]. A micro-CHP, boiler and TES are available to satisfy the energy demand. Different design scenarios were developed and analysed to assess TES integration. They are, design with and without TES, and design with/without incentives. Since TES is not a stand-alone technology, fiscal incentives such as the FIT to support the deployment of micro-CHP is adapted; where the value for production and export are 13.45 and 4.91 p/kWh respectively [36]. Another instrument explored is the exemption from the $CO_2$ levy (current central value is 0.063 £/kg) [37], especially when the $CO_2$ of the system designed is less than the BAU.

**Table 1.** Energy demand and house characteristics.

|  |  | Detached | Semi-Detached | Terrace | Flat |
|---|---|---|---|---|---|
| Peak demand (kW) | Space heating | 7.65 | 6.89 | 4.42 | 5.11 |
|  | Hot water | 27.2 | 34.9 | 34.6 | 34.7 |
| Total demand (kWh/year) | Space heating | 9904 | 5400 | 2480 | 1940 |
|  | Hot water | 1712 | 1590 | 1351 | 1304 |
| Floor area (m²) | - | 121–152 | 74–93 | 66–83 | 45–57 |

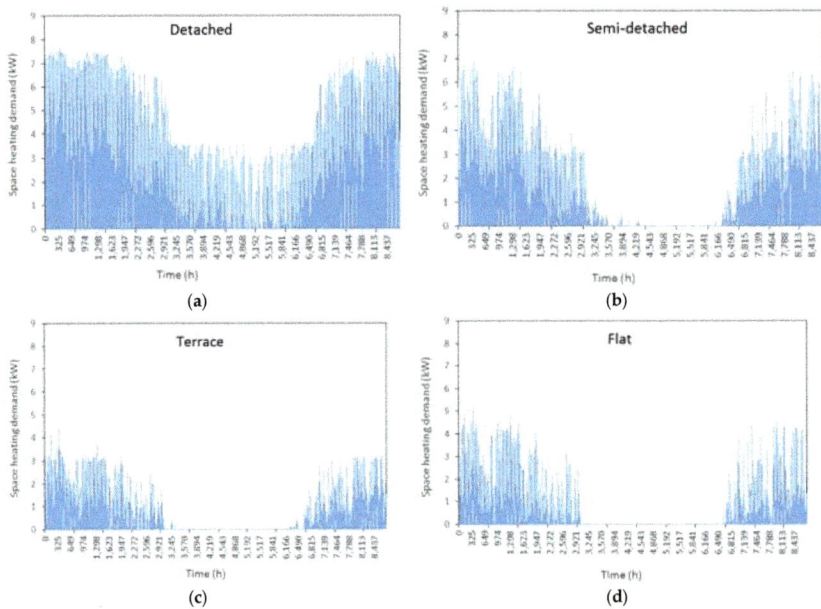

**Figure 2.** Space heating demand for all house types: (**a**) Detached; (**b**) Semi-detached; (**c**) Terrace; (**d**) Flat.

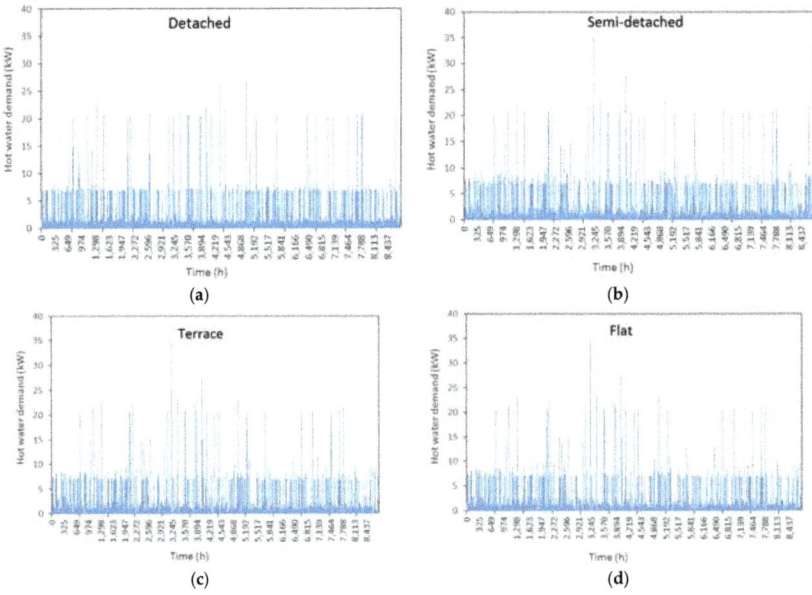

**Figure 3.** Hot water demand for all house types: (**a**) Detached; (**b**) Semi-detached; (**c**) Terrace; (**d**) Flat.

## 3.2. Results and Discussion

The contribution of each technology to heat produced under different scenarios are presented in Figures 4–7. In the absence of incentives, and for design without TES, the micro-CHP contributes more to the heat supply compared to the auxiliary boiler. The contribution depends on the house type, increasing from detached to flats. A micro-CHP is better suited to a house with high demands, hence the EAI is highest for a detached house as shown in Table 2. The optimiser maximises the use of the micro-CHP for the semi-detached, terrace and flat, in order to improve their economics. For design with TES, the boiler contribution reduces to 0% for a flat. The majority of the heat in the store is supplied from the micro-CHP. The TES size for a semi-detached, terrace and flat is higher than for a detached house, because since the design for the other houses are uneconomic, the optimiser will maximise the use of storage to improve the economic viability. This will ensure the micro-CHP does not operate all the time resulting in a decrease in operational costs. Hence the heat diverted to storage increases from detached to flat (Figure 6). The heat diverted to storage is more for a non-incentivised design compared to the design with incentives.

Other benefits of TES integration are reduction in net $CO_2$ emissions, TDE, and in this case more income for the home owner (Tables 2 and 3). The benefits are dependent on house type, increasing based on the total space heating and hot water demand.

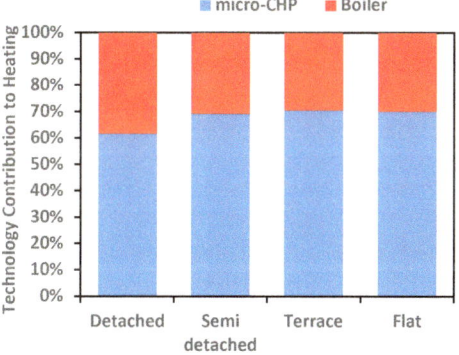

**Figure 4.** Non-incentivised design without thermal energy storage (TES).

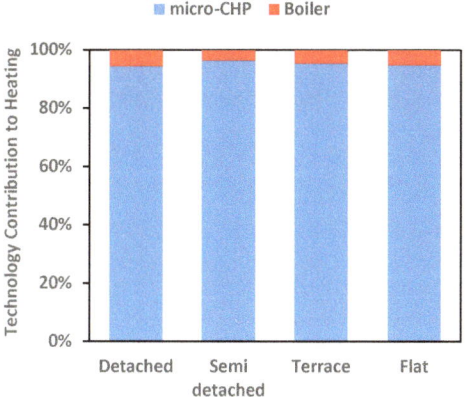

**Figure 5.** Incentivised design (Feed in Tariff (FIT)) without TES.

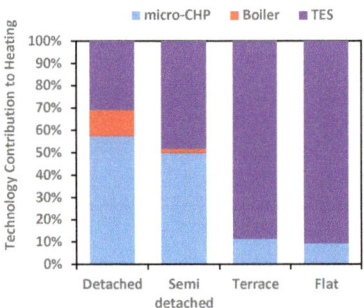

**Figure 6.** Non-incentivised design with TES.

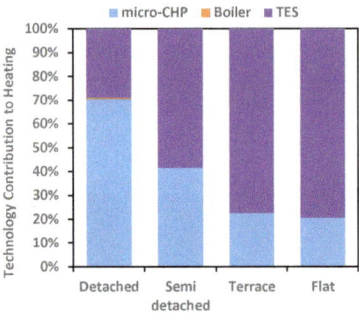

**Figure 7.** Incentivised design (FIT) with TES.

A major barrier to implementing micro generation is cost, when the FIT is introduced in the optimisation, the income for the home owner increases (Table 3). A further increase is possible when TES is integrated. However, for this design the net $CO_2$ emissions and TDE is slightly higher (Table 3) due to an increase in the micro-CHP fuel consumption. For the design without TES, over 90% of the heat is supplied from the micro-CHP. With TES, the boiler's contribution is negligible for a detached house, and nil for the remaining house types (Figure 7). Grid electricity import increases from detached to flat for the cases, that is, without and with incentives. The contribution from the grid is more when thermal storage is not integrated. Contribution from the grid is less with TES and incentives, however, there is no change for terraces and flats.

**Table 2.** Non-incentivised design results.

|  |  | Detached | Semi-Detached | Terrace | Flat |
|---|---|---|---|---|---|
| EAI (£/year) | Design with TES | 100.9 | −19.2 | −133.84 | −132.48 |
|  | Design without TES | 41.22 | −52.5 | −138.5 | −156 |
| System payback (year) | Design with TES | 14 | 20 | 36 | 36 |
|  | Design without TES | 16 | 24 | 43 | 51 |
| System NPV (£) | Design with TES | −1348 | −2680 | −3860 | −3830 |
|  | Design without TES | −1716 | −2690 | −3580 | −3760 |
| NTCO$_2$ (kg/year) | Design with TES | 6233 | 5416 | 4967 | 4876 |
|  | Design without TES | 6379 | 5514 | 4984 | 4889 |
| TDE (kWh/year) | Design with TES | 29,286 | 25,609 | 23,640 | 23,242 |
|  | Design without TES | 30,078 | 26,161 | 23,788 | 23,367 |
| TES size | - | 26 | 34 | 34 | 32 |

**Table 3.** Incentivised (FIT) design results.

|  |  | Detached | Semi-Detached | Terrace | Flat |
|---|---|---|---|---|---|
| EAI (£/year) | Design with TES | 751 | 414 | 125 | 91 |
|  | Design without TES | 610 | 296 | 51 | 4 |
| System Payback (year) | Design with TES | 4.9 | 7.4 | 12.8 | 14.0 |
|  | Design without TES | 5.5 | 8.7 | 15.8 | 18.7 |
| System NPV (£) | Design with TES | 5525 | 2019 | −928 | −1266 |
|  | Design without TES | 4186 | 933 | −1616 | −2103 |
| NTCO$_2$ (kg/year) | Design with TES | 6009 | 5407 | 4954 | 4864 |
|  | Design without TES | 5966 | 5308 | 4881 | 4802 |
| TDE (kWh/year) | Design with TES | 28,169 | 25,551 | 23,579 | 23,191 |
|  | Design without TES | 28,025 | 25,138 | 23,275 | 22,933 |
| TES size |  | - | 13 | 13 | 9 | 7 |

Without incentives, the payback of the accompanying system is greater than 10 and has a negative return on investment. A higher payback and lower NPV is observed for design without TES. Additionally, the TES size is greater compared to the incentivised design in Table 3. The TES is also discharged more efficiently (Figures 8 and 9). A higher TES size is selected in order to maximise the use of the micro-CHP for off-setting the increased costs when electricity import tariff is high. Therefore, more heat is diverted to storage (Figure 8) compared to Figure 9. For the incentivised design, it is economic to integrate the micro-CHP, hence the need to reduce the cost becomes less and the TES size reduces (Table 3 and Figure 9). The x-axis in Figures 8 and 9 are the temporal precision of the first 10 days in the year.

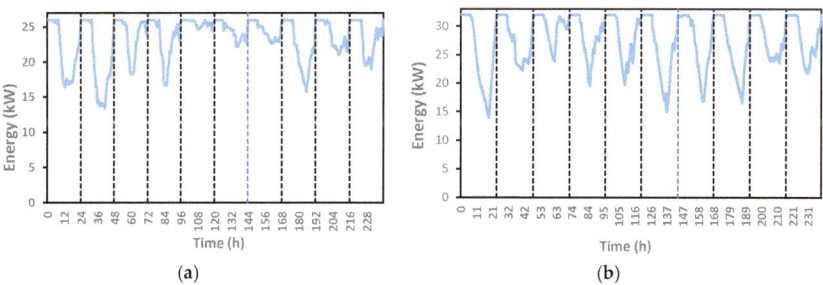

**Figure 8.** TES discharge profile for non-incentivised design: (**a**) detached house; (**b**) flat.

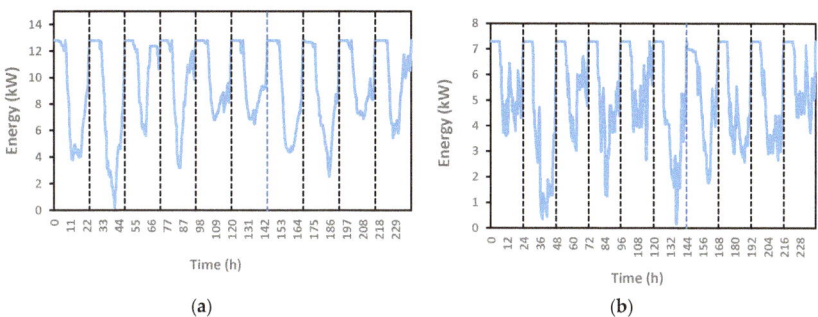

**Figure 9.** TES discharge profile for incentivised design: (**a**) detached house; (**b**) flat.

In the non-incentivised scenario with TES, the system payback is 14 years for a detached house; additional benefits from TES integration (i.e., increase in the EAI), can pay for the additional investment in TES within the first 6.24 years, with a positive NPV of £368 (Figure 10a). The TES payback and NPV is different for each house type. Whilst in the incentivised scenario (based on the FIT) TES investment can be paid for during the first 1.78 years for a system payback of 4.9 years. The NPV is £1339 (Figure 10b). Another way to incentivise is savings based on reduced $CO_2$ emissions when TES is integrated (Table 2). In 2016, the $CO_2$ central non-traded levy is 0.063 £/kg [36]. Using the $CO_2$ reduced, the savings for a detached, semi-detached, terrace and flat are £70, £57, £37 and £32. This has potential to reduce the additional investment in TES. When incentivised using the $CO_2$ levy, the TES investment is paid back in the first 5.65 years, with NPV of £463.60 (Figure 10c). There is a clear relationship with the type of house; the payback increases and NPV reduces (negative in some houses). Hence, even when the system is not economically viable, TES integration is economically viable.

The benefit of micro-CHP and TES in a single dwelling has been established. Even though the NPV in system investment is negative (Table 2), the NPV for TES investment is positive even in the absence of incentives (Figure 10a).

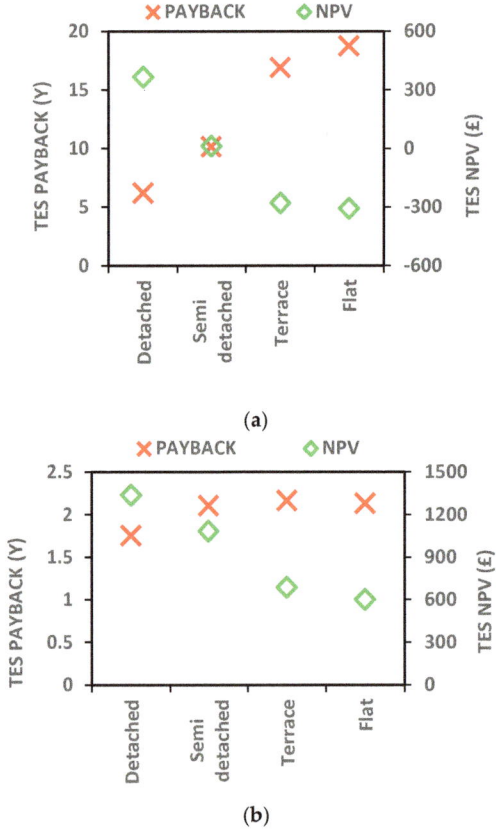

Figure 10. *Cont.*

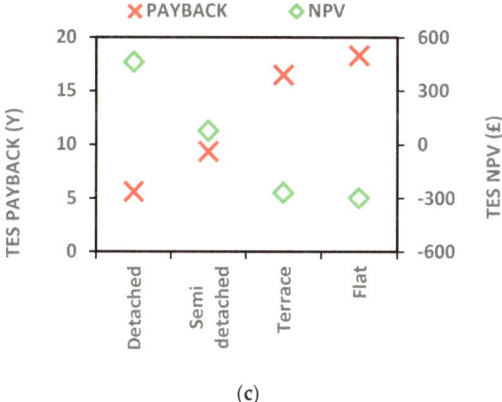

(c)

**Figure 10.** Payback and Net Present Value (NPV) associated with TES investment for (**a**) non-incentivised design; (**b**) incentivised design (FIT); and (**c**) incentivised design ($CO_2$ levy exemption).

*3.3. Limitations of This Work*

The findings of this work need to be viewed alongside the assumptions on demand, energy prices, and $CO_2$ emission factors. Whilst the assumptions have been taken from reputable sources, they are subject to variations. Furthermore, the support mechanisms mentioned in the report are currently in operation and are also dynamic.

**4. Conclusions**

A multi-period MILP model is presented and used to provide a case to support incentivising TES. The accuracy of the model is improved through the use of less temporal precision to represent demand and technology characteristics. The introduction of additional binary variables for charging/discharging the store, and import/export of electricity also reduces the model complexity.

The case to support incentivising TES was established based on a techno-economic assessment of the design obtained with and without TES, and with and without incentives. Assessment metrics used were the EAI for a homeowner, net $CO_2$ emissions, the TDE, and the system payback and NPV. A new assessment based on the payback and NPV associated with additional investment in TES was also introduced. Results show the benefit from micro generation and TES integration depend on the house type; in this case highest for a detached house. For a detached house, TES integration reduces the TDE by 792 kWh/r, $CO_2$ by 146 kg/year and the homeowner makes 60 £/year. These additional incomes means TES investment can be paid for in the first 6.24 years during the 14 years of system payback. This provides evidence to support incentivisation. When value is added to $CO_2$ reduction based on $CO_2$ levy, the payback for TES reduces to 5.65 years and NPV also increases. With the FIT scheme, for a system payback of 4.9 years, TES investment can be paid back during the first 1.759 years. In the non-incentivised case, even though investment in micro-CHP is not economically viable, the additional investment in TES is economically viable. Therefore, TES incentivisation needs to receive more attention. The future uptake of LZC technologies will be influenced by financial incentives to encourage end-user investments.

There is need to establish which of the accompanying technologies is more beneficial. This forms the basis of future work. Another area worth investigating is establishing the flexibility potential of thermal storage in domestic houses.

**Author Contributions:** Gbemi Oluleye and Adam Hawkes defined the research questions, and contextualized the problem statement; Gbemi Oluleye wrote the GAMS code and ran the optimization model; John Allison and

Nicolas Kelly developed the simulation models for the four key UK house types considered. Gbemi Oluleye wrote the paper with input from all co-authors.

**Acknowledgments:** The authors would like to thank the Engineering and Physical Sciences Research Council (EPSRC) for funding the research reported in this paper under grant EP/N021479/1 as part of its Thermal Energy Challenge programme.

**Conflicts of Interest:** The authors declare no conflict of interest.

## Nomenclature

| | |
|---|---|
| Sets | |
| $r \in R$ | Set of all time periods |
| Independent variables | |
| $Q^{CHP}$ | Heat produced from micro-CHP (kW) |
| $Q^{BOI}$ | Heat produced from boiler (kW) |
| $Q^{TES,IN}$ | Heat diverted into storage, (kW) |
| $Q^{TES,OUT}$ | Heat delivered from storage (kW) |
| $Size^{TES}$ | Total storage capacity (kWh) |
| $W^{IMP}$ | Electricity imported to satisfy demand (kW) |
| $W^{EXP}$ | Electricity exported (kW) |
| Dependent variables | |
| ACC | Annualised capital cost (£/year) |
| EAC | Equivalent Annual Cost (£/year) |
| FC | Fuel cost (£/year) |
| MC | Maintenance cost (£/year) |
| $W_{GRID,NET}$ | Net electricity cost (£/year) |
| Binary variables | |
| $Y^{CHP}$ | Binary variable for technology operation |
| $Y^{EI}$ | Binary variable for export/import of electricity |
| $Y^{TES}$ | Binary variable for charging/discharging the store |
| $Z^{CHP}$ | Binary variable for technology existence |
| Parameters | |
| FEF | Fuel emission factor (kg/kWh) |
| GIMP | Grid electricity import price (£/kWh) |
| GEXP | Grid electricity export price (£/kWh) |
| GEE | Grid energy efficiency (%) |
| GEF | Grid emission factor (kg/kWh) |
| $IC^{CHP}$ | Micro-CHP installed capital (£/kW) |
| $IC^{TES}$ | TES installed capital (£/kW) |
| ISH | Initial store heat (%) |
| IR | Interest rate |
| $\theta$ | Storage losses (%) |
| Lo | Lower limit |
| $MC^{CHP}$ | Micro-CHP maintenance cost (£/kWh) |
| $MC^{TES}$ | TES maintenance cost (£/kWh) |
| n | Technology lifetime (years) |
| $Perf^{BOI}$ | Boiler performance |
| $Perf^{CHP}$ | Micro-CHP performance |
| Qdemand | Heat demand (both hot water and space heating) (kW) |
| $Size^{CHP}$ | Micro-CHP capacity (kW) |
| ts | Time precision |
| $U_{EI}$ | Upper limit for electricity export and import |
| Uo | Upper limit for micro-CHP power |
| Wdemand | Electricity demand (kW) |
| $\eta_{charge}$ | TES charge efficiency (%) |
| $\eta_{discharge}$ | TES discharge efficiency (%) |

## Appendix A

**Table A1.** Technical attributes of technologies in superset [7,29].

| Technical Attributes | Natural Gas Boiler | Micro-CHP (1 kW) | Micro-CHP (2 kW) | Micro-CHP (4 kW) | Hot Water Tank |
|---|---|---|---|---|---|
| Turnkey cost (£/kW) | 163 | 3110 | 2400 | 1900 | 20 |
| Maintenance cost (£/kWh) | 0.001 | 0.01 | 0.01 | 0.01 | 0.001 |
| Performance | 0.895 | 0.9 | 0.9 | 0.9 | 2.75 |
| Charge efficiency (%) | - | - | - | - | 90 |
| Discharge efficiency (%) | - | - | - | - | 90 |
| Initial store heat (%) | - | - | - | - | 100 |

The micro-CHP heat to power ratio is 0.385. The design year is 2016. Off-peak and peak electricity import tariffs are 5.5 and 15.29 p/kWh, and fuel price is 3.48 p/kWh [36]. The average CO2 factor associated with natural gas and electricity was assumed to be 0.185 and 0.519 kg/kWh [36]. Technology lifetime is 15 years and discount rate 5%. CEPCI cost factors were applied to adjust the costs from the year provided to 2016.

## References

1. International Energy Agency; Energy Technology Systems Analysis Program (ETSAP); International Renewable Energy Agency. Thermal Energy Storage, Technology Brief E17. January 2013. Available online: https://www.irena.org/DocumentDownloads/Publications/IRENA-ETSAP%20Tech%20Brief%20E17%20Thermal%20Energy%20Storage.pdf (accessed on 23 March 2017).
2. Balcombe, P.; Rigby, D.; Azapagic, A. Investigating the importance of motivations and barriers related to microgeneration uptake in the UK. *Appl. Energy* **2014**, *130*, 403–418. [CrossRef]
3. Cherrington, R.; Goodship, V.; Longfield, A.; Kirwan, K. The feed-in tariff in the UK: A case study focus on domestic photovoltaic systems. *Renew. Energy* **2013**, *50*, 421–426. [CrossRef]
4. EASE/EERA. European Energy Storage Technology Development Roadmap Toward 2030. 2013. Available online: http://www.eera-set.eu/wp-content/uploads/148885-EASE-recommendations-Roadmap-04.pdf (accessed on 8 December 2016).
5. Spataru, C.; Kok, Y.C.; Barrett, M.; Sweetnam, T. Techno-economic assessment for optimal energy storage mix. *Energy Procedia* **2015**, *83*, 515–524. [CrossRef]
6. Oluleye, G.; Hawkes, A.D.; Allison, J.; Kelly, N.; Clarke, J. An optimisation study on integrating and incentivising Thermal Energy Storage (TES) in a dwelling energy system. In Proceedings of the Sustainable Development of Energy, Water and Environment Systems Conference, Dubrovnik, Croatia, 4–8 October 2017.
7. Brandoni, C.; Arteconi, A.; Ciriachi, G.; Polonara, F. Assessing the impact of micro-generation technologies on local sustainability. *Energy Convers. Manag.* **2014**, *87*, 1281–1290. [CrossRef]
8. Sorace, M.; Gandiglio, M.; Santarelli, M. Modelling and techno-economic analysis of the integration of a FC-based micro-CHP system for residential application with a heat pump. *Energy* **2015**, *120*, 275–2015.
9. Díaz, G.; Moreno, B. Valuation under uncertain energy prices and load demands of micro-CHP plants supplemented by optimally switched thermal energy storage. *Appl. Energy* **2016**, *177*, 553–569. [CrossRef]
10. Fubara, T. K.; Cecelja, F.; Yang, A. Modelling and selection of micro-CHP systems for domestic energy supply: The dimension of network-Wide primary energy consumption. *Appl. Energy* **2014**, *114*, 327–334. [CrossRef]
11. Murugan, S.; Horák, B. A review of micro combined heat and power systems for residential applications. *Renew. Sustain. Energy Rev.* **2016**, *64*, 144–162. [CrossRef]
12. Eames, P.; Loveday, D.; Haines, V.; Romanos, P. *The Future Role of Thermal Energy Storage in the UK Energy System: An Assessment of the Technical Feasibility and Factors Influencing Adoption—Research Report*; REF UKERC/RR/ED/2014/001; UKERC: London, UK, 2014.
13. Barbieri, E.S.; Melino, F.; Morini, M. Influence of the thermal energy storage on the profitability of micro-CHP systems for residential building applications. *Appl. Energy* **2012**, *97*, 714–722. [CrossRef]
14. Arteconi, A.; Hewitt, N.; Polonara, F. State of the art of thermal storage for demand-side management. *Appl. Energy* **2012**, *93*, 371–389. [CrossRef]
15. Bogdan, Z.; Kopjar, D. Improvement of the cogeneration plant economy by using heat accumulator. *Energy* **2006**, *31*, 2285–2292. [CrossRef]

16. Oluleye, G.; Vasquez, L.; Smith, R.; Jobson, M. A multi-period mixed integer linear program for design of residential distributed energy centres with thermal demand data discretisation'. *Sustain. Prod. Consum.* **2016**, *5*, 16–28. [CrossRef]

17. Committee on Climate Change 2013. Fourth Carbon Budget Review-Part 2: The Cost-Effective Path to the 2050 Target. Committee on Climate Change. Available online: http://www.theccc.org.uk/wp-content/uploads/2013/12/1785a-CCC_AdviceRep_Singles_1.pdf (accessed on 25 July 2017).

18. Kelly, N.J.; Tuohy, P.G.; Hawkes, A.D. Performance assessment of tariff-based air source heat pump load shifting in a UK detached dwelling featuring phase change-enhanced buffering. *Appl. Therm. Eng.* **2014**, *71*, 809–820. [CrossRef]

19. Salata, F.; Golasi, I.; Domestico, U.; Banditelli, M.; Lo Basso, G.; Nastasi, B.; de Lieto Vollaro, A. Heading towards the nZEB through CHP+HP systems. A comparison between retrofit solutions able to increase the energy performance for the heating and domestic hot water production in residential buildings. *Energy Convers. Manag.* **2017**, *138*, 61–76. [CrossRef]

20. Mongibello, L.; Capezzuto, M.; Graditi, G. Technical and cost analyses of two different heat storage systems for residential micro-CHP plants. *Appl. Therm. Eng.* **2014**, *71*, 636–642. [CrossRef]

21. Vögelin, P.; Koch, B.; Georges, G.; Boulouchos, K. Heuristic approach for the economic optimisation of combined heat and power (CHP) plants: Operating strategy, heat storage and power. *Energy* **2017**, *121*, 66–77. [CrossRef]

22. Shaneb, O.; Coates, G.; Taylor, P. Sizing of residential μCHP systems. *Energy Build.* **2011**, *43*, 1991–2001. [CrossRef]

23. Hawkes, A.; Leach, M. Modelling high level system design and unit commitment for a microgrid. *Appl. Energy* **2009**, *86*, 1253–1265. [CrossRef]

24. Ren, H.; Gao, W.; Ruan, Y. Optimal Sizing for Residential CHP System. *Chall. Power Eng. Environ.* **2007**, 73–79. [CrossRef]

25. Merkel, E.; Mckenna, R.; Fichtner, W. Optimisation of the capacity and the dispatch of decentralised micro-CHP systems: A case study for the UK. *Appl. Energy* **2015**, *140*, 120–134. [CrossRef]

26. Mckenna, R.; Merkel, E.; Fichtner, W. Energy autonomy in residential buildings: A techno-economic model-based analysis of the scale effects. *Appl. Energy* **2017**, *189*, 800–815. [CrossRef]

27. Pruitt, K.A.; Braun, R.J.; Newman, A.M. Evaluating shortfalls in mixed-integer programming approaches for the optimal design and dispatch of distributed generation systems. *Appl. Energy* **2013**, *102*, 386–398. [CrossRef]

28. Hawkes, A.; Leach, M. Impacts of temporal precision in optimisation modelling of micro-Combined Heat and Power. *Energy* **2005**, *30*, 1759–1779. [CrossRef]

29. Baeten, B.; Rogiers, F.; Helsen, L. Reduction of heat pump induced peak electricity use and required generation capacity through thermal energy storage and demand response. *Appl. Energy* **2017**, *195*, 184–195. [CrossRef]

30. Fragaki, A.; Andersen, A.N.; Toke, D. Exploration of economical sizing of gas engine and thermal store for combined heat and power plants in the UK. *Energy* **2008**, *33*, 1659–1670. [CrossRef]

31. Renaldi, R.; Kiprakis,, A.; Friedrich, D. An optimisation framework for thermal energy storage integration in a residential heat pump heating system. *Appl. Energy* **2017**, *186*, 520–529. [CrossRef]

32. Hussain, A.; Bui, V.; Kim, H.; Im, Y.; Lee, J. Optimal Energy Management of Combined Cooling, Heat and Power in Different Demand Type Buildings Considering Seasonal Demand Variations. *Energies* **2017**, *10*, 789. [CrossRef]

33. Bianco, V.; Scarpa, F.; Tagliafico, L. Estimation of primary energy savings by using heat pumps for heating purposes in the residential sector. *Appl. Therm. Eng.* **2017**, *114*, 938–947. [CrossRef]

34. Allison, J.; Bell, K.; Clarke, J.; Cowie, A.; Elsayed, A.; Flett, G.; Oluleye, G.; Hawkes, A.; Hawker, G.; Kelly, N.; et al. Assessing domestic heat storage requirements for energy flexibility over varying timescales. *Appl. Therm. Eng.* **2018**, *136*, 602–616. [CrossRef]

35. Standard European Electrical Profiles. IEA/ECBCS Annex 42 The Simulation of Building-Integrated Fuel Cell and Other Cogeneration Systems. 2006. Available online: http://www.buildup.eu/en/practices/publications/iea-ecbes-annex-42-simulation-building-integrated-fuel-cell-and-other (accessed on 19 October 2016).

36. Office of Gas and Electricity Markets (Ofgem) Feed-In Tariff (FIT) Rates. 2017. Available online: https://www.ofgem.gov.uk/environmental-programmes/fit/fit-tariff-rates (accessed 10 December 2017.

37. Department of Business, Energy and Industrial Strategy, Data tables 1–20 supporting the toolkit and the guidance. Available online: https://www.gov.uk/government/publications/valuation-of-energy-use-and-greenhouse-gas-emissions-for-appraisal (accessed on 15 September 2016).

*Article*

# Numerical Simulation of Flow and Heat Transfer in Structured Packed Beds with Smooth or Dimpled Spheres at Low Channel to Particle Diameter Ratio

**Shiyang Li, Lang Zhou, Jian Yang and Qiuwang Wang ***

Key Laboratory of Thermo-Fluid Science and Engineering, Ministry of Education, School of Energy and Power Engineering, Xi'an Jiaotong University, Xi'an 710049, China; lishiyang@stu.xjtu.edu.cn (S.L.); zhoulangqi@126.com (L.Z.); yangjian81@mail.xjtu.edu.cn (J.Y.)
* Correspondence: wangqw@mail.xjtu.edu.cn; Tel.: +86-29-8266-5539

Received: 1 March 2018; Accepted: 10 April 2018; Published: 15 April 2018

**Abstract:** Packed beds are widely used in catalytic reactors or nuclear reactors. Reducing the pressure drop and improving the heat transfer performance of a packed bed is a common research aim. The dimpled structure has a complex influence on the flow and heat transfer characteristics. In the present study, the flow and heat transfer characteristics in structured packed beds with smooth or dimpled spheres are numerically investigated, where two different low channel to particle diameter ratios ($N = 1.00$ and $N = 1.15$) are considered. The pressure drop and the Nusselt number are obtained. The results show that, for $N = 1.00$, compared with the structured packed bed with smooth spheres, the structured packed bed with dimpled spheres has a lower pressure drop and little higher Nusselt number at $1500 < Re_H < 14{,}000$, exhibiting an improved overall heat transfer performance. However, for $N = 1.15$, the structured packed bed with dimpled spheres shows a much higher pressure drop, which dominantly affects the overall heat transfer performance, causing it to be weaker. Comparing the different channel to particle diameter ratios, we find that different configurations can result in: (i) completely different drag reduction effect; and (ii) relatively less influence on heat transfer enhancement.

**Keywords:** dimpled sphere; structured packed bed; low channel to particle diameter ratio; numerical simulation; pressure drop; Nusselt number

---

## 1. Introduction

Packed beds are widely used in industrial applications, such as catalytic reactors or nuclear reactors. Reducing the pressure drop and improving the heat transfer performance a packed bed is a common research aim. Traditionally, a randomly packed bed is utilized due to the low cost and ease of use. The pressure drop of randomly packed beds, which is calculated by Ergun empirical correlation [1], is usually much higher than that of other packed beds, such as the structured packed bed. A novel type of structured catalytic reactor packing with a very low channel to particle diameter ratio (between 1.0 and 2.0) called composite structured packing (CSP) is reported by Strangio et al. [2]. This is a feasible way to achieve the structured packing form. Calis et al. [3] studied the flow characteristics of these CSP by computational fluid dynamics (CFD) and experiment. Five different channel to particle diameter ratio ($N$) packing forms were studied, as shown in Figure 1a. Their results showed that the pressure drop of these structured packed beds was lower than that of randomly packed beds. At the same time, they fitted a two-parameter pressure drop correlation for different $N$ by using the CFD results. Rokmes et al. [4] numerically and experimentally investigated the heat transfer performance of these structured packed beds and gave correlations of Nusselt numbers by fitting the CFD results. Palle and Aliabadi [5] used direct numerical simulation for a structured packed bed in simple cubic

(SC) configuration. They studied the friction factor for both infinite and wall bounded structured packed beds and proposed correlations for modified friction factor. Lin et al. [6] numerically studied two dimensional transient turbulent fluid flow and heat transfer in a packed sphere bed used in a regenerator furnace. In our group, Yang et al. [7] investigated flow and forced convective heat transfer in structured packed beds with spherical or ellipsoidal particles with symmetry boundary conditions. Wang et al. [8] designed a grille-sphere composite structured packed bed (GSCSPB), aiming to achieve the SC structured packing configuration easily, which is similar to $N = 1.00$ packing form, as shown in Figure 1b. Pressure drop and heat transfer performance of GSCSPB have been experimentally and numerically studied. The results showed that the GSCSPB can lower the pressure drop of the randomly packed bed and improve the heat transfer coefficient of the structured packed bed. Hu et al. [9] numerically studied the GSCSPB design parameters by using Taguchi method. All these works focus on packed beds with smooth particles at low channel to particle diameter ratios.

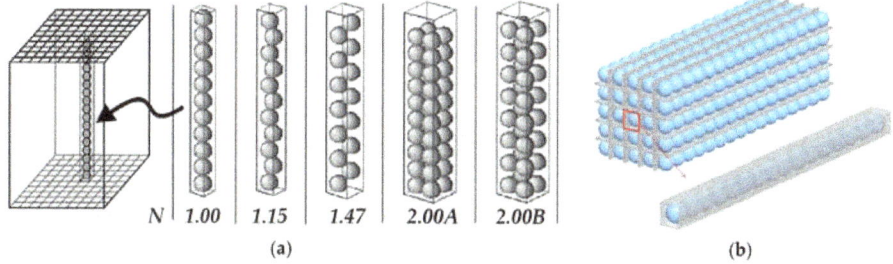

Figure 1. Schematic of (a) composite structured packing (CSP) by Calis et al. [3] and Rokmes et al. [4]; and (b) grille-sphere composite structured packed bed (GSCSPB) by Wang et al. [8].

With different aerodynamic characteristics, a single dimpled sphere has been investigated by many researchers. Some studies show that the dimples on the sphere surface can reduce drag significantly at high Reynolds number. Jin et al. [10] measured the streamwise velocity above the dimpled surface to get a detailed mechanism of drag reduction by dimples on the surface. They considered that dimples caused local flow separation and large turbulence intensity. As a result, dimples delay the main separation and reduce drag significantly. Aoki et al. [11] applied the oil film method and particle image velocimetry (PIV) technology to obtain the flow pattern of the stationary and rotating dimple balls. Their results showed that, as the number of dimples became larger and the depth became deeper, the critical region shifted toward the lower Reynolds number range. Smith et al. [12] used direct numerical simulation to investigate the flow over a golf ball in the subcritical and supercritical regimes. Prediction of the drag coefficient was in reasonable agreement with measurements. At the same time, some researchers applied the dimpled structure to a rectangular channel [13,14] or shell and helically-coiled tube heat exchangers [15]. They all concluded that the dimpled structure could enhance heat transfer, while Kim et al. [15] found that the inline and staggered dimples showed the highest pressure drop.

Motivated by these researches, we wonder the influence of a series of dimpled spheres in wall bounded structured packed bed on flow and heat transfer performance, while the reference is rare. Since the dimpled surface can be approximately treated as roughness surface, we find an experimentally study about the influence of surface roughness on resistance to flow through randomly packed beds by Crawford and Plumb [16]. In their work, glass microspheres have been glued to smooth surfaces to obtain specific roughness. The results showed that the pressure drop was substantially increased by the presence of surface roughness. In 2017, Yang et al. [17] have initially studied the flow and heat transfer in infinite structured packed beds of dimple-particles with SC configuration by numerical simulation.

The results showed that, the overall heat transfer efficiency can be improved by up to about 7% compared with the structured packed bed with smooth particles at the same inlet velocity condition.

In the present study, the flow and heat transfer characteristics in wall bounded structured packed beds with smooth or dimpled spheres are numerically investigated. Two different low channel to particle diameter ratios ($N = 1.00$ and $N = 1.15$) are considered to investigate the influence of different configuration. The pressure drop and the Nusselt number are obtained by CFD and the overall heat transfer performance of packed bed with smooth or dimpled spheres is compared.

## 2. Computational Model and Method

### 2.1. Physical Model

The schematic of wall bounded structured packed beds are shown in Figure 2. The spheres are packed in the square cross-section channel. The channel to particle diameter ratio $N$ is defined as the ratio of the sphere diameter ($d_p$) to the width of square channel ($H$). The diameter of the spheres is 42.87 mm, while the widths of square channels are 42.87 mm and 49.30 mm respectively, obtained two packing configuration: $N = 1.00$ and $N = 1.15$. The packing with $N = 1.00$ is the simplest structured packed bed and the porosity $\varepsilon$ equals 0.48. The packing with $N = 1.15$ is chosen to investigate the influence of small deviation of $N$ and the porosity equals 0.60.

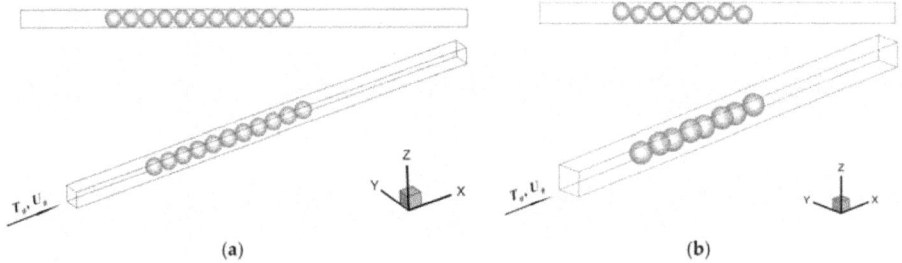

(a)  (b)

**Figure 2.** Schematic of wall bounded structured packed beds for (a) $N = 1.00$; and (b) $N = 1.15$.

A uniform freestream velocity $U_0$ at different Reynolds numbers with 0.5% turbulence intensity and constant temperature of air $T_0 = 293$ K are applied to the domain inlet. The pressure outlet condition with zero gauge pressure is used at the domain outlet. The walls of the square channel are no-slip and adiabatic boundaries. The spheres surfaces are no-slip, constant temperature boundaries at $T_p = 303$ K, 10 K higher than the inlet temperature.

The dimpled sphere used in this study is shown in Figure 3. The size and shape of dimples refer to the experiment test balls by Aoki et al. [11], as shown in Table 1. We rebuilt this geometry of the dimpled sphere [17]. This dimpled sphere with 184 dimples on the surface is similar with the golf balls.

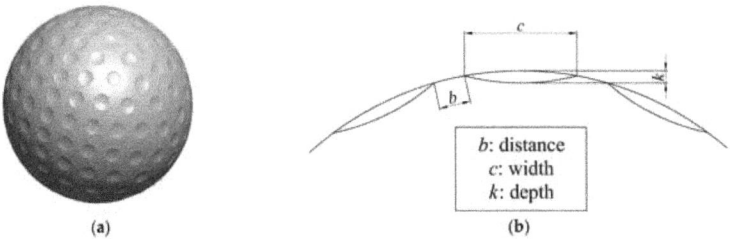

| b: distance |
| c: width |
| k: depth |

(a)  (b)

**Figure 3.** (a) The dimpled sphere; and (b) the detail view of the dimples.

**Table 1.** The geometry parameters of dimples.

| $N_D$ | $b$ (mm) | $c$ (mm) | $k$ (mm) |
|-------|----------|----------|----------|
| 184   | 2.043    | 3.528    | 0.338    |

*2.2. Computational Method*

The hydraulic Reynolds number for packed bed in this study is based on the interstitial velocity ($U_i$) and hydraulic diameter ($d_H$):

$$Re_H = \frac{\rho U_i d_H}{\mu},$$ (1)

where the interstitial velocity ($U_i$) and hydraulic diameter ($d_H$) are defined as follows:

$$U_i = \frac{U_0}{\varepsilon},$$ (2)

$$d_H = \frac{4V_f}{A_p + A_w},$$ (3)

where $U_0$ is the superficial velocity and $\varepsilon$ is the porosity of the packed part. $V_f$ is the volume of fluid, $A_p$ and $A_w$ represent the surface area of particles and walls.

In order to analyse the hydrodynamics of the packed bed, we investigate the pressure drop through a unit length and friction factor $f$, defined as follows:

$$\frac{\Delta p}{L} = f \cdot \frac{1}{2}\rho U_i^2 \frac{1}{d_H},$$ (4)

At the same time, we study the heat transfer performance by obtaining the heat transfer coefficient and the average Nusselt number of one sphere in developed section of the packed bed:

$$h = \frac{q}{A_p \cdot (T_p - T_f)},$$ (5)

$$Nu = \frac{h d_p}{\lambda},$$ (6)

where $d_p$ is the diameter of particle, $\lambda$ is the thermal conductivity. $q$ is the particle-to-fluid heat transfer rate, $T_p$ is the temperature of particle and $T_f$ is the average of mass averaged temperature on the upstream and downstream cross-sections.

Considering the unit pressure drop and the heat transfer coefficient, we get overall heat transfer efficiency:

$$\gamma = \frac{h}{\Delta p / L},$$ (7)

In the numerical simulation, the geometric models are built by Pro/ENGINEER WILDFIRE 5.0 (5.0 Parametric Technology Corporation, Needham, MA, USA) and the unstructured grids are generated by ANSYS ICEM (14.5, ANSYS Inc., Cecil Township, PA, USA). To avoid the poor quality grids on the contact point, the diameters of the packed spheres are shrunk by 1%, same as the treatment in Ref. [3]. The hydraulic diameter, porosity and other parameters are calculated with 0.99 $d_p$. For the same inlet velocity, the $Re_H$ has a little difference between the structured packed bed with smooth and dimpled spheres, due to a difference of 0.66% in surface area and a difference of 0.77% in volume of sphere. Since the difference is tiny, we assume the $Re_H$ of the structured packed bed with smooth and dimpled spheres to be the same for the convenience of the comparison. For the structured packed beds with smooth spheres (SPBS) and the structured packed beds with dimpled spheres (SPBD), four different grids are used for grid the independence test, respectively. The total elements numbers vary from 1,002,223 to 3,352,966 for SPBS, and from 1,276,351 to 3,774,227 for SPBD. As shown in Figure 4, the differences of the unit pressure drop between the largest two grids are about 0.83% and 0.28% for SPBS and SPBD. So, the most refined grids are chosen to get credible results. The mesh near the contact point of packed bed with dimple sphere is shown in Figure 5.

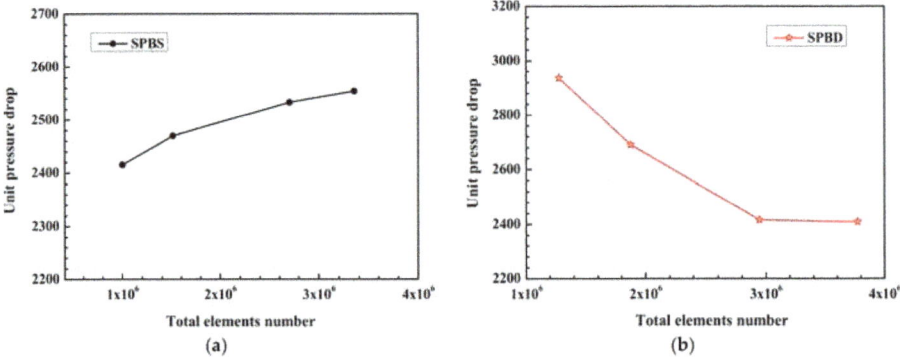

**Figure 4.** The unit pressure drop for (**a**) structured packed beds with smooth spheres (SPBS); and (**b**) structured packed beds with dimpled spheres (SPBD) with different grids.

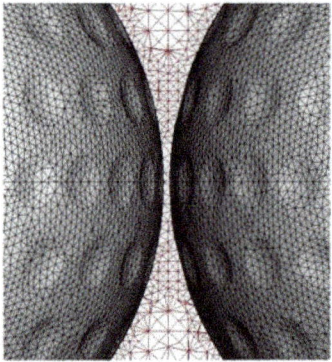

**Figure 5.** The mesh near the contact point of packed bed with dimple sphere.

The three-dimensional Navier–Stokes and energy equations for steady state incompressible flow and heat transfer are used. The hydraulic Reynolds number $Re_H$ in this study is from 1500 to 14,000. The RNG $k$-$\varepsilon$ model with scalable wall function is used as the turbulence model. The effect of swirl on turbulence is included in the RNG model, enhancing accuracy for swirling flows. Scalable wall functions avoid the deterioration of standard wall functions under grid refinement below $y^* < 11$ [18]. The computations are performed using the commercial software ANSYS FLUENT (14.5, ANSYS Inc., Cecil Township, PA, USA). The SIMPLE algorithm is used for the pressure-velocity coupling in the Navier-Stokes equations. The Green-Gauss cell based scheme is used for the gradient and the second order upwind scheme is applied for the momentum, turbulence kinetic energy and dissipation rate, and energy equations.

The unit pressure drop and Nusselt number of the packed bed with smooth spheres at $N = 1.00$ are validated with the correlations by Calis et al. [3] and Rokmes et al. [4], as shown in Figure 6. The maximal and average deviations of friction factor are 9.1% and 5.4%, respectively. The maximal and average deviations of Nusselt number are 11.7% and 7.1%, respectively. Our results are in good agreement with the prediction results by these correlations.

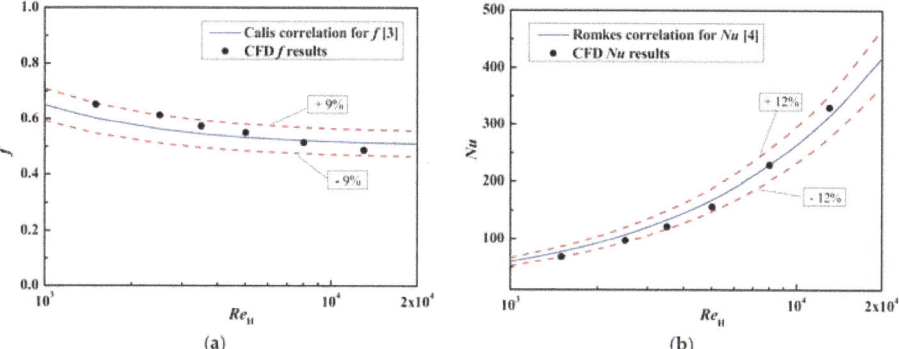

**Figure 6.** Validation of (**a**) friction factor; and (**b**) Nusselt number of the packed bed with smooth spheres at $N = 1.00$.

## 3. Results and Discussion

### 3.1. N = 1.00 Packing

3.1.1. The Flow Characteristics

The dimpled structure has a complex influence on the flow characteristics. Firstly, the unit pressure drop and friction factor are discussed. The unit pressure drop is obtained by fitting the pressures and the location of the middle planes between adjacent spheres in packed bed, as shown in Table 2. In Table 2, it can be seen that the unit pressure drop of SPBD is smaller than that of SPBS at $N = 1.00$ in the range of simulated Reynolds number, with maximal and average differences about 7.4% and 4%, respectively. It means that the change from smooth spheres to dimpled spheres in packed bed could slightly reduce the drag of packed bed. The SPBD has a drag reduction effect at this configuration. The friction factor of SPBS and SPBD at $N = 1.00$, computed with the pressure drop and hydraulic diameter, is shown in Figure 7. We can see the apparent drag reduction in SPBD, especially when $Re_H > 4000$.

**Table 2.** The unit pressure drop of the structured packed beds with smooth or dimpled spheres at $N = 1.00$ (Pa·m$^{-1}$).

| $Re_H$ | SPBS | SPBD | Unit Pressure Drop Reduction |
|---|---|---|---|
| 1504 | 114.0 | 106.4 | 6.63% |
| 2506 | 297.6 | 296.8 | 0.27% |
| 3509 | 544.7 | 540.9 | 0.68% |
| 5013 | 1066.8 | 987.6 | 7.42% |
| 8021 | 2554.0 | 2409.5 | 5.66% |
| 13,033 | 6388.8 | 6204.7 | 2.88% |

In order to investigate why SPBD has a drag reduction effect at $N = 1.00$, we illustrate the streamlines of SPBS and SPBD as well as the streamwise velocity in the cross section near the 7th sphere, where the flow has been fully developed. In Figure 8, although there is a difference of about 7% between the friction factors of SPBS and SPBD at $Re_H = 5013$, we could hardly find obvious distinctions between the streamlines. In a way, the streamlines are more continuous in the corner zone in SPBD.

From Figure 9 we can see that the streamwise velocity in the corner zone on plane 1 and plane 3 of SPBD is a little larger than that of SPBS. The maximum streamwise velocity appears on plane 2 of SPBD is larger as well. We infer that the wake zone behind or near the dimpled spheres is larger than that of smooth spheres, resulting in more fluid flowing through the corner zone. This may cause the reductions of pressure drop and friction factor in SPBD.

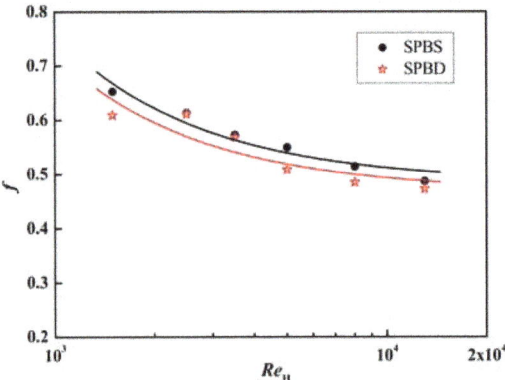

**Figure 7.** The friction factor of the structured packed bed with smooth or dimpled spheres at $N = 1.00$.

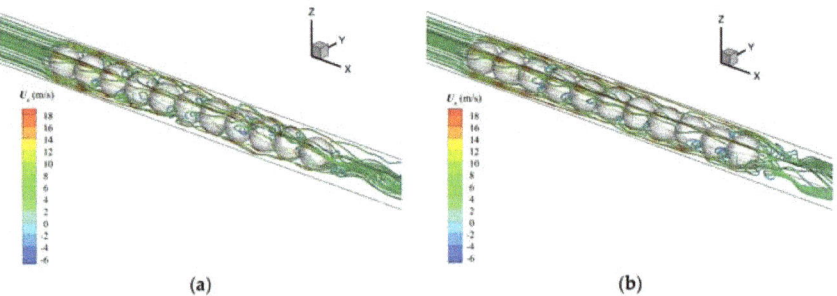

**Figure 8.** The streamlines of (a) SPBS; and (b) SPBD at $N = 1.00$ ($Re_H = 5013$).

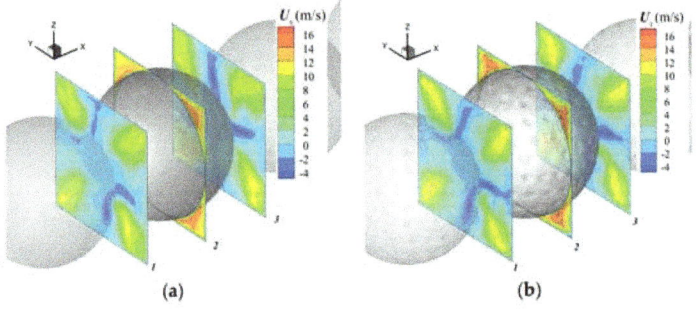

**Figure 9.** The streamwise velocity in the cross sections of (a) SPBS and (b) SPBD at $N = 1.00$ ($Re_H = 5013$).

3.1.2. The Heat Transfer Characteristics

The Nusselt number and heat transfer coefficient on the surface of one sphere in full development section in packed bed are computed. At $N = 1.00$, these characteristics are obtained on the 7th sphere of the 11 spheres packing, as shown in Table 3 and Figure 10. It can be seen that the heat transfer enhancement of SPBD is negative at low and high Reynolds number, namely that the heat transfer performance of SPBD is worse than that of SPBS in this situation. However, it increases with the $Re_H$ first and then decreases. The maximum of heat transfer enhancement of SPBD reaches to 6% at $Re_H$ about 3500.

**Table 3.** The Nusselt number of the structured packed beds with smooth or dimpled spheres at $N = 1.00$.

| $Re_H$ | SPBS | SPBD | Heat Transfer Enhancement |
|--------|------|------|---------------------------|
| 1504 | 68.2 | 64.9 | −4.94% |
| 2506 | 97.1 | 100.9 | 3.96% |
| 3509 | 121.6 | 129.5 | 6.46% |
| 5013 | 156.4 | 160.5 | 2.65% |
| 8021 | 228.8 | 226.8 | −0.87% |
| 13,033 | 330.3 | 325.2 | −1.54% |

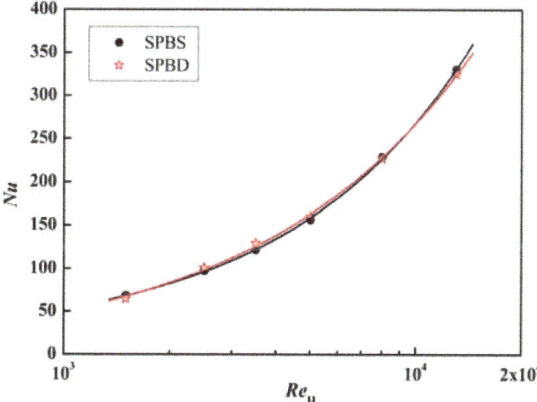

**Figure 10.** The Nusselt number of the structured packed bed with smooth or dimpled spheres at $N = 1.00$.

Figure 11 shows the surface heat transfer coefficient of SPBS and SPBD at $N = 1.00$ ($Re_H = 3509$). The zone A, which depicts the stronger heat transfer performance, is larger for the dimpled sphere than that for the smooth one. Similarly, the heat transfer weaker zone (zone B) in the back part of the dimpled spheres is smaller. However, it is found that surface heat transfer coefficient in each dimple is weaker than that in the surroundings. It means that the stagnation fluid in each dimple deteriorates the heat transfer from sphere to the main flow area. So, the enlargement of heat transfer enhanced area of SPBD is the dominant reason for the increase in the average heat transfer coefficient and Nusselt number.

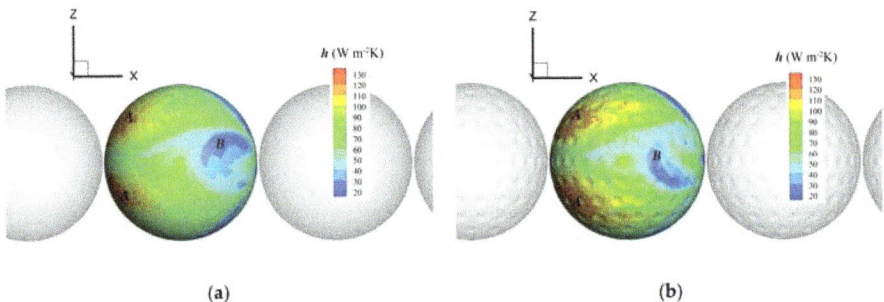

**Figure 11.** The surface heat transfer coefficient on a smooth or dimpled sphere in packed bed at $N = 1.00$ ($Re_H = 3509$) for (**a**) smooth spheres packed bed and (**b**) dimpled spheres packed bed.

Combining the effects of unit pressure drop and heat transfer coefficient, we calculated the overall heat transfer efficiency of SPBS and SPBD, as shown in Figure 12. The difference in overall heat transfer efficiency between SPBD and SPBS increases with the $Re_H$ first and then decreases, while it is always positive indicating that the overall heat transfer efficiency of SPBD can be improved in the range of simulated Reynolds number.

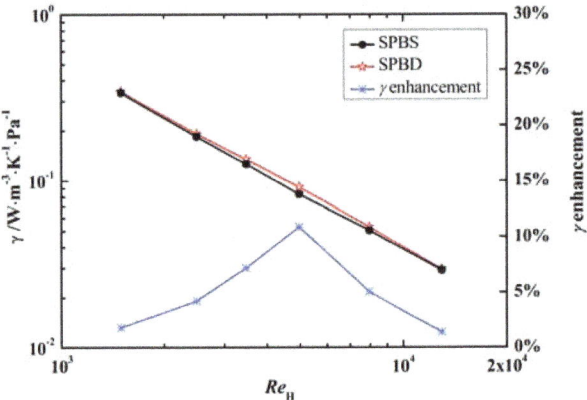

**Figure 12.** The overall heat transfer efficiency of the structured packed bed with smooth or dimpled spheres at $N = 1.00$.

By analysing the flow and heat transfer characteristics of SPBS and SPBD at $N = 1.00$, it is concluded that, compared with SPBS, the SPBD shows a drag reduction about 4%, while the heat transfer enhancement increases with the $Re_H$ first and then decreases and the Nusselt number is slightly higher in a certain range. So, the overall heat transfer efficiency will be improved at the configuration of $N = 1.00$ packing.

### 3.2. N = 1.15 Packing

#### 3.2.1. The Flow Characteristics

The pressure drop and friction factor of SPBS and SPBD at $N = 1.15$ are shown in Table 4 and Figure 13. Completely different from the $N = 1.00$ packing, the unit pressure drop of SPBD is much larger than that of SPBS. The negative sign of unit pressure drop reduction means that the SPBD has no drag reduction effect anymore. Averagely, the unit pressure drop increases by about 26.3%.

**Table 4.** The unit pressure drop of the structured packed beds with smooth or dimpled spheres at $N = 1.15$ (Pa·m$^{-1}$).

| Re$_H$ | SPBS | SPBD | Unit Pressure Drop Reduction |
|---|---|---|---|
| 1840 | 31.4 | 39.5 | −25.45% |
| 3219 | 91.0 | 117.9 | −29.62% |
| 4599 | 183.4 | 232.9 | −26.98% |
| 13,797 | 1528.0 | 1883.0 | −23.23% |

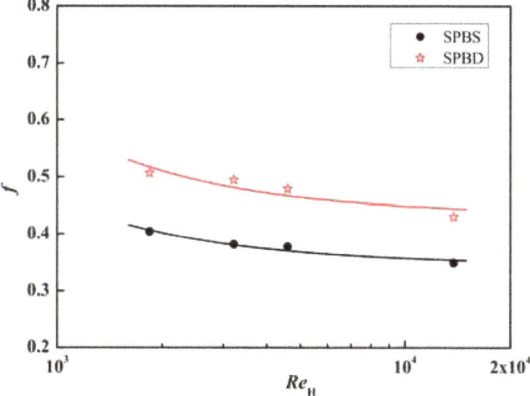

**Figure 13.** The friction factor of the structured packed bed with smooth or dimpled spheres at $N = 1.15$.

The streamlines of SPBS and SPBD at $N = 1.15$ are exhibited in Figure 14. Unlike the situation at $N = 1.00$, an obvious distinction in the streamlines between SPBS and SPBD at $N = 1.15$ is found. More streamlines are interrupted in the SPBD. We suppose that the stagger arrangement of the dimpled spheres and the existence of channel walls result in a more chaotic flow so that the backward flow behind every dimple spheres of SPBD is stronger. From Figure 15 we can see that, the streamwise velocity in the corner zones B and D of SPBD is larger than that of SPBS. However, the streamwise velocity in the corner zone A of SPBD is slower, especially on plane 2. So, we guess that, the more chaotic flow in SPBD brings greater viscous resistance. This may cause the increase in the pressure drop of the SPBD.

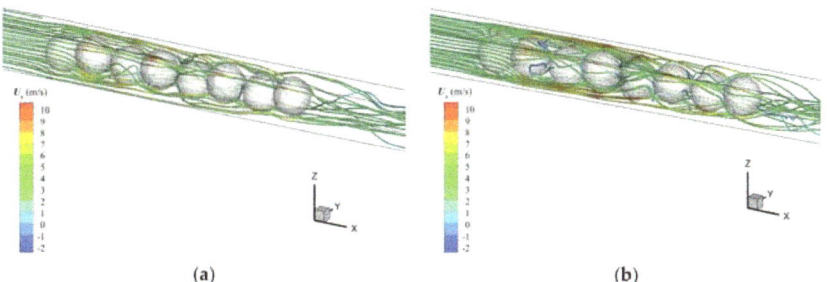

(a)                                       (b)

**Figure 14.** The streamlines of the structured packed bed with smooth or dimpled spheres at $N = 1.15$ ($Re_H = 4599$) for (**a**) smooth spheres packed bed and (**b**) dimpled spheres packed bed.

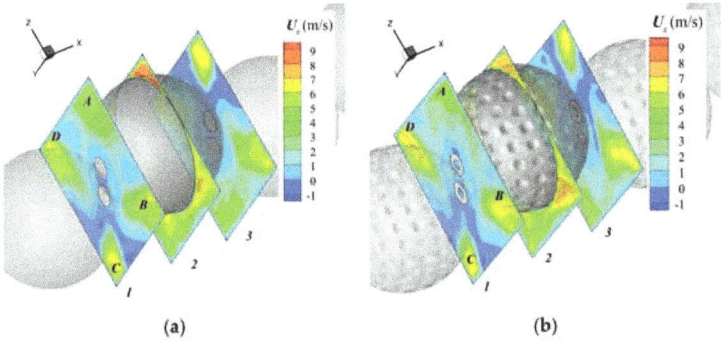

**Figure 15.** The streamwise velocity in the cross sections of (**a**) SPBS and (**b**) SPBD at $N = 1.15$ ($Re_H = 4599$).

### 3.2.2. The Heat Transfer Characteristics

The Nusselt number and heat transfer coefficient of SPBS and SPBD are obtained on the 6th sphere of the 8 spheres packing at $N = 1.15$, where the flow has been fully developed, as shown in Table 5 and Figure 16. The heat transfer enhancement of SPBD is distinctly improved, by about 10%, compared with SPBS. Similarity to $N = 1.00$, it increases with the $Re_H$ first and then decreases.

**Table 5.** The Nusselt number of the structured packed beds with smooth or dimpled spheres at $N = 1.15$.

| $Re_H$ | Smooth Spheres Packed Bed | Dimpled Spheres Packed Bed | Heat Transfer Enhancement |
|---|---|---|---|
| 1840 | 47.6 | 52.4 | 10.09% |
| 3219 | 72.1 | 81.1 | 12.51% |
| 4599 | 93.1 | 105.6 | 13.36% |
| 13,797 | 208.0 | 217.5 | 4.56% |

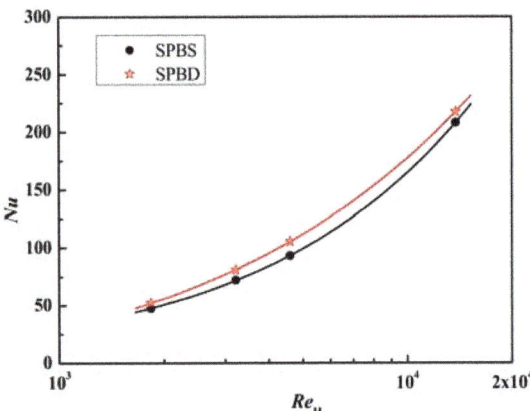

**Figure 16.** The Nusselt number of the structured packed bed with smooth or dimpled spheres at $N = 1.15$.

Figure 17 shows the surface heat transfer coefficient of SPBS and SPBD at $N = 1.15$ ($Re_H = 4599$). In the flow direction($x$ axis), the part of surface covered by the front sphere has a stronger heat transfer performance (zone A). The area of this part on the dimpled sphere surface is obviously larger than that on the smooth sphere. According to the conclusion in Section 3.1.2, the enlargement of heat transfer enhanced area of SPBD is the dominant reason for the increase in the average heat transfer coefficient. We also think that the enlargement of heat transfer enhanced area of SPBD might be the results of chaotic flow and wake zone enlargement.

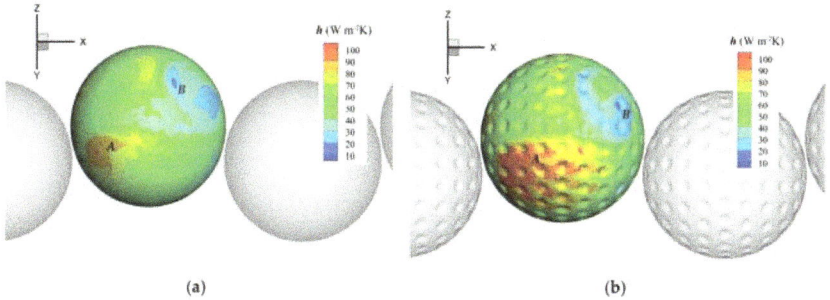

(a)                                                   (b)

**Figure 17.** The surface heat transfer coefficient on a smooth or dimpled sphere in packed bed at $N = 1.15$ ($Re_H = 4599$) for (a) smooth spheres packed bed and (b) dimpled spheres packed bed.

The overall heat transfer efficiency of SPBS and SPBD at $N = 1.15$ is shown in Figure 18. Although the heat transfer coefficient of SPBD is improved by about 10%, the drag of SPBD largely increases resulting in a worse overall heat transfer efficiency compared with SPBS in the range of simulated Reynolds number.

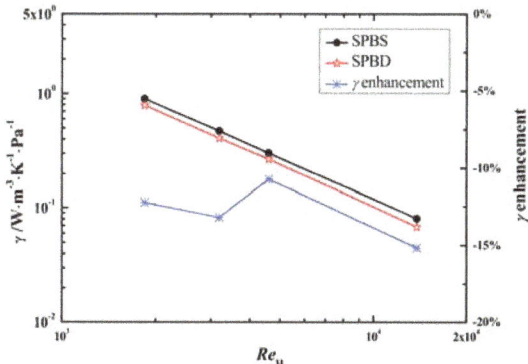

**Figure 18.** The overall heat transfer efficiency of the structured packed bed with smooth or dimpled spheres at $N = 1.15$.

From the analysis and discussion of the comparison of flow and heat transfer characteristics between SPBS and SPBD at $N = 1.15$, we have a brief summary. Compared with SPBS, the SPBD does not exhibit the drag reduction effect and the friction factor of SPBD increases by about 26.3%. Heat transfer enhancement distinctly improves by about 10%. However, the overall heat transfer performance of SPBD would be weaker at the configuration of $N = 1.15$ packing.

### 3.3. The Effect of Packing Configuration

Comparing the flow and heat transfer characteristics of different channel to particle diameter ratios, we find that, due to the combined effect of channel walls and dimpled surfaces, the flow in the wall bounded structured packed bed is quite complex. The drag reduction effect of SPBD turns into drag increasing effect as the channel to particle diameter ratio changes from $N = 1.00$ to $N = 1.15$. Therefore, we can also deduce that the drag coefficient of randomly packed beds with dimpled spheres may be larger than that with smooth spheres. At the same time, the heat transfer enhancement of SPBD strengthens with the channel to particle diameter ratio changing from $N = 1.00$ to $N = 1.15$. We think that the difference in the channel to particle diameter ratio has relatively less effect on the heat transfer enhancement.

## 4. Conclusions

The present study numerically investigated the flow and heat transfer characteristics in structured packed beds with smooth or dimpled spheres. Two different low channel to particle diameter ratios ($N = 1.00$ and $N = 1.15$) are simulated. The friction factor, Nusselt number and overall heat transfer efficiency are obtained. The conclusions are given as follows:

(1) For $N = 1.00$, the packed bed with dimpled spheres shows a little lower pressure drop and a slightly higher Nusselt number than those in the packed bed with smooth spheres, the overall heat transfer performance of packed bed with dimpled spheres would be improved. Therefore, the packed bed with dimpled spheres should be used in real applications to obtain a better overall performance at this configuration.

(2) For $N = 1.15$, the structured packed bed with dimpled spheres shows a much higher pressure drop, while the Nusselt number is still higher the overall heat transfer performance of a packed bed with dimpled spheres would be weaker. Therefore, the packed bed with dimpled spheres should be used when a better heat transfer performance is pursued in spite of the higher pressure drop in this configuration.

(3) Different channel to particle diameter ratios packing configuration can result in totally different drag reduction effect, while the heat transfer enhancement is relatively less affected.

**Acknowledgments:** We would like to acknowledge financial supports for this work provided by National Natural Science Foundation of China (No. 51536007, 51476124).

**Author Contributions:** Shiyang Li performed the numerical simulation and wrote the paper; Lang Zhou generated the geometry; Jian Yang contributed to writing and revising the paper; Qiuwang Wang supervised the work and revising the paper. All authors contributed to this work.

**Conflicts of Interest:** The authors declare no conflict of interest.

## Nomenclature

| | |
|---|---|
| $A$ | area (m$^2$) |
| $b$ | distance between dimples (m) |
| $c$ | width of dimple (m) |
| $d$ | diameter (m) |
| $f$ | friction factor |
| $h$ | heat transfer coefficient (W·m$^{-2}$·K) |
| $H$ | widths of the square channel (m) |
| $k$ | depth of dimple (m) |
| $L$ | length of the square channel |
| $N$ | channel to particle diameter ratio |
| $N_\mathrm{D}$ | the number of dimples on the sphere |
| $Nu$ | Nusselt number |
| $\Delta p$ | pressure drop (Pa) |
| $q$ | heat transfer rate (J·s$^{-1}$) |
| $Re_\mathrm{H}$ | hydraulic Reynolds number |
| $T$ | temperature (K) |
| $U$ | velocity (m·s$^{-1}$) |
| $V$ | volume (m$^3$) |

*Greek letters*

| | |
|---|---|
| $\varepsilon$ | porosity |
| $\rho$ | density (kg·m$^{-3}$) |
| $\mu$ | dynamic viscosity (kg·m$^{-1}$·s$^{-1}$) |
| $\lambda$ | thermal conductivity (W·m$^{-1}$·K$^{-1}$) |
| $\gamma$ | overall heat transfer efficiency (W·m$^{-3}$·K$^{-1}$·Pa$^{-1}$) |

*Subscripts*

| | |
|---|---|
| 0 | initial state at inlet |
| f | fluid |
| i | interstitial |
| H | hydraulic |
| p | particle |
| w | wall |

## References

1. Ergun, S. Fluid flow through packed columns. *Chem. Eng. Prog.* **1952**, *48*, 89–94.
2. Strangio, V.A.; Dautzenberg, F.M.; Calis, H.P.A.; Gupta, A. Fixed Catalytic Bed Reactor. International Patent Application PCT/US99/06242, 23 March 1998.
3. Calis, H.P.A.; Nijenhuis, J.; Paikert, B.C.; Bleek, C.M.V.D. CFD modelling and experimental validation of pressure drop and flow profile in a novel structured catalytic reactor packing. *Chem. Eng. Sci.* **2001**, *56*, 1713–1720. [CrossRef]
4. Romkes, S.J.P.; Dautzenberg, F.M.; Bleek, C.M.V.D.; Calis, H.P.A. CFD modelling and experimental validation of particle-to-fluid mass and heat transfer in a packed bed at very low channel to particle diameter ratio. *Chem. Eng. J.* **2003**, *96*, 3–13. [CrossRef]
5. Palle, S.; Aliabadi, S. Direct simulation of structured wall bounded packed beds using hybrid fe/fv methods. *Comput. Fluids* **2013**, *88*, 730–742. [CrossRef]
6. Lin, C.N.; Jang, J.Y.; Lai, Y.S. Two Dimensional Thermal-Hydraulic Analysis for a Packed Bed Regenerator Used in a Reheating Furnace. *Energies* **2016**, *9*, 995. [CrossRef]
7. Yang, J.; Wang, Q.; Zeng, M.; Nakayama, A. Computational study of forced convective heat transfer in structured packed beds with spherical or ellipsoidal particles. *Chem. Eng. Sci.* **2010**, *65*, 726–738. [CrossRef]
8. Wang, J.; Yang, J.; Cheng, Z.; Liu, Y.; Chen, Y.; Wang, Q. Experimental and numerical study on pressure drop and heat transfer performance of grille-sphere composite structured packed bed. *Appl. Energ.* **2017**, in press. [CrossRef]
9. Hu, Y.; Yang, J.; Wang, J.; Zhou, L.; Li, S.; Wang, Q. Numerical study of forced convective heat transfer in grille-sphere composite packed bed with Taguchi-CFD method. *Chem. Eng. Trans.* **2017**, *61*, 313–318.
10. Jin, C.; Jeon, W.P.; Choi, H. Mechanism of drag reduction by dimples on a sphere. *Phys. Fluids* **2006**, *18*, 112. [CrossRef]
11. Aoki, K.; Ohike, A.; Yamaguchi, K.; Nakayama, Y. Flying characteristics and flow pattern of a sphere with dimples. *J. Visual-Japan* **2003**, *6*, 67–76. [CrossRef]
12. Smith, C.E.; Beratlis, N.; Balaras, E.; Squires, K.; Tsunoda, M. Numerical investigation of the flow over a golf ball in the subcritical and supercritical regimes. *Int. J. Heat Fluid Flow* **2010**, *31*, 262–273. [CrossRef]
13. Burgess, N.K.; Ligrani, P.M. Effects of Dimple Depth on Channel Nusselt Numbers and Friction Factors. *J. Heat Trans.-T. ASME* **2005**, *127*, 839–847. [CrossRef]
14. Samad, A.; Lee, K.D.; Kim, K.Y. Shape Optimization of a Dimpled Channel to Enhance Heat Transfer Using a Weighted-Average Surrogate Model. *Heat Transf. Eng.* **2010**, *31*, 1114–1124. [CrossRef]
15. Kim, S.M.; Jo, J.H.; Lee, Y.E.; Yoo, Y.S. Comparative Study of Shell and Helically-Coiled Tube Heat Exchangers with Various Dimple Arrangements in Condensers for Odor Control in a Pyrolysis System. *Energies* **2016**, *9*, 1027. [CrossRef]
16. Crawford, C.W.; Plumb, O.A. The influence of surface roughness on resistance to flow through packed beds. *J. Fluid Eng.-T. ASME* **1986**, *108*, 343. [CrossRef]

17. Yang, J.; Zhou, L.; Hu, Y.; Li, S.; Wang, Q. Numerical study of forced convective heat transfer in structured packed beds of dimple-particles. *Heat Transf. Eng.* **2017**, in press. [CrossRef]

18. ANSYS Inc. *ANSYS User and Theory Guide, ANSYS Fluent, Release 14.5*; ANSYS Inc.: Cecil Township, PA, USA, 2012.

*Article*

# From Problems to Potentials—The Urban Energy Transition of Gruž, Dubrovnik

Andy van den Dobbelsteen [1,*] , Craig Lee Martin [1], Greg Keeffe [2], Riccardo Maria Pulselli [3] and Han Vandevyvere [4]

[1]  Faculty of Architecture and the Built Environment, Delft University of Technology, Julianalaan 134, 2628 BL Delft, The Netherlands; c.l.martin@tudelft.nl
[2]  School of Natural and Built Environment, Queens University Belfast, University Road, Belfast BT7 1NN, UK; g.keeffe@qub.ac.uk
[3]  Ecodynamics Group, University of Siena, Italy INDACO$_2$, via Roma 21B int.3, 53034 Colle Val d'Elsa (Siena), Italy; riccardo.pulselli@indaco2.it
[4]  Unit Smart Energy & Built Environment, EnergyVille, Thor Park 8310, 3600 Genk, Belgium; han.vandevyvere@energyville.be
*  Correspondence: a.a.j.f.vandendobbelsteen@tudelft.nl; Tel.: +31-6-3925-1421

Received: 4 March 2018; Accepted: 5 April 2018; Published: 13 April 2018

**Abstract:** In the challenge for a sustainable society, carbon-neutrality is a critical objective for all cities in the coming decades. In the EU City-zen project, academic partners collaborate to develop an urban energy transition methodology, which supports cities in making the energy transition to sustainable lifestyles and carbon neutrality. As part of the project, so-called Roadshows are organised in cities that wish to take the first step toward zero-energy living. Each Roadshow is methodologically composed to allow sustainability experts from across Europe to co-create designs, strategies and timelines with local stakeholders in order to reach this vital goal. Following a precursory investigative student workshop (the SWAT Studio), Dubrovnik was the third city to host the Roadshow in November 2016. During these events the characteristics of Dubrovnik, and the district of Gruž in particular, were systematically analysed, leading to useful insights into the current problems and potentials of the city. In close collaboration with local stakeholders, the team proposed a series of interventions, validated by the calculation of carbon emission, to help make Gruž, and in its wake the whole city of Dubrovnik, net zero energy and zero carbon. The vision presented to the inhabitants and its key city decision makers encompassed a path towards an attainable sustainable future. The strategies and solutions proposed for the Dubrovnik district of Gruž were able to reduce the current carbon sequestration compensation of 1200 hectares of forestland to only 67 hectares, an area achievable by urban reforestation projects. This paper presents the City-zen methodology of urban energy transition and that of the City-zen Roadshow, the analysis of the city of Dubrovnik, proposed interventions and the carbon impact, as calculated by means of the carbon accounting method discussed in the paper.

**Keywords:** sustainability; carbon neutrality; zero carbon cities; net zero energy; urban energy transition; energy renovation; neighbourhoods; SWAT Studio; City-zen; Roadshow

---

## 1. Introduction

### 1.1. Urgency

Climate change, depletion of fossil fuels, and scarcity of resources demand a sustainable transition of the world as we know it. Cities are increasingly dependent on energy supplies from politically volatile delta areas. Their heat-accumulating mass and high density are inherently vulnerable to excessive heat. The call for resilience is clearly evident on many fronts and is now without any need for further justification. The European Union underlines this, as can be seen from its sustainability programmes and targets [1,2].

### 1.2. City-zen EU

City-zen is an FP7 project of the EU [3], in which 34 public, private and academic partners collaborate to develop a method, tools and solutions for urban energy transition, to initiate and monitor demonstrator projects and to disseminate results throughout Europe. In one of the nine City-zen work packages, academic partners collaborate to develop an urban energy transition methodology, apply to the two partner cities, Amsterdam and Grenoble, and help them to develop a Roadmap towards the year 2050. In another work package, Roadshows are organised.

### 1.3. City-zen Roadshows

As part of the dissemination work package of City-zen, so-called Roadshows are organised. The aims of the City-zen Roadshow are to engage city stakeholders with innovative technologies and their applications, and to facilitate the development of a sustainable city agenda. These aims are met through multidisciplinary working groups and co-creative interactive sessions. The City-zen Roadshows take place in cities that do not form part of the project but that have expressed their desire to become carbon neutral in due time. For the City-zen project they also function to introduce, test and enhance the urban energy transition methodology presented.

The Roadshow is novel, in that its aim is not that of a typical urban design project usually undertaken by city governance bodies and their agencies. The Roadshow is intended as a challenge, one that inspires cities to meet environmental issues head on. It achieves this by using a design methodology to future-cast a possible solution for a neighbourhood that is not only achievable in terms of a time-based intervention and process but is also carbon accounted in a way that is didactic and accessible for a range of stakeholders.

### 1.4. Effectiveness of the Roadshow

Market failures usually prevent environmental policies or measures from being effective. Planning considerations are therefore critical in correctly implementing policies, as discussed by Deal et al. [4], Deal & Pan [5] and even earlier by Klosterman in 1985 [6]. In order to make plans or policies viable, understanding how planning information should be used is essential.

Obviously, spending one week in a neighbourhood, with a small team of academic designers and carbon-accountants, cannot be seen as a sustained intervention, nor can it be judged as community engagement in the normal sense, but the Roadshow is not intended to replace these sustained interventions. More focused and long-term engagement is necessary to make real change (see [4]). However, it does offer a new way of seeing the challenges of urban sustainability as issues of the stakeholders' imagination, rather than just a matter of technical know-how. Inspiring stakeholders' imagination as to what is possible and introducing a simple process-based methodology could unlock stasis in a socially-orientated context such as a West-European city.

## 2. Methodology

### 2.1. The Urban Energy Transition Roadmap

Put in a time-frame and within specific non-technical boundary conditions, Figure 1 illustrates the process for moving towards the desired future sustainable state of a city, as applied in the City-zen project. Starting from the analysis of the (non-sustainable) situation at present, the urban development already planned (2) and the stakeholders involved (3), the desires of these stakeholders together with scenarios for the future (4) define a future (sustainable) vision of the city, which is then translated back into interventions on the way towards this desired future state (6), the urban energy transition roadmap. This methodology in fact, is a more elaborate process of backcasting, as presented by Robinson [7] and Dreborg [8], and as proposed by Zhivov et al. [9] for communities and campuses.

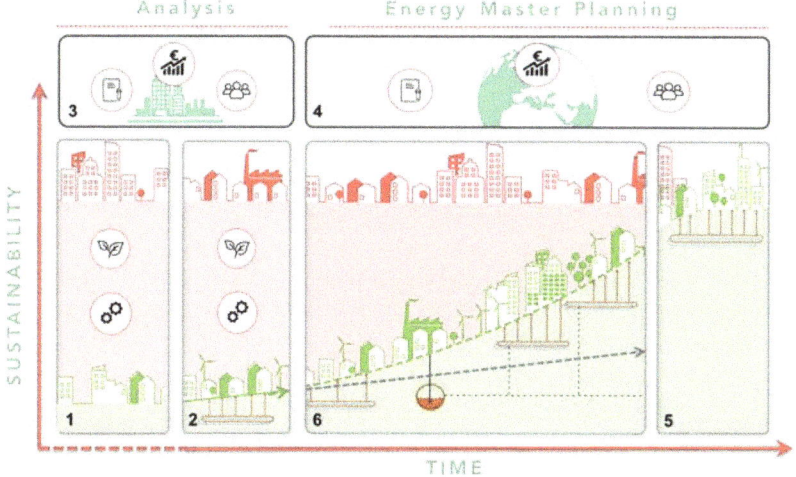

**Figure 1.** City-zen Road mapping scheme, from the present to the desired future state, taking into account future scenarios and using a catalogue of measures and book of inspiration.

For the green curved route in Figure 2, acceleration is needed along a transitional S-curve; otherwise, business-as-usual will ensure that very little progress is made, that is, the energy transition and future carbon emissions will remain on the grey line.

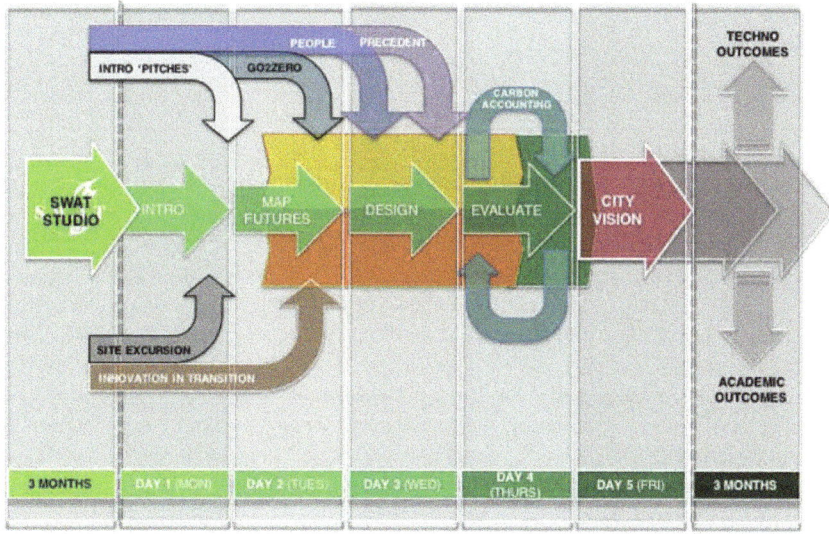

**Figure 2.** Organisational schedule of the City-zen Roadshow, starting with input from the SWAT Studio at left and with the resulting City Vision at right (scheme by Craig Lee Martin, TU Delft).

## 2.2. The City-zen Roadshow Methodology

The City-zen Roadshow methodology is a thoroughly structured timeline of events designed to create meaningful partnerships, knowledge sharing and rigorously prepared Sustainable City Vision (Figure 2).

The City-zen Roadshow is preceded by the SWAT Studio, an explorative two-week collaborative workshop with students from TU Delft, which typically occurs two months in advance of the specialist five-day Roadshow workshop. The outputs of the SWAT Studio consist of group proposals for sustainable interventions in the city, which the professionals of City-zen develop further during the Roadshow itself.

The Roadshow itself is an intensive and impactful co-creative 5-day model that has evolved over a -year period of live onsite implementation (see Figure 3). It strategically places Pecha Kucha pitches, and organizes interactive field visits, stakeholder interviews, Energy Potential Mapping [10], design workshops, serious gaming, mini-masterclasses (social and technical), seminars and interactive Carbon Accounting. The five days culminate with a Sustainable City Vision that shows how the city can achieve carbon neutrality through the implementation of variously scaled strategic and design interventions and measure.

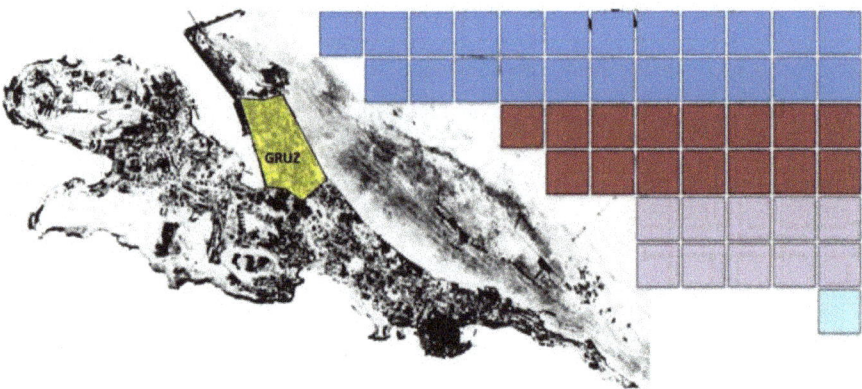

**Figure 3.** Gruž (the area depicted in yellow) and its estimated forestland grabbing (in 25-hectare square blocks) with details on emission sources: electricity (blue); waste (brown); mobility (purple); water (cyan) (image by Riccardo Pulselli, University of Siena).

## 2.3. Roadshow Week Activities

The largest part of the work of the Roadshow is done in interactive sessions between the researchers and stakeholders, creative work informed by hard data from the city and validated by energy and carbon calculations. The Roadshow team analyses the use of energy and emission of carbon dioxide, in total numbers and per household. In one specific workshop, the explanation of carbon usage and its measurement is described through the individual daily experiences of each of the attending stakeholders. This is an exciting component that has evolved incrementally with each Roadshow. The approach builds confidence in the math by supporting it with a visual carbon narrative and graphics. Once an understanding of individual carbon usage is gained it is up-scaled to include the street, neighbourhood, district, city and depending on location, an entire island. Carbon data is coupled to the forest area that needs to be planted to compensate for the emissions, but the eventual goal is to avoid this principle of compensation. The designs, calculations and proposed implementation eventually leads to a city vision on different scales—at city, neighbourhood and building levels—which is presented to the decision makers of the city on the final morning of the Roadshow.

## 2.4. The Dubrovnik Roadshow

Under the initiative of the City of Dubrovnik Development Agency (DURA) and by a mandate from the Mayor, the City was host to both the SWAT Studio and the specialist City-zen 'Dubrovnik' Roadshow, in fall 2016.

This paper presents the process and outcomes of that Roadshow, which are made specific by the carbon emission assessment of proposed interventions. The authors do not intend to discuss urban planning theory or to put the City-zen Roadshow in perspective using different urban planning methods. Although extensive work was done on sustainable urban planning in the overarching project, the focus of this paper is on strategies and measures of carbon emission reductions within the urban context of Dubrovnik.

## 3. Introducing Dubrovnik and Gruž

Before and during the Roadshow, the characteristics of Dubrovnik, the district of Gruž in particular, were systematically analysed, leading to in-depth understanding of the current problems and potentials of the city.

### 3.1. Dubrovnik

The city of Dubrovnik is best known for its historical UNESCO protected old town centre overlooking the Adriatic Sea—home-base for the imaginary city called King's Landing in the popular TV series Game of Thrones. However, the reason Dubrovnik was selected to host a Roadshow was Gruž, the port town of Dubrovnik, which lies to the north-west of the city.

DURA (the city's development agency), EE Info (Energy Efficiency office) and the University of Zagreb, as well as other local stakeholders collaborated closely with the Roadshow leader to prepare both the SWAT Studio and the City-zen Gruž Roadshow. The DURA offices on 15 Braniteja Dubrovnika was the studio base of the Roadshow during its co-creative efforts to develop a Sustainable City Vision for Dubrovnik.

### 3.2. Problems

A comprehensive analysis was undertaken to document the environmental challenges of Gruž. According to the Dubrovnik Tourist Board, in 2016 more than 1 million tourists arrived in the city with approximately 3.7 million overnight stays [11]. This means that on average ten thousand visitors descend into the city daily, creating a considerable impact on the daily routines of local life. During the height of the tourist summertime season this number is manifold. The majority of tourists arrive by cruise ship, which are moored in the Gruž harbour, and then transported by a continuous fleet of coaches to the gate of the old town.

Considering this cruise ship activity and the projected flow of tourists, a 2012 traffic study of Dubrovnik revealed the following problems: a lack of adequate parking space, bicycle transportation is difficult, the public transportation system is in need of improvement, and there is a pedestrian traffic problem in the city centre during the tourist season. As Peručić and Puh [12] found, the citizens of Dubrovnik experience many problems related to cruise ship tourism. This particularly holds true for Gruž, the district adjacent to Dubrovnik's port.

### 3.3. Cruise Ships

The environmental impact of cruise ships, essentially floating cities of 2000–2500 inhabitants, is enormous. The scope of the Roadshow and this paper cannot present a full study of the impact of cruise ship tourism, but we want to mention a few figures. Per week, 600–1000 $m^3$ of sewage is discharged into the sea. This is an environmental problem at sea, but the way ports accept mooring ships plays an important role in the sustainable transition of cruise ships, as will be discussed later. The annual carbon emission of one ship is comparable to the $CO_2$ absorbed by 11,900 hectares of forestland, approximately ten times larger than a town the same size. On a daily basis, a cruise ship expels the same amount of sulphur dioxide fumes as 13.1 million cars, which is generated from the most polluting and harmful form of diesel. Part of this emission is released in harbours such as Gruž.

### 3.4. Gruž

Gruž suffers severely from the fumes and soot emitted by the cruise ships. Gruž is a relatively poor community with minimal public transport and mobility partly due to the topography; buildings are also generally energy-inefficient. All these factors are exacerbated by the fact that the area has a high proportion of elderly people.

The economically beneficial, but equally environmentally adverse relationship between the cruise ships and the district was a major driver during the development of the City Vision for Dubrovnik/Gruž.

## 4. Analysis of Carbon Emission

### 4.1. Carbon Emission Accountancy

The carbon accounting developed here estimates the greenhouse gas emission in a city or neighbourhood. The emission is given in t $CO_2$-eq corresponding to the quantity of the main greenhouse gases released into the atmosphere i.e., $CO_2$, $CH_4$ and $N_2O$, multiplied by their Global Warming Potentials (GWP), see Table 1. The main references for this procedure are the Global Protocol for Community-Scale Greenhouse Gas (GHG) Emission Inventories [13], other carbon assessments (such as [14]), some previous experience and case studies. In particular, the monitoring of GHG emissions and impact mitigation scenarios has been demonstrated to be a powerful tool to address choices and hinder global warming and climate change at the regional [15] and urban level [16] or for specific sectors such as urban waste [17].

**Table 1.** Emission factors used for the carbon accounting and main references (overview by Riccardo Pulselli, University of Siena).

| Emission Factors | EF | Unit | References |
|---|---|---|---|
| Electricity grid | 0.341 | kg $CO_2$eq/kWh$_e$ | This study |
| Car driving (diesel) | 0.169 | kg $CO_2$eq/km | IPCC2006 |
| Waste to energy | 0.652 | kg $CO_2$eq/kg | IPCC Waste model |
| Organic to compost | 0.091 | kg $CO_2$eq/kg | IPCC Waste model |
| Landfill | 1.160 | kg $CO_2$eq/kg | IPCC Waste model |
| Water management | 0.585 | kg $CO_2$eq/m$^3$ | IPCC2006 |

### 4.2. Carbon Emissions in Gruž

A simplified but reliable carbon assessment was carried out for an average Dubrovnik-Gruž household, through the analysis of the domestic energy demand, mobility, waste management and water management. Statistical data (2015) was provided by the Municipality of Dubrovnik during the co-working sessions of the Roadshow. The emission factors used for the assessment refer to the IPCC standards [18], as reported in Table 1, except for the site-specific emission factor of electricity, which was estimated and based on the Croatian grid mix (Table 2).

Table 3 shows results for the single household and neighbourhood (almost 8000 inhabitants). Outcomes show that the average emission per household in Dubrovnik (on average 2.7 people on 100 m$^2$) is around 5.70 tonne $CO_2$eq/a mostly depending on the use of electricity for lighting, appliances and cooling, and methane for heating, water heating and cooking (energy demand: 47%). Other sources of emission are the use of private cars (mobility: 20%), the collection and treatment of domestic waste, considering that most of it is landfilled (waste management: 30%) and water withdrawal from the aqueduct (water management: 3%).

**Table 2.** Emission factor of electricity in Croatia, based on the national electricity grid mix (overview by Riccardo Pulselli, University of Siena).

| Croatia | LCA-Based EF | Data | | GHG Emission | Notes |
|---|---|---|---|---|---|
| | kg CO$_2$eq/kWh | kWh | % | kt CO$_2$eq/a | References |
| **General Data** | | | | | |
| Electricity demand | - | $1.75 \times 10^{10}$ | 100.0 | - | |
| Electricity production | - | $1.36 \times 10^{10}$ | 77.4 | - | |
| Import | 0.712 | $3.95 \times 10^9$ | 22.6 | $2.81 \times 10^9$ | [19] Table 8. Various generators/turbine types |
| Thermo-electricity | | $3.50 \times 10^9$ | 20.0 | $3.03 \times 10^9$ | |
| Natural gas | 0.443 | $1.00 \times 10^9$ | 5.7 | $4.44 \times 10^8$ | [19] Table 8. Various combined cycle turbines |
| Petroleum products | 0.778 | $1.29 \times 10^8$ | 0.7 | $1.00 \times 10^8$ | [19] Table 8. Various generators/turbine types |
| Coal | 1.050 | $2.37 \times 10^9$ | 13.5 | $2.49 \times 10^9$ | [19] Table 8. Various generator types |
| Renewables | | $1.01 \times 10^{10}$ | 57.4 | $1.23 \times 10^8$ | |
| Solar PV | 0.032 | $3.50 \times 10^7$ | 0.2 | $1.12 \times 10^6$ | [19] Table 8. 80 MW, parabolic trough |
| Wind | 0.010 | $7.30 \times 10^8$ | 4.2 | $7.30 \times 10^6$ | [19] Table 8. 1.5 MW, onshore, 10 g CO$_2$-eq/kWh |
| Hydro | 0.012 | $9.13 \times 10^9$ | 52.1 | $1.10 \times 10^8$ | [19] Table 8. 3.1 MW, reservoir, 10 g CO$_2$-eq/kWh; 300 kW, river run-off, 13 g CO$_2$/kWh |
| Biomass | 0.028 | $1.65 \times 10^8$ | 0.9 | $4.62 \times 10^6$ | [19] Table 8. Short rotation forestry steam turbine |
| Total | 0.341 | $1.75 \times 10^{10}$ | 100.0 | $5.97 \times 10^9$ | |

**Table 3.** Carbon footprint of a typical household and the Gruž neighbourhood (8000 inhabitants; 2900 households) (overview by Riccardo Pulselli, University of Siena).

| Carbon Footprint (State of the Art) | | Gruž Household | | | | Gruž Neighbourhood | | |
|---|---|---|---|---|---|---|---|---|
| Emission Sources | Unit/a | Raw Data | % | kg CO$_2$eq | % | Raw Data | t CO$_2$eq | % |
| Electricity | kWh$_e$ | 7930 | 100 | 2703 | 47.4 | 22,997,000 | 7839 | 47.4 |
| Lighting + appliances | kWh$_e$ | 2450 | 31 | 835 | 14.6 | 7,105,000 | 2422 | 14.6 |
| Cooling | kWh$_e$ | 1850 | 23 | 631 | 11.1 | 5,365,000 | 1829 | 11.1 |
| Cooking | kWh$_e$ | 520 | 7 | 177 | 3.1 | 1,508,000 | 514 | 3.1 |
| Water heating | kWh$_e$ | 2210 | 28 | 753 | 13.2 | 6,409,000 | 2185 | 13.2 |
| Heating | kWh$_e$ | 900 | 11 | 307 | 5.4 | 2,610,000 | 890 | 5.4 |
| Mobility | km | 6600 | - | 1113 | 19.5 | 19,140,000 | 3227 | 19.5 |
| Private car use | km | 6600 | 100 | 1113 | 19.5 | 19,140,000 | 3227 | 19.5 |
| Waste | kg | 1670.35 | 90 | 1726 | 30.3 | 4,359,614 | 5005 | 30.3 |
| % waste-to-energy | kg | 0.00 | 0 | 0 | 0.0 | 0 | 0 | 0.0 |
| % organic | kg | 16.70 | 1 | 2 | 0.0 | 48,440 | 4 | 0.0 |
| % landfill | kg | 1486.61 | 89 | 1724 | 30.2 | 4,311,173 | 5001 | 30.2 |
| Water | m$^3$ | 280 | - | 164 | .9 | 811,696 | 475 | 2.9 |
| m$^3$ per annum (house) | m$^3$ | 279.90 | 100 | 164 | 2.9 | 811,696 | 475 | 2.9 |
| Carbon uptake | ha | 0.00 | 100 | 0 | 0.0 | 0 | 0 | 0.0 |
| Total | | | | 5706 | 100 | | 16,546 | 100 |

## 4.3. Forestland Sequestration of Carbon

As calculated according to the carbon accounting method described above, from the single household profile (average of 80 m$^2$) up to the neighbourhood scale (2900 households), the greenhouse gas emission of the neighbourhood corresponds to more than 16,000 tonne CO$_2$eq. Compensation for this amount of carbon emission would require a hypothetical forestland of about 1200 hectares, as illustrated by Figure 3, using the same spatial scale of the neighbourhood. This carbon sequestration area has been divided by emission sources i.e., energy demand (blue), mobility (purple), waste (brown) and water management (cyan). The graphical representation is used to communicate and visualise the state of the art to the wider public, which was the initial challenge to be faced in planning new strategies. The primary goal was to propose a coherent and clear plan for a carbon-neutral Gruž, as well as to develop a full energy transition strategy for Dubrovnik.

The carbon footprint per family corresponds to 0.42 hectare of forestland (considering an average carbon uptake of 1.35 kg $CO_2eq/m^2$ of forest), which is equivalent to the size of football pitch in area. The area of football pitch is familiar to many and is independent of cultural or demographical background.

## 5. Analysis of Energy Potential

### 5.1. The Local Climate of Dubrovnik

The climate of Dubrovnik seems quite favourable (Figure 4): the temperature ranges between 5 and 30 °C in winter and summer, with an average of 16 °C, which means that the soil will have a temperature of around 16 °C, ideal for cooling in the summertime and pre-heating in wintertime. At the moment, there are no soil energy systems operating in Dubrovnik.

Annually, Dubrovnik receives more than 1000 mm of precipitation. With this quantity, all the water consumption for toilet flushing, washing and watering of plants could be served with collected and stored rain water. However, such energy-efficient solutions are not commonly practiced in Dubrovnik.

In spite of the high rainfall, Dubrovnik has high average daily sunlight hours, making the city—not least for its good slope orientation towards the south—very suited for active and passive solar technologies. As with water recycling, such solar applications are rarely in use in the area, the exception being the intermittent uptake of solar heat collectors for hot water.

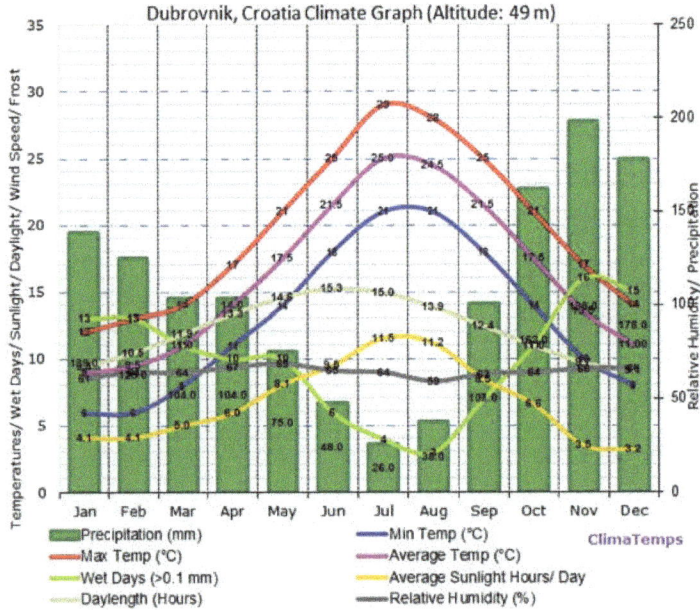

**Figure 4.** Climate chart for Dubrovnik, with precipitation in green bars, and minimum, average and maximum temperatures represented by the blue, pink and red line [20], respectively.

## 5.2. Energy Potentials

In order to determine the renewable energy potentials of Dubrovnik, the method of Energy Potential Mapping [10] was used.

Based on the climatic analysis undertaken, it was very apparent that the solar potential of the entire city is significant: Dubrovnik has 2480 h of sunshine per year, and an average solar irradiation of 1810 kWh/m² (SW orientation). In comparison, the solar irradiation of North-Western Europe is around 1000 kWh/m².

Further, wind turbines could be potentially effective when located on top of the coastal ridges, where wind from the sea accelerates them. The sea near the city, including the harbour of Gruž, also has the potential to be used as a source of heat and cold, assuming that the wave energy is not too complicated to harness. Finally, biomass could be obtained from bio-organic waste and from forest maintenance around the city's hinterland.

## 6. Strategies for Improvement

### 6.1. Energy Master Planning

In line with other energy master planning methods (e.g., [9]), the urban energy transition methodology of City-zen is based on incremental approaches such as REAP [21,22] and LES [23]. The basis of these methods lies in the New Stepped Strategy (NSS [24]): reduce (the demand for energy), reuse (waste energy flows), produce (energy from renewable sources). These steps are then translated to the urban level.

Figure 5 depicts an urban energy metabolism scheme from the Energy Master Plan [25], which is indebted to the 'Eco-device' concept of Wirdum [26] and Tjallingii [27,28], and includes a stepped approach based on REAP and the NSS. Basic preparative work includes attaining insight into the energy characteristics of a city and identifying energy sources and sinks (0, the blue arrow above). In the globalised world, provisions for cities and their regions are now commonly established by supply from outside, with discharge of waste flows (waste heat, waste water, waste materials, nutrients) to outside the city (the grey arrows left and right). The first step towards a more sustainable energy system is to reduce the energy demand in the urban environment (1), followed by solving the reuse of waste flows within and outside the city boundaries (2a and 2b), and the production of energy from renewable sources (3a and 3b). These technical-spatial strategies can be used for any urban sustainable transition process.

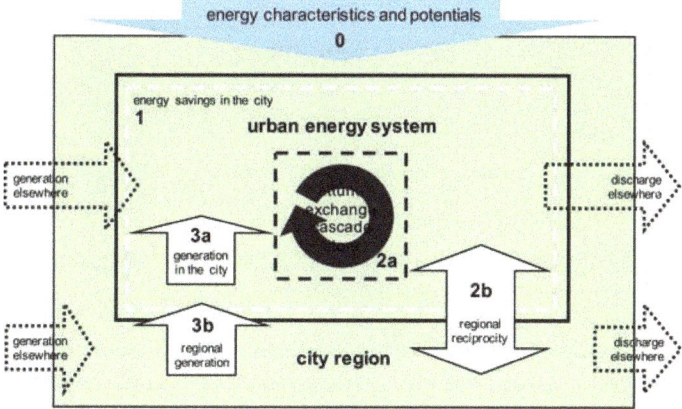

**Figure 5.** Metabolism urban energy scheme, including steps to transition to a more sustainable system, as part of the Energy Master Plan (adaptation of [25]).

In close collaboration with local stakeholders, the Roadshow team co-created a proposal for a series of variously scaled interventions to make Gruž, and thus, the entire city of Dubrovnik, net zero energy and zero carbon.

*6.2. Reduce: Smart Bioclimatic Design and Energy Renovation*

Energy conservation is probably the largest factor in becoming carbon neutral. This includes for instance, saving energy for heating, cooling, ventilating and lighting, in newly constructed buildings and existing building stock. The greatest challenge lies in reducing the cooling demand of existing buildings.

The local analysis made clear that Gruž' position on a slope next to the harbour makes it suited for cooling breezes passing through the streets and alleys, powered by thermal daytime drafts, and by downward night-time mountain drafts (Figure 6). This would save mechanical cooling needs and hence save energy. To calculate the exact impact of these cooling breezes, advanced simulations with CFD models would be needed, which was not possible within the time-frame of the Roadshow. Therefore, this particular planning proposal was not included in the calculated carbon savings. Nonetheless, it could be an additional factor.

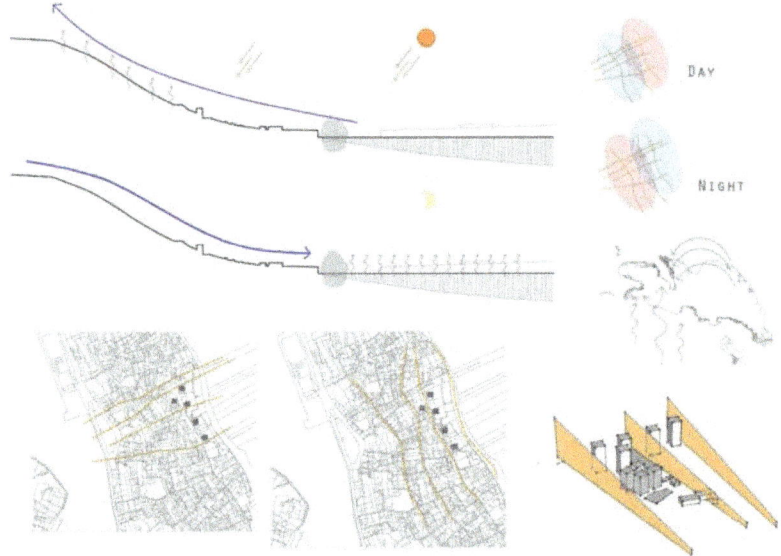

**Figure 6.** Creating cooling breezes via thermal drafts from the harbour during the day and mountain drafts at night (images by Andy van den Dobbelsteen and QUB students).

An energy renovation scheme was proposed for Gruž's building stock, starting with the larger apartment blocks, which could be turned into energy-neutral buildings with post-insulation and photovoltaics (PV), algae or greenhouse facades (Figure 7).

After studying the situation in situ and using the numbers provided by the Dubrovnik Development Agency and University of Zagreb, the effects of the proposed measures of energy renovation were calculated and later translated to carbon emission reductions. The energy calculations were informed by the expert judgement of the Roadshow team members.

Roof and façade shading—applicable to 50% of all households—could lead to a cooling demand reduction of 10%. Greening buildings could have the same effect and is applicable to 60% of Gruž households; greening the street blocks further helps to reduce demand by 5%. The greatest effect is achieved by post-insulation of roofs and walls and replacing windows with low-emittance alternatives, applicable to 80% of the households and resulting in a 35% reduction in heating demand and 5% in cooling demand. The impact of switching to heat pump systems, was calculated as adding a further saving of 26%.

**Figure 7.** Energy renovation by post-insulation and PV (photovoltaic) facades (**left**) or algae facades (**right**) (images by Andy van den Dobbelsteen and QUB students0.

*6.3. Reuse: The Green Port*

With a clear understanding of the contextual challenges ahead and their hierarchy in terms of importance, need and resolution, the strategy was to seriously address the issues related to cruise ship tourism. Requesting that the cruise ships transport their waste water to Gruž, instead of accepting that they release it directly into the sea, as is the present case, would be a step toward energy efficiency and creating a greener, cleaner economy and environment. The use of waste water treatment by algae enables the production of biodiesel and nutrients that are used to grow food, both of which can be sold back to the cruise ships. This green port solution (Figure 8) would create a win-win for the local community of Gruž.

**Figure 8.** An impression of algae arrays for waste water treatment for cruise ships [29].

### 6.4. Produce: Golf Course Wind Park

For Gruž, the City-zen Roadshow proposed a set of renewable energy production opportunities within the district, use of photovoltaics in particular, and these were calculated as savings of fossil fuels and carbon emissions. We want to highlight one unusual proposal that was made.

The Roadshow method ensures that not only past and present situations are analysed but also that future plans for the built environment influence the final outcomes. In the case of Gruž, an expansive and locally controversial golf resort has been proposed on the ridge overlooking the city. Considering the loss of biodiversity, use of pesticides, energy required for frequent grass trimming, material use of golf balls lost, etc., golf courses are seldom sustainable (for instance see data accessible through the Golf Course Superintendents Association of America [30]).

The City-zen Roadshow suggested that the strategic placement of wind turbines between the greens could add to the productive renewable energy capacity of Gruž (Figure 9). The effect of the wind turbines was included in the calculation as a saving in the use of non-sustainable electricity. A side effect would be that communal wind turbines would improve the acceptability of the controversial golf course.

**Figure 9.** Wind turbines on the golf course green (image by QUB students).

### 6.5. Sustainable and Healthy Transportation

"A developed country is not a place where the poor have cars, it is where the rich use public transportation", said Petro Gustavo, Mayor of Bogotá. Or bicycles, we could add. Dubrovnik transportation has a large share of private cars and other fossil fuel powered vehicles. Cruise ship tourists converge on the city centre in diesel buses, contributing further to fossil energy consumption, carbon emissions and air pollution. Based on carbon figures for various forms of transportation (see below), the modal split of Dubrovnik entails a relatively large emission of $CO_2$ as compared to other European cities. The City-zen Roadshow tried to reduce the carbon emissions of a default modal split of transportation by a shift to more use of public electric transportation, cycling and walking. The impact of this was included in the calculations.

For example, the Roadshow proposed a new tramway linking the Gruž port and the historic centre of Dubrovnik. Bicycle lanes and walkways for more health conscious citizens and tourists would complement the logical green corridor scheme. Figure 10 illustrates this. It is a good example of the design-driven workshop (a form of research by design [31]), which includes interactive, co-creative processes, including science (most notably, impact assessments), technology (knowledge of latest technical developments) and design skills (the ability to imagine new solutions and draw them in present-day situations). The traffic solution shown in Figure 10 was hand-drawn by an architect with input from local stakeholders and assessed for its energy and carbon impact by scientists.

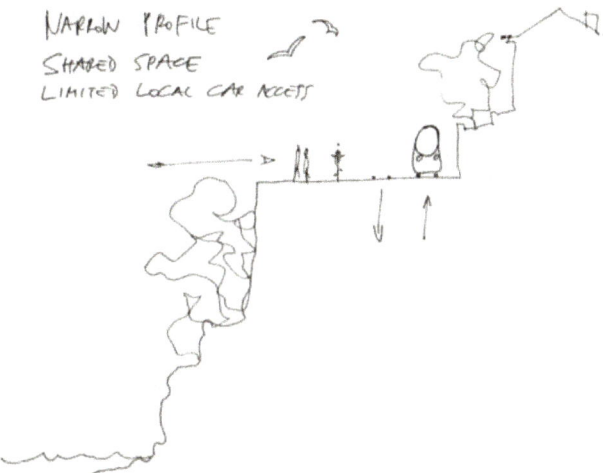

**Figure 10.** New cross-section of a combined pedestrian-bicycle-tramway track on the cliff stretch between Gruž and Dubrovnik city centre, where at present cars and diesel buses are dominant (drawing by Han Vandevyvere, EnergyVille).

It should be noted that a minority of tourists do attempt the journey by foot today, but the 20-min journey is unattractive, exhaust fumed and precarious. The lack of appropriately safe, wide and level footpaths causes coaches and buses to stop abruptly to prevent their wing mirrors striking unaware, single-filed and city-unfamiliar pedestrians.

## 7. Impact Assessment

The Sustainable City Vision presented to stakeholders of the city incorporated a path towards an attainable sustainable future, validated by calculations of the carbon emission effects. Table 4 shows a sequence of possible measures that can be potentially applied to progressively decrease the impacts on the Gruž neighbourhood. In particular, ten measures are selected as follows:

1.  Building greenery and shading. This combines solutions for shading building roofs and facades from direct solar irradiation. It also includes nature-based solutions such as roof gardens, vertical greenery systems or common vegetation and trees in squares and streets to avoid the Urban Heat Island (UHI) effect.
2.  Heat pumps. This concerns the replacement of old electric systems that are diffusely used in houses. Based on a coefficient of performance (COP) of 4, this would provide an energy saving for heating and cooling.
3.  Building envelope insulation. Considering the current state of dwellings, energy retrofitting of buildings in the neighbourhood would be a strategic choice.

4.  Solar PV and solar thermal for domestic hot water was estimated based on real availability of flat roofs and well-oriented walls. Considering 2480 h of sunshine per year, an average solar irradiation of 1810 kWh/m$^2$ and the orientation of buildings (6% South-East; 6% South; 61% South-West out of 750 available roofs per 80 m$^2$ each), the potential of electricity production has been estimated to be around 4300 MWh/a.

5.  Mini wind towers. It is estimated that 50 towers would produce up to 350 MWh/a of generated electricity for the neighbourhood.

6.  Increase in public transport. This measure is highly desirable in Gruž and would potentially involve 60% of households avoiding the use of private cars.

7.  Transition to electric mobility. This is a medium- or long-term scenario hypothesising full electrification of the mobility system with an additional electricity demand estimated as 1.4 GWh.

8.  Wind turbines. This hypothesises the installation of 12 wind turbines of 1 MW each on the planned golf course, which would support the electricity demand of the neighbourhood once the transition to electric mobility has been achieved.

9.  Avoid waste to landfill. Increase the fraction of differentiated waste (most of it is just landfilled) by redirecting it to composting (40% organic waste), incineration (10%) and recycling (up to 30%).

10. Carbon uptake by urban forestry. The final assessment reveals that urban (re)forestation of 67 hectares would compensate the residual emission mostly due to waste and water management that cannot be avoided anyhow.

**Table 4.** Carbon footprint mitigation measures (long-term scenario) from the state of the art to the carbon-neutral neighbourhood (Riccardo Pulselli, University of Siena).

| | Measure | Electricity | Cooling | Heating | Mobility | Waste | Carbon Uptake | CF | Notes |
|---|---|---|---|---|---|---|---|---|---|
| | | | | | | | | t CO$_2$eq/a | |
| | Single household | | | | | | | 5.706 | Based on the typical household profiling |
| | | MWh/a | MWh/a | MWh/a | 10$^3$ km/a | t/a | ha | t CO$_2$eq/a | |
| | State of the Art Gruž | | | | | | | 16,546 | Referring to 8000 inhabitants and 2900 households |
| 1 | Building greenery + shading | −563 | −563 | | | | | −192 | Estimated households involved: 70%; estimated cooling energy saving: −15% |
| 2 | Heat pumps | −2424 | −858 | −1566 | | | | −826 | Estimated households involved: 80%; cooling energy saving: −20%; heating energy saving: −75% (CoP = 4) |
| 3 | Building envelope insulation | −1588 | −805 | −783 | | | | −541 | Estimated households involved: 50%; cooling energy saving: −30%; heating energy saving: −60% |
| 4 | Solar PV + solar thermal | −4300 | | | | | | −1466 | Based on the real estimate of solar energy potential in Gruž (available surface of roofs and well-oriented walls) |
| 5 | Mini wind turbines | −350 | | | | | | −111 | Based on the real estimate of wind energy potential in the neighbourhood (very small, for 50 mini-wind turbines) |
| 6 | Public transport increase | | | | −11,484 | | | −1936 | Estimated 60% avoided private car use |
| 7 | Transition to electric mobility | 1421 | | | −7656 | | | −806 | Estimated 40% fully electrified private cars or bikes with an increase of electricity demand of 1.4 GWh/a |
| 8 | Wind turbine (up to 7 MWh) | −14,948 | | | | | | −5095 | Estimated 12 towers with 1 MW turbine each |

**Table 4.** *Cont.*

| | Measure | Electricity | Cooling | Heating | Mobility | Waste | Carbon Uptake | CF | Notes |
|---|---|---|---|---|---|---|---|---|---|
| 9 | Avoid waste to landfill | | | | | −3880 | | −4499 | Estimated differentiated fractions: waste-to-energy: 20%; organic waste-to-compost: 40%; recycling: 30%; landfill 10% |
| 10 | Carbon uptake by urban forestry | | | | | | 67 | −905 | Estimated area needed to compensate the remaining emission (mainly due to waste and water management) |

Realisation of the proposed measures would ensure that near all of Gruž's greenhouse gas emissions incurred by energy consumption and other sources would be reduced to zero. Of the 1200 hectares of forestland initially required, only 67 remained, and these relate to the emission from the waste-to-energy plant, landfills and water resource use. This emission cannot be completely avoided but could be potentially compensated by the reforestation of 36 hectares of disused land or brownfields. Integrated measures for energy saving and renewable energy generation in buildings (e.g., greenery and shading systems, envelope insulation, PV panels and micro-wind turbines, etc.) as well as new installations (e.g., a wind farm) would increase the differentiated fraction of domestic waste addressed by an integrated waste management system (including recycling, a waste-to-energy plant, organic waste composting and just a little fraction sent to landfill). Enhanced public transport (to replace at least 60% of private cars in the short term) and the provision of a new mobility system, based on electric public and private vehicles in the medium term, are among the possible solutions to achieve the goal of carbon neutrality.

## 8. Conclusions, Discussion and Outlook

The City-zen Roadshow Dubrovnik demonstrated that through a meaningful merger of co-creative inputs from global experts and local stakeholders that cities can become carbon neutral and energy neutral. The measures proposed for Gruž are of course theoretical and not presently in operation. The zero-energy challenge must now move to the realisation phase in which the lead is taken by the City to formulate an action plan towards a desired future, for which they can take inspiration from the City Vision as presented at the City-zen Roadshow.

It should be mentioned that during the Roadshow more technical, spatial and management proposals were made than could be discussed and elaborated in this paper. The examples given are the most important carbon measures, a timely illustration of the science- and information-based creativity that the City-zen Roadshow entails, without proposing unrealistic solutions.

Although efforts were made by the Roadshow coordinator to invite a wide representation of Dubrovnik stakeholders and although many of these attended the workshop on one or more occasions, it is unknown to us if this set of stakeholders is sufficient to represent the city or district of Gruž, and if the measures proposed will work effectively. It is possible that participants who showed interest are biased while others with influence, or citizens themselves, may not understand the importance of a sustainable transition and measures of carbon emission reduction and renewable energy production.

The endorsement of the mayor of Dubrovnik and active involvement of the spatial planning department however, gives some confidence that the findings and proposals made for Gruž will be processed beyond the Roadshow week. Moreover, the Roadshow revealed pressing issues regarding threats to the health of citizens and to the environment, which directly affect the local population. Turning these problems into potentials for sustainable development and economic prospects, as proposed by the Roadshow, clearly appealed to the local community. Related or not, shortly after the Roadshow the City of Dubrovnik announced a limit to the number of cruise ships mooring in Gruž. The coming years will indicate whether effective change will occur, but the signs are good.

As stated earlier, the Roadshow is not intended to be a substitute for socially-engaged urban planning, and indeed, many of the participants in the cities have roles, that engage directly in this way. What the Roadshow offers them is the chance to play and imagine solutions with experts, that empower them to be bolder in the offerings they present to stakeholders.

Therefore, in addition to the design outcomes and the calculated energy and carbon performance (though vital to the EU and the rest of the world), perhaps the greatest value of the City-zen Roadshow is the collaboration and co-creation it establishes between people with different expertise and between academics, students, experts and local stakeholders [32]. The parties involved identified and acknowledged this as something making the dissemination of knowledge and strategies to the cities and their citizens much more effective than earlier methods, such as conferences, lectures, meetings, guidelines and reports.

Before Dubrovnik, the City-zen Roadshow visited Belfast (winter 2016) and Izmir (spring 2016); both Roadshows were successful in that they outlined the scale of the energy challenge, the timeline needed to meet it and the location-specific design interventions that could facilitate it both socially and spatially. Lessons were learned from earlier versions of the Roadshow in Florence and Sarajevo. Since Dubrovnik, Menorca was visited in spring 2017, followed by Seville (fall 2017) and Roeselare (spring 2018). The Roadshow has recently been described by its overseeing European Policy Commissioner as "remarkable" for its method and societal contribution. The Roadshow has now been recognised as an example of an EU 'Best Practice'. The City-zen Roadshows will continue until the end of 2019.

**Acknowledgments:** City-zen has received funding from the European Union's Seventh Programme for research, technological development and demonstration under grant agreement No 608702. The Dubrovnik Roadshow would not have been possible without the commitment and energy of two individuals, Andrea Novaković (Director, City of Dubrovnik Development Agency DURA) and Goran Krajacic (Assistant Professor, University of Zagreb). Both Andrea's and Goran's unwavering support during the preparations for the SWAT Studio and later Roadshow were pivotal in their joint success. The authors would also like to thank Andrea's colleagues at DURA: Marko Cosmai, Ana Marija Pilato and Tomislav Matković, all three major factors in the success of the event. We wish them every success in taking the zero-energy outcomes of the Dubrovnik (Gruž) Roadshow to the next level of realisation. The DURA staff worked tirelessly to ensure that the objectives of the Roadshow were met. A special mention must also go to Viktorija Dobravec (PhD candidate, University of Zagreb) whose daily support during the 2-week intensive SWAT Studio was highly appreciated by both TU Delft staff and students. Finally, a mention for Siir Kilkis (Scientific and Technological Research Council of Turkey) who originally nominated the City of Dubrovnik to be a Roadshow city.

**Author Contributions:** Andy van den Dobbelsteen, Craig Lee Martin, Greg Keeffe and Han Vandevyvere conceived and designed the strategies and solutions; Riccardo Maria Pulselli analysed the data; all authors wrote the paper.

**Conflicts of Interest:** The authors declare no conflict of interest.

## References

1. EU. Horizon 2020 Programme Secure Clean and Efficient Energy. Available online: http://ec.europa.eu/programmes/horizon2020/en/h2020-section/secure-clean-and-efficient-energy (accessed on 4 March 2018).
2. EU. Energy Strategy and Energy Union. Available online: http://ec.europa.eu/energy/en/topics/energy-strategy-and-energy-union (accessed on 4 March 2018).
3. City-zen, EU FP7 Project. Available online: http://www.cityzen-smartcity.eu (accessed on 4 March 2018).
4. Deal, B.; Pan, H.; Pallathucheril, V.; Fulton, G. Urban Resilience and Planning Support Systems: The Need for Sentience. *J. Urban Technol.* **2017**, *24*, 29–45. [CrossRef]
5. Deal, B.; Pan, H. Discerning and Addressing Environmental Failures in Policy Scenarios Using Planning Support System (PSS) Technologies. *Sustainability* **2017**, *9*, 13. [CrossRef]
6. Klosterman, R.E. Arguments for and against planning. *Town Plan. Rev.* **1985**, *56*, 5–20. [CrossRef]
7. Dreborg, K.H. Essence of Backcasting. *Futures* **1996**, *28*, 813–828. [CrossRef]
8. Robinson, J.B. Energy Backcasting: A Proposed Method of Policy Analysis. *Energy Policy* **1982**, *10*, 337–345. [CrossRef]
9. Zhivov, A.M.; Case, M.; Liesen, R.; Kimman, J.; Broers, W. Energy Master Planning Towards Net-Zero Energy Communities/Campuses. *ASHRAE Trans.* **2014**, *120*, 114–129.

10. Broersma, S.; Fremouw, M.; van den Dobbelsteen, A. Energy Potential Mapping—Visualising Energy Characteristics for the Exergetic Optimisation of the Built Environment. *Entropy* **2013**, *2*, 490–510. [CrossRef]

11. Dubrovnik Tourist Board (TZ Dubrovnik), Record-Breaking Tourism Results: Dubrovnik Is a Living and Eventful City with a Quality Tourism Product. Available online: http://www.tzdubrovnik.hr/ (accessed on 5 April 2018).

12. Peričić, D.; Puh, B. Attitudes of Citizen of Dubrovnik towards the Impact of Cruise Tourism on Dubrovnik. *Tour. Hosp. Manag.* **2012**, *18*, 213–228.

13. World Resources Institute (WRI). GHG Protocol Summary—Global Protocol for Community-Scale Greenhouse Gas Emissions Inventories, Executive Summary 12. 2014. Available online: https://www.wri.org/sites/default/files/global_protocol_for_community_scale_greenhouse_gas_emissions_inventory_executive_summary.pdf (accessed on 4 April 2018).

14. Lin, J.; Liu, Y.; Meng, F.; Cui, S.; Xu, L. Using hybrid method to evaluate carbon footprint of Xiamen City, China. *Energy Policy* **2013**, *58*, 220–227. [CrossRef]

15. Marchi, M.; Pulselli, F.M.; Mangiavacchi, S.; Menghetti, F.; Marchettini, N.; Bastianoni, S. The greenhouse gas inventory as a tool for planning integrated waste management systems: A case study in central Italy. *J. Clean. Prod.* **2017**, *142*, 351–359. [CrossRef]

16. Marchi, M.; Niccolucci, N.; Pulselli, R.M.; Marchettini, N. Environmental policies for GHG emissions reduction and energy transition in the medieval historic centre of Siena (Italy): The role of solar energy. *J. Clean. Prod.* **2018**, *185*, 829–840. [CrossRef]

17. Bastianoni, S.; Marchi, M.; Caro, D.; Casprini, P.; Pulselli, F.M. The connection between 2006 IPCC GHG inventory methodology and ISO 14064-1 certification standard e a reference point for the environmental policies at sub-national scale. *Environ. Sci. Policy* **2014**, *44*, 97–107. [CrossRef]

18. IPCC. *IPCC Guideline for National Greenhouse Gas Inventories*; Eggleston, H.S., Buendia, L., Miwa, K., Ngara, T., Tanabe, K., Eds.; IGES: Tsukuba, Japan, 2006.

19. Sovacool, B. Valuating the greenhouse emissions from nuclear power: A critical survey. *Energy Policy* **2008**, *36*, 2950–2963. [CrossRef]

20. Dubrovnik Climate Maps. Available online: www.dubrovnik.climatemps.com (accessed on 4 March 2018).

21. Tillie, N.; van den Dobbelsteen, A.; Doepel, D.; de Jager, W.; Joubert, M.; Mayenburg, D. Towards $CO_2$ Neutral Urban Planning—Introducing the Rotterdam Energy Approach and Planning (REAP). *J. Green Build.* **2009**, *4*, 103–112. [CrossRef]

22. Van den Dobbelsteen, A.; Keeffe, G.; Tillie, N. Cities Ready for Energy Crises—Building Urban Energy Resilience. In Proceedings of the SASBE2012, Sao Paulo, Brazil, 27–29 June 2012.

23. Van den Dobbelsteen, A.; Tillie, N.; Kurschner, J.; Mantel, B.; Hakfoort, L. The Amsterdam Guide to Energetic Urban Planning. In Proceedings of the MISBE2011, Amsterdam, The Netherlands, 19–23 June 2011.

24. Van den Dobbelsteen, A. Towards closed cycles—New strategy steps inspired by the Cradle to Cradle approach. In Proceedings of the PLEA2008—25th Conference on Passive and Low Energy Architecture, Dublin, Ireland, 22–24 October 2008.

25. Van den Dobbelsteen, A.; Tillie, N.; Broersma, S.; Fremouw, M. The Energy Master Plan: Transition to self-sufficient city regions by means of an approach to local energy potentials. In Proceedings of the PLEA2014, Ahmedabad, India, 16–18 December 2014.

26. Van Wirdum, G. *Ecoterminologie en Grondwaterregime, Mededeling Werkgemeenschap Landschapsoecologisch Onderzoek 6*; WLO: Utrecht, The Netherlands, 1979; pp. 19–24.

27. Tjallingii, S. *Ecologisch Verantwoorde Stedelijke Ontwikkeling*; IBN-DLO Rapport nr 706; Universiteit Wageningen: Wageningen, The Netherlands, 1992.

28. Tjallingii, S. *Ecopolis—Strategies for Ecologically Sound Urban Development*; Backhuys Publishers: Kerkwerve, The Netherlands, 1995.

29. Keeffe, G. Synergetic City: Urban algae production as a regenerative tool for a post-industrial city. In Proceedings of the IFHP 09, Berlin, Germany, 7–9 May 2009.

30. Golf Course Superintendents Association of America (GCSAA). Golf Course Environmental Profile—Growing Awareness about the Environmental Impact of Golf Courses; Phase 2 Energy Use Survey Results Now. 2017. Available online: https://www.gcsaa.org/environment/golf-course-environmental-profile (accessed on 5 April 2018).

*Energies* **2018**, *11*, 922

31.  De Jong, T.M.; van der Voordt, D.J.M. (Eds.) *Ways to Study and Research—Urban, Architectural and Technological Design*; Delft University Press: Delft, The Netherlands, 2002.
32.  Martin, C.L.; van den Dobbelsteen, A.; Keeffe, G. The Societal Impact Methodology—Connecting Citizens, Sustainability Awareness, Technological Interventions & Co-creative City Visions. In Proceedings of the International Conference on Passive and Low Energy, Edinburgh, UK, 3–5 July 2017; Volume II, pp. 2791–2798.

*Article*

# Comparison of the Energy Conversion Efficiency of a Solar Chimney and a Solar PV-Powered Fan for Ventilation Applications

Lubomír Klimeš [1,*] , Pavel Charvát [2] and Jiří Hejčík [2]

1   Sustainable Process Integration Laboratory—SPIL, NETME Centre, Brno University of Technology, Brno 61669, Czech Republic
2   Energy Institute, Faculty of Mechanical Engineering, Brno University of Technology, Brno 61669, Czech  Republic; charvat@fme.vutbr.cz (P.C.); hejcik@fme.vutbr.cz (J.H.)
*   Correspondence: klimes@fme.vutbr.cz; Tel.: +420-54114-3241

Received: 16 February 2018; Accepted: 8 April 2018; Published: 12 April 2018

**Abstract:** A study into the performance of a solar chimney and a solar photovoltaic (PV)-powered fan for ventilation applications was carried out using numerical simulations. The performance of the solar chimney was compared with that of a direct current (DC) fan powered by a solar PV panel. The comparison was carried out using the same area of the irradiated surface—the area of the solar absorber plate in the case of the solar chimney and the area of the solar panel in the case of the photovoltaic-powered fan. The two studied cases were compared under various solar radiation intensities of incident solar radiation. The results indicate that the PV-powered fans significantly outperform solar chimneys in terms of converting solar energy into the kinetic energy of air motion. Moreover, ventilation with PV-powered fans offers more flexibility in the arrangement of the ventilation system and also better control of the air flow rates in the case of battery storage.

**Keywords:** building ventilation; solar energy; solar chimney; solar photovoltaics; DC fan; energy conversion

---

## 1. Introduction

The stack effect induced by solar radiation can be employed for various purposes, including building ventilation. Interest in solar-driven ventilation in general and solar chimneys in particular has risen in the last two decades as utilization of renewable energy sources has become one of the main approaches to reducing the carbon footprint. A solar chimney (sometimes also called a solar stack or a solar updraft tower) is a device that converts the thermal energy of solar radiation into the kinetic energy of air motion. The interest in the utilization of solar chimneys has significantly increased over the last three decades, as can be seen from the number of publications on this topic.

Another area of solar energy utilization that has experienced significant development in the last several decades is solar photovoltaics. The decreasing prices of photovoltaic (PV) panels and developments in the area of direct current (DC) motors have brought about an alternative option for solar ventilation—fans powered by solar photovoltaics. Similar to solar chimneys, PV-powered fans do not require access to the power grid. The aim of the theoretical study presented in this paper was to compare the performance of a solar chimney with the performance of a DC fan powered by solar photovoltaics. A research question to be answered in this study can be formulated as follows: From an energy conversion point of view, when the area of the solar absorber plate of the chimney is the same as the area of the PV panel, is it more efficient to use the solar chimney or a DC fan powered by a PV panel? Computer models of both systems were implemented in MATLAB in order to perform simulations and make a comparison.

## 1.1. Solar Chimneys

The conversion of solar heat into air motion by means of solar chimneys can be used not only for building ventilation but also for power generation. One analysis of solar chimney efficiency was presented by Mullett et al. [1]. The author considered a solar chimney for power generation consisting of a horizontal solar collector and a vertical tower (chimney) with a uniform circular cross-section. The author used a simplified approach to address the overall efficiency of the system. In terms of the overall efficiency, the efficiency of conversion of solar radiation into thermal energy, the efficiency of conversion of thermal energy into air motion, and the efficiency of conversion of air motion into the shaft power of the turbine were taken into account. The overall efficiency was less than 1 percent for most of the investigated cases.

Suárez-López et al. [2] built a comprehensive three-dimensional (3D) CFD model of a solar chimney, validated by means of experimental data from the literature. The model was used for the investigation of the conditions in the solar chimney (such as the fluid flow patterns and the temperature distribution), and for the identification of parameters that directly influence the exergetic efficiency. The authors reported that their numerical results indicated a small efficiency in case of solar chimneys used for natural ventilation systems. The thermal exergetic efficiency for the studied case was 0.55%, and the useful exergetic efficiency was 0.0006%. The improved performance of the chimney was shown by the minimization of fluid dynamic losses, optimization of the channel geometry, and by the reduction of losses through the glass cover. Naraghi and Blanchard [3] developed a mathematical model of a solar chimney for the dynamic analysis. The model was based on the implicit finite differences applied to the energy balance equations for components of the solar chimneys. The Newton–Raphson numerical method was used for the solution of the discretized equations in time and a clear sky model was utilized for the determination of solar irradiance. A parametric study into the performance of a solar chimney was carried out. The authors reported that solar chimneys with a relatively large thermal mass can provide a reasonable airflow during the night and early morning with no solar irradiance. A high thermal mass of the absorbing plate also helped to reduce fluctuations of the airflow rate.

A numerical model of an inclined solar chimney was presented by Imran et al. [4]. The model solved two-dimensional (2D) steady-state turbulent flow induced by natural convection, and its functionality was validated with experimental data. The model was implemented in FORTRAN with the use of the finite volume method. The authors reported that the optimum inclination angle was 60°, providing an about 20% higher rate of ventilation. The results indicated that the flow rate between 50 $m^3/h$ and 425 $m^3/h$ could be achieved for solar irradiation between 150 $W/m^2$ and 750 $W/m^2$, corresponding to between 4 and 35 air changes per hour for a room with a volume of 12 $m^3$. An analytical iterative model of the airflow in solar chimneys based on the thermal boundary layer was reported by He et al. [5]. The model takes into account the spatial variation of the density which is ignored in most solar chimney models. The model is based on energy balance equations and the theory of thermal boundary layers. The author stated that their model outperformed existing analytical models in terms of the predicted airflow rates, and it allowed for the identification of the optimal chimney gap that would provide maximum air flow rate.

Sudprasert et al. [6] numerically investigated the influence of moist air on the solar chimney performance. An ANSYS Fluent model of heat transfer and fluid flow was assembled for the investigation of the chimney operation with dry and moist air (humidity between 30–80%). The authors reported that the airflow rate with the moist air was about 15–25% lower and the air temperature was higher than in case of the dry air. The authors recommended that the optimal aspect ratio should be about 14:1, with a limited opening height for the maximization of ventilation. An empirical model for solar chimneys assembled on the basis of experimental data acquired from the literature for various test rigs and configurations of solar chimneys was presented by Shi et al. [7]. The relationship between the airflow rate, cavity height and width, and solar radiation was derived. The model was then used in a parametric analysis addressing the various parameters of the solar chimneys. Saleem et al. [8]

developed a mathematical model for the solar chimney which was based on the overall energy balances. The model was used for the determination of optimum parameters for a specified set of conditions. Further, the authors assembled a CFD model of the solar chimney which included the solution of the k-omega turbulence model for the fluid flow applied to the solution of the mass and energy equations. A good agreement between the simple model and the CFD model was reported which, in the authors' opinion, justified the use of the simple model in optimization processes.

### 1.2. Solar-Powered Electric Motors

The application of solar photovoltaics in ventilation is not as widely addressed in the literature as the application of solar chimneys. Nonetheless, several authors have addressed the connection between PV cells and the electric motors. Badescu [9] developed a mathematical model for the system, including photovoltaic cells, an electric battery, an electric motor, and a water pump. The system was proposed for water storage instead of direct storage of electrical energy. The model of the system, developed in FORTRAN, consisted of a set of ordinary differential equations with the time derivative, and it allowed for the time-dependent simulations under various conditions. A submodel for the meteorological and actinometric data was implemented. The author presented a parametric analysis carried out with the use of the model for two modes of PV cell operation. At sufficient solar irradiance, the PV cells supplied electricity to both the battery and the motor. At insufficient solar irradiance the motor was powered from the battery.

Atlam and Kolhe [10] developed a mathematical model of the solar PV-powered direct current permanent magnet motor–propeller system. They used the model for the selection of a proper type of the motor and its parameters. The model developed using MATLAB/Simulink consisted of a set of algebraic equations and it allowed for steady-state simulations. The authors observed that a properly selected motor had operating points that closely matched the maximum power points of the PV array. The authors pointed out that the motor and its parameters had to be selected specifically according to a particular application. Gupta et al. [11] investigated a stand-alone system of the PV array and batteries powering a refrigerator. The authors assembled a computer model in the TRNSYS simulation tool and they validated the model with experimental data. Standard TRNSYS components were utilized for the PV panel, battery, inverter, and refrigerator. The model was then used in a parametric study with the aim to properly design the optimal parameters of the PV cells, batteries and of the insulation level of the refrigerator. The authors concluded that a refrigerator with an insulation layer of 50 mm thickness can feasibly be operated by means of a 200 W photovoltaic array coupled with a battery with a capacity of 50 Ah. The authors reported that such a configuration enables the stand-alone operation of the refrigerator over two cloudy days.

### 1.3. Scope of the Study

Solar chimneys are often proposed as an energy efficient way of building ventilation (see e.g., [12–14]). However, studies providing an analysis of energy performance of a solar chimney and its comparison to other ways of solar energy utilization in building ventilation are rarely found. With the decreasing prices of photovoltaic solar panels and advancements in the design of direct current motors for ventilation fans, there appears to be a more efficient, more flexible, and probably also cheaper option for building ventilation in comparison to solar chimneys.

The present study deals with the comparison of the energy conversion efficiency of a solar chimney and a PV-powered fan. In case of the solar chimney, a theoretical case, non-achievable in real life, was considered. The models of the PV panel and the DC fan were based on the data provided by manufacturers. The accuracy of the model in this case was assessed by comparison of the results with the data from the fan manufacturer. Though the conducted study is rather simple, the comparison of a solar chimney to a ventilation fan powered by solar photovoltaics (as far as we know) has not been published yet.

The main focus of the study was to assess how effective a solar chimney was in building ventilation in comparison to a fan powered by a solar PV panel. The solar chimneys are typically used for air exhaust and thus the thermal energy contained in the air is not utilized in the building. The solar chimneys basically substitute air exhaust fans as the drivers for air movement. The energy efficiency was therefore expressed in terms of conversion of the incident solar radiation to the kinetic energy of air motion.

## 2. Solar Chimney Scenario

There are many possible designs of solar chimneys. The solar chimneys intended for power generation usually consist of a vertical tower (tube) and a horizontal solar collector. The power-generating turbines are usually located at the base of the tower. Thermal storage in the ground under the solar collector or other forms of thermal storage mass can extend the operation of the solar chimney power plants to night-time hours. The solar chimneys used for ventilation are usually rectangular cavities with the glass cover on one side of the cavity and the solar absorber plate on the other side. The chimney can either be vertical or inclined (for better irradiation). This design of the solar chimney, which is very similar to to that of a solar collector, was considered in the present study.

The calculation of the airflow rate through the solar chimney can be rather complex depending on the level of details that are taken into account. Solar radiation-induced buoyancy is not the only driving force for airflow in solar chimneys. The effect of wind in this respect can be significant and under certain circumstance it can become a dominant driving force. Also, in the case of solar chimneys installed in buildings, the buoyancy force due to the difference between the indoor and outdoor air temperature can very much influence the air flow rate through the solar chimney. Neither the effect of wind nor the influence of the temperature difference between the indoors and outdoors was considered in the present study. These influences are also present when a fan powered by photovoltaics is used. The aim of the study was to compare solar-driven air flow in the case of two forms of exploitation of solar energy and not to investigate the complex performance of a solar chimney or a PV-powered fan for a particular installation.

## 3. Computer Model of Solar Chimney

### 3.1. Basic Considerations and Governing Equations

A simple iterative model for a solar chimney was derived and implemented in MATLAB. A chimney in the form of the rectangular glass-covered cavity, as shown in Figure 1, was considered. An idealized theoretical case was considered in the study. Only the optical loss (transmittance and absorptance) was taken into account in case of the solar chimney. The heat loss of the solar chimney to the ambient environment was neglected. The friction and local losses for the air flow in the solar chimney were also neglected. It would be relatively easy to account for transmission and radiation heat loss in the solar chimney model, but the heat loss depends on the actual materials used in the design of the chimney. As the aim of the study was to compare the principles of solar energy utilization for ventilation rather than comparing the actual designs, the neglect of the heat loss seemed justifiable.

The solar radiation incident on the solar chimney warms up the air in the chimney cavity and the induced buoyancy force causes the air movement in the cavity. Solar heat (thermal energy) is thus converted into air motion (kinetic energy of air). It was assumed that the solar radiation absorbed by the solar absorber plate was entirely transferred to the air in the cavity of the solar chimney. The energy balance of the air passing through the solar chimney is then

$$I_g S_g \tau \alpha = \dot{m}_{\mathrm{sc}} c_p (T_{\mathrm{out}} - T_{\mathrm{in}}) = S_A \varrho_{\mathrm{sc}} w_{\mathrm{sc}} c_p (T_{\mathrm{out}} - T_{\mathrm{in}}) \tag{1}$$

where $I_g$ is the intensity of solar radiation incident on the glazing of the chimney, $S_g$ is the area of the glazing, $\tau$ is the transmissivity of the glazing, $\alpha$ is the absorptivity of the solar absorber plate, $\dot{m}_{\mathrm{sc}}$ is the mass flow rate through the chimney, $c_p$ is the specific heat of the air at a constant pressure, $T_{in}$ and $T_{out}$ are the inlet and outlet air temperatures, respectively, $S_A$ is the the cross-section area of the chimney, and $\varrho_{sc}$ and $w_{sc}$ are the mean density and velocity of the air in the chimney, respectively. The stack effect in the solar chimney can be expressed as

$$gh(\varrho_a - \varrho_{sc}) = \left(1 + \frac{fh}{D} + \sum_i \check{\zeta}_i\right)\frac{1}{2}\varrho_a w_a^2 = K\frac{1}{2}\varrho_a w_a^2 \tag{2}$$

where $g$ is the standard gravity, $h$ is the height of the chimney, $\varrho_a$ is the density of the ambient air, $K$ is overall resistance coefficient, $f$ is the friction factor, $D$ is the equivalent diameter, $\check{\zeta}_i$ represents local loss coefficients, and $w_a$ is the inlet air velocity to the chimney. The case with $K = 1$ corresponds to the situation without both the local losses and the friction loss. Since the mass flow rate of the air depended on the buoyancy force and the buoyancy force induced by the difference of densities $\varrho_{sc}$ and $\varrho_a$ is dependent on the mass flow rate, the solution of the model for the solar chimney is performed iteratively in a loop until a balance between the two mentioned terms is achieved. The values of the parameters and inputs to the computer model of the solar chimney are summarized in Table 1.

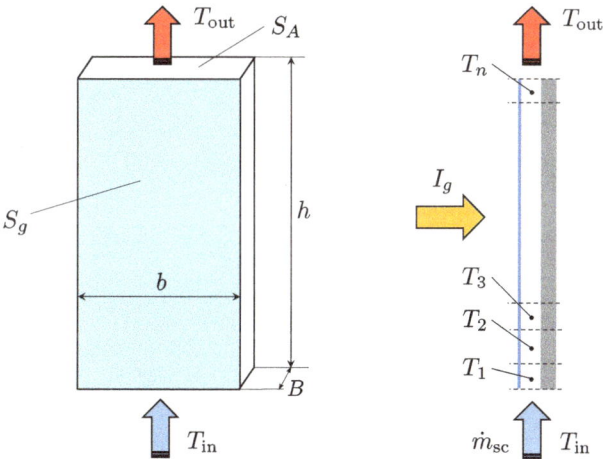

**Figure 1.** Schematic of the solar chimney (**left**) and the numerical discretization (**right**).

**Table 1.** Parameters and inputs to the model of the solar chimney.

| Parameter | Value | Description |
|---|---|---|
| $\alpha$ | 0.95 | absorptivity of the solar absorber plate |
| $B$ | 0.2 m | depth of the chimney cavity |
| $c_p$ | 1005 J/kg·K | specific heat of air at constant pressure |
| $\varrho_a$ | 1.19 kg/m³ | density of ambient (inlet) air |
| $S_g$ | 1.3 m² | area of glazing |
| $T_{in}$ | 20 °C | inlet air temperature |
| $\tau$ | 0.91 | transmissivity of glazing |

*3.2. Numerical Discretization*

A one-dimensional (1D) discretization of the solar chimney was considered. A schematic of the discretization is shown in Figure 1. The solar chimney was vertically divided into $n$ segments in the direction of the air flow. The heat flux $\dot{Q}_i$ transferred from the solar radiation into the air is

$$\dot{Q}_i = I_g S_{g,i} \tau \alpha \tag{3}$$

where $S_{g,i}$ is the area of glazing in the $i$-th section. The mass flow rate of the air through the chimney can be determined from the parameters of the air at the inlet as

$$\dot{m}_{sc} = S_A \varrho_a w_a. \tag{4}$$

The temperature of the air in the $(i+1)$-th section is then given as

$$T_{i+1} = T_i + \frac{\dot{Q}_i}{\dot{m}_{sc} c_p} \tag{5}$$

where $T_i$ is the temperature of the air in the $i$-th section. The mean temperature of the air in the solar chimney is the mean of the temperatures $T_i$. The inlet air velocity to the chimney is determined as

$$w_a = \sqrt{\frac{gh(\varrho_a - \varrho_{sc})}{K_\frac{1}{2}\varrho_{sc}}} \tag{6}$$

and the pressure difference in the solar chimney induced by the buoyancy was calculated as

$$\Delta p = gh(\varrho_a - \varrho_{sc}) - \frac{1}{2}\varrho_a w_a^2. \tag{7}$$

**4. Scenario with the Fan Powered by Solar PV**

The simulated scenario with the direct current (DC) fan powered by solar PV was similar to the scenario with the solar chimney. A solar panel with the same area as the area of the solar chimney absorber plate was considered. A DC fan was assumed to be connected to a PV panel through a control unit without battery energy storage. The control unit of the fan was not modeled in detail. It was only assumed that the control unit would prevent damage of the fan when the solar panel provided more power than the fan could handle. The fan air flow rates were calculated for a range of incident solar radiation intensities with the use of the parameters of the solar panel and the characteristics of the DC fan. As mentioned earlier, the goal of the study was to compare two ways of converting solar energy into air motion rather than comparing a particular design of a solar chimney to a particular configuration of a PV-powered fan. Nonetheless, the parameters of both the solar panel and the DC fan were adopted from data sheets of real product in order to consider a realistic performance of these devices in the study.

*4.1. Solar Panel*

A solar panel with the area of $1.3\,\text{m}^2$ was considered in the study. The panel had the maximum power of 210 W, its maximum power current was 5.1 A, its maximum power voltage was 41.3 V, and the open circuit voltage was 50.9 V. The short circuit current of the solar panel was 5.57 A, the temperature coefficient of the open-circuit voltage in the standard test conditions (STCs) was $-0.127\,\text{V}/^\circ\text{C}$, and the short-circuit current in STC was $0.00167\,\text{A}/^\circ\text{C}$.

## 4.2. DC Fan

An axial DC fan with the operation voltage between 30 V and 57 V was considered in the study. The maximum speed of the fan was 5300 rpm, corresponding to a maximum power of 66 W, a maximum flow rate of 10 m$^3$/min, and a maximum pressure of 410 Pa. The *I-V* characteristic slope of 0.017 A/V and the $\omega$-*V* characteristic slope of 83.3 rpm/V were considered. The pressure difference-air flow rate ($\Delta p$-*Q*) curves from a data sheet were approximated with the use of exponential functions with the $R^2$-value of 0.9993. The $\Delta p$-*Q* relationships of the fan are shown in Figure 2 for various speeds of the fan. Similar to the solar chimney scenario, the pressure loss of the connected ventilation ductwork was not considered and thus the fan operated at the maximum flow rate for the supply voltage.

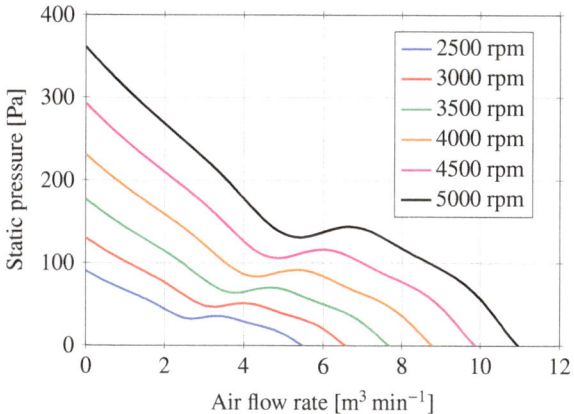

**Figure 2.** $\Delta p$-*Q* characteristics of the considered direct current (DC) fan for various speeds.

## 5. Computer Model of the PV Panel with the DC Fan

Two separate models for the PV panel and for the DC fan were implemented in MATLAB. The models were coupled together by means of the *I-V* characteristics of the PV panel and the *I-V* and $\omega$-*V* characteristics of the electric motor of the fan. This approach was already verified by other investigators (see e.g., the work by Odeh et al. [15]).

### 5.1. Model of PV Panels

Though some simple models of PV panels can be utilized, a more detailed model proposed by Sera et al. [16] was adopted in the simulations. The implemented model of the PV panel is based on an electric circuit of a photovoltaic cell using the single exponential model. The performance of the PV panel is given by its current–voltage (*I-V*) characteristics

$$I = I_{ph} - I_o \left( e^{\frac{q(v+IR_s)}{n_s A k T_{stc}}} - 1 \right) - \frac{V + IR_s}{R_{sh}} \qquad (8)$$

where $I_{ph}$ is the photo-generated current in STC, $I_o$ is the dark saturation current in STC, $R_s$ is the panel series resistance, $R_{sh}$ is the panel shunt resistance, $A$ is the ideality factor, $k$ is the Boltzmann constant, $q$ is the charge of the electron, $n_s$ is the number of cells in the panel connected in a series, and $T_{stc}$ is the temperature of the cell in STC. The *I-V* characteristics has to be calculated iteratively since Equation (8) is implicit for the electric current. The quantities $I_{ph}$ and $I_o$ are determined according to incident solar irradiation, cell temperature, and datasheet parameters of the PV panel. Since the paper is not concerned with the detailed description of the adopted model of the PV panel, the reader is referred to [16] for further details. A typical output of the model in the form of the current-voltage *I-V* (solid lines) and power-voltage *P-V* (dashed lines) characteristics is shown in Figure 3. Figure 4 demonstrates the dependence of the current-voltage ratio for various values of the cell temperatures of the PV panel. The results in Figures 3 and 4 were calculated for the PV panel defined in the foregoing section. The system consisting of such a PV panel coupled with a DC fan is considered below in the comparison between the solar chimney and the PV panel–fan system.

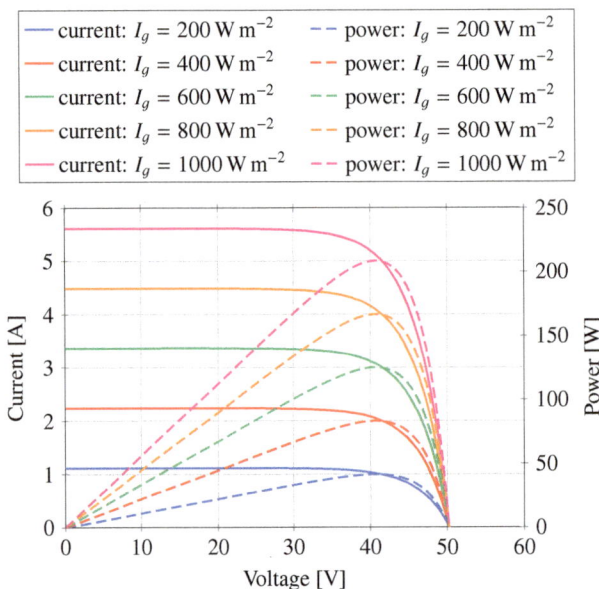

**Figure 3.** *I-V* (solid lines) and *P-V* (dashed lines) characteristics of the considered photovoltaic (PV) panel for various solar irradiations and for a cell temperature of 30 °C.

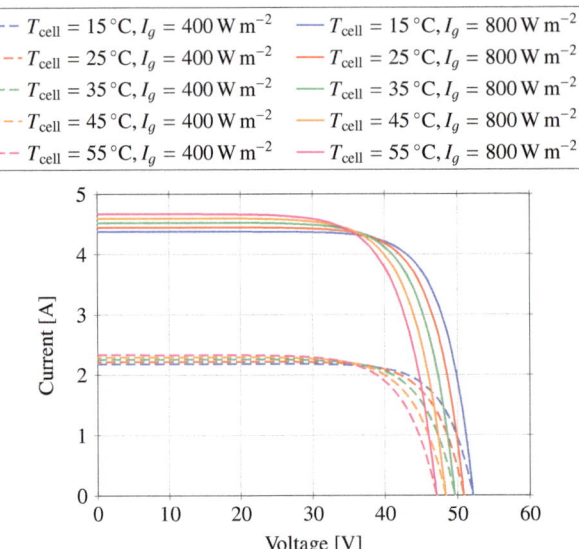

**Figure 4.** *I-V* characteristics of the considered PV panel for various cell temperatures and for a solar irradiation of 400 W/m$^2$.

*5.2. Model of the DC Fan*

The current-voltage (*I-V*) and speed-voltage (*ω-V*) characteristics of the DC fan can be considered as linear and they can be used for the determination of the speed of the fan with the assumption that those are valid for a constant torque [17]. Such assumption was already used by other investigators (see e.g., [15]). The *I-V* characteristics of the PV panel and of the DC fan are therefore solved simultaneously and the intersection of the two *I-V* characteristics determines the actual current and voltage transferred from the PV panel to the fan. The speed of the fan is then determined from the *ω-V* characteristics. The performance of the fan is defined by its Δ*p*-*Q* relationship (see Figure 2). Such dependence, usually provided by the manufacturer, is given for a defined speed, air pressure, and temperature. The air flow rate *Q* for the actual speed of the fan *ω* determined from the *ω-V* characteristics for actual air pressure *p* and air temperature *T* was calculated by means of the affine relationship [15]

$$Q = Q_{ref} \left( \frac{\omega}{\omega_{ref}} \right) \left( \frac{p}{p_{ref}} \right) \left( \frac{T_{ref}}{T} \right) \tag{9}$$

where $Q_{ref}$, $\omega_{ref}$, $p_{ref}$, and $T_{ref}$ are the reference flow rate, the reference speed of the fan, the reference air pressure, and the reference air temperature, respectively.

## 6. Comparison of Performance and Discussion

The performance of the solar chimney and the PV-powered fan was compared in terms of the achieved air flow rates and energy conversion efficiencies. As shown before, the simplified model of the solar chimney neglected many factors that would decrease the air flow rate through the chimney in real life operation (heat loss, friction loss, constraints on heat transfer between the solar absorber plate and the passing air).

*6.1. Mass Flow Rates*

Figure 5 shows the simulation results for the air mass flow rate through the solar chimney for various heights of the chimney, denoted *h* in Figures 1 and 5. The width of the chimney, denoted *b*, was determined in such a way that the area of the chimney was the same as the area of the solar panel ($1.3\,m^2$). The depth of the chimney, denoted *B* in Figure 1, was constant and equal to 0.2 m in all cases. The transmissivity values of the glass and the absorptivity of the solar absorber plate were set to 0.91 and 0.95, respectively. As can be seen seen in Figure 5 the air mass flow rate decreases with the increasing height of the chimney. As the irradiated area of the solar chimney and the chimney gap remain the same, the cross-section area of the chimney cavity decreases with the increasing height of the chimney. As a result, higher velocity is needed for the same air flow rate.

Figure 6 presents the simulation results for the mass flow rate through the DC fan powered by the PV panel. The characteristics of the PV panel and the DC fan have been described in detail in the foregoing sections. The ideality factor of the PV panel was set to 1.5 and the series resistance of the cells in the PV panel of 0.004 Ω was considered. The results are shown for various temperatures of the PV panel ($T_{PV\,cell}$).

The comparison of the results depicted in Figures 5 and 6 indicates that the DC fan powered by a solar panel provides higher mass flow rates of air than the solar chimney with the same irradiated area. The air mass flow rate of the fan is almost constant for radiation intensity above $400\,W/m^2$. The reason for this is the power input of the fan. The fan has the maximum power input of 66 W while the solar panel provides almost 200 W at a solar radiation intensity of $1000\,W/m^2$ (as can be seen in Figure 3) and thus it would be able to power three fans at that level of irradiation. For the solar radiation intensity below $200\,W/m^2$ the air mass flow rate falls sharply as the solar panel does not provide enough power for the nominal performance of the fan.

The accuracy of the model of the DC fan was assessed by the comparison of the flow rates obtained from Equation (9) (Figure 6) and the data from the fan manufacturer shown in Figure 2. The density of air entering the fan was considered constant and equal to $1.19 \, \text{kg/m}^3$. The pressure drop of the connected ductwork was neglected and as a result $\frac{p}{p_{ref}} = 1$ in Equation (9). The fan speed $\omega$ depended on the current and voltage of the PV panel, as explained in Section 4.2. It was thus possible to compare the flow rate obtained from Equation (9) for a certain fan speed with the flow rate for the same fan speed from the data of the fan manufacturer. The difference between the flow rates was less than 1% in the entire range of the flow rates.

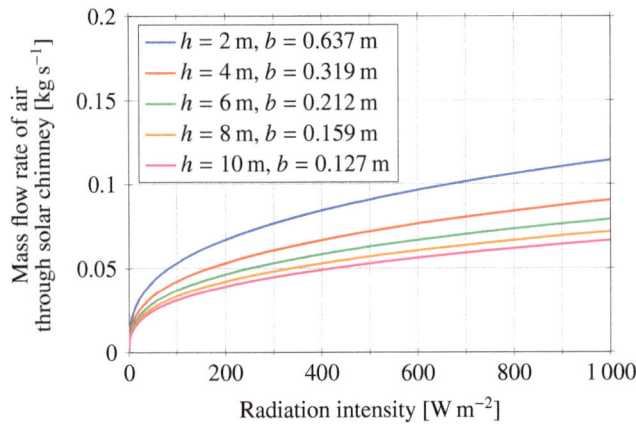

**Figure 5.** Dependence of the mass flow rate through the solar chimney for various heights.

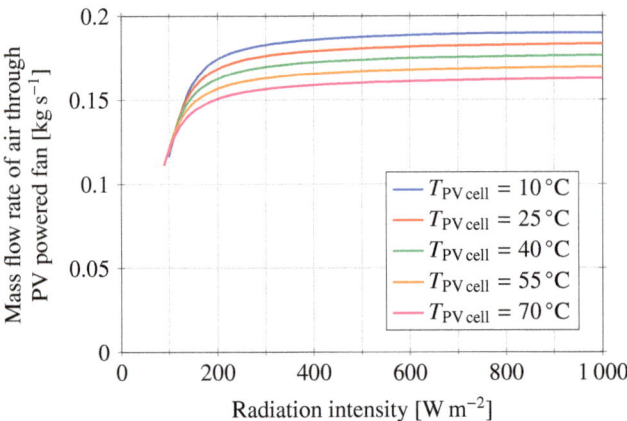

**Figure 6.** Mass flow rate of the PV-powered fan for various solar cell temperatures.

### 6.2. Energy Conversion Efficiency

Since the pressure drop of the connected ductwork was not considered in the study, the energy conversion efficiency can be expressed as the ratio between the kinetic energy of the moving air and the total solar radiation incident on the area of the solar chimney and the solar panel, respectively. The air mass flow rates were already presented in the previous section. The kinetic energy $E_k$ of the air flowing through the chimney and through the fan can be obtained as

$$E_k = \frac{1}{2} \dot{m} w^2 \tag{10}$$

where $\dot{m}$ and $w$ are the mass flow rate and the mean air velocity, respectively. The energy conversion efficiency $\eta$ can then be expressed as follows

$$\eta = \frac{\frac{1}{2}\dot{m}w^2}{I_g S} \tag{11}$$

where $I_g$ is the solar radiation incident on the area $S$ of the solar chimney or the PV panel.

Figure 7 shows the energy conversion efficiency for the solar chimney (denoted as SC) and the PV-powered fan (denoted as PV) depending on the radiation intensity. The energy conversion efficiency for several temperatures of the PV panel and several heights of the solar chimney are plotted in Figure 7. Note that the logarithmic scale for the energy conversion efficiency is used in Figure 7. As can be seen, the energy conversion efficiency of a solar chimney is very small for all the considered heights. The value of the efficiency ranges between 0.0025% and 0.013% for the chimneys with heights of 2 m and 10 m, respectively. This is in agreement with conclusions of other investigators (see e.g., [1]). Moreover, the energy conversion efficiency for the solar chimney is virtually independent of the radiation intensity. This is due to the neglected heat loss and friction losses.

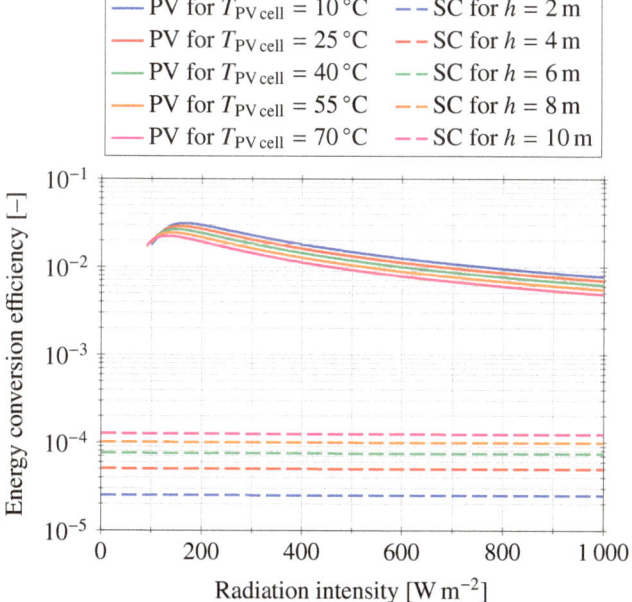

**Figure 7.** Energy conversion efficiency for the solar chimney and PV-powered fan.

As for the PV-powered fan, the conversion efficiency is about two orders of magnitude higher than in the case of the solar chimney. As expected, the efficiency decreases with the increasing cell temperature of the PV panel since its efficiency decreases as well. Considering the cell temperature of the PV panel of 40 °C, the total conversion efficiency peaks at approximately 2.7% for about 150 W/m² of the radiation intensity. As the radiation intensity increases, the conversion efficiency decreases to about 0.8% for the radiation intensity of 800 W/m². The reason for such behavior is clear in Figure 6. The air mass flow rate of the fan does not increase very much for a radiation intensity above 400 W/m². As a result, the energy conversion efficiency decreases with increasing radiation intensity, as can be deduced from Equation (11). The main reason for low energy conversion efficiency at high radiation intensities is the maximum power input of the DC fan considered in the study, which was about 66 W.

At a solar radiation intensity of $1000\,W/m^2$ the solar panel would be able to power three considered DC fans and both the air mass flow rate and overall conversion efficiency would increase significantly. The reason for choosing a DC fan with much smaller power than the nominal output (at $1000\,W/m^2$) of the photovoltaic solar panel was the considered application in building ventilation. The DC fan considered in the study is able to provide a high air mass flow rate at a relatively small radiation intensity (from about $100\,W/m^2$). The performance curves of a more powerful fan would shift towards higher radiation intensities with a significant operation gap for small solar radiation intensities.

### 6.3. Economic Considerations

Both the solar chimney and the PV-powered fan can operate without access to the power grid. The potential cost savings of these two ways of exploiting solar energy are twofold; capital cost savings and operating cost savings. In situations where solar chimneys or PV-powered fans are used instead of grid-connected fans, the potential capital cost savings are related to the installation of the power supply for the grid-connected fans. These costs can be significant in case of retrofits of existing passive stack ventilation systems when the performance of the passive stack ventilation is to be extended by the use of a grid-connected fan. The installation of power supply for the fan mounted on top of an existing passive stack usually exceeds the cost of the fan itself. The project of the power supply installation needs to be prepared (usually by an electrical engineer), the power supply with safety features needs to be installed by an electrician, and the installation of power supply needs to be commissioned by an authorized specialist. The costs associated with the installation of power supply for a fan can reach several hundred euros. On the other hand, a solar chimney and a PV-powered fan in particular can often be installed on top of an existing passive stack with minimum additional costs.

The operating cost savings mainly depend on the electricity prices. The average price of electricity in 28 member states of the EU in the first half of 2017 was 0.204 EUR/kWh in case of household use and 0.114 EUR/kWh in case of non-household use [18]. The considered 66 W fan operating at full capacity 5 h a day would consume 120 kWh/year of electricity. The potential operating cost savings are thus 25 EUR/year (14 EUR/year in case of non-household use) assuming that maintenance costs are the same in case of grid-connected fan, solar PV-powered fan, and the solar chimney. Both the solar chimney and the PV-powered fan are intended to assist passive stack ventilation during unfavorable conditions (high outdoor temperature, low wind speed). These conditions depend very much on the climatic region, but the year-average of 5 h a day seems to be a safe figure in the case of Central Europe. As was shown, the operating cost savings are relatively small. Therefore, the total capital cost will be the main factor when deciding between the grid-connected fan and the solar chimney or the solar PV-powered fan. The PV-powered fan can be an attractive option in countries with high electricity prices and high capital costs for installation of power supply (high wages of involved specialists).

As far as the capital costs of the solar chimney and the solar PV-powered fan are concerned, the conducted study indicates that at the same ventilation capacity the PV-powered fan can be a less expensive option. The acquisition cost of the considered fan was about 90 EUR and the cost of the solar PV panel was about 520 EUR. The capital costs of PV-powered fans can decrease when compact units containing the fan, the solar panel, and the control unit are used. There already are some manufacturers of such units and more suppliers will likely emerge with the increasing demand.

Since the overall efficiency of the considered PV-powered fan was about two orders of magnitude higher than that of the considered solar chimney, the area of the solar chimney would need to be much larger to provide the same air flow rates (induced by solar irradiation). Solar chimneys are not consumer products; therefore, they need to be custom built for a given application. That significantly increases per-unit cost in comparison to mass market products such as DC fans and solar PV panels. The materials for the construction of solar chimneys are generally inexpensive. The retail price of ordinary clear glass usable as a glass cover can be less than $20\,EUR/m^2$. However, toughened or even safety glass may be required or recommended for most solar chimneys. The clear safety glass can cost more than $50\,EUR/m^2$. The frame of the chimney can be welded from metal profiles. The sheet metal can be used for solar absorber plate and also for external surfaces of the chimney. Thermal insulation (i.e., styrofoam, mineral wool, or polyurethane) that is needed in most climates contributes to both better energy efficiency and higher capital costs. The most expensive part of the capital cost is the cost of labor (manufacturing) and in some cases also the transportation and installation costs. A crane may be needed to put the solar chimney on top of a building. Figure 8 shows an experimental solar chimney being installed on the roof of a demonstration house with hybrid ventilation on the campus of Brno University of Technology in November 2003. Unfortunately, in this case the solar chimney was the only part of the house installed by a crane, and the cost of renting a crane contributed significantly to the capital cost of the solar chimney.

**Figure 8.** Installation of an experimental solar chimney on the roof of a demonstration house with a hybrid ventilation system.

## 7. Conclusions

A performance comparison between a solar chimney and a DC fan powered by solar photovoltaics was undertaken using various solar radiation intensities. Solar radiation was the only source of energy for the motion of air considered in the study. The results shown in Figures 5 and 6 demonstrate that a PV-powered DC fan can provide up to three times higher air mass flow rates as compared to a solar chimney for the same solar incidence area (the area of solar absorber plate in the case of the solar chimney and the area of the PV panel in the case of the PV-powered fan). Many factors that would decrease the air mass flow rate through the solar chimney in real-life operation were not taken into account in the study. The results indicate that the fans powered by solar photovoltaics provide two orders of magnitude higher conversion efficiency of solar radiation to air motion as compared to solar chimneys. The difference in the efficiency will further increase as more efficient PV panels, DC motors, and fan impellers are developed.

PV-powered fans can be used for both air supply and air exhaust, while the use of solar chimneys is limited to air exhaust. Another advantage of a PV-powered fan is the much easier positioning of the solar panel in terms of the slope and azimuth in comparison to a solar chimney. Moreover, the electricity from the PV panels that is not immediately consumed by the fans can be stored in the batteries or it can be used for other purposes. A steady-state operation of the fan at constant irradiation was considered in the present study. Future work should focus on transient operation of the fan for the solar irradiation of a location and on the possibilities of flow rate control and electric storage.

**Acknowledgments:** This paper has been supported by the project Sustainable Process Integration Laboratory—SPIL, No. CZ.02.1.01/0.0/0.0/15_003/0000456, funded by European Research Development Fund,

Czech Republic Operational Programme Research, Development and Education; Priority 1: Strengthening capacity for quality research.

**Author Contributions:** P.C. and L.K. conceived and designed the computer model for the solar chimney. L.K. and J.H. conceived and designed the computer model for the PV-powered fan. L.K. performed the computer simulations. L.K., P.C. and J.H. assessed and evaluated the computational results. L.K. and P.C. wrote the paper.

**Conflicts of Interest:** The authors declare no conflict of interest.

## Nomenclature

| | |
|---|---|
| $A$ | ideality factor |
| $b$ | width |
| $B$ | depth |
| $c_p$ | specific heat at constant pressure |
| $D$ | equivalent diameter |
| $f$ | friction factor |
| $h$ | height |
| $g$ | standard gravity |
| $I$ | current |
| $I_g$ | incident radiation |
| $k$ | Boltzmann constant |
| $K$ | overall resistance coefficient |
| $\dot{m}$ | mass flow rate |
| $n_s$ | number of cells |
| $p$ | pressure |
| $q$ | charge of electron |
| $Q$ | volume flow rate |
| $\dot{Q}$ | heat flux |
| $r$ | gas constant |
| $R$ | electric resistance |
| $S_g$ | area of glazing |
| $S_A$ | cross-section area of chimney |
| $T$ | temperature |
| $V$ | voltage |
| $w$ | air velocity |
| $\alpha$ | absorptivity |
| $\xi$ | local loss coefficient |
| $\varrho$ | density |
| $\tau$ | transmissivity |
| $\omega$ | speed |

## References

1. Mullett, L.B. The solar chimney—Overall efficiency, design and performance. *Int. J. Ambient Energy* **1987**, *8*, 35–40, doi:10.1080/01430750.1987.9675512.
2. Suárez-López, M.J.; Blanco-Marigorta, A.M.; Gutiérrez-Trashorras, A.J.; Pistono-Favero, J.; Blanco-Marigorta, E. Numerical simulation and exergetic analysis of building ventilation solar chimneys. *Energy Convers. Manag.* **2015**, *96*, 1–11, doi:10.1016/j.enconman.2015.02.049.
3. Naraghi, M.H.; Blanchard, S. Twenty-four hour simulation of solar chimneys. *Energy Build.* **2015**, *94*, 218–226, doi:10.1016/j.enbuild.2015.03.001.
4. Imran, A.A.; Jalil, J.M.; Ahmed, S.T. Induced flow for ventilation and cooling by a solar chimney. *Renew. Energy* **2015**, *78*, 236–244, doi:10.1016/j.renene.2015.01.019.
5. He, G.; Zhang, J.; Hong, S. A new analytical model for airflow in solar chimneys based on thermal boundary layers. *Sol. Energy* **2016**, *136*, 614–621, doi:10.1016/j.solener.2016.07.041.

6. Sudprasert, S.; Chinsorranant, C.; Rattanadecho, P. Numerical study of vertical solar chimneys with moist air in a hot and humid climate. *Int. J. Heat Mass Transf.* **2016**, *102*, 645–656, doi:10.1016/j.ijheatmasstransfer.2016.06.054.

7. Shi, L.; Zhang, G.; Cheng, X.; Guo, Y.; Wang, J.; Chew, M.Y.L. Developing an empirical model for roof solar chimney based on experimental data from various test rigs. *Build. Environ.* **2016**, *110*, 115–128, doi:10.1016/j.buildenv.2016.10.002.

8. Saleem, A.A.; Bady, M.; Ookawara, S.; Abdel-Rahman, A.K. Achieving standard natural ventilation rate of dwellings in a hot-arid climate using solar chimney. *Energy Build.* **2016**, *133*, 360–370, doi:10.1016/j.enbuild.2016.10.001.

9. Badescu, V. Dynamic model of a complex system including PV cells, electric battery, electrical motor and water pump. *Energy* **2003**, *28*, 1165–1181, doi:10.1016/j.enbuild.2016.10.001.

10. Atlam, O.; Kolhe, M. Performance evaluation of directly photovoltaic powered DC PM (direct current permanent magnet) motor—Propeller thrust system. *Energy* **2013**, *57*, 692–698, doi:10.1016/j.energy.2013.05.052.

11. Gupta, B.L.; Bhatnagar, M.; Marhur, J. Optimum sizing of PV panel, battery capacity and insulation thickness for a photovoltaic operated domestic refrigerator. *Sustain. Energy Technol. Assess.* **2014**, *7*, 55–67, doi:10.1016/j.seta.2014.03.005.

12. Hosien, M.A.; Selim, S.M. Effects of the geometrical and operational parameters and alternative outer cover materials on the performance of solar chimney used for natural ventilation. *Energy Build.* **2017** *138*, 355–367, doi:10.1016/j.enbuild.2016.12.041.

13. Lei, Y.; Zhang, Y.; Wang, F.; Wang, X. Enhancement of natural ventilation of a novel roof solar chimney with perforated absorber plate for building energy conservation. *Appl. Therm. Eng.* **2016**, *107*, 653–661, doi:10.1016/j.applthermaleng.2016.06.090.

14. Khanal, R.; Lei, C. An experimental investigation of an inclined passive wall solar chimney for natural ventilation. *Sol. Energy* **2014**, *107*, 461–474, doi:10.1016/j.solener.2014.05.032.

15. Odeh, N.; Grassie, T.; Henderson, D.; Muneer, T. Modelling of flow rate in a photovoltaic-driven roof slate-based solar ventilation air preheating system. *Energy Convers. Manag.* **2006**, *47*, 909–925, doi:10.1016/j.enconman.2005.06.005.

16. Sera, D.; Teodorescu, R.; Rodriguez, P. PV panel model based on datasheet values. In Proceedings of the IEEE International Symposium on Industrial Electronics, Vigo, Spain, 4–7 June 2007; pp. 2392–2396. doi:10.1109/ISIE.2007.4374981.

17. Hadj Arab, A.; Chenlo, F.; Benghanem, M. Loss-of-load probability of photovoltaic water pumping systems. *Sol. Energy* **2004**, *76*, 713–723, doi:10.1016/j.solener.2004.01.006.

18. Eurostat—Statistics Explained, Electricity Price Statistics. Available online: http://ec.europa.eu/eurostat/statistics-explained/index.php/Electricity_price_statistics (accessed on 3 April 2018).

Article

# On the Effects of Variation of Thermal Conductivity in Buildings in the Italian Construction Sector

Umberto Berardi [1], Lamberto Tronchin [2,*], Massimiliano Manfren [3] and Benedetto Nastasi [4]

[1] Department of Architectural Science, Ryerson University, 350 Victoria Street, Toronto, ON M5B2K3, Canada; uberardi@ryerson.ca

[2] Department of Architecture (DA), University of Bologna, Viale Europa 596, 47521 Cesena, Italy

[3] Faculty of Engineering and the Environment, University of Southampton, Highfield, Southampton SO17 1BJ, UK; M.Manfren@soton.ac.uk

[4] Department of Architectural Engineering & Technology, TU Delft University of Technology, Julianalaan 134, 2628BL Delft, The Netherlands; benedetto.nastasi@outlook.com

* Correspondence: lamberto.tronchin@unibo.it; Tel.: +39-051-2090542

Received: 10 February 2018; Accepted: 5 April 2018; Published: 9 April 2018

**Abstract:** Stationary and dynamic heat and mass transfer analyses of building components are an essential part of energy efficient design of new and retrofitted buildings. Generally, a single constant thermal conductivity value is assumed for each material layer in construction components. However, the variability of thermal conductivity may depend on many factors; temperature and moisture content are among the most relevant ones. A linear temperature dependence of thermal conductivity has been found experimentally for materials made of inorganic fibers such as rockwool or fiberglass, showing lower thermal conductivities at lower temperatures. On the contrary, a nonlinear temperature dependence has been found for foamed insulation materials like polyisocyanurate, with a significant deviation from linear behavior. For this reason, thermal conductivity assumptions used in thermal calculations of construction components and in whole-building performance simulations have to be critically questioned. This study aims to evaluate how temperature affects thermal conductivity of materials in building components such as exterior walls and flat roofs in different climate conditions. Therefore, experimental conductivities measured for four common insulation materials have been used as a basis to simulate the behavior of typical construction components in three different Italian climate conditions, corresponding to the cities of Turin, Rome, and Palermo.

**Keywords:** insulation materials; thermal conductivity; building energy consumption; temperature dependence; high-performance buildings

## 1. Introduction

In 2010, buildings accounted for 32% of total global final energy use, 19% of energy-related Greenhouse Gas (GHG) emissions, 51% of global electricity consumption, 33% of carbon emissions, and an eighth to a third of F-gases emissions [1]. In residential buildings, space heating shows the highest share of total primary energy consumption, equal to 32%. In commercial buildings as well, space heating dominated consumption with a 33% share of the total primary energy consumption [1]. In the European Union (EU), important efforts have been put into energy policies and different directives have resulted in recent years. Among them, the most important ones are the Energy Performance of Buildings Directive [2,3] and the Energy Efficiency Directive [4]. Further, there is much evidence that improving energy efficiency practices in the existing building stock will be crucial for energy sustainability at the EU level [5]. This strategy is even defined as the "new start" for the new EU economy [6] since the finance of energy efficiency can be unlocked by public and private partnership and not rely only on EU funds [7]. Considering the problem of space heating demand reduction, heat losses can be decreased by improving envelope performance with increased levels of insulation. This measure is the most effective way to drastically reduce heating demand, considering, of course, dependence on climate conditions [8]. However, in the existing building stock, this measure is much more costly than the replacement of boilers in heating systems [9,10]. Nonetheless, there are evident synergies between building envelope performance enhancement and sizing and operation of technical systems [11], even in the case of advanced energy conversion systems [12]. Following this evidence, many research efforts have been concentrated on the definition of methodologies [13] for the determination of cost-optimal levels of energy performance [13] in new and retrofitted buildings [14,15], and the impact of insulation can be extremely relevant in modelling [16]. Clearly, a reasonably robust performance estimate [17] is necessary to evaluate project feasibility. In this sense, uncertainty of energy performance represents an issue in techno-economic assessment methodologies and relevant sources of uncertainty have to be considered to limit as much as possible the "performance gap" [18], or side effects such as "re-bound" [19], "pre-bound" [20], and overheating risk [21]. These effects could potentially undermine the credibility of energy efficiency practices and, for these reasons, appropriate methodological tools are needed to account for uncertainty in building applications—for example, at the energy performance contracting level [22].

This paper focuses on one particular aspect that may affect the performance of building insulation materials (and, consequently, overall building performance)—temperature dependence of thermal conductivity—and how the approximations used in calculation tools may affect performance estimates. So far, this aspect is normally neglected and has not been considered in a number of scientific publications about energy behavior of buildings. In particular, the potential uncertainty introduced by constant and linear temperature dependence approximations is addressed by combining experimental analysis and thermohygrometric simulations for selected case studies in three Italian climate conditions.

## 2. Impact of Thermal Insulation Variability in Building Energy Performance

In common design practice, the starting point for building energy performance analysis is the thermal analysis of building construction components and fabric. A more accurate analysis requires the use of thermohygrometric modelling tools, based on heat, air, and moisture transfer (HAMT) algorithms [23]. This type of analysis is not generally integrated into energy simulation and is conducted in a separate way according to specific standards [24], considered for building code compliance checking.

Heat transfer in building components occurs through three modes: conduction, convection, and radiation [25]. In dynamic building energy modelling tools, monodimensional heat transfer and zonal energy and mass balances are generally considered [26,27]. For these reasons, several approximations are introduced in the modelling process. These approximations will be briefly summarized hereafter, showing the impact of thermal conductivity at multiple scales of analysis:

1. single layers of material;
2. construction components;
3. overall building fabric.

First, the thermal conductivity concept was introduced by Fourier with his phenomenological law of heat [11]. This law states that heat transfer rate through a material is proportional to the negative temperature gradient and to the area through which heat flows. By considering a monodimensional heat transfer problem through a single material layer in stationary conditions, we can write

$$q = -\frac{\lambda}{s}(\theta_e - \theta_i)A \tag{1}$$

$$Q = \frac{\lambda}{s}(\theta_i - \theta_e)A\Delta t \tag{2}$$

where

$q$ is heat transfer rate;
$Q$ is thermal energy transfer;
$\lambda$ is thermal conductivity;
$s$ is depth of the material layer;
$A$ is area;
$\Delta t$ is time interval;
$\theta_i$ is internal side temperature;
$\theta_e$ is external side temperature.

Building construction components generally constitute multiple layers of material and we can account for stationary heat transfer by introducing thermal transmittance $U$ [28]. Limits on $U$ (or $R$) values of individual construction components are generally imposed by building regulations. Convective and radiative heat transfer taking place respectively on the internal and external surfaces of components are accounted for by specific resistances in the calculation [28]. The stationary thermal energy transfer for a multilayered construction component can be calculated as follows:

$$U = \frac{1}{R} = \frac{1}{\left(R_{si} + \sum_i \frac{s_i}{\lambda_i} + R_{se}\right)} = \frac{1}{\left(\frac{1}{h_{si}} + \sum_j \frac{s_i}{\lambda_i} + \frac{1}{h_{se}}\right)} \tag{3}$$

$$Q = UA(\theta_e - \theta_i)\Delta t \tag{4}$$

where

$U$ is thermal transmittance;
$R$ is thermal resistance;
$\lambda_i$ is thermal conductivity of layer $i$;
$s_i$ is depth of material layer $i$;
$R_{si}$ is thermal resistance on internal side;
$R_{se}$ is thermal resistance on external side;
$h_i$ is thermal heat transfer coefficient on internal side, accounting for convection and radiation;
$h_e$ is thermal heat transfer coefficient on external side, accounting for convection and radiation;
$Q$ is the thermal energy transfer.

In order to easily account for the overall heat transfer performance of the building fabric in stationary conditions, heat transfer coefficient $H$ can be introduced [29]. Similar to the $U$ value, in building regulations, limits can be set for the overall fabric performance using $H$. The heat transfer coefficient can be calculated as follows:

$$H_{tr} = \sum_i U_i \cdot A_i + \sum_j \psi_j \cdot l_j + \sum_k \chi_k \tag{5}$$

$$Q_{tr} = H_{tr}(\theta_i - \theta_e)\Delta t \tag{6}$$

where

$H_{tr}$ is the thermal heat transfer coefficient for the building fabric;
$U_i$ is the thermal transmittance of construction component $i$;
$A_i$ is the surface area of construction component $i$;
$\psi_j$ is the heat transmission coefficient for two-dimensional thermal bridge $j$;
$l_j$ is the length of the two-dimensional thermal bridge $j$;

$\chi_k$ is the heat transmission coefficient for three-dimensional thermal bridge $k$;

$H_{tr}$ is the heat transfer coefficient for envelope transmission;

$Q_{tr}$ is the heat transfer of the overall building envelope in stationary conditions.

In Equation (5), quantities are introduced to account for bidimensional and tridimensional heat transfer happening in thermal bridges. Even in this case, simplifications are possible, for example, by performing regression analysis on thermal bridges calculated with detailed bidimensional stationary heat transfer models in multiple conditions [30]. Further, moving from stationary to dynamic behavior of building construction components and fabric [26,27], conceptual models can be introduced [31] to simplify performance assessment [32,33]; however, the estimation of aggregated (lumped) thermal properties has to be considered carefully [34], as the potential advantages themselves depend on the operational strategy adopted in buildings [35].

The impact of insulation materials in building components is determined by the low conductivity value. Consequently, the use of insulation sensibly affects all the aggregated physical properties (i.e., $U$ or $R$, $H$, but also dynamic parameters) from individual components to the overall building fabric. For this reason, it is very important to give boundaries to assumptions and simplifications introduced in calculations. They depend on the specific context of application, i.e., type of construction technologies, climate conditions, and operation strategies. Further, it is also necessary to synthetically visualize the impact of design choices at multiple scales, from individual components to the whole building fabric [36].

Insulation materials generally used in buildings are characterized by a thermal conductivity lower than 0.10 W/mK, although the most common insulation materials used in the construction industry have thermal conductivities ranging from 0.03 to 0.05 W/mK. Further, materials with thermal conductivity values lower than 0.03 W/mK are becoming more popular today in building applications.

The way an insulation material resists heat flux depends on the microscopic cells where air or other gaseous substances are locked up. In closed-cell materials, the insulation effect is guaranteed by the fact that air or gas contained in the cells is prevented from moving and, therefore, convective heat transfer is significantly suppressed. For example, cells in plastic foam insulation (e.g., polystyrene and polyurethane) contain fluorocarbon gas instead of air, obtaining lower thermal conductivity than air. A different behavior can be observed for blown foam insulation materials like polyisocyanurate. This kind of insulation generally shows low conductivity values since the blowing agent is locked up within the pores, and this condition permits a lower heat transfer by convection and conduction with respect to air. However, if temperature decreases under the condensation value of the aforementioned agent, condensation occurs in the pores and, subsequently, the thermal conductivity value rises due to the higher conductive property of the liquid phase compared to the gaseous phase of the agent [37].

Therefore, thermal conductivity is not directly linked to the insulation material but it is associated with the thermal resistance of the gas used within the material which determines the theoretical value limit [25,38]. As such, conduction properties are heavily influenced by the characteristics of raw material, insulation material density, nature and microscopic structure of the solid component, and moisture content and temperature [38–40].

Although commonly used insulation materials show variations in their thermal conductivity determined by temperature, their physical properties are measured at 23.8 °C, the standard test conditions. However, when insulation materials experience realistic temperature oscillations, their thermal conductivities vary significantly with respect to the values obtained in standard test conditions. This effect will be investigated experimentally and by means of simulations in the following sections.

## 3. Research Methodology

In order to evaluate the effects of temperature-dependent thermal conductivities of insulation material in typical Italian construction components, thermohygrometric models were created and a series of simulations were run in different climatic conditions. Experimentally measured data from conductivity tests were used as a basis for these simulations. The hygrothermal modelling tool chosen was Wufi® Pro (Fraunhofer Institut für Bauphysik, Stuttgart, Germany) [23], a well-established and tested software used for both research and design purposes. Wufi is used to simulate heat and moisture transfer in multilayered building components and its calculation engine can account for temperature dependence of thermal conductivity of materials. Simulations are used to assess the impact of temperature-dependent thermal conductivities on external walls and flat roofs, representing typical components used in the Italian construction sector. Four insulation materials were considered for experimental analysis: fiberglass, rockwool, polyisocyanurate (PIR), and extruded polystyrene (XPS). These materials were chosen based on their high market share among the conventional solutions [41].

Three different thermal conductivity profiles were used in the simulations for each insulation material:

1. a constant value based on the value measured at standard temperature, i.e., 23.8 °C;
2. a linear temperature-dependent function;
3. experimental values which represent the temperature-dependent thermal conductivities, measured in the laboratory.

In order to demonstrate the effect of temperature-dependent thermal conductivity, simulations were run in three different Italian climate conditions corresponding to three cities: Turin (northern Italy), Rome (central Italy), and Palermo (southern Italy). According to Köppen climate classification [42], Turin has a humid temperate climate, classified as Cfa, with cold, foggy winters and hot summers; Rome has a warm temperate climate, classified as Csa, with warm winters and hot summers; and Palermo has a semi-arid climate, classified as BSk, with moderately cold winters and hot and dry summers. Figure 1 depicts the average monthly temperature trends.

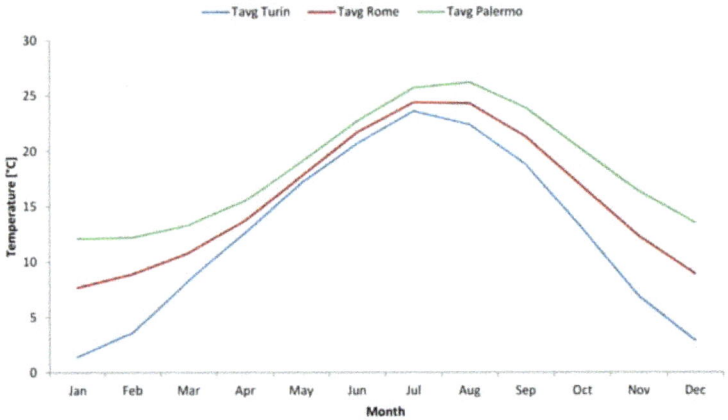

**Figure 1.** Average monthly temperature (Tavg) trends plotted for Turin, Rome, and Palermo.

In terms of typical construction components, a brick cavity structure for walls and a hollow brick-cement structure for flat roofs were selected for the simulations, as shown in Figure 2.

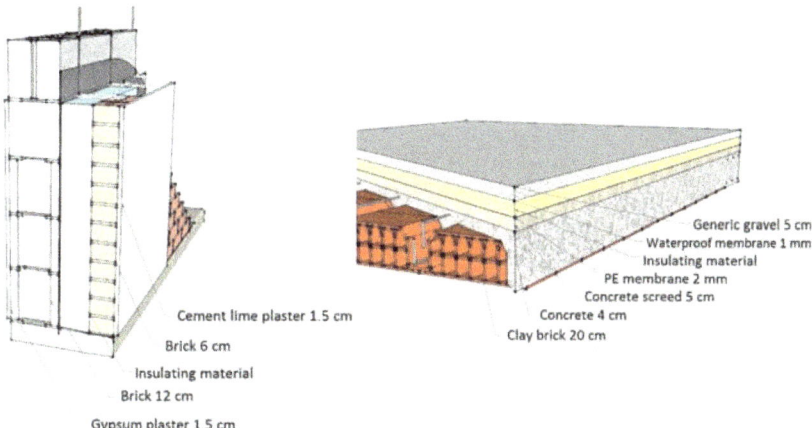

**Figure 2.** Type of construction components simulated, brick-cavity structure for walls (**on the left**), hollow brick–cement structure for flat roofs (**on the right**).

Thermal simulations were run to obtain hourly time-series data of thermal conductivities and heat transfer. In terms of boundary conditions, all the calculations were conducted with 80% initial relative humidity (RH) and an initial temperature of 20 °C. The initial RH accounts for materials exposed to the open air before being included in the building envelope. A moisture source was taken into account within the wall assembly for each simulation performed, derived from a percentage (1%) of exterior driving rain [43]. In order to account for worst-case conditions, external walls considered were the ones oriented towards the north direction. For each city, a specific maximum *U*-value has been determined for vertical walls and roofs according to current technical standards [28] and Italian national regulations, as reported in Tables 1 and 2. Building indoor temperature was supposed to be constantly 21 °C in the winter period and 26 °C in the summer period.

**Table 1.** Maximum *U*-value data prescribed in Italy used for the selection of the building systems and insulation thickness for constant, linear, and measured conductivities—external walls.

| Location | Max *U*-Value for Vertical Wall (W/(m²·K)) | Typology | Insulation Thickness for Constant and Linear Conductivities/for Measured Conductivities | | | |
|---|---|---|---|---|---|---|
| | | | Fiberglass (cm) | Rockwool (cm) | PIR (cm) | XPS (cm) |
| Turin | 0.30 | Wall (brick cavity structure) | 8.0/7.5 | 6.5/6.0 | 5.0/5.5 | 5.5/5.0 |
| Rome | 0.34 | Wall (brick cavity structure) | 6.0/6.0 | 5.0/4.5 | 4.0/4.0 | 4.5/4.0 |
| Palermo | 0.45 | Wall (brick cavity structure) | 3.0/3.0 | 2.5/2.5 | 2.0/2.0 | 2.0/2.0 |

**Table 2.** Maximum *U*-value data prescribed in Italy used for the selection of the building systems and insulation thickness for constant, linear, and measured conductivities—horizontal roofs.

| Location | Max *U*-Value for Roofs(W/(m²·K)) | Typology | Insulation Thickness for Constant and Linear Conductivities/for Measured Conductivities | | |
|---|---|---|---|---|---|
| | | | Rockwool (cm) | PIR (cm) | XPS (cm) |
| Turin | 0.25 | Flat roof (hollow brick–cement) | 8.5/8.0 | 6.5/ 6.5 | 7.0/6.5 |
| Rome | 0.30 | Flat roof (hollow brick–cement) | 6.0/5.5 | 4.5/5.0 | 5.0/4.5 |
| Palermo | 0.38 | Flat roof (hollow brick–cement) | 3.5/3.5 | 2.5/3.0 | 3.0/3.0 |

Constant and linear temperature-dependent thermal conductivities considered for calculations are reported in Figure 3 for the four insulation materials, in comparison with experimental data. The thermal conductivity of the samples was measured at 23.8 °C using the heat flow meter apparatus HFM 436/3/1E Lambda produced by Netzsch (Selb, Germany). The apparatus was calibrated with a standard fiberglass board, supplied by the National Institute of Standards and Technology. The measurement accuracy of the apparatus was set to ±1% (so an error of ±1% can be expected in the measurement results). Specimens sized with variable thicknesses were placed between the hot and cold plates and the thermal conductivity was measured by the heat flux sensor upon reaching the thermal equilibrium at a defined temperature difference and for a uniform temperature gradient throughout the sample. The sample size was $305 \times 305$ mm$^2$, although the heat flow was measured in the central $100 \times 100$ mm$^2$ area of the sample. The large sample size compared to the measurement area ensured steady-state thermal conditions for the measuring area so that the surrounding area of the transducer acted as an effective guard against lateral heat flow. To keep track of the moisture content, samples were weighed before and after thermal measurement using a digital scale with 0.1 g accuracy.

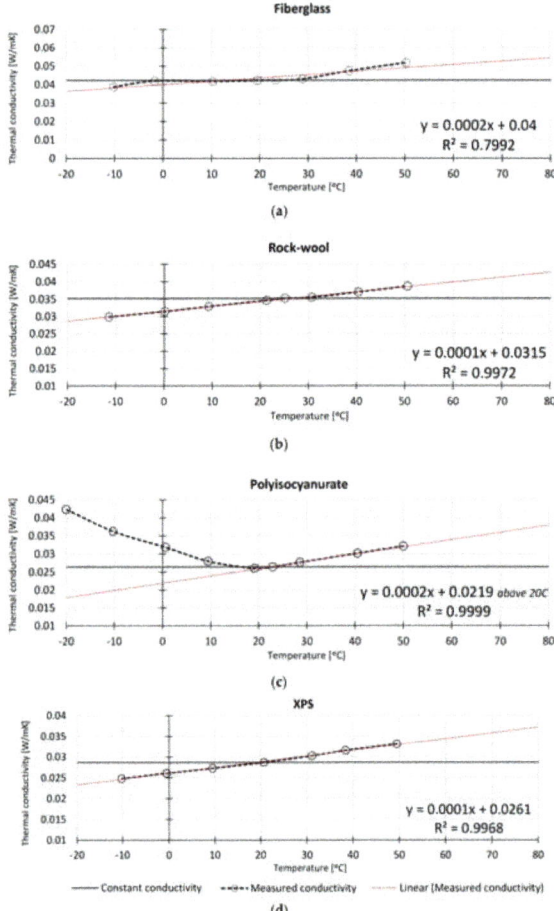

**Figure 3.** Temperature dependence of the thermal conductivity for the four analyzed materials, comparing constant, linear, and measured values. (**a**) Fiberglass; (**b**) Rockwool; (**c**) Polyisocyanurate; (**d**) Extruded polystyrene.

Conductivities were measured experimentally multiple times in the range from $-20\,^{\circ}$C to $+60\,^{\circ}$C and the final results of the test are reported in Figure 3; data points are between $-10\,^{\circ}$C and $+50\,^{\circ}$C for all materials except for polyisocyanurate, for which the nonlinear behavior is reported up to $-20\,^{\circ}$C. The trend lines, supported by the equation reported in the legend, depict the data behavior of materials. It is noteworthy that the thermal conductivity values for rockwool, fiberglass, and extruded polystyrene (XPS) are nearly linear with respect to temperature while polyisocyanurate exhibits a nonlinear behavior with higher conductivity values for colder temperatures. In particular, the deviation from linear behavior begins at $20\,^{\circ}$C and the main change is reached around $10\,^{\circ}$C; this is when the blowing agent within the polyisocyanurate microstructure starts condensing, determining the rise of conductivity. This effect has recently been extensively discussed by Berardi and Naldi [44].

## 4. Temperature-Dependent Thermal Conductivity in Building Components

This section concerns the study of variations in thermal conductivity determined by variability of climate conditions. Mid-thickness temperature of the insulation materials has been investigated for the months of January and July, which represent the two extreme conditions, following the methodology described in the previous section for the different locations.

Figures 4 and 5 present the data results for vertical walls and flat roofs, respectively, and are organized by material for each location. The thermal conductivities of exterior walls with fiberglass, rockwool, and extruded polystyrene (XPS) insulation in Figure 4 exhibit better performance at lower temperatures, as shown in the winter time series for Turin and Palermo. They represent the two extreme conditions, Turin corresponding to northern Italy and Palermo to southern Italy. On the contrary, polyisocyanurate (PIR) conductivity increases at lower temperatures with a nonlinear trend. The deviation from linear temperature dependence begins at $20\,^{\circ}$C and the value remains nearly constant up to $10\,^{\circ}$C. This fact depends on the gas condensation process inside the polyisocyanurate when temperature falls below $10\,^{\circ}$C, as explained before. The case of Rome in winter presents a larger variability because of the combination of low temperatures with high solar radiation that determines larger oscillations.

Fiberglass is not considered for roof insulation because it is not a suitable material for flat roof components. Analyzing the plots relative to flat roofs in Figure 5, we can see how simulations confirm the different behavior of polyisocyanurate with respect to the other insulation materials—rockwool and XPS. What appears to be evident from Figure 1 is the fact that temperatures in Turin are much lower in winter then those in Rome and Palermo and are below $10\,^{\circ}$C, the point of condensation of the blowing agent in the polyisocyanurate. Again, Rome in winter presents larger conductivity oscillations because of the combination of low temperatures with high solar radiation.

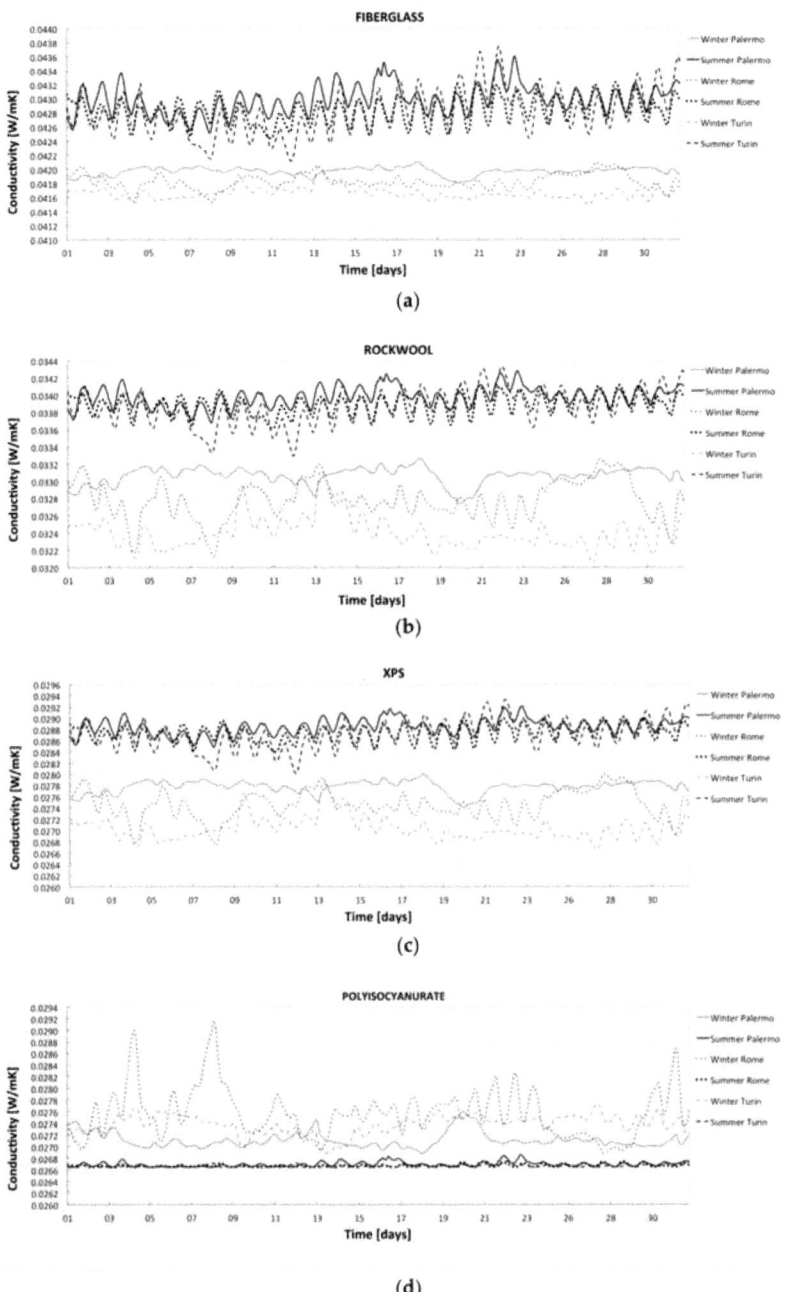

**Figure 4.** Time series of thermal conductivity values of insulation materials in external walls.
(**a**) Fiberglass; (**b**) Rockwool; (**c**) Extruded polystyrene; (**d**) Polyisocyanurate.

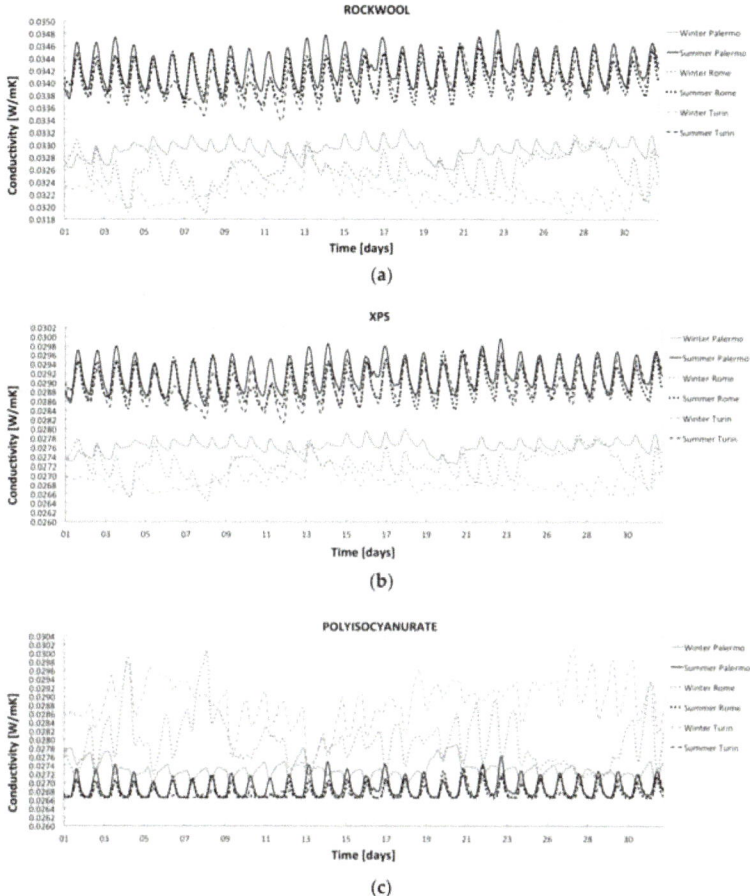

**Figure 5.** Time series of thermal conductivity values of insulation materials in flat roofs. (a) Rockwool; (b) Extruded polystyrene; (c) Polyisocyanurate.

## 5. Seasonal Average Heat Flux through Building Components

The definition of normative requirements for building efficiency accounts for climate conditions (and possibly for climate change effects in the medium–long term [45,46]); however, it is difficult to account for complex effects, such as temperature dependency, without using detailed simulation models. Clearly, underestimations or overestimations of building thermal performance have an impact on energy supply design and operation [47]. For this reason, average heat flux values have been analyzed in the subsequent section, comparing percentage deviations.

In order to evaluate how each assumed behavior (in terms of thermal conductivity) for the different materials affects energy consumption in buildings, respective hourly simulated time series of heat losses and gains were analyzed for winter and summer selected periods. The winter period analyzed goes from 1 December to 31 March, while the summer period analyzed goes from 1 June to 30 September. Results for the exterior walls and flat roofs are reported in Figures 6–9 for the two cities Turin and Palermo, representing the two extreme conditions, corresponding respectively to the northern and southern Italian climates. The data presented in Figures 6–9 are average heat fluxes through construction components obtained with the three different assumptions:

1. constant thermal conductivity;
2. linear temperature dependence of thermal conductivity;
3. experimentally measured thermal conductivity.

Negative values of average heat fluxes represent an average heat loss (heat exiting from thermal zone), while positive values indicate an average heat gain (heat entering thermal zone). As explained in Section 2, the energy balance of construction components depends on its characteristics and on the dynamic conditions of temperature, as well as other effects such as thermal gains due to incident solar radiation. In the case study considered, following Italian normative requirements, the differences between the *U* values for the components in Turin and Palermo are very large, as reported in Tables 1 and 2.

In the Turin cases, heat fluxes are negative in winter conditions both for external walls and horizontal roofs, while in summer conditions, heat fluxes are negative for external walls and positive for horizontal roofs. In Palermo cases, heat fluxes are negative in winter conditions both for external walls and horizontal roofs, while in summer conditions, heat fluxes are positive both for external walls and horizontal roofs.

The aim of the research in this case was to highlight differences among performance estimates obtained using the different assumptions about thermal conductivity reported previously. In particular, our goal was to show the difference obtained with measured data with respect to constant and linear assumptions. For this reason, Tables 3 and 4 report the percentage difference of average heat fluxes for all the components and locations, comparing

1. experimentally measured with respect to constant thermal conductivity;
2. experimentally measured with respect to linear temperature dependence of thermal conductivity.

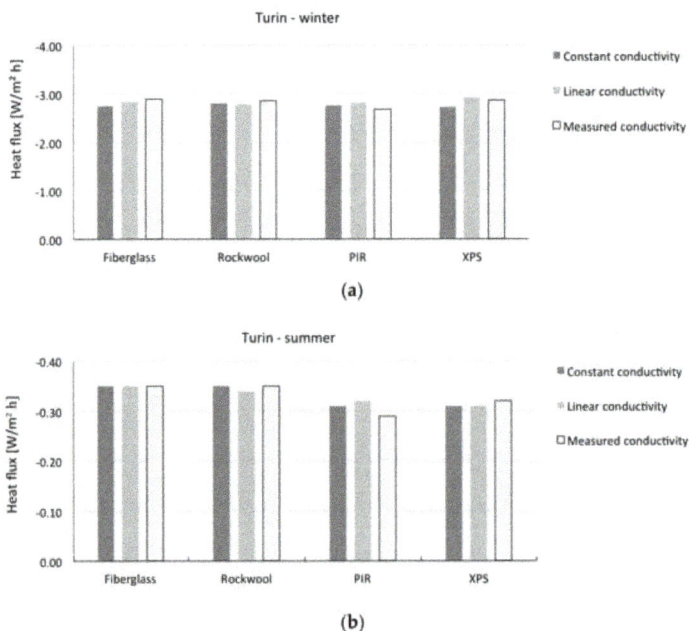

**Figure 6.** Heat flux average in exterior walls in Turin. (**a**) Turin—winter; (**b**) Turin—summer.

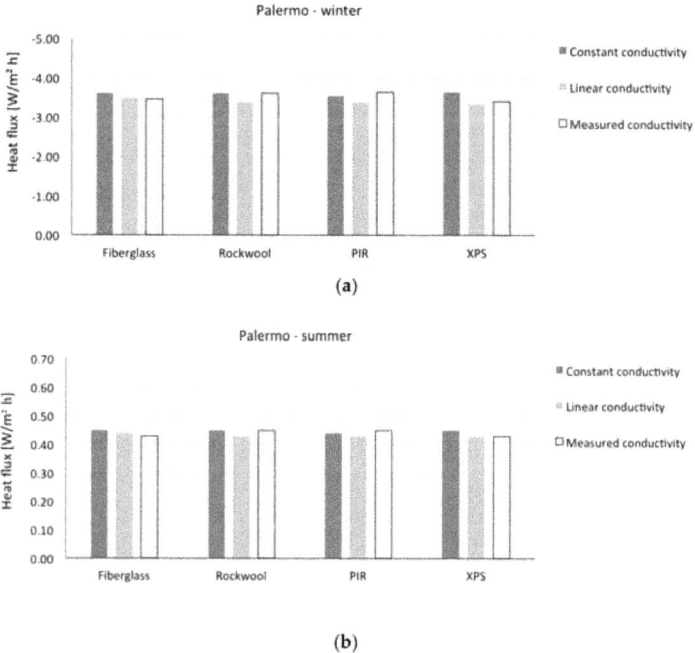

**Figure 7.** Heat flux average in exterior walls in Palermo. (**a**) Palermo—winter; (**b**) Palermo—summer.

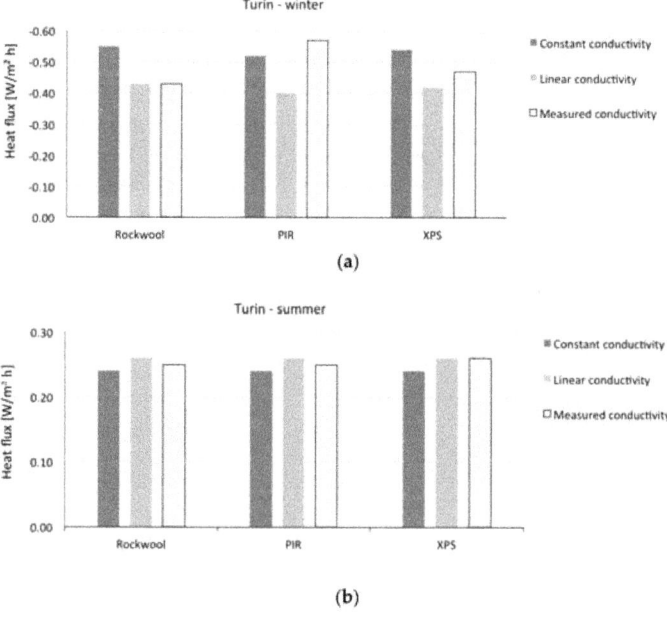

**Figure 8.** Heat flux average in flat roofs in Turin. (**a**) Turin—winter; (**b**) Turin—summer.

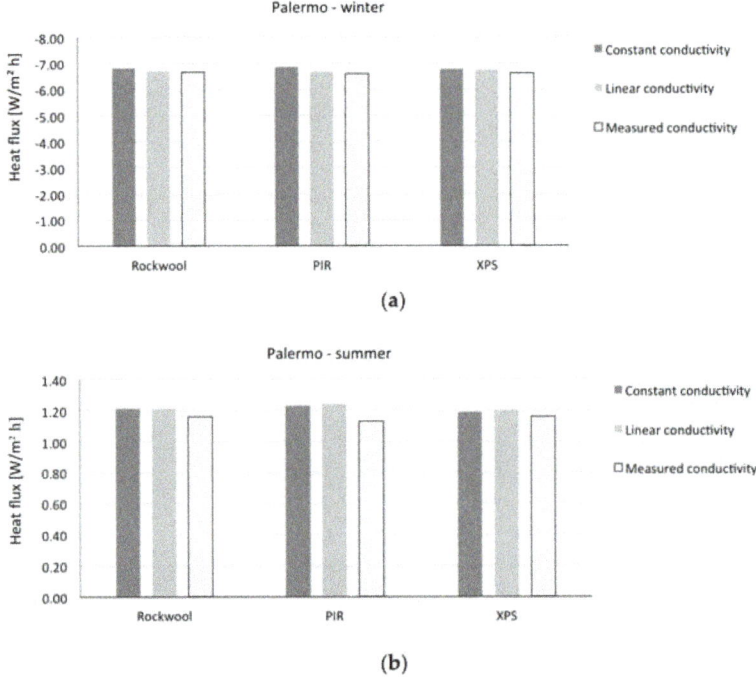

**Figure 9.** Heat flux average in flat roofs in Palermo. (**a**) Palermo—winter; (**b**) Palermo—summer.

**Table 3.** Percentage difference of heat flux for exterior walls.

| Location | Material | Measured vs. Constant | | Measured vs. Linear | |
|---|---|---|---|---|---|
| | | **Winter** | **Summer** | **Winter** | **Summer** |
| Turin | Fiberglass | 5.5% | 0.0% | 2.1% | 0.0% |
| | Rockwool | 1.8% | 0.0% | 2.5% | 2.9% |
| | PIR | −2.5% | −6.4% | −4.6% | −9.4% |
| | XPS | 5.5% | 3.2% | −1.7% | 3.2% |
| Rome | Fiberglass | −3.2% | −3.8% | −1.3% | −3.8% |
| | Rockwool | −0.5% | 0.0% | 1.0% | 0.0% |
| | PIR | −1.3% | 0.0% | −1.5% | 0.0% |
| | XPS | 1.3% | −4.0% | −2.5% | −7.7% |
| Palermo | Fiberglass | −4.4% | −4.4% | −1.0% | −2.3% |
| | Rockwool | 0.0% | 0.0% | 6.5% | 4.6% |
| | PIR | 2.5% | 2.3% | 7.7% | 4.6% |
| | XPS | −6.3% | −4.4% | 2.1% | 0.0% |

Highlighted: positive values.

**Table 4.** Percentage difference of heat flux for flat roofs.

| Location | Material | Heat Flux Difference Percentage | | | |
|---|---|---|---|---|---|
| | | Measured vs. Constant | | Measured vs. Linear | |
| | | Winter | Summer | Winter | Summer |
| Turin | Rockwool | −21.8% | 4.2% | 0.0% | −3.8% |
| | PIR | 9.6% | 4.2% | 42.5% | −3.8% |
| | XPS | −13.0% | 8.3% | 12.0% | 0.0% |
| Rome | Rockwool | −1.9% | 2.9% | 10.6% | 0.0% |
| | PIR | −5.1% | −2.9% | −2.2% | −8.1% |
| | XPS | −0.5% | 8.8% | 2.2% | 0.0% |
| Palermo | Rockwool | −2.3% | −4.1% | −0.7% | −4.1% |
| | PIR | −3.6% | −8.1% | −1.0% | −8.9% |
| | XPS | −2.1% | −2.5% | −1.9% | −3.3% |

Highlighted: positive values.

A positive value of the percentage difference indicates that the heat flux obtained from simulations with experimentally measured conductivity is higher, in absolute value, than the one obtained with either constant or linear assumptions. This implies that in the winter period an underestimation of heat losses and, consequently, an underestimation of thermal demand for heating flux is negative in all cases. On the other hand, the summer results depend on the sign of the heat flux: when the heat flux is negative, a negative percentage difference implies an underestimation of heat losses (e.g., external walls for Turin) or, vice versa, an underestimation of gains when the heat flux is positive (e.g., external walls for Palermo, horizontal roofs for Turin and Palermo). The differences found by means of simulation remain in most of the cases lower than 10% in absolute value. The higher values encountered are for polyisocyanurate (PIR) and extruded polystyrene (XPS) in the flat roof in Turin for winter conditions, when comparing linear temperature dependence with respect to measured. The percentage differences in these cases are respectively 42.5% and 12%. Therefore, we can state that for the cases considered, most of the time a sensitivity analysis assuming constant thermal conductivity would be sufficient to address the issue of accounting for performance variability. This implies the possibility of using conventional simulation tools, although considering a spectrum of input data variability. Clearly, in order to be able to derive general conclusions (e.g., practical indications for sensitivity analyses using conventional stationary and dynamic simulation tools), more extensive experimental and simulation analysis would be necessary, employing statistical reference buildings [48].

## 6. Conclusions

Insulation materials are some of the most crucial elements in energy efficient building design, especially in buildings where heating represents the predominant demand. Standard thermal conductivities assumed in most of the cases in building performance simulations are measured in standard conditions at 23.8 °C. However, a variation of thermal conductivity is clearly to be expected in actual building operation.

The relation between temperature and thermal conductivity can be approximated with a linear function for many materials. For example, inorganic fiber insulations, e.g., rockwool and fiberglass, and some petrochemical-based ones, e.g., extruded polystyrene, are characterized by lower conductivity values at lower temperatures. However, there are materials such as refrigerant blown foam insulation, e.g., polyisocyanurate, that have a nonlinear relation between temperature and conductivity, determined by the condensation of the blowing agent below certain temperatures.

This paper presented experimental analysis of the thermal conductivity of four materials, namely, rockwool, fiberglass, extruded polystyrene, and polyisocyanurate. Temperatures ranging from −10 °C to +50 °C were used to measure conductivities. The outcomes of laboratory tests were the basis for creating thermal conductivity functions (temperature dependent) to represent the behavior within a dynamic thermohygrometric modelling tool, WUFI. By means of this tool, simulations were run for typical construction components such as exterior walls and flat roofs in different Italian climatic conditions, corresponding to the north (Turin), center (Rome), and south (Palermo) of the country.

According to the results obtained, polyisocyanurate determines the larger performance variability with respect to other materials. This becomes more evident in colder climates, as the blowing agent inside the material begins condensing at 10 °C. The research shows how assumptions about thermal conductivity in simulations affect performance estimates. While for Turin the results with measured conductivities implies in many cases a higher simulated thermal demand for heating with respect to constant and linear assumptions, in Rome and in Palermo this effect may not be particularly relevant because of the temperature ranges. In summer conditions, the results depend on the sign of the average heat flux leading to an underestimation of heat losses when the flux is negative or, vice versa, an underestimation of gains when the heat flux is positive. In the last case, the effect may be negative in terms of increased cooling demand or overheating.

This research represents an evaluation of the potential inaccuracies in building performance estimation introduced by assumptions about thermal conductivity of insulation materials. The impact of these inaccuracies when using simulation tools for building design and code compliance checking is relevant because it affects design choices that will influence performance for the whole building life cycle. Further, climate change issues play an important role as well, requiring the designer to consider not only present but also future climate conditions. Finally, at the policy-making level, there should be awareness of the potential faults related to simplified assumptions in the thermohygrometric behavior of construction components and overall building performance.

**Acknowledgments:** The authors would like to acknowledge Matteo Naldi for the initial analysis on the case study building. The first author wishes to thank the financial support of the Ontario Center for Excellence, Program, the Natural Sciences and Engineering Research Council of Canada (NSERC) and the Ontario Ministry of Research, Innovation and Science.

**Author Contributions:** Umberto Berardi and Lamberto Tronchin conceived, designed and performed the experiments; Massimiliano Manfren and Benedetto Nastasi analyzed and commented the data; All the Authors equally contributed to write the paper.

**Conflicts of Interest:** The authors declare no conflict of interest.

## References

1. Berardi, U. A cross-country comparison of the building energy consumptions and their trends. *Resour. Conserv. Recycl.* **2017**, *123*, 230–241. [CrossRef]
2. Europa. *Directive 2002/91/EC of the European Parliament and of the Council of 16 December 2002 on the Energy Performance of Buildings*; Europa: Brussels, Belgium, 2003.
3. Europa. *Directive 2010/31/EU of the European Parliament and of the Council of 19 May 2010 on the Energy Performance of Buildings (EPBD Recast)*; Europa: Brussels, Belgium, 2010.
4. Europa. *Directive 2012/27/EU of the European Parliament and of the Counci of 25 October 2012 on Energy Efficiency*; Europa: Brussels, Belgium, 2012.
5. Buildings Performance Institute Europe (BPIE). *Europe's Buildings under the Microscope*; BPIE: Brussels, Belgium, 2011.
6. Saheb, Y.; Bodis, K.; Szabo, S.; Ossenbrink, H.; Panev, S. *Energy Renovation: The Trump Card for the New Start for Europe*; JRC EU Commission: Brussels, Belgium, 2015.
7. Energy Efficiency Financial Institutions Group (EEFIG). Energy Efficiency—The First Fuel for the EU Economy, How to Drive New Finance for Energy Efficiency Investments. 2015. Available online: https://www.db.com/cr/en/concrete-EEFIG-report--Energy-Efficiency-the-first-fuel-for-the-EU-Economy.htm (accessed on 7 April 2018).
8. Asdrubali, F.; D'Alessandro, F.; Schiavoni, S. A review of unconventional sustainable building insulation materials. *Sustain. Mater. Technol.* **2015**, *4*, 1–17. [CrossRef]

9.    Noussan, M.; Nastasi, B. Data Analysis of Heating Systems for Buildings—A Tool for Energy Planning, Policies and Systems Simulation. *Energies* **2018**, *11*, 233. [CrossRef]

10.   Aste, N.; Buzzetti, M.; Caputo, P.; Manfren, M. Local energy efficiency programs: A monitoring methodology for heating systems. *Sustain. Cities Soc.* **2014**, *13*, 69–77. [CrossRef]

11.   Papadopoulos, A.M. State of the art in thermal insulation materials and aims for future developments. *Energy Build.* **2005**, *37*, 77–86. [CrossRef]

12.   Lo Basso, G.; Nastasi, B.; Salata, F.; Golasi, I. Energy retrofitting of residential buildings—How to couple Combined Heat and Power (CHP) and Heat Pump (HP) for thermal management and off-design operation. *Energy Build.* **2017**, *151*, 293–305. [CrossRef]

13.   Tronchin, L.; Tommasino, M.C.; Fabbri, K. On the "cost-optimal levels" of energy performance requirements and its economic evaluation in Italy. *Int. J. Sustain. Energy Plan. Manag.* **2014**, *3*. [CrossRef]

14.   Aste, N.; Adhikari, R.S.; Manfren, M. Cost optimal analysis of heat pump technology adoption in residential reference buildings. *Renew. Energy* **2013**, *60*, 615–624. [CrossRef]

15.   EU Project ENTRANZE. Available online: http://www.entranze.eu/pub/pub-optimality (accessed on 7 February 2018).

16.   Kaynakli, O. A review of the economical and optimum thermal insulation thickness for building applications. *Renew. Sustain. Energy Rev.* **2012**, *16*, 415–425. [CrossRef]

17.   Kotireddy, R.; Hoes, P.-J.; Hensen, J.L.M. A methodology for performance robustness assessment of low-energy buildings using scenario analysis. *Appl. Energy* **2018**, *212*, 428–442. [CrossRef]

18.   Imam, S.; Coley, D.A.; Walker, I. The building performance gap: Are modellers literate? *Build. Serv. Eng. Res. Technol.* **2017**, *38*, 351–375. [CrossRef]

19.   Herring, H.; Roy, R. Technological innovation, energy efficient design and the rebound effect. *Technovation* **2007**, *27*, 194–203. [CrossRef]

20.   Sunikka-Blank, M.; Galvin, R. Introducing the prebound effect: The gap between performance and actual energy consumption. *Build. Res. Inf.* **2012**, *40*, 260–273. [CrossRef]

21.   Taylor, J.; Davies, M.; Mavrogianni, A.; Chalabi, Z.; Biddulph, P.; Oikonomou, E.; Das, P.; Jones, B. The relative importance of input weather data for indoor overheating risk assessment in dwellings. *Build. Environ.* **2014**, *76*, 81–91. [CrossRef]

22.   Ligier, S.; Robillart, M.; Schalbart, P.; Peuportier, B. Energy performance contracting methodology based upon simulation and measurement. In Proceedings of the Building Simulation 2017, San Francisco, CA, USA, 7–9 August 2017.

23.   Wufi Software. Available online: https://wufi.de/en/ (accessed on 7 February 2018).

24.   Ente Nazionale Italiano di Unificazione—ISO. *Hygrothermal Performance of Building Components and Building Elements—Internal Surface Temperature to Avoid Critical Surface Humidity and Interstitial Condensation—Calculation Methods*; UNI EN ISO 13788; UNI: Milan, Italy, 2008.

25.   Pfundstein, M.; Gellert, R.; Spitzner, M.; Rudolphi, A. *Insulating Materials: Principles, Materials, Applications*; Walter de Gruyter: Berlin, Germany, 2008.

26.   Ente Nazionale Italiano di Unificazione—ISO. *Energy Performance of Buildings—Calculation of Energy Use for Space Heating and Cooling*; UNI EN ISO 13790; UNI: Milan, Italy, 2008.

27.   International Organization for Standardization (ISO). *Energy Performance of Buildings—Overarching EPB Assessment—Part 1: General Framework and Procedures (Draft)*; ISO/DIS 52000-1; ISO: Geneva, Switzerland, 2007.

28.   Ente Nazionale Italiano di Unificazione—ISO. *Building Components and Building Elements—Thermal Resistance and Thermal Transmittance—Calculation Method*; UNI EN ISO 6946; UNI: Milan, Italy, 2008.

29.   Ente Nazionale Italiano di Unificazione—ISO. *Energy Performance of Buildings—Transmission and Ventilation Heat Transfer Coefficients—Calculation Method*; UNI EN ISO 13789; UNI: Milan, Italy, 2008.

30.   Capozzoli, A.; Gorrino, A.; Corrado, V. A building thermal bridges sensitivity analysis. *Appl. Energy* **2013**, *107*, 229–243. [CrossRef]

31.   Ente Nazionale Italiano di Unificazione—ISO. *Thermal Performance of Building Components—Dynamic Thermal Characteristics—Calculation Methods*; UNI EN ISO 13786; UNI: Milan, Italy, 2008.

32.   Karlsson, J.; Wadsö, L.; Öberg, M. A conceptual model that simulates the influence of thermal inertia in building structures. *Energy Build.* **2013**, *60*, 146–151. [CrossRef]

33. Martin, K.; Erkoreka, A.; Flores, I.; Odriozola, M.; Sala, J.M. Problems in the calculation of thermal bridges in dynamic conditions. *Energy Build.* **2011**, *43*, 529–535. [CrossRef]

34. Mantesi, E.; Hopfe, C.J.; Cook, M.J.; Glass, J.; Strachan, P. The modelling gap: Quantifying the discrepancy in the representation of thermal mass in building simulation. *Build. Environ.* **2018**, *131*, 74–98. [CrossRef]

35. Aste, N.; Leonforte, F.; Manfren, M.; Mazzon, M. Thermal inertia and energy efficiency—Parametric simulation assessment on a calibrated case study. *Appl. Energy* **2015**, *145*, 111–123. [CrossRef]

36. Tronchin, L.; Manfren, M.; Tagliabue, L.C. Optimization of building energy performance by means of multi-scale analysis—Lessons learned from case studies. *Sustain. Cities Soc.* **2016**, *27*, 296–306. [CrossRef]

37. Lepage, R.; Schumacher, C.; Straube, J.; Luxachko, A. *The Implications of Temperature Dependent Thermal Conductivity of Exterior Wall Using Insulated Sheathing*; Building Science Consulting: Waterloo, ON, Canada, 2013.

38. Al-Homoud, D.M.S. Performance characteristics and practical applications of common building thermal insulation materials. *Build. Environ.* **2005**, *40*, 353–366. [CrossRef]

39. Budaiwi, I.; Abdou, A.; Al-Homoud, M. Variations of thermal conductivity of insulation materials under different operating temperatures: Impact on envelope-induced cooling load. *J. Archit. Eng.* **2002**, *8*, 125–132. [CrossRef]

40. Budaiwi, I.; Abdou, A. The impact of thermal conductivity change of moist fibrous insulation on energy performance of buildings under hot–humid conditions. *Energy Build.* **2013**, *60*, 388–399. [CrossRef]

41. Cuce, E.; Cuce, P.M.; Wood, C.J.; Riffat, S.B. Toward aerogel based thermal superinsulation in buildings: A comprehensive review. *Renew. Sustain. Energy Rev.* **2014**, *34*, 273–299. [CrossRef]

42. Arnfield, A.J. Köppen Climate Classification. 2016. Available online: https://www.britannica.com/science/Koppen-climate-classification (accessed on 7 February 2018).

43. Kontoleon, K.J.; Giarma, C. Dynamic thermal response of building material layers in aspect of their moisture content. *Appl. Energy* **2016**, *170*, 76–91. [CrossRef]

44. Berardi, U.; Naldi, M. The impact of the temperature dependent thermal conductivity of insulating materials on the effective building envelope performance. *Energy Build.* **2017**, *144*, 262–275. [CrossRef]

45. Jentsch, M.F.; Bahaj, A.S.; James, P.A.B. Climate change future proofing of buildings—Generation and assessment of building simulation weather files. *Energy Build.* **2008**, *40*, 2148–2168. [CrossRef]

46. Jentsch, M.F.; James, P.A.B.; Bourikas, L.; Bahaj, A.S. Transforming existing weather data for worldwide locations to enable energy and building performance simulation under future climates. *Renew. Energy* **2013**, *55*, 514–524. [CrossRef]

47. Noussan, M.; Jarre, M. Multicarrier energy systems: Optimization model based on real data and application to a case study. *Int. J. Energy Res.* **2018**, *42*, 1338–1351. [CrossRef]

48. Corgnati, S.P.; Fabrizio, E.; Filippi, M.; Monetti, V. Reference buildings for cost optimal analysis: Method of definition and application. *Appl. Energy* **2013**, *102*, 983–993. [CrossRef]

*Article*

# Construction of Biodigesters to Optimize the Production of Biogas from Anaerobic Co-Digestion of Food Waste and Sewage

**Claudinei de Souza Guimarães [1],\*, David Rodrigues da Silva Maia [1] and Eduardo Gonçalves Serra [2]**

[1] Department of Biochemical Engineering, School of Chemistry, Federal University of Rio de Janeiro, 21941909 Rio de Janeiro, Brazil; davidrsmaia@gmail.com

[2] Department of Marine Engineering, Polytechnic School, Federal University of Rio de Janeiro, 21941909 Rio de Janeiro, Brazil; serra@poli.ufrj.br

\* Correspondence: claudinei@eq.ufrj.br; Tel.: +55-21-3938-7572

Received: 22 February 2018; Accepted: 27 March 2018; Published: 9 April 2018

**Abstract:** The objective of this study was to build and develop anaerobic biodigesters for optimization of biogas production using food waste (FW) and sewage (S) co-digestion from a wastewater treatment plant (WWTP). The biodigesters operated with different mixtures and in mesophilic phase (37 °C). During the 60 days of experiments, all control and monitoring parameters of the biodigesters necessary for biogas production were tested and evaluated. The biodigester containing FW, S and anaerobic sludge presented the biggest reduction of organic matter, expressed with removal of 88.3% TVS (total volatile solid) and 84.7% COD (chemical oxygen demand) the biggest biogas production (63 L) and the highest methane percentage (95%). Specific methane production was 0.299 l.CH$_4$/gVS and removed. The use of biodigesters to produce biogas through anaerobic digestion may play an important role in local economies due to the opportunity to produce a renewable fuel from organic waste and also as an alternative to waste treatment. Finally, the embedded control and automation system was simple, effective, and robust, and the supervisory software was efficient in all aspects defined at its conception.

**Keywords:** development of biodigesters; biogas production; anaerobic co-digestion; sewage; food waste; renewable fuel

## 1. Introduction

Nowadays, one of the biggest problems faced by many countries, particularly developing ones, is the final disposal of municipal solid waste (MSW), mainly due to environmental, social, and economic problems caused by its poor management. Costs related to collection, transport and treatment still make it difficult for adequate waste management, as its disposal is made in inappropriate areas, such as dumps, ditches, and other places devoid of adequate infra-structure [1,2]. Besides the problems associated with MSW, pollution of water resources and access to energy sources have historically represented challenges for economic growth, human health, and environmental preservation all over the world [3,4].

In Brazil, MSW presents organic matter as its biggest share, coming mainly from restaurants and households [5]. Anaerobic digestion is one of the solutions to reduce these problems and also an attempt to reuse MSW. It is known that anaerobic digestion is a process through which organic waste is biologically converted using a microbial consortium in the absence of oxygen [6]. In addition to stabilizing the organic load of waste, it generates products such as biogas, rich in methane, and digestate, which can be used as soil conditioner and is historically utilized for stabilizing sludge originated from sewage treatment, although it is a feasible application for any organic matter treatment [7]. Besides the potential for renewable energy generation, anaerobic digestion has become increasingly studied and more popular due to many factors, such as reduction of waste disposal in sanitary landfills and provision of energy to small communities situated away from urban centers. Another considerably evident advantage is the smaller generation of sludge. In anaerobic digestion about 10% of organic waste is turned into sludge, and the remaining 90% is used as biogas. Also important to highlight is the application of anaerobic processes at both small and large scale, with low implementation costs, low demand of area and good tolerance for high organic loads [8]. Therefore, biogas production and development of technologies for biomethane generation have been encouraged by many countries as an alternative for electricity generation or cogeneration of internal power engines [9–12].

Initially, mesophilic anaerobic co-digestion of a mixture consisting of a college restaurant food waste and sewage from a treatment plant were evaluated, in different proportions, with a view to obtaining a better use of organic waste and bigger methane production. The use of anaerobic digestion for treatment of the organic share of municipal solid waste, as well as food and street market waste, is largely mentioned in literature [13–16]. Consequently, anaerobic digestion has revealed itself as a promising treatment for urban solid waste produced in Brazil. Besides biogas, anaerobic digestion produces a digestate rich in nutrients which, depending on their characteristics, can be used as fertilizer or soil corrective. Environmentally speaking, its application on soil represents a more attractive option, as it allows nutrients to be recovered and reduces loss of organic matter suffered by soils under agricultural exploitation [17].

However, for power generation viability, it is necessary the use of anaerobic biodigesters, built at low cost and with high technology, along with control of all parameters for biogas production optimization. In this sense, the main objective of this study was to develop anaerobic biodigesters with control and automation systems, and for that purpose this project was divided into the following phases: bioreactor development (anaerobic biodigesters), heating system construction, agitation system, gases monitoring system, using initially, methane sensors ($CH_4$), carbon dioxide ($CO_2$), and hydrogen sulfide ($H_2S$), and finally software development to monitor and control the most important parameters during the process. The main bases were use of low cost materials, current monitoring technologies and simple control and methods to use the system with a view to large-scale reproduction in future.

## 2. Materials and Methods

### 2.1. Sample Collection and Description: Food Waste, Sewage and Anaerobic Sludge (Inoculum)

Raw sewage was collected from a wastewater treatment plant located in the city of Rio de Janeiro after removing sand from the preliminary treatment. Description was made according to Chemical Oxygen Demand (COD), moisture, pH, Total Solids (Volatile and Fixed), Total Kjedahl Nitrogen (TKN) and Total Phosphorus (TP). The anaerobic sludge used as inoculum in the experiments was collected from Upflow Anaerobic Sludge Blanket (UASB) reactor in operation at a local industry, making the same descriptions of raw sewage. After withdrawing an aliquot for characterization, sludge was stored under refrigeration (4 °C) until time of use. All analyses were determined according to Standard Methods for the Examination of Water and Wastewater [18]. For food waste, the same characterization as the one used for raw sewage and sludge was made, as well as total carbon test (CT). Collection of food waste was executed after meal time, when the rest removed from plates and utensils was submitted to screening, for separation from organic fraction, and homogenized through quartering, according to Brazilian norm [19]. Next, the homogenized material was ground with distilled water in the appropriate proportions, and part of the ground material (called food waste) was stored under refrigeration (4 °C) until time of use, and part preserved in a freezer (−20 °C).

Initially many experiments were carried out at small scale, using 100 mL penicillin bottles to find the ideal proportions for ultimate biogas production. In experiments with penicillin bottles the ground food waste was mixed with raw sewage to obtain different proportions of Dry Weight (DW): 20% 15%, 10% e 5%, with a view to evaluating moisture effect. Experiments were conducted with and without anaerobic sludge addition (inoculum) at the proportion of 10% $v/v$, for evaluation of seeding effect. The selected moisture was also used as basis for waste mixture in benchtop digesters experiments. Mixtures had their pH corrected to values between 7 and 8 using 1 M sodium bicarbonate solution (NaHCO$_3$), and were supplemented with potassium phosphate monobasic to correct phosphorus concentration, according to a relation COD:P of 350:1. The bottles were closed with rubber bungs and aluminum seals, and incubated at $37 \pm 2\,^{\circ}$C until stabilization of biogas production was reached, for 30 days, and the volume was measured through displacement of 60 mL plastic syringe plunger connected to rubber seals. Final volume used was 75% of total bottle volume, thus providing a 25% safety margin from middle surface to mouth bottle. This margin was important for, during the process, biogas blisters disturb the medium and transport it upwards, which can block the syringes. Every time stabilization of biogas volume in the syringes was reached, the biogas produced and conditioned in syringes was stored at $-5\,^{\circ}$C until analysis was made using gas chromatography. In this initial optimization phase of experiments, analyses of COD, Total Solids (Volatile and Fixed) and pH were carried out before and after anaerobic digestion process. Anaerobic sludge used as inoculum was collected in anaerobic reactor from poultry slaughter industry, being characterized in terms of volatile total solids. Each condition was evaluated with triplicates, and average values of biogas production were considered. Following this experimental phase, the ideal fraction was used in anaerobic biodigesters, at benchtop scale, built and developed in laboratory with total volume of 7 L. The experiments carried out followed the phases described in flow chart of Figure 1.

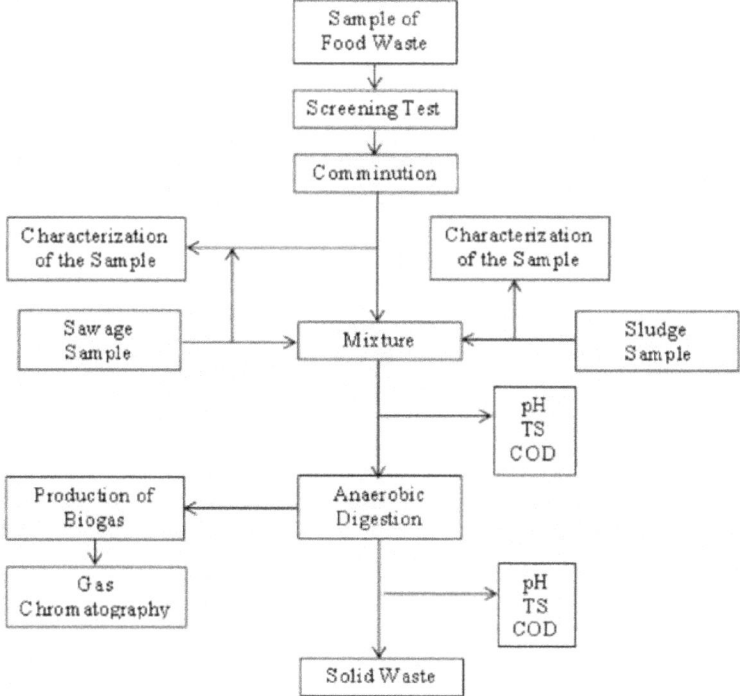

**Figure 1.** Flow chart of experimental phases carried out in this study.

### 2.2. Ideal Dry Weight Determination (DW) for Use in Biodigesters

The choice of ideal DW quantity to be put in biodigesters at benchtop scale was made from three experiments, in penicillin bottles, described as follows: Experiment 1 with different mixtures of food waste, with and without seeding. Mixtures were made to obtain 20%, 15% and 10% in DW; in Experiment 2 triplicates with different mixtures of sewage and food waste were made, with and without seeding. The objective was to obtain 20%, 15%, 10% and 5% in DW; and Experiment 3 used triplicates of different mixtures of 10% and 5% DW of sewage and food waste with anaerobic sludge (inoculum), and pH adjustment at the beginning of the experiment. All experiments were performed at $37 \pm 2\,^{\circ}$C and atmospheric pressure, for 30 days.

### 2.3. Development and Construction of Anaerobic Biodigesters for Biogas Production

After optimization of proportions made in penicillin bottles, experiments in bioreactors (anaerobic biodigesters) were initialized. Following the tendency to use simple materials, firstly the bioreactor was conceived and a glass container with cylindrical geometry was manufactured, involved with acrylic with total volume of 7 L, hermetically closed with a polyurethane cap. Seven holes were done on this cap, with a central hole for the stirring rod passage and the others for: biogas exit, temperature sensors accommodation, $CH_4$, $CO_2$, $H_2S$, pH, entrance and withdrawal of material, and the remaining closed and reserved for future application. The heating system was based on a thermostatic bath, consisting of one recipient manufactured in acrylic and dimensioned so that it could accommodate the bioreactor and the bath liquid, an electrical resistance vulgarly known as "body heater", and a water pump, used in car windshield wiper nozzle, for the circulation of the bath fluid. Agitation system was developed using an engine, commonly used for spinning microwave plates. This kind of engine was chosen as it has a low cost and functions in low rotation, which is preferable when it comes to anaerobic biodigesters. A stirring rod specially developed for these biodigesters was attached to the engine. The assembly scheme and addition of materials used in the experiments are shown in Figure 2; after that pH was adjusted before sealing to block oxygen entrance.

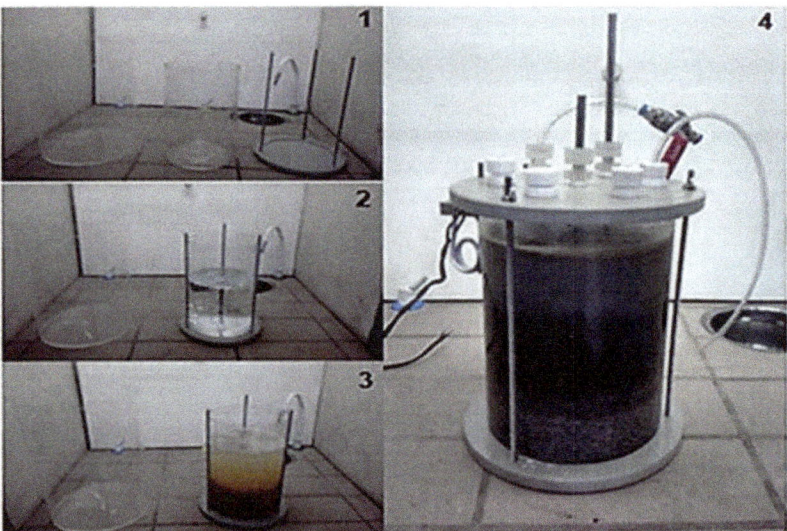

**Figure 2.** Biodigesters assembly phases for experiments execution. (1) The glass bioreactor and the support; (2) The bioreactor coupled to the support; (3) The bioreactor with solid waste; and (4) The bioreactor after addition of the inoculum and closed.

## 2.3.1. Software Development for Control and Monitoring

The developed control and automation unit consisted of a system furnished with control and Programmable Logic Controller (PLC), which communicates with a supervisory software. This system was completely developed: supervisory hardware, firmware and software [20]. The projected hardware can be divided into following main parts: microcontroller for the Central Processing Unit (CPU), communication between microcomputer-microcontroller, power modules and sensors signal conditioning modules aimed at providing for the system's needs, digital-analog converters (DAC) for the sensors, high power actuators, agitation engine and system for communication with the microcomputer. The control and automation software was developed through Microsoft Visual Studio 2012 using programming oriented to objects and Visual Basic language, having as requirements, defined at its conception, pH online monitoring, temperature, water pump activation control, agitation engine, PID temperature control, collected data storage, and analysis and visualization of parameters monitored by remote access. The complete biodigester functioning scheme is shown in Figure 3.

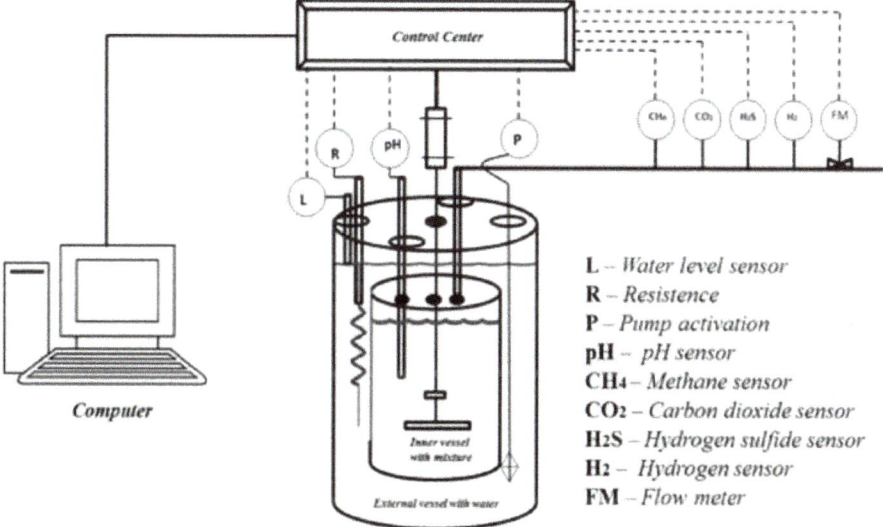

**Figure 3.** Complete representation of anaerobic biodigester functioning, developed and used on this study experiments.

## 2.3.2. Analysis and Calibration of Sensors in Biogas Produced Compounds

Initially biogas produced in biodigesters was stored in Sigma Aldrich Tedlar bags (St. Louis, MO, USA), specially prepared for gases. Biogas characterization was made in a gas chromatograph (Agilent Technologies, model 7820A, Santa Clara, CA, USA), with thermal conductivity detectors (TCD) and flame ionization detectors (FID). Samples were analyzed in triplicates, and to calculate concentrations, patterns of $CO_2$, $CH_4$, $H_2S$ (White Martins) were used. The gas sensors used were models MQ, Hanwei Electronics Co., Ltd., (Zhengzhou, China), which were pattern-calibrated. Results were compared with chromatographic analyses to verify detection efficiency and quantification of online monitoring system made by the sensors installed and the software developed.

## 3. Results and Discussion

### 3.1. Description of Food Waste Used before Blending

During this optimization phase, with a view to verifying the best proportions to be used in the biodigesters, three collections of secondary sludge, food waste and sewage were carried out, and only one test of Total Carbon (CT) was performed for food waste of 68.6 mg/L; the remaining descriptions are presented on Table 1.

**Table 1.** Description of residues used in mixtures.

| Parameters | Food Waste | Sewage | Sludge |
|---|---|---|---|
| | (Average ± SD) | (Average ± SD) | (Average ± SD) |
| Moisture % | 75 ± 8.2 | 97 ± 2.1 | 93 ± 4.1 |
| pH | 5.4 ± 0.2 | 6.2 ± 0.6 | 6.5 ± 0.3 |
| TFS (mg/g) | 8.4 ± 4.6 | 0.7 ± 0.2 | 22.4 ± 3.1 |
| TVS (mg/g) | 95.4 ± 34.1 | 0.6 ± 0.2 | 31.6 ± 4.4 |
| TKN (mg/L) | 18.5 ± 3.6 | 42.4 ± 5.4 | 26.5 ± 4.8 |
| TP (mg/L) | 0.05 ± 0.02 | 8.2 ± 0.9 | 15.4 ± 2.3 |

Food waste presented moisture lower than sludge and sewage, but after being blended it presented moisture close to values obtained for sewage and sludge. Residues presented acid pH, below 7, showing that its final mixture will have acid pH, and that the addition of an alkalizer to adjust pH at the beginning of the experiments will be necessary. Initially pH adjustment was done with $NaHCO_3$, which could hinder anaerobic digestion as it would increase sodium concentration at inhibiting levels. Food waste low C/N ratio (3.8) indicated presence of high nitrogen concentrations on residue which, in this case, can be converted into ammonia, promoting pH increase and inhibiting methanogenic microorganisms [21]. High TVS value in food waste suggests the possibility of biogas production increase, specially methane, when blended with sewage sludge. Work by Kim et al. [22] evaluated the blend of food waste with sewage sludge and noted that the addition of organic waste improves anaerobic digestion. Based on the results obtained in the initial phase, the best proportion was 10% in dry weight (DW).

### 3.2. Biodigesters Construction and Efficiency of Food Waste and Sewage Co-Digestion

In this second phase of the study, after optimization of ideal proportion, three biodigesters were built in the following situations: Biodigester B1 containing only a mixture of food waste and Sewage; Biodigester B2 containing a mixture of Food Waste, Sewage and Anaerobic Sludge (inoculum); and Biodigester 3 with a blend of Food Waste, Water and Anaerobic Sludge. Biodigester B1 was used as a standard experiment, that is, an experiment with the objective of verifying co-digestion viability without the presence of inoculum. For the mixtures, tests of Chemical Oxygen Demand (COD), Total Solids (TS) and Total Volatile Solids (TVS) were performed at the beginning and monitored during the 60 days of the experiment, besides pH, yield, biogas production and all biodigesters operability parameters. A complete scheme of the biodigesters in operation during the experiments is shown in Figure 4.

Efficiency profiles of anaerobic co-digestion bioprocess of food waste and sewage for Biodigesters B2 and B3 are presented in Figure 5 and, as it can be noted, TS, TVS and COD concentrations were similar at the beginning. For Biodigester B3, a low removal of TVS occurred, compared with B2, but with stable removal since the beginning of the experiment, obtaining removal of total solids and total volatile solids of 53.8% TS, 75.1% TVS and 46.2 COD at the end of the 60 days of experiment. Biodigester B2 showed high elimination of organic waste, expressed with the removal of 88.3% TSV and 84.7% COD. These results can be associated with different types of residues present in biodigesters, specially co-digestion of sewage with food waste, which probably contributed for the increase of organic matter degradation and the higher biogas production such as presented on Table 2. These results showed that the specific methane production was 0.299 $LCH_4/gVS$ removed to Biodigester B2, and 0.116 $LCH_4/gVS$ for Biodigester B3.

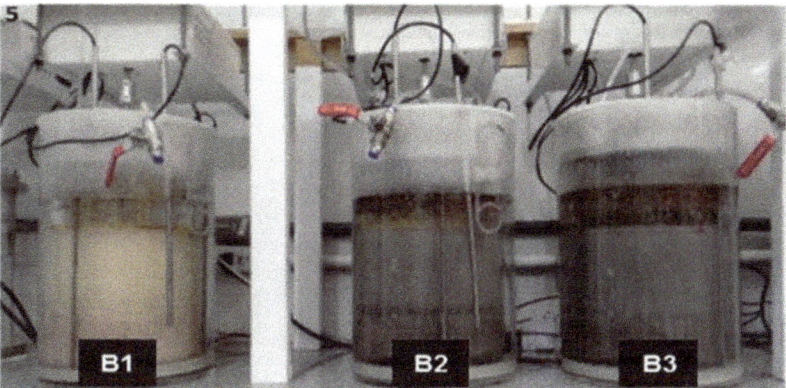

**Figure 4.** Operational and efficiency experiments on biodigesters constructed using different mixtures and proportions. B1: Food residue + Sewage; B2: Food residue + Sewage + Sludge (inoculum); and B3: Food residue + Water + Sludge (inoculum).

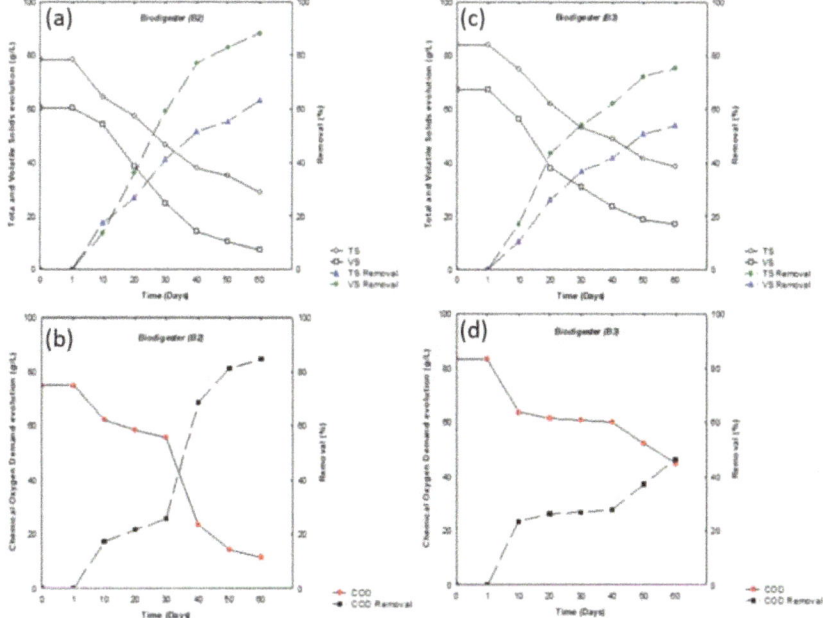

**Figure 5.** Efficiency profiles of anaerobic co-digestion bioprocess of food waste and sewage for Biodigesters B2 and B3 during the 60 days of experiments. (**a**) removal of TS and VS from B2; (**b**) removal COD from B2; (**c**) removal of TS and VS from B3; and (**d**) removal COD from B3.

**Table 2.** Production of removed organic waste and methane yield in Biodigesters B2 and B3 after 60 days of experiments.

| 60 Days | Organic Matter Removal (%) | | | Biogas Production (L/day) | Accumulative (L) | |
|---|---|---|---|---|---|---|
| | TVS | COD | TS | | Biogas | Methane |
| Biodigester (B2) | 88.3 | 84.7 | 63.3 | 1.1 | 63 | 59.9 |
| Biodigester (B1) | 75.1 | 46.2 | 53.8 | 0.5 | 28 | 23.8 |

To evaluate detection level and quantification of gases sensors ($CH_4$, $CO_2$ and $H_2S$), a system which could condition biogas in a closed system was used. Every two days this sample passed through the sensors and the result was then compared with the ones from chromatographic analyses. In all samples errors below 5% were found, implying a high reliability on the monitoring system developed in the biodigesters. Efficiency profile of methane production associated with pH variation in Biogas is shown in Figure 6. Although the results of biogas total volume production were different for Biodigesters B2 and B3, gases behavior ($CH_4$, $CO_2$ and $H_2S$) produced by biodigesters, in percentage, had similar values, that is, initially a high $CO_2$ production (acetogenesis phase) occurred, and as time went by concentration was decreasing and concentration of methane with basic pH was increasing. $H_2S$ behavior was similar for both biodigesters. A pH correction during the first 10 days in Biodigester B2 became necessary, the same occurring in Biodigester B3 in the first 20 days. In Biodigester B2 necessary pH control was performed with addition of sodium hydroxide (6N).

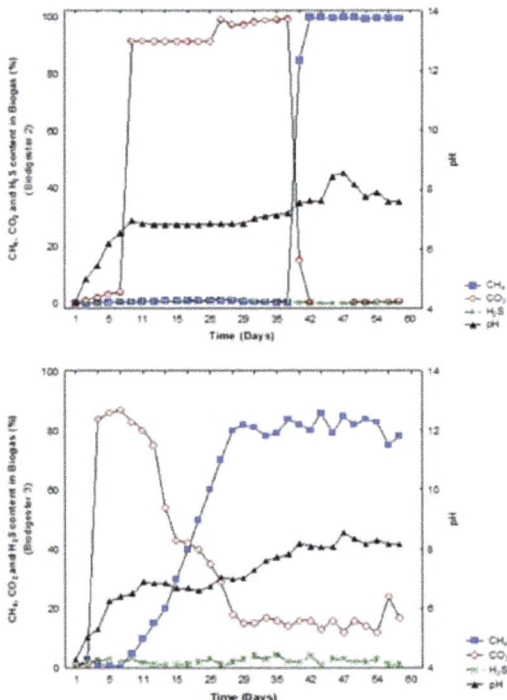

**Figure 6.** Efficiency profile of methane production associated with pH variation in Biogas. The Biodigester 2 presented higher methane production from the thirty-seventh day of experiment and the Biodigester 3 started the production of methane in a shorter period of time (from the tenth day of experiment).

## 4. Conclusions

It was possible to evaluate the influence of food waste and sewage co-digestion in biogas production for anaerobic digestion processes. A mixture of Biodigester B2 presented the highest biogas production (63 L), the biggest methane return (95%) and the best TS, TVS and COD reductions on the 60 days of experiments. The better efficiency of Biodigester B2, compared to B3, cannot be attributed only to sewage addition to the mixture, since the concentration of organic matter in sewage is usually low. Such result might be associated with composition of substances present in sewage which complement anaerobic microorganisms needs and contributed to better conditions during biodegradation of food waste components. Another important factor is that probably food waste does not contain representatives of all microbial population necessary for its complete degradation; consequently, it might have influenced biogas production in Biodigester B3, making it slower and lower. Regarding the developed biodigesters, the bioreactor physical system proved adequate and allowed us a clear visualization of the fermenting medium, as well as checking whether the agitation system was effective or not. Besides, it guaranteed a low $H_2S$ production in biogas, due to a complete system sealing, preventing oxygen entrance in the medium. Finally, with these and other results we can conclude that the biodigester developed with the automation and control system was satisfactory for biogas production. The control parameters of pH, temperature, agitation, and the software developed for biogas monitoring worked well. The use of current technologies and low-cost materials was enough for the experimental purpose and can be used at larger scale in future.

**Acknowledgments:** The authors would like to acknowledge financial support from CNPq (Brazil).

**Author Contributions:** Claudinei de Souza Guimarães conceived and designed the experiments, corrected and revised the manuscript; Eduardo Gonçalves Serra made a great contribution to the revision of the manuscript, analysis and interpretation of the experimental results; David Rodrigues da Silva Maia carried out the assembly and calibration of the experiments, developed the automation and control system for biodigesters. All authors read, corrected, and approved the manuscript.

**Conflicts of Interest:** The authors declare no conflict of interest.

## Abbreviations

| | |
|---|---|
| CPU | central processing unit |
| DAC | digital-analog converters |
| FID | flame ionization detectors |
| FW | food waste |
| MSW | municipal solid waste |
| PID | proportional–integral–derivative |
| PLC | programmable logic controller |
| S | sewage |
| TCD | thermal conductivity detectors |
| WWTP | wastewater treatment plant |

## References

1. Carlini, M.; Mosconi, E.M.; Castelluci, S.; Villarini, M.; Colantoni, A. An Economical Evaluation of Anaerobic Digestion Plants Fed with Organic Agro-Industrial Waste. *Energies* **2017**, *10*, 1165. [CrossRef]
2. Apetato, M.M.; Nobre, A.M.; Alves, J.C.; Robalo, G.S.; Ferreira, F. Taxa de Resíduos Urbanos: Deficiências e Soluções. In Proceedings of the 6a Conferência Nacional sobre Qualidade do Ambiente, Lisboa, Portugal, 20 October 1999; Editora Plátano: Lisboa, Portugal, 1999; Volume 3, pp. 363–369.
3. Lindkvist, E.; Johansson, M.T.; Rosenqvist, J. Methodology for Analysing Energy Demand in Biogas Production Plants—A Comparative Study of Two Biogas Plants. *Energies* **2017**, *10*, 1822. [CrossRef]
4. Lansing, S.; Botero, R.B.; Martin, J.F. Waste treatment and biogas quality in small-scale agricultural digesters. *Bioresour. Technol.* **2008**, *99*, 5881–5890. [CrossRef] [PubMed]

5. Plano Nacional de Resíduos Sólidos. Lei n° 12.305/10. Available online: http://www.mma.gov.br/estruturas/253/_publicacao/253_publicacao02022012041757.pdf (accessed on 30 March 2017).

6. Li, Y.; Park, S.Y.; Zhu, J. Solid-state anaerobic digestion for methane production from organic waste. *Renew. Sustain. Energy Rev.* **2011**, *15*, 821–826. [CrossRef]

7. Cecchi, F.; Pavan, P.; Alvarez, J.M.; Bassetti, A.; Cozzolino, C. Anaerobic digestion of municipal solid waste: Thermophilic vs. mesophilic performance at high solids. *Waste Manag. Res.* **1991**, *9*, 305–315. [CrossRef]

8. Chernicaro, C.A.L. *Princípios do Tratamento Biológico de águas Residuárias. Reatores Anaeróbios*; Universidade Federal de Minas Gerais: Belo Horizonte, Brazil, 1997.

9. Budzianowski, W.M.; Budzianowska, D.A. Economic analysis of biomethane and bioelectricity generation from biogas using different support schemes and plant configurations. *Energy* **2015**, *88*, 658–666. [CrossRef]

10. Patterson, T.; Esteves, S.; Dinsdale, R.; Guwy, A. Life cycle assessment of biogas infrastructure options on a regional scale. *Bioresour. Technol.* **2011**, *102*, 7313–7323. [CrossRef] [PubMed]

11. Jha, A.K.; Li, J.; Zhang, L.; Ban, Q.; Jin, Y. Comparison between Wet and Dry Anaerobic Digestions of Cow Dung under Mesophilic and Thermophilic Conditions. *Adv. Water Resour. Prot.* **2013**, *1*, 28–38.

12. Venkatesh, G.; Elmi, R.A. Economic-environmental analysis of handling biogas from sewage sludge digesters in WWTPs (wastewater treatment plants) for energy recovery: Case study of Bekkelaget WWTP in Oslo (Norway). *Energy* **2013**, *58*, 220–235. [CrossRef]

13. Ward, A.J.; Hobbis, P.J.; Holliman, P.J.; Jones, D.L. Optimization of the Anaerobic Digestion of Agricultural Resources. *Bioresour. Technol.* **2008**, *99*, 7928–7940. [CrossRef]

14. Fernandez, J.; Perez, M.; Romero, L.I. Kinetics of mesophilic anaerobic digestion of the organic fraction of the municipal solid waste: Influence of initial total solid concentration. *Bioresour. Technol.* **2010**, *101*, 6322–6328. [CrossRef] [PubMed]

15. Kim, D.H.; OH, S.E. Continuous high-solids anaerobic co-digestion of organic solid wastes under mesophilic conditions. *Waste Manag.* **2011**, *31*, 1943–1948. [CrossRef] [PubMed]

16. Charles, W.; Walker, L.; Cord-Ruwisch, R. Effect of pre-aeration and inoculum on the start-up of batch termophilic anaerobic digestion of municipal solid waste. *Bioresour. Technol.* **2009**, *100*, 2329–2335. [CrossRef] [PubMed]

17. Mata-Alvarez, J.; Macé, S.; Llabrés, P. Anaerobic Digestion of Organic Solid Wastes. An Overview of Research Achievements and Perspectives. *Bioresour. Technol.* **2000**, *74*, 3–16. [CrossRef]

18. American Public Health Association (APHA). *Standard Methods for the Examination of Water and Wastewater*, 21st ed.; American Public Health Association: Washington, DC, USA, 2005.

19. Brazilian Association of Technical Standards. *NBR 10007. Sampling of Solid Waste*, 2nd ed.; Brazilian Association of Technical Standards: Rio de Janeiro, Brazil, 2004.

20. Guimarães, C.S.; Maia, D.R.S.; Serra, E.G. Control and Monitoring Software Website. GitHub. 2018. Available online: https://github.com/DavidRSMaia/Software-de-Controle-e-Monitoramento (accessed on 22 February 2018).

21. Appels, L.; Baeyens, J.; Degrève, J.; Dewil, R. Principles and potential of the anaerobic digestion of waste-activated sludge. *Prog. Energy Combust. Sci.* **2008**, *34*, 755–781. [CrossRef]

22. Kim, H.-W.; Han, S.-K.; Shin, H.-S. The optimisation of food waste addition as a co-substrate in anaerobic digestion of sewage sludge. *Waste Manag. Res.* **2003**, *21*, 515–526. [CrossRef] [PubMed]

*Article*

# The Key Role of the Vector Optimization Algorithm and Robust Design Approach for the Design of Polygeneration Systems

**Alfredo Gimelli * and Massimiliano Muccillo**

DII—Department of industrial engineering, University of Naples Federico II, 80125 Napoli, Italy;
massimiliano.muccillo@unina.it
* Correspondence: gimelli@unina.it; Tel.: +39-081-768-3271

Received: 19 February 2018; Accepted: 30 March 2018; Published: 2 April 2018

**Abstract:** In recent decades, growing concerns about global warming and climate change effects have led to specific directives, especially in Europe, promoting the use of primary energy-saving techniques and renewable energy systems. The increasingly stringent requirements for carbon dioxide reduction have led to a more widespread adoption of distributed energy systems. In particular, besides renewable energy systems for power generation, one of the most effective techniques used to face the energy-saving challenges has been the adoption of polygeneration plants for combined heating, cooling, and electricity generation. This technique offers the possibility to achieve a considerable enhancement in energy and cost savings as well as a simultaneous reduction of greenhouse gas emissions. However, the use of small-scale polygeneration systems does not ensure the achievement of mandatory, but sometimes conflicting, aims without the proper sizing and operation of the plant. This paper is focused on a methodology based on vector optimization algorithms and developed by the authors for the identification of optimal polygeneration plant solutions. To this aim, a specific calculation algorithm for the study of cogeneration systems has also been developed. This paper provides, after a detailed description of the proposed methodology, some specific applications to the study of combined heat and power (CHP) and organic Rankine cycle (ORC) plants, thus highlighting the potential of the proposed techniques and the main results achieved.

**Keywords:** vector optimization; evolutionary genetic algorithm; robust design optimization; combined heat and power; ORC systems

## 1. Introduction

In recent decades, the world energy consumption has continuously increased, especially because of the strong economic growth of non-OECD (Organisation for Economic Co-operation and Development) countries [1,2]. Fossil fuels have mostly been used to cope with this increased energy demand, leading to considerable concerns about climate change. Therefore, an increasing use of renewable energy sources and a more efficient exploitation of primary energy sources are indispensable to reduce carbon dioxide emissions and limit global warming effects, as suggested by the chart in Figure 1 [3]. In this figure, the first column represents the world's total primary energy supply in 2011. The second bar represents the outlook of the primary energy supply at 2050 in the current energy scenario, for which the average global temperature rise is projected to be 6 °C. The third column represents the outlook at 2050 in the 450 Scenario, for which the average global temperature increase should be limited to 2 °C. The figure highlights how the share of renewable energy sources should reach about 40%, while the primary energy consumption should decrease about 25% compared to the 6 °C Scenario (i.e., the energy efficiency should increase by about 25%) to limit the average temperature rise.

Due to improvements in energy efficiency and to an increasing use of low-carbon energy and renewable energy sources, energy-related $CO_2$ emissions stalled in 2015, as reported in [4]. However, with the development of innovative technologies in the future [5] in addition to traditional engines [6–11], increasing exploitation of energy sources and increasing share of renewable energies in final energy consumption will still be fundamental if we are to cope with the challenges due to the world energy balance and recognized by the Paris Agreement on climate change. In particular, the World Energy Outlook of 2016 (WEO-2016) highlighted how a key player in further emissions reductions is recognized to be the wider use of renewable energy sources in the power sector. The WEO-2016 also highlighted how in the industrial sector alone, an additional investment of about 300 billion dollars could reduce the 2040 global electricity demand by about 5%, thus avoiding investments of about 450 billion dollars in power generation.

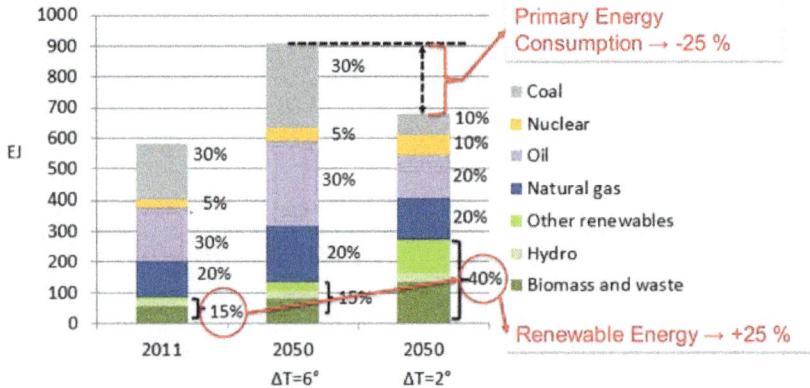

**Figure 1.** World total primary energy supply detailed by fuel [3].

In this scenario, a key role for primary energy saving and greenhouse gas emission reduction could be played by polygeneration systems [12–19]. These are mainly cogeneration systems delivering useful electric (or mechanical) and heat output making use of a single primary energy source (often referred to as combined heat and power (CHP) or combined cooling, heat, and power (CCHP) systems) or more complex integrated energy systems providing heat and power from a combination of renewable and non-renewable power plants and equipment (e.g., solar PV systems, wind turbines, biomass boilers, and CHP systems). More precisely, as stated by Serra et al. in [12] and Song et al. in [20], polygeneration could be defined as the combined supply of two or more energy services and/or manufactured products with the aim of maximizing the exploitation of the energy source that supplies the plant. Examples of polygeneration systems are cogeneration and trigeneration plants, dual-purpose power, and desalination plants [12]. In particular, combined heat and power generation can allow for considerable decreases in primary energy consumption, $CO_2$ emissions, and costs [21–26].

The strategic role of polygeneration systems in the achievement of the Paris Agreement goals involves leading the transition from centralized energy generation to mature, distributed, small- and medium-scale energy generation. However, due to the decrease in thermal efficiency and the increase in specific investment costs with the reduction in plant size, the actual utilization of the thermal energy provided by a cogeneration plant is essential to optimize energetic and economic performance (for example, fuel utilization factor, net present value, and $CO_2$ emission). Figure 2 clearly shows this concept applied to a CHP plant based on the internal combustion engine. In particular, it shows the primary energy saving (PES) as a function of the plant's electrical efficiency (i.e., engine size if ON/OFF operation is assumed) for $\eta_{eREF} = 0.46$, $\eta_b = 0.9$, and the different ratios of the available nominal thermal power actually exploited by the final user.

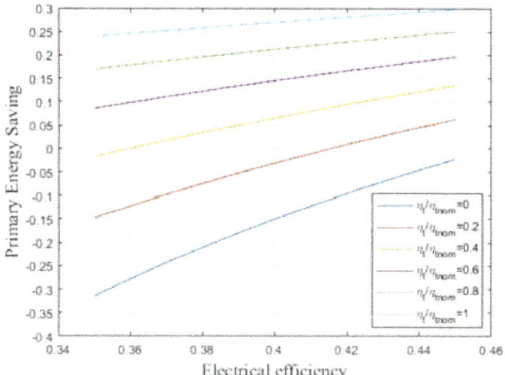

**Figure 2.** Primary energy saving (PES) as function of the combined heat and power (CHP) plant's electrical efficiency for different ratios of the nominal thermal power actually exploited (the chart is representative of CHP plants adopting a reciprocating internal combustion engine).

As for the possible energy subsystem of polygeneration plants, an increasingly important role is played by organic Rankine cycle (ORC) plants [27] thanks to the possibility of exploiting low-enthalpy heat sources. ORC plants convert waste heat into electrical energy. However, they often integrate complex polygeneration plants to improve the energetic performance or to adjust the electricity to the thermal energy output ratio of the whole system, as demonstrated in [28–30]. In this context, unlike centralized power plants [31,32], ORC technology encourages distributed power generation [33,34]. However, the efficiency and cost optimization of ORC plants is a key issue, due to their high specific investment cost and low thermal efficiency [35]. Therefore, the goal of this research work is to highlight the key role that advanced mathematical methods could have for the optimal configuration of polygeneration plants, also focusing on the optimization of ORC plants given their increasing use within complex polygeneration systems. In particular, evolutionary genetic optimization algorithms could be useful in identifying optimal solutions, even when conflicting goals are pursued. For this purpose, and with reference to the load profiles of an Italian hospital facility, in the first part of this research, the energetic and economic advantages achievable with the use of optimized cogeneration plants ensuring electricity, sanitary hot water, and space heating are addressed. Vector optimization techniques were adopted by coupling an evaluation algorithm developed by the authors with an evolutionary optimization algorithm.

The calculation algorithm includes variable energy demands, variability for the specific investment costs, revenue from selling the exceeding electrical energy to the grid taking into consideration different time periods, different pricing periods based on an Italian three-tier tariff; nominal efficiencies depending on CHP engine size, and all of the main elements of complexity discussed in research activities already published [36,37]. Moreover, the methodology includes other topics that have not yet been addressed, enabling the design of a CHP plant when energetic, economic, or legislative scenarios change. In fact, many research works have ignored uncertainties that could affect the expected results, as stated in [38]. In [39], the development of an operation optimization model for CHP plants is addressed, and the energy prices are forecasted. However, most of the studies have assumed fixed values for tariffs and other quantities, while these values are variable during the plant's life span. Furthermore, most of the developed methodologies do not identify technical solutions actually available in the market, as stated in [40].

Therefore, the following two key problems were analyzed in this article:

- the instability of the results due to mismatches between the marketed CHP engines and those provided by the calculation procedure;

- the instability of the results due to variations in the reference energetic and electricity tariff scenarios.

In this study, firstly, optimal energetic and economic solutions (i.e., engine number and size) were found. Subsequently, a multi-objective robust design optimization approach was used to find the most stable plant solutions. Although robust design techniques have been rarely adopted within the field of energy systems [41], this research proposes a novel and valid application of such techniques to polygeneration plants. Robust optimization is a methodology addressing uncertainty in the input data of an optimization problem; further details can be found in [42–44].

Lastly, a vector optimization problem concerning an ORC plant supplied by biomass [45] was solved. Objectives of the optimization problem were the maximization of the global electric efficiency and the minimization of the overall heat exchangers area, which could be related to the cost of the plant and its size. Most of the research works on ORC power plants focus on the thermodynamic optimization and fluid selection [46–48]. However, few studies address the vector optimization of an ORC plant according to specific thermodynamic and economic objective functions, which is a more reasonable approach than single-objective optimization [49].

## 2. The Proposed Methodology

### 2.1. The Vector Optimization Approach for CHP System Optimization

Starting from data concerning thermal and electric power required by the reference hospital facility [1], the energetic and economic performance of the plant were calculated over the plant's life span, which is estimated to be 10 years. In particular, the heat demand curve represented in Figure 3 was obtained by reordering the annual thermal load to ensure its contemporaneity with the duration curve of the electrical load reported in the same figure. A vector optimization process was performed to find the optimized modular plant configurations. Similar methodologies were already adopted in [50–53]. Vector optimization [54–57] can be useful to conduct predictive investigations on a high number of plant solutions, also highlighting eventual tradeoffs between energetic and economic results. The vector optimization problem is generally formalized as follows [58]:

$$minF(x) = min(F_1(x), F_2(x), \ldots, F_k(x)) \tag{1}$$

where $x \in X$, $F_i : R^n \rightarrow R$, $i = 1, \ldots, k$, and $k \geq 2$, where $F_1(x)$, $F_2(x), \ldots, F_k(x)$ are conflicting functions, $R^k$ is the objectives space, and $R^n$ is the decision variable space. For this reason, vector $x \in R^n$ is a decision variable, while $y = F(x) \in R^k$ is a vector of objectives. Table 1 shows decision variables and objective functions of the problem. The optimal solutions are identified from the notion of partial ordering. The minimum problem of Equation (1) is based on the Pareto dominance concept and usually provides a set of optimal solutions.

The authors developed a specific evaluation algorithm. This algorithm, which will be described in detail in the evaluation algorithm, was coded and coupled to the genetic algorithm MOGA II, according to the logic scheme shown in Figure 4.

**Table 1.** Decision variables and objective functions of the optimization problem. SPB: simple payback period.

| DECISION VARIABLES | | OBJECTIVE FUNCTIONS | |
|---|---|---|---|
| VARIABLE | RANGE | OBJECTIVE | TARGET |
| Electric CHP power | 150–1000 kW | PES | maximize |
| CHP engine number | 1–9 | SPB | minimize |

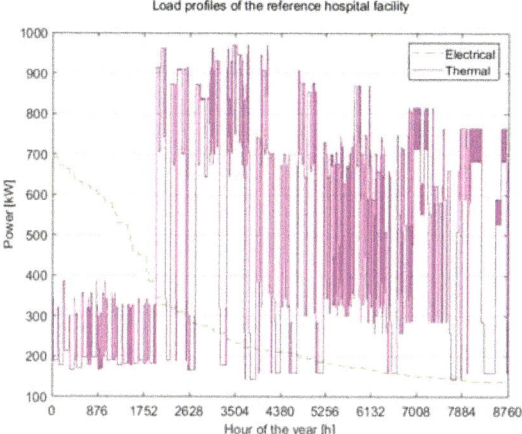

**Figure 3.** Annual duration curve of the electrical load and contemporary thermal load profiles of the analyzed hospital facility.

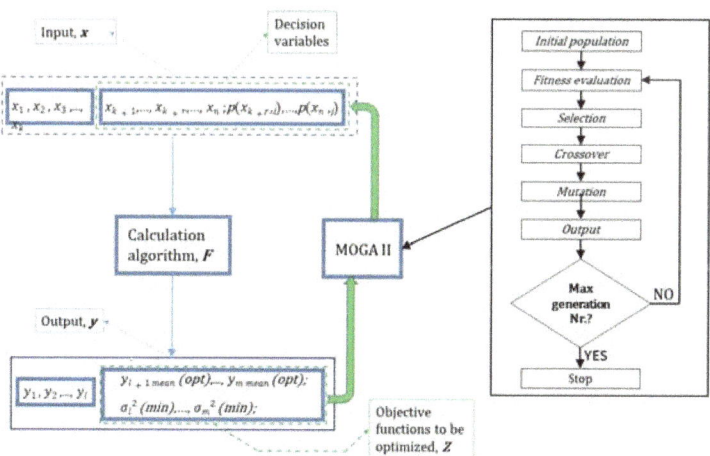

**Figure 4.** Workflow of the multi-objective optimization process.

Although each evolutionary algorithm proposes a different approach, each of them simulates the evolution of populations through the application of the genetic operators of selection, mutation, and crossover. Genetic algorithms are iteratively executed on a set of coded chromosomes, called the population. Each genetic operator is applied to the current individual according to predefined probabilities. MOGA II is a version of MOGA improved by Poloni [59,60].

However, several inputs to the algorithm can be characterized by uncertain values or can vary during the plant's life span. Therefore, vector optimization problems were also solved to evaluate the stability of the results to mismatches between calculated and marketed solutions. Possible variations in the energetic and tariff scenarios have also been considered. In particular, a robust design methodology was used. For this reason, some specific decision variables or economic and energetic input quantities to the proposed evaluation algorithm were redefined using a probability distribution. The vector optimization processes were then performed. A stable solution is characterized by a lower sensitivity to fluctuations of unknown variables. However, the most stable solutions may not include the solutions on the Pareto Front. The discrete formulation for the MORDO (multi-objective robust design optimization) problem can be written as [61]

$$minF(x, \sigma)$$
$$p(x_j): P(x_j) = \sum_{j=1}^{j} p(x_j) \in [0, 1]$$
$$max\ F_{mean},\ where\ F_{mean} = \overline{F} = \sum_{j=1}^{q} \frac{F_j}{q} \tag{2}$$
$$min\ \sigma_F^2,\ where:\ \sigma_F^2 = \sum_{j=1}^{q} \frac{(F_j - F_{mean})^2}{q-1};\ x_j \in R\ and\ F: R \to R.$$

In Equation (2), $\sigma$ represents the stochastic description of the variable $x$, $p(x_j)$ represents the probability density function, and $P(x_j)$ represents the cumulative distribution function. The mean value and the variance for the random variable $x$ can be calculated as follows:

$$\overline{x} = \sum_{j=1}^{q} \frac{x_j}{q} \tag{3}$$

$$\sigma^2 = \sum_{j=1}^{q} \frac{(x_j - \overline{x})^2}{q-1}. \tag{4}$$

The uniform and the normal distributions were adopted in this study. More details concerning these probability density functions can be found in [36].

The Evaluation Algorithm

Cogeneration plants based only on natural gas internal combustion engines (ICEs) were considered because of the dominance of ICEs in small- and medium-scale applications. The main input variables to the developed algorithm are as follows:

- annual electrical ($P_e$) and thermal load ($P_t$) of the user;
- nominal electrical power of the CHP gas engine ($P_{enom}$) and their number ($N_{CHP}$);
- reference efficiency for thermo-electric power generation ($\eta_{eref}$);
- average boiler efficiency ($\eta_b$);
- CHP plant maintenance costs for kWh of generated electrical energy ($M$);
- fuel lower heating value ($HV_L$);
- electrical energy price based on a three-tier time-of-use (TOU) tariff, not including VAT ($C_{ueref,Fi}$);
- electricity taxation $I_{ue}$;
- electrical energy selling price in the billing periods $Fi$ ($P_{ue,Fi}$);
- the selling price of energy efficiency certificates (EECs) for cogeneration plants to be recognized as highly efficient ($P_{EEC}$);
- natural gas tariff, not including VAT ($T$);
- natural gas taxation ($I_{um}$);
- peak demand charge ($C_{uepeak}$);
- discount rate ($a$);
- lifetime of the plant ($n$).

The annual electrical ($P_e$) and thermal loads ($P_t$) of the user were determined via hourly average values, providing the possibility to simulate the hourly operation of the entire CHP-user system over an entire year. Further details are discussed in [1].

Total PES (TPES) and simple payback period (SPB) are the main output of the evaluation algorithm, which depend on the specific operation strategies adopted for the CHP engines. An ON/OFF operation was imposed to each cogeneration plant (i.e., the electrical power delivered by the plant depends on the engine size). The engines are switched ON when their operation results in a positive contribution to the PES. Otherwise, when no energetic advantage is achieved, the engines are switched OFF. In particular, according to the management strategy adopted, any hour of the year that is characterized by positive values of the primary energy saving index is included in the operating range of the CHP plant. Variations in the nominal electrical ($\eta_{enom}$) and thermal ($\eta_{tnom}$) efficiency with the engine size were imposed, according to the curves represented in Figure 5. The efficiency curves in Figure 5 are based on the rated values of some cogeneration natural gas engines currently on the market. Figure 5. also depicts the regression curve adopted to evaluate the specific investment cost of each single cogeneration unit ($C_{uCHP}$) as a function of its size. The nominal thermal power of the CHP engine is then evaluated as follows:

$$P_{tnom} = (P_{enom}/\eta_{enom}) \cdot \eta_{tnom}. \tag{5}$$

To calculate the operating costs of the entire CHP-user system, the electrical load profile has been characterized according to a three-tier Italian tariff [1].

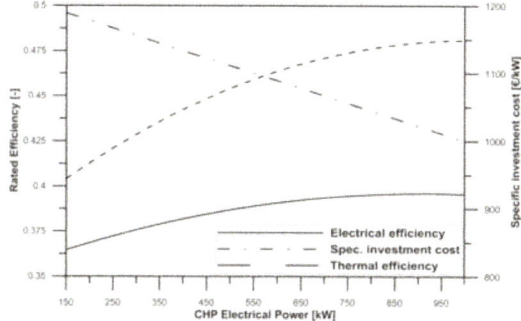

**Figure 5.** Variation in the nominal efficiencies and specific investment cost with the cogeneration plant size.

Analyses are based on the estimation of the hourly average thermal power actually exploited ($P_{tCHP}$), as calculated from thermodynamic considerations. This estimation is then used to define the operating range of the cogeneration system through an hourly primary energy calculation that is performed as follows by assuming the CHP plant is ON over the whole year:

$$TPES(t) = \frac{E_{p,RS}(t) - E_{p,PS}(t)}{E_{p,RS}(t)} \cdot 100 \tag{6}$$

where

$$E_{p,RS}(t) = E_{pb,RS}(t) + E_{pe,RS}(t) + E_{eexc}(t)/\eta_{eref} \tag{7}$$

$$E_{p,PS}(t) = E_{pb,PS}(t) + E_{pCHP}(t) + E_{peint}(t). \tag{8}$$

The $TPES(t)$ in Equation (6) compares the primary energy consumption that characterizes the interaction among the cogeneration system, the user, the electrical grid, and the auxiliary boilers in the analyzed energy system to the primary energy consumption, which characterizes the separate production of the same amount of energy. Therefore, $TPES(t)$ also considers the thermal energy provided by auxiliary boilers and the electrical energy imported from the grid. Moreover, when considering the separate production, the electrical energy provided by the cogeneration system and exceeding the user load demand must be considered as generated with the average efficiency of thermoelectric power generation ($\eta_{eref} = 0.46$ in Italy).

Once the vector including the number of CHP engines that are switched ON hourly (N(t)), and thus the actual CHP engine number ($N_{CHP}$), has been redefined according to the adopted operating strategy, it is possible to evaluate the hourly average electric ($P_{eCHP}(t)$) and thermal ($P_{tCHP}(t)$) power provided by the CHP plant. Moreover, to perform detailed energetic and economic analyses, the electric loads of the user ($P_e(t)$), the CHP electric power ($P_{eCHP}(t)$), and therefore the rate of $P_{eCHP}(t)$ self-consumed by the user ($P_{eself}(t)$) have been characterized according to the considered TOU tariff, allowing for the definition of vectors $P_{e,Fi}$, $P_{eCHP,Fi}$, and $P_{eself,Fi}$. The individual contributions in Equations (6) and (7) are calculated through the following equations:

$$E_{pb,PS}(t) = (P_{tb,PS}(t) \cdot \Delta t)/\eta_b \tag{9}$$

$$E_{pCHP}(t) = (P_{eCHP}(t) \cdot \Delta t)/\eta_{enom} \tag{10}$$

$$E_{peint}(t) = (P_{eint}(t) \cdot \Delta t)/\eta_{eref} \tag{11}$$

$$E_{pb,RS}(t) = (P_t(t) \cdot \Delta t)/\eta_b \tag{12}$$

$$E_{pe,RS}(t) = (P_e(t) \cdot \Delta t)/\eta_{eref} \tag{13}$$

given that the following definitions are assumed:

$$P_{tb,PS}(t) = P_t(t) - P_{tCHP}(t) \tag{14}$$

$$P_{eCHP}(t) = N(t) \cdot P_{enom} \tag{15}$$

$$P_{tnom}(t) = N(t) \cdot P_{tnom} \tag{16}$$

$$P_{eexc}(t) = P_{eCHP}(t) - P_{eself}(t). \tag{17}$$

Once the self-consumed electrical power ($P_{eself}(t)$) is determined, the electrical power imported from the grid $P_{eint}(t)$ is given by:

$$P_{eint}(t) = P_e(t) - P_{eself}(t) \tag{18}$$

where $P_e(t)$ is the average power requested from the final user. Once $P_{eint}(t)$ has been evaluated with Equation (18), $P_{eexc,Fi}(t)$ and $P_{eint,Fi}(t)$ can be recalculated before the energetic balance of the whole reference year is performed, and the resulting TPES can finally be determined according to the following equation:

$$TPES = \left(1 - \frac{E_{PCHP} + \frac{E_{eint}}{\eta_{eref} \cdot P_{grid}} + \frac{E_{tint}}{\eta_b}}{\frac{E_e}{\eta_{eref} \cdot P_{grid}} + \frac{E_t}{\eta_b} + \frac{E_{eexc}}{\eta_{eref} \cdot P_{grid}}}\right) \tag{19}$$

where $P_{grid}$ accounts for transmission and transformation losses on the electrical grid.

The TPES in Equation (19) is the total primary energy savings, which considers all energy flows between user, cogeneration system, and the grid. Finally, detailed economic analyses can be performed, and the ability to comply with the conditions required to be recognized as a high-efficiency cogeneration plant can be verified. Further details are reported in [36].

## 2.2. The Multi-Objective Approach for ORC System Optimization

The optimization approach described in this paragraph involves the coupling of the thermodynamic model of the ORC system with the evolutionary algorithm MOGA II. Starting from the assigned values of the input parameters and the ORC system model deeply discussed in [35], the optimization process enabled the identification of a set of Pareto dominant solutions for the specific system configuration and application. With reference to the general scheme of the optimization process represented above in Figure 4, analyses were conducted by selecting the following two objective functions:

$$\text{global electric efficiency (to be maximized)}: \eta_{el} = \frac{P_{el}}{\dot{Q}_{eva}} \tag{20}$$

$$\text{total area of the heat exchangers (to be minimized)}: A_{tot} = A_{eva} + A_{cond} + A_{reg}. \tag{21}$$

Although there is a shortage of reliable cost data for ORC plants already installed [62], the energetic and economic optimization of these plants is a fundamental issue to be addressed. Therefore, to overcome this limitation, the plant investment cost was related to the total exchange area of the heat exchangers as defined in Equation (21). The total heat transfer area was then set as the objective function (to be minimized) to indirectly optimize the economic performance of the ORC. Actually, it should be noted that the investment cost of these systems is dominated by the cost of the heat exchangers rather than that of the pump and turbine [63]. Wang et al. [63], instead, considered the ratio between the net power output and the total heat transfer area as a single objective function to achieve both thermodynamic and economic optimization. Important information about the cost for ORC plants is reported in [64]. At any rate, if the type and technological level of the heat exchanger are defined through a fixed value of the surface/volume ratio, and the cost of the heat exchanger is proportional to its mass, then the ORC investment cost increases proportionally to the overall exchange area. This indirect approach to the cost estimation also enables more general results that are independent of the technology maturity level, which characterizes each specific application.

The following parameters were selected as decision variables of the optimization problem: minimum and maximum pressure ($p_{min}$ and $p_{max}$) of the thermodynamic cycle, regenerator efficiency ($\varepsilon_{reg}$), superheating at the evaporator outlet ($\Delta T_{super}$), and sub-cooling at the condenser outlet ($\Delta T_{sub}$). The range of definition for the pressure in the decision variable space (Table 2) was limited according to the thermodynamic restrictions imposed by the hot and cool sources. More details are reported in [35].

**Table 2.** Range of definition for the decision variables.

| Decision Variable | Range of Definition |
|---|---|
| $p_{min}$ | $(10 \div 50)$ [kPa] |
| $p_{max}$ | $(600 \div 950)$ [kPa] |
| $\varepsilon_{reg}$ | $(0.55 \div 0.85)$ |
| $\Delta T_{super}$ | $(0 \div 15)$ [°C] |
| $\Delta T_{sub}$ | $(0 \div 10)$ |

The overall heat exchangers area and the global electric efficiency were evaluated via a 0D model of the ORC system, which became the calculation algorithm coupled with the optimization algorithm MOGA II according to the scheme shown in Figure 4.

## 3. Analyses and Results

### 3.1. CHP Plant Configuration Optimization

Figure 6 shows, on the objective function plane, the distribution of the calculated solutions obtained through the maximization of the TPES and the minimization of the payback period (SPB). Specifically, the bubble chart includes, for each numerical solution, details concerning the number CHP engine adopted and the electrical power output provided by each CHP engine (i.e., CHP engine size). The Pareto optimal front was also depicted, highlighting how solutions that maximize the total energy savings are characterized by an increased payback period. Moreover, this result is in agreement with other analyses available in the literature [59,65].

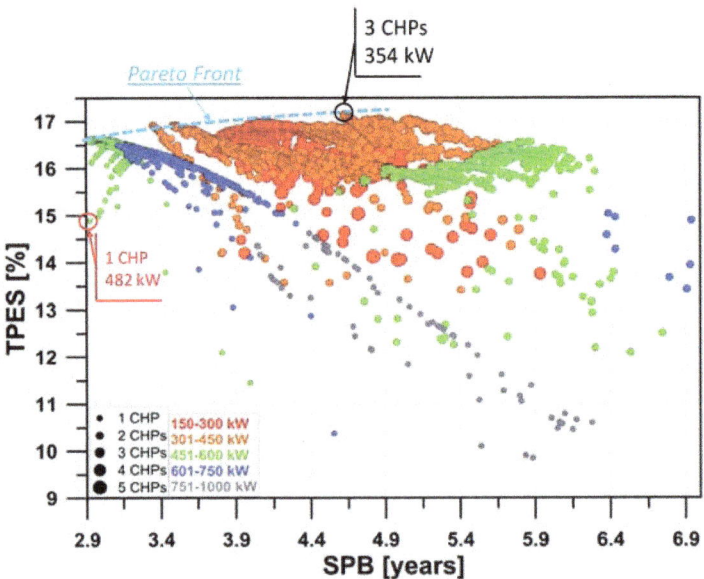

**Figure 6.** Bubble chart with details concerning CHP engine number and size.

It should also be noted that low TPES increases can be achieved with a high worsening of the payback period. Dominant solutions show primary energy saving exceeding 16.5%, payback periods of 2.9–4.6 years, and the adoption of one to three CHP engines with sizes in the range of 260 to 570 kW for each engine. In particular, the plant solution ensuring the minimum payback period is characterized by one CHP engine providing 554 kW of electric power while the maximum PES solution (highlighted by a black circle in Figure 6.) consists of three CHP engines providing about 350 kW of electric power. This solution allows for a TPES of over 17% and a SPB of just over 4.5 years. Figure 6. also shows how modular plants consisting of two or three CHP engines ensure a reasonable compromise between energetic and economic objectives. Figure 7 shows how, unlike heuristic or random techniques, the statistical genetic optimization algorithm MOGA II provides fast convergence toward global optimum solutions starting from the initial set of solutions belonging to the DoE (Design of Experiment). Specifically, solutions in the objective functions space are labeled according to their iteration number, while DoE solutions are depicted with green circles. After DoE methods were performed and solution number 67 identified, it should be noted that only three subsequent iterations were required by MOGA II to find a solution reasonably close to the Pareto dominant solutions (i.e., close to the minimum SPB solution).

Finally, Figure 8 shows the electric power delivered by the CHP plant and the electrical load demand of the user with reference to the minimum SPB solution (i.e., one CHP engine providing 554 kW of electric power output) and a specific day of the year. Figure 9 shows the difference between the thermal power provided by the same system configuration and the thermal power required by the user.

To estimate the robustness of the results to the possible unavailability on the market of CHP systems whose sizes are quite close to the calculated optimal plants, a second vector optimization problem was solved. The cogeneration engine size was defined through a stochastic decision variable described by a uniform distribution. A set of 25 sample solutions was adopted to describe this distribution. The sample designs are spread over an interval of 60 kW. Moreover, they were centered around the mean value currently evaluated by the optimization algorithm. Figure 10 shows, in the

$\sigma_F/\overline{F}$ (SPB) $- \sigma_F/\overline{F}$ (TPES) plane, the obtained dominant solutions, where $\sigma_F$ is the standard deviation of the quantity under consideration, and $\overline{F}$ is its mean value.

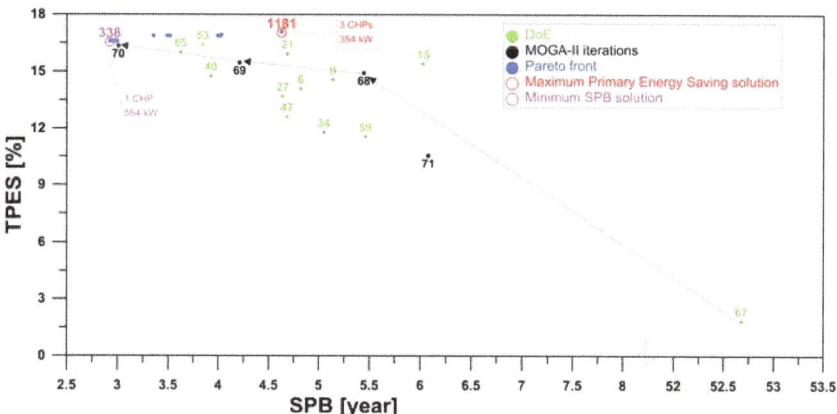

**Figure 7.** MOGA II's fast convergence toward Pareto dominant solutions.

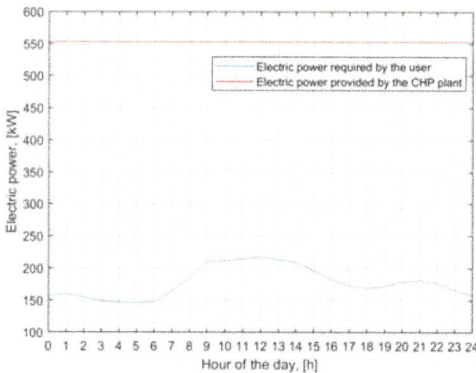

**Figure 8.** The electric power required by the user and that provided by the CHP plant.

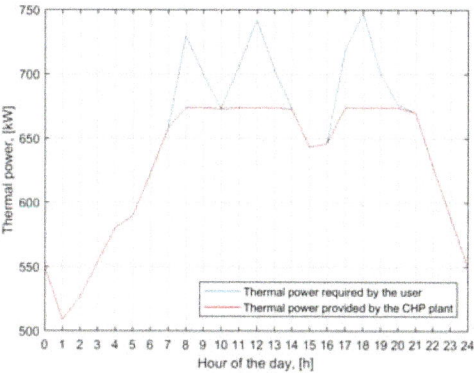

**Figure 9.** The thermal power provided by the CHP plant and that required by the user.

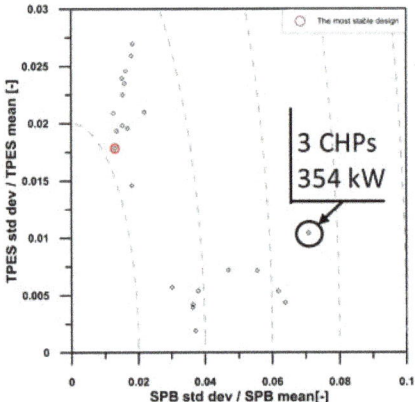

**Figure 10.** Pareto dominant solutions obtained through the first robust design analysis.

Figure 10 shows that the standard deviation accounts for up to 7% of its mean value for the payback period. The ratio $\sigma_F/\overline{F}$ for the TPES is always under 3%. Solutions were also ranked according to the arcs of circumference, defined as equal-stability curves, and are represented in Figure 10. The most stable solution is highlighted in red and its main characteristics are summarized in Table 3. It should be noted that this Pareto solution obtained through the robust design approach becomes a dominated solution if a deterministic approach is adopted for the optimization process, as demonstrated in Figure 6, where this solution is highlighted with a red circle. Figure 10 also shows that the most stable solutions include the best energetic performance solution obtained through the deterministic approach (i.e., three CHP engines—354 kW of electric power highlighted in Figure 6). In this last configuration, the uncertainties related to the actual commercial availability of the CHP engine size under consideration have a greater effect on the economic sensitivity, while the ratio $\sigma_F/\overline{F}$ for the TPES is around 1%, showing high energetic stability.

**Table 3.** The main characteristics of the most stable solution.

| The Most Stable Solution | | | | | | | |
|---|---|---|---|---|---|---|---|
| CHP engine number | Electrical power (mean) | TPES min | TPES mean | TPES max | SPB min | SPB mean | SPB max |
| [-] | [kW] | [%] | [%] | [%] | [years] | [years] | [years] |
| 1 | 482.19 | 14.32 | 14.87 | 15.23 | 2.91 | 2.98 | 3.05 |

Figure 11 demonstrates that the results, achievable when a deterministic definition of the decision variables is adopted, may lead to an overestimation of the objective functions if they are compared to the values achievable through the MORDO.

In fact, the red circle highlights the maximum TPES solution (three engines with 354 kW of electric power output) represented in Figure 6. The blue crosses represent 25 sample solutions that belong to a single statistical distribution for the CHP engine size (i.e., the same robust design solution) whose mean value is just 354 kW, which characterizes the maximum TPES solution represented in red. Figure 11 shows that the SPB can range from 4.3 to 5.3 years, while the TPES could vary in the range 16.7–17.1 when the robust design approach is adopted. These energetic and economic fluctuations are due to the uncertainties related to the actual commercial availability of the considered engine size. To estimate the performance fluctuations due to eventual variations in the energetic and economic scenarios, a further MORDO problem was solved. Specifically, the selling price of the electricity in the three price periods, the efficiency of the Italian thermoelectric generation, and the selling price of the EECs granted in Italy to high-efficiency CHP plants were represented through normal probability distributions (Table 4).

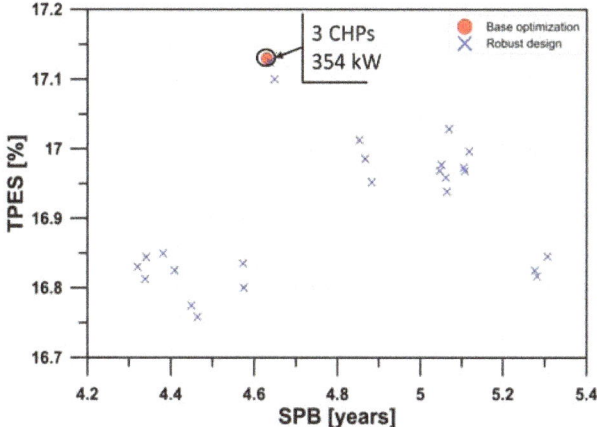

**Figure 11.** Deterministic and probabilistic approaches to vector optimization.

**Table 4.** Probabilistic decision variables adopted in the second MORDO problem.

| Input Decision Variable | Range | Unit | Distribution | Standard Dev. |
|---|---|---|---|---|
| Selling price in time period F1 | 0.10–0.14 | €/kWh | Normal | 0.003 |
| Selling price in time period F2 | 0.076–0.116 | €/kWh | Normal | 0.003 |
| Selling price in time period F3 | 0.045–0.085 | €/kWh | Normal | 0.003 |
| Reference efficiency | 43.5–48.5 | % | Normal | 1 |
| Selling price for the EEC | 90–110 | €/certificate | Normal | 3 |

Figure 12 summarizes, in the $\sigma_F/\overline{F}$ (SPB) $- \sigma_F/\overline{F}$ (TPES) plane, the expected energetic and economic stability for the dominant solutions.

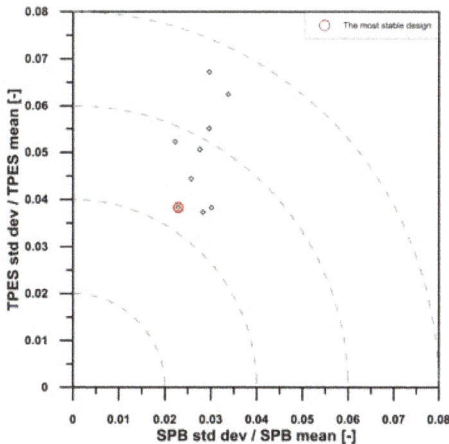

**Figure 12.** Pareto dominant solutions obtained through the second robust design analysis.

The standard deviation for the payback period is always lower than 3.5% of its mean value. This percentage, that provides an estimation of the variations of the objective functions during the plant's life span, assumes values up to 7% for the total primary energy saving. The most stable solution is highlighted with a red circle in Figure 12, and its characteristics are summarized in Table 5.

**Table 5.** Main characteristics of the most stable energetic and economic solution.

| The Most Stable Solution | | | | | | | |
|---|---|---|---|---|---|---|---|
| CHP number | Electrical power (mean) | TPES min | TPES mean | TPES max | SPB min | SPB mean | SPB Max |
| [-] | [kW] | [%] | [%] | [%] | [years] | [years] | [years] |
| 3 | 331 | 17.12 | 17.69 | 19.25 | 4.45 | 4.58 | 4.79 |

This plant configuration is somehow similar to the best TPES solution represented in Figure 6. For this reason, further analysis will be conducted in future works to investigate if the maximum energy-saving solutions belonging to the Pareto frontier and calculated with a deterministic approach to the definition of the decision variables generally lead to more stable economic and energetic results, in comparison to the other dominant solutions.

### 3.2. ORC System Optimization

The vector optimization process generated over 3500 different solutions. Figure 13 shows, among Pareto dominant solutions, a clear tradeoff between the two objectives functions. Therefore, an increase in the global electric efficiency (due to the improvement of the thermal efficiency) was associated with an increased overall heat exchange area, and thus with an increase in the investment cost, as also confirmed in [65]. The increase of the electrical efficiency is mainly correlated with increased values of the degree of regeneration, as highlighted in Figure 14.

Pareto optimal solutions are characterized by electric efficiencies between 14.1 and 18.9% and an overall heat exchangers area from 446 to 1079 m$^2$. Solutions characterized by higher values of the global electric efficiency show higher values of the regenerator efficiency $\varepsilon_{reg}$ and higher values of the thermal power recovered by the regenerator.

Figure 15 shows, for the same Pareto dominant solutions, the history chart for the minimum and maximum pressures of the thermodynamic cycle. Minimum pressure values are concentrated in a small range between 12.7 (Solution 6) and 20.7 kPa (Solution 0), whose saturation temperatures are 88.1 °C and 100.9 °C, respectively.

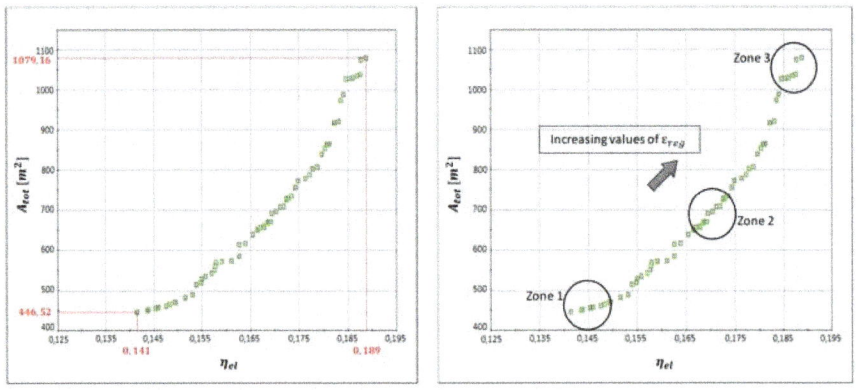

**Figure 13.** Pareto optimal front and clustering of the calculated solutions. In this figure, the comma is used as a symbol to separate the integer part from the fractional part of a number.

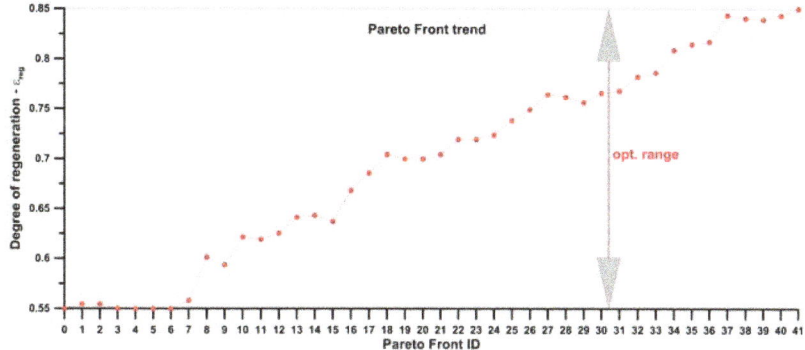

**Figure 14.** Pareto history chart of the degree of regeneration.

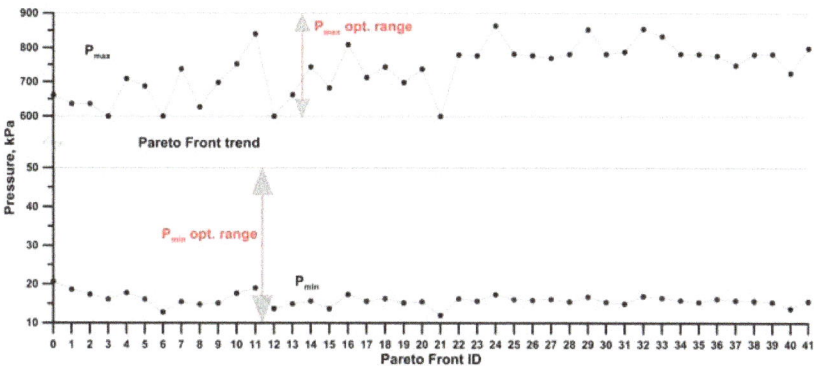

**Figure 15.** History chart for the minimum and maximum pressures of the thermodynamic cycle.

Almost all of the optimal solutions show a superheating phase (Figure 16) that is mostly under 5 °C, except for the solutions identified by the numbers 13, 15, 21, and 36. However, due to the negligible superheating phase, the maximum temperature of the thermodynamic cycle mainly coincides with the saturation temperature at the maximum cycle pressure for all of the Pareto solutions. Figure 16 also shows that the sub-cooling phase can be neglected.

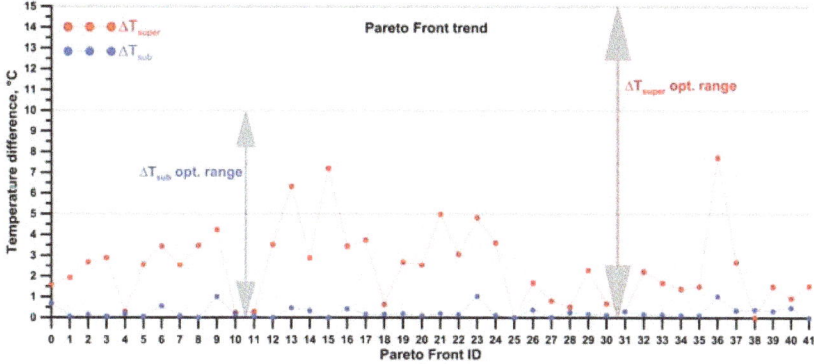

**Figure 16.** History chart of the superheating and sub-cooling temperatures.

Table 6 summarizes some important details concerning three specific solutions belonging to the three zones highlighted in Figure 13.

**Table 6.** Characteristics of three solutions belonging to the three zones identified in Figure 13.

| Solution–Zone | Decision Variables | | | | | Objective Functions | |
|---|---|---|---|---|---|---|---|
| | $p_{min}$ [kPa] | $p_{max}$ [kPa] | $\varepsilon_{reg}$ | $\Delta T_{super}$ [°C] | $\Delta T_{sub}$ [°C] | $A_{tot}$ [m²] | $\eta_{el}$ |
| Solution 0–Zone 1 | 20.70 | 661.05 | 0.55 | 1.60 | 0.70 | 446.52 | 0.141 |
| Solution 20–Zone 2 | 16.26 | 687.71 | 0.64 | 7.19 | 0.16 | 570.78 | 0.158 |
| Solution 41–Zone 3 | 16.04 | 771.18 | 0.74 | 0.00 | 0.00 | 727.00 | 0.172 |

Finally, with reference to Solution 0 and Solution 41, Figure 17 provides more detailed information for comparison. It should be noted that Solution 41 is characterized by a higher heat recovery through the regenerator. This higher thermal power is due to an increased heat exchange area for the regenerator, which was indirectly related to an increased investment cost in this study. These conditions enable the increase of the electric efficiency of the ORC power plant.

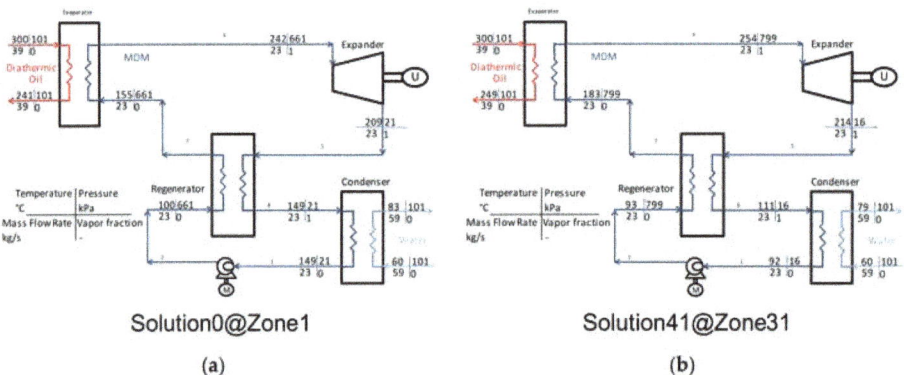

Solution0@Zone1

(a)

Solution41@Zone31

(b)

**Figure 17.** Scheme of the studied ORC power plant with details of the thermodynamic conditions for Solution 0 and Solution 41. (a) Solution 0@Zone 1; (b) Solution 41@Zone 31.

## 4. Conclusions

The increasingly stringent requirements for carbon dioxide reduction have led to a more widespread adoption of distributed energy systems. One of the most effective techniques employed to face the energy-saving challenges is the adoption of polygeneraton systems. These plants can provide a relevant increase in overall efficiency and cost savings. For this reason, a simultaneous reduction of greenhouse gas emissions can be also achieved. However, the use of small-scale polygeneration systems does not ensure the achievement of mandatory, but sometimes conflicting, aims without the proper sizing and operation of the plant. Advanced mathematical techniques such as vector optimization based on evolutionary genetic algorithms and the robust design approach could play a key role in identifying optimal solutions. After a detailed description of these techniques, some specific applications to the study of CHP and ORC systems were presented in this research paper to highlight the potential of these methods and the main results achieved. In particular, the stability of the results to possible mismatches between the cogeneration engine size actually marketed and that calculated was evaluated. Then, the robustness of the achievable results to the eventual variation in the reference energetic scenario and electricity tariffs was also analyzed.

Finally, a vector optimization technique was adopted to simultaneously optimize the electric efficiency and the plant investment cost of a specific ORC system. The genetic optimization algorithm

*Energies* **2018**, *11*, 821

MOGA II and an indirect approach to the evaluation of the plant cost were adopted. This study clearly highlighted how vector optimization techniques based on evolutionary genetic algorithms and the robust design approach provide effective mathematical tools that can support and promote original investigations concerning polygeneration and energy systems in general. In particular, the insertion of robust design procedures in the vector optimization methodology proposed by the authors enables the minimization of the effects of uncertainties on the expected results provided by the energy system under investigation.

**Author Contributions:** Both authors played an equal role in the development of the methodology and its applications, the analyses of the results, and the writing of the manuscript in the present research work.

**Conflicts of Interest:** The authors declare no conflict of interest.

## Nomenclature

| | |
|---|---|
| $a$ | discount rate |
| $A$ | area [m²] |
| $A_{tot}$ | total area for the heat exchange [m²] |
| $E_{e,self}$ | hourly electrical energy delivered by the CHP plant and self-consumed by the user [kWh] |
| $E_P$ | primary energy [kWh] |
| $E_e$ | yearly electrical energy supplied by the CHP plant [kWh] |
| $E_{e,exc}$ | yearly electrical energy supplied by the CHP plant exceeding the user needs [kWh] |
| $E_{e,int}$ | yearly electrical energy integrated by the electrical grid [kWh] |
| $E_{P,CHP}$ | yearly primary energy supplied to the CHP plant [kWh] |
| $E_t$ | yearly thermal energy supplied by the CHP plant [kWh] |
| $E_{t,int}$ | yearly thermal energy integrated by auxiliary boilers [kWh] |
| $p_{grid}$ | factor representative of transmission and transformation losses on the electrical grid [-] |
| $P_{tCHP}$ | actual thermal power provided by the CHP plant [kW] |
| $P_{tnom}$ | nominal thermal power of a single CHP gas engine [kW] |
| $P_{eCHP}(t)$ | average power output of the CHP during the $t$th time interval |
| $P_e(t)$ | average electrical load of the user during the $t$th time interval of the year |
| $P_{eint}(t)$ | electrical power to be integrated from the electrical grid during the $t$th time interval |
| $P_{enom}$ | nominal electrical power of each CHP gas engine [kW] |
| *std dev* | standard deviation of the considered quantity |
| CHP | combined heat and power |
| CCHP | combined cooling, heat, and power |
| DoE | design of experiment |
| EEC | energy efficiency certificates |
| ICE | internal combustion engine |
| MOGA | multi-objective genetic algorithm |
| MORDO | multi-objective robust design optimization |
| PES | primary energy savings |
| PS | proposed system (CHP) |
| PV | photovoltaic |
| RDO | robust design optimization |
| RS | reference system (separate production of electrical and thermal energy demand) |
| SPB | simple payback period |
| TPES | total (or technical) primary energy savings |
| $\Delta t$ | duration of the time interval (1 h in this paper) |
| $\Delta T_{sub}$ | superheating at the evaporator outlet [°C] |
| $\Delta T_{super}$ | subcooling at the condenser outlet [°C] |
| $\varepsilon_{reg}$ | regenerator efficiency or degree of regeneration [-] |
| $\eta_b$ | average boiler efficiency [-] |
| $\eta_{enom}$ | nominal electrical efficiency of the CHP gas engine [-] |
| $\eta_{el}$ | global electric efficiency [-] |

| $\eta_{e,ref}$ | reference efficiency for thermo-electric power generation [-] |
| $\eta_{eREF}$ | average efficiency of the Italian thermoelectric power generation [-] |
| $\eta_t$ | actual thermal efficiency of the cogeneration plant ($\eta_t = P_{tCHP}/P_{P,CHP}$) [-] |
| $\eta_{tnom}$ | nominal thermal efficiency of the cogeneration plant [-] |
| $\eta_{t,ref}$ | reference efficiency for thermal energy production [-] |
| *cond* | condenser |
| *eva* | evaporator |
| *reg* | regenerator |

## References

1. Gimelli, A.; Muccillo, M. Optimization Criteria for Cogeneration Systems: Multi-Objective Approach and Application in a Hospital Facility. *Appl. Energy* **2013**, *104*, 910–923. [CrossRef]
2. International Energy Agency. *Key World Energy Statistics*; International Energy Agency: Paris, France, 2016.
3. International Energy Agency. *Energy Technology Perspectives*; International Energy Agency: Paris, France, 2014; ISBN 978-92-64-20800-1.
4. International Energy Agency. *World Energy Outlook*; International Energy Agency: Paris, France, 2016.
5. Stambouli, A.B.; Traversa, E. Solid oxide fuel cells (SOFCs): A review of an environmentally clean and efficient source of energy. *Renew. Sustain. Energy Rev.* **2002**, *6*, 433–455. [CrossRef]
6. Fontanesi, S.; Severi, E.; Bozza, F.; Gimelli, A. Investigation of Scavenging, Combustion and Knock in a Two-Stroke SI Engine Operated with Gasoline and CNG. *Int. J. Automot. Eng.* **2012**, *3*, 97–105.
7. De Simio, L.; Gambino, M.; Iannaccone, S.; Borrelli, L.; Gimelli, A.; Muccillo, M. Experimental Analysis of a Natural Gas Fueled Engine and 1-D Simulation of VVT and VVA Strategies. In Proceedings of the ICE2013—11th International Conference on Engines & Vehicles, Capri, Italy, 15–19 September 2013. SAE Technical Paper number 2013-24-0111. [CrossRef]
8. Gimelli, A.; Muccillo, M.; Pennacchia, O. The Study of a New Mechanical VVA System. Part I: Valve Train Design and Friction Modeling. *Int. J. Res. Engines* **2015**, *16*, 750–761. [CrossRef]
9. Gimelli, A.; Muccillo, M.; Pennacchia, O. Study of a New Mechanical VVA System. Part II: Estimation of the Actual Fuel Consumption Improvement through 1D Fluid Dynamic Analysis and Valve Train Friction Estimation. *Int. J. Engine Res.* **2015**, *16*, 762–772. [CrossRef]
10. De Nola, F.; Giardiello, G.; Gimelli, A.; Molteni, A.; Muccillo, M.; Picariello, R. A Model-Based Computer Aided Calibration Methodology Enhancing Accuracy, Time and Experimental Effort Savings through Regression Techniques and Neural Networks. In Proceedings of the ICE2017—13th International Conference on Engines and Vehicles, Capri, Italy, 10–14 September 2017. SAE Technical Paper # 2017-24-0054. [CrossRef]
11. De Bellis, V.; Gimelli, A.; Muccillo, M. Effects of Pre-Lift Intake Valve Strategies on the Performance of a DISI VVA Turbocharged Engine at Part and Full Load Operation. *Energy Procedia* **2015**, *81*, 874–882. [CrossRef]
12. Serra, L.M.; Lozano, M.A.; Ramos, J.; Ensinas, A.V.; Nebra, S.A. Polygeneration and efficient use of natural resources. *Energy* **2009**, *34*, 575–586. [CrossRef]
13. Calise, F.; Figaj, R.D.; Massarotti, N.; Mauro, A.; Vanoli, L. Polygeneration system based on PEMFC, CPVT and electrolyzer: Dynamic simulation and energetic and economic analysis. *Appl. Energy* **2017**, *192*, 530–542. [CrossRef]
14. Calise, F.; Cipollina, A.; Dentice D'Accadia, M.; Piacentino, A. A novel renewable polygeneration system for a small Mediterranean volcanic island for the combined production of energy and water: Dynamic simulation and economic assessment. *Appl. Energy* **2014**, *135*, 675–693. [CrossRef]
15. Guo, Z.; Wang, Q.; Fang, M.; Luo, Z.; Cen, K. Thermodynamic and economic analysis of polygeneration system integrating atmospheric pressure coal pyrolysis technology with circulating fluidized bed power plant. *Appl. Energy* **2014**, *113*, 1301–1314. [CrossRef]
16. Buonomano, A.; Calise, F.; Ferruzzi, G.; Vanoli, L. A novel renewable polygeneration system for hospital buildings: Design, simulation and thermo-economic optimization. *Appl. Therm. Eng.* **2014**, *67*, 43–60. [CrossRef]
17. Bracco, S.; Delfino, F.; Pampararo, F.; Robba, M.; Rossi, M. A dynamic optimization-based architecture for polygeneration microgrids with tri-generation, renewables, storage systems and electrical vehicles. *Energy Convers. Manag.* **2015**, *96*, 511–520. [CrossRef]

18. El-Emam, R.S.; Dincer, I. Assessment and Evolutionary Based Multi-Objective Optimization of a Novel Renewable-Based Polygeneration Energy System, ASME. *J. Energy Resour. Technol.* **2016**, *139*, 012003. [CrossRef]

19. Calise, F.; D'Accadia, M.D.; Libertini, L.; Quiriti, E.; Vicidomini, M. A novel tool for thermoeconomic analysis and optimization of trigeneration systems: A case study for a hospital building in Italy. *Energy* **2017**, *126*, 64–87. [CrossRef]

20. Song, H.; Starfelt, F.; Daianova, L.; Yan, J. Influence of drying process on the biomass-based polygeneration system of bioethanol, power and heat. *Appl. Energy* **2012**, *90*, 32–37. [CrossRef]

21. D'Accadia, M.D.; Sasso, M.; Sibilio, S.; Vanoli, L. Micro-combined heat and power in residential and light commercial applications. *Appl. Therm. Eng.* **2003**, *23*, 1247–1259. [CrossRef]

22. Muccillo, M.; Gimelli, A. Experimental Development, 1D CFD Simulation and Energetic Analysis of a 15 kW Micro-CHP Unit based on Reciprocating Internal Combustion Engine. *Appl. Therm. Eng.* **2014**, *71*, 760–770. [CrossRef]

23. Merkel, E.; McKenna, R. Wolf Fichtner, Optimisation of the capacity and the dispatch of decentralised micro-CHP systems: A case study for the UK. *Appl. Energy* **2015**, *140*, 120–134. [CrossRef]

24. Monteiro, E.; Moreira, N.A.; Ferreira, S. Planning of micro-combined heat and power systems in the Portuguese scenario. *Appl. Energy* **2009**, *86*, 290–298. [CrossRef]

25. Sannino, R. Thermal characterization of CHP-User Needs interaction and optimized choice of the Internal Combustion Engines in the CHP plants. *Energy Procedia* **2015**, *82*, 929–935. [CrossRef]

26. Xie, D.; Lu, Y.; Sun, J.; Gu, C.; Yu, J. Optimal Operation of Network-Connected Combined Heat and Powers for Customer Profit Maximization. *Energies* **2016**, *9*, 442. [CrossRef]

27. Ahmadi, P.; Marc Rosen, A.; Dincer, I. Multi-objective exergy-based optimization of a polygeneration energy system using an evolutionary algorithm. *Energy* **2012**, *46*, 21–31. [CrossRef]

28. Calise, F.; D'Accadia, M.D.; Macaluso, A.; Piacentino, A.; Vanoli, L. Exergetic and exergoeconomic analysis of a novel hybrid solar-geothermal polygeneration system producing energy and water. *Energy Convers. Manag.* **2016**, *115*, 200–220. [CrossRef]

29. Maraver, D.; Uche, J.; Royo, J. Assessment of high temperature organic Rankine cycle engine for polygeneration with MED desalination: A preliminary approach. *Energy Convers. Manag.* **2012**, *53*, 108–117. [CrossRef]

30. Fang, F.; Wei, L.; Liu, J.; Zhang, J.; Hou, G. Complementary configuration and operation of a CCHP-ORC system. *Energy* **2012**, *46*, 211–220. [CrossRef]

31. Gimelli, A.; Muccillo, M. Regulation Problems of Combined Cycle Gas-Steam Turbine Power Plant in a Liberalized Market: Part I—Experimental Investigation and Energetic Analysis. *Int. Rev. Modell. Simul.* **2016**, *9*, 1974–9821. [CrossRef]

32. Gimelli, A.; Muccillo, M. Regulation Problems of Combined Cycle Gas-Steam Turbine Power Plant in a Liberalized Market: Part II—Thermodynamic Analysis. *Int. Rev. Modell. Simul.* **2016**, *9*, 348–354. [CrossRef]

33. Cameretti, M.C.; Muccillo, M. Combined MGT-ORC solar—Hybrid system. PART A: Plant optimization. *Energy Procedia* **2015**, *81*, 368–378. [CrossRef]

34. Cameretti, M.C.; Ferrara, F.; Gimelli, A.; Tuccillo, R. Combined MGT-ORC solar-hybrid system. PART B: Component analysis and prime mover selection. *Energy Procedia* **2015**, *81*, 379–389. [CrossRef]

35. Gimelli, A.; Luongo, A.; Muccillo, M. Efficiency and cost optimization of a regenerative Organic Rankine Cycle power plant through the multi-objective approach. *Appl. Therm. Eng.* **2017**, *114*, 601–610. [CrossRef]

36. Gimelli, A.; Muccillo, M.; Sannino, R. Optimal design of modular cogeneration plants for hospital facilities and robustness evaluation of the results. *Energy Convers. Manag.* **2017**, *134*, 20–31. [CrossRef]

37. Gimelli, A.; Muccillo, M.; Sannino, R. Effects of uncertainties on the stability of the results of an optimal sized modular cogeneration plant. *Energy Procedia* **2017**, *126*, 369–376. [CrossRef]

38. Akbari, K.; Nasiri, M.M.; Jolai, F.; Ghaderi, S.F. Optimal investment and unit sizing of distributed energy systems under uncertainty: A robust optimization approach. *Energy Build* **2014**, *85*, 275–286. [CrossRef]

39. Gu, C.-H.; Xie, D.; Sun, G.-B.; Wang, X.-T.; Ai, Q. Optimal operation of combined heat and power system based on forecasted energy prices in real-time markets. *Energies* **2015**, *8*, 14330–14345. [CrossRef]

40. Alvarado, D.C.; Acha, S.; Shah, N.; Markides, C.N. A Technology Selection and Operation (TSO) optimisation model for distributed energy systems: Mathematical formulation and case study. *Appl. Energy* **2016**, *180*, 491–503. [CrossRef]

41. Piacentino, A.; Cardona, F. EABOT—Energetic analysis as a basis for robust optimization of trigeneration systems by linear programming. *Energy Convers. Manag.* **2008**, *49*, 3006–3016. [CrossRef]
42. Bertsimas, D.; Sim, M. The Price of Robustness. *Oper. Res.* **2004**, *52*, 35–53. [CrossRef]
43. Büsing, C.; D'Andreagiovanni, F. New Results about Multi-band Uncertainty in Robust Optimization. In *International Symposium on Experimental Algorithms*; Klasing, R., Ed.; Springer: Berlin/Heidelberg, Germany, 2012; Volume 7276, ISBN 978-3-642-30849-9.
44. Yu, X.; Jin, Y.; Tang, K.; Yao, X. Robust optimization over time—A new perspective on dynamic optimization problems. In Proceedings of the IEEE Congress on Evolutionary Computation, Barcelona, Spain, 18–23 July 2010; pp. 1–6.
45. Ferrara, F.; Gimelli, A.; Luongo, A. Small-scale concentrated solar power (CSP) plant: ORCs comparison for different organic fluids. *Energy Procedia* **2013**, *45*, 217–226. [CrossRef]
46. Tchanche, B.F.; Lambrinos, G.; Frangoudakis, A.; Papadakis, G. Low-grade heat conversion into power using organic Rankine cycles—A review of various applications. *Renew. Sustain. Energy Rev.* **2011**, *15*, 3963–3979. [CrossRef]
47. Schuster, A.; Karellas, S.; Kakaras, E.; Spliethoff, H. Energetic and economic investigation of Organic Rankine Cycle applications. *Appl. Therm. Eng.* **2009**, *29*, 1809–1817. [CrossRef]
48. Özkaraca, O.; Keçebas, P.; Demircan, C.; Keçebas, A. Thermodynamic Optimization of a Geothermal-Based Organic Rankine Cycle System Using an Artificial Bee Colony Algorithm. *Energies* **2017**, *10*, 1691. [CrossRef]
49. Pezzuolo, A.; Benato, A.; Stoppato, A.; Mirandola, A. The ORC-PD: A versatile tool for fluid selection and Organic Rankine Cycle unit design. *Energy* **2016**, *102*, 605–620. [CrossRef]
50. Sayyaadi, H. Multi-objective approach in thermoenvironomic optimization of a benchmark cogeneration system. *Appl. Energy* **2009**, *86*, 867–879. [CrossRef]
51. Wang, J.-J.; Jing, Y.-Y.; Zhang, C.-F. Optimization of capacity and operation for CCHP system by genetic algorithm. *Appl. Energy* **2010**, *87*, 1325–1335. [CrossRef]
52. Gimelli, A.; Sannino, R. A multi-variable multi-objective methodology for experimental data and thermodynamic analysis validation: An application to micro gas turbines. *Appl. Therm. Eng.* **2018**, *134*, 501–512. [CrossRef]
53. Toffolo, A.; Lazzaretto, A. Evolutionary algorithms for multi-objective energetic and economic optimization in thermal system design. *Energy* **2002**, *27*, 549–567. [CrossRef]
54. Das, I.; Dennis, J. Normal boundary intersection, Alternate method for generating Pareto optimal points in multicriteria optimization problems. Available online: http://www.dtic.mil/dtic/tr/fulltext/u2/a320782.pdf (accessed on 18 February 2018).
55. Coello, C.A.; Van Veldhuizen, D.A.; Lamont, G.B. *Evolutionary Algorithms for Solving Multi-Objective Problems*; Springer: New York, NY, USA, 2007.
56. Lotov, A.V.; Bushenkov, V.A.; Kamenev, G.K. *Interactive Decision Maps: Approximation and Visualization of Pareto Frontier*; Kluwer Academic Publishers: Boston, MA, USA, 2004.
57. Ngatchou, P.; Zarei, A.; El-Sharkawi, M.A. Pareto multi-objective optimization. In Proceedings of the 13th International Conference on Intelligent Systems Application to Power Systems, Arlington, VA, USA, 6–10 November 2005; pp. 84–91.
58. Branke, J.; Deb, K.; Miettinen, K.; Slowinski, R. *Multiobjective Optimization. Interactive and Evolutionary Approaches*; Springer: Berlin, Germany, 2008; ISBN 978-3-540-88908-3.
59. Poloni, C.; Giurgevich, A.; Onesti, L.; Pediroda, V. Hybridization of a multi-objective genetic algorithm, a neural network and a classical optimizer for a complex design problem in fluid dynamics. *Comput. Methods Appl. Mech. Engrgy* **2000**, *186*, 403–420. [CrossRef]
60. Poloni, C.; Pediroda, V. GA coupled with computationally expensive simulations: Tools to improve efficiency. In *Genetic Algorithms and Evolution Strategies in Engineering and Computer Science*; John Wiley and Sons: Chichester, NH, USA, 1997; pp. 267–288. ISBN 0471977101.
61. Padovan, L.; Pediroda, V.; Poloni, C. Multi objective robust design optimization of airfoils in transonic field (M.O.R.D.O.). In Proceedings of the International Congress on Evolutionary Methods for Design, Optimization and Control with Applications to Industrial Problems EUROGEN 2003, Barcelona, Spain, 15–17 September 2003.

62.  Vélez, F.; Segovia, J.J.; Martín, M.C.; Antolín, G.; Chejne, F.; Quijano, A. A technical, economical and market review of organic Rankine cycles for the conversion of low-grade heat for power generation. *Renew. Sustain. Energy Rev.* **2012**, *16*, 4175–4189. [CrossRef]

63.  Wang, J.; Yan, Z.; Wang, M.; Ma, S.; Dai, Y. Thermodynamic analysis and optimization of an (organic Rankine cycle) ORC using low grade heat source. *Energy* **2013**, *49*, 356–365. [CrossRef]

64.  Lemmens, S. Cost Engineering Techniques and Their Applicability for Cost Estimation of Organic Rankine Cycle Systems. *Energies* **2016**, *9*, 485. [CrossRef]

65.  Wang, J.; Yan, Z.; Wang, M.; Li, M.; Dai, Y. Multi-objective optimization of an organic Rankine cycle (ORC) for low grade waste heat recovery using evolutionary algorithm. *Energy Convers. Manag.* **2013**, *71*, 146–158. [CrossRef]

*Article*

# Technical Aspects and Energy Effects of Waste Heat Recovery from District Heating Boiler Slag

**Mariusz Tańczuk *** , **Maciej Masiukiewicz, Stanisław Anweiler and Robert Junga**

Faculty of Mechanical Engineering, Opole University of Technology, ul. Mikołajczyka 5, 45-271 Opole, Poland; m.masiukiewicz@po.opole.pl (M.M.); s.anweiler@po.opole.pl (S.A.); r.junga@po.opole.pl (R.J.)
* Correspondence: m.tanczuk@po.opole.pl; Tel.: +48-664-475-355

Received: 18 February 2018; Accepted: 27 March 2018; Published: 30 March 2018

**Abstract:** Coal continues to dominate in the structure of the heat production system in some European countries. Coal-fired boilers in district heating and power generation systems are accompanied by the formation of large quantities of slag and ash. Due to considerable high temperature, slag may be used as a source of waste energy. In this study, the technical possibilities of recovery slag's physical enthalpy from grate-fired district heating boiler of 45 MW thermal capacity are analyzed. The aim of the work is to estimate the waste energy potential of the slag in analyzed boiler and proposition of the heat recovery system. The construction and design of the existing deslagging system was examined. Studies have shown that high water temperature accelerates system wear. Recovering heat from this system decreases the water temperature, which extends the trouble-free working time. The slag parameters were determined, including the temperature at the outlet of the boiler and the temperature after leaving the slag water tub. The annual amount of heat regenerative potential was estimated. On the basis of the research, the authors propose a waste heat recovery facility with high temperature R134a heat pump system. The result of the conducted research is that the proposed heat pump provides energy savings that are worth considering by recovering from 58.8% to 88.0% of energy slag potential.

**Keywords:** waste energy; coal slag; thermodynamic analysis; heat recovery; high temperature heat pump

---

## 1. Introduction

A huge interest in modern energy production, conversion, transmission and storage is observed in the international scientific and industrial fields [1]. Energy demand is rapidly rising, and the need for sustainable management of our environment drives us to search for new energy solutions [2]. One of the solutions is energy savings and recovery. Energy that is saved from utilization or recovered from the wastes or from the energy conversion processes can be treated as another useful form of energy, which gives measureable financial effects. Sustainable energy management including energy recovery has become a challenge in the engineering community. A significant technology shift will be necessary to enable efficient energy recovery and advanced energy conversion and management is becoming a primary focus of this technology shift [3]. Some of the important elements of this action are the reduction of the consumption of energy and natural resources. Much attention is being focused on the recycling of wastes and by-products to achieve more sustainable development, including processes, devices and materials [4,5]. There is an interesting approach, which claims that significant energy efficiency can be gained in zones with concentrated energy activity. This gain from energy intensive industries can often be achieved by recovering and reusing waste heat between processing plants [6] or by creating energy sharing networks inside industrial areas [7] and optimizing them to obtain the highest possible techno-economic efficiency [8].

One of the promising options of recovering energy is the physical enthalpy usage of the combustion solid products, such as slag and ash, which are generated from different technological processes. Although the waste heat recovery from furnace slags can be treated as a whole process in national heavy industry [9], most of the cases are extremely individual and have to be treated separately.

According to [10], physical enthalpy of molten slag constitutes from 3.5% to 25% of total energy released during combustion. The promising area for slag waste energy recovery is high temperature slags [11]. In the case of furnace solid slag, mostly produced in power stoker-boilers as well as in heat generating stoker-boilers, its potential is much lower—from 0.1% to 1.5% of total energy. Recovering energy from solid slag in such installations is rather occasional. Regardless of the low share of energy contained in slag in the total energy balance of the furnaces, it is still worth highlighting that the amount of waste energy should not be rejected from both technological and economic concerns.

Boilers fired with solid fuels generate a significant amount of hot solid slag, which is in most cases unproductively cooled and removed. Unfortunately, one of the characteristic features of most silicate slags is their low heat conductivity, which creates difficulties in the energy recovery [12]. According to [13] heat conductivity of coal slags stays in the range of 1–2 W/mK and is very dependent on the slag temperature [14]. That feature enforces some additional activities to be done, in order to efficiently recover the energy from the slag. The temperature of the slag discharging from district heating boilers with grate furnaces is in the range 250–450 °C. There is a need for the exact energy balance to be performed, to determine how much waste heat can be extracted. The energy balance should include the energy demand of additional facilities needed for slag energy recovery.

Apart from waste energy issue of the slag, another operating problem can be noticed in solid fuels boiler systems. It is the tribocorrosion of deslagging facility elements. This problem is particularly distinctive and inconvenient in water slag traps where triborrosion phenomena leads to material damages that significantly reduces the lifetime of the device. In general, tribocorrosion involve mechanical and chemical or electrochemical interactions between surfaces in relative motion in presence of a corrosive environment. However, the different mechanisms of tribocorrosion are not entirely understood yet, as they involve properties of contacting material surfaces, mechanics of the contact and corrosion conditions [15]. Mechanisms for the synergistic effect are also generally not well understood and there has been little effort done so far in modelling synergistic processes [16].

In water type deslagging facilities installed in solid fuels boilers systems, tribocorrosion is reported to be more intensive due to the high temperature of water cooling the hot slag. There is not sufficient research on tribocorrosion dependent on temperature in aqueous solutions, although triboelectrochemical techniques are widely used for the investigation of tribocorrosion wear in aqueous solutions of still temperature [17]. There are some investigations on fretting wear behavior in water under various temperatures which state that, in water, the main wear mechanisms were delamination and abrasive wear, and the delamination wear increased as the temperature of water increased. Tests were conducted in different temperatures and solutions: three different temperatures 20, 60, 90 °C [18] or in two temperatures 2 and 23 °C. Moreover, the corrosion resistance of the steels is higher in lower temperatures [19]. Increasing the temperature from 32 to 52 °C leads to an increase in the corrosion rate from 50 to 100% depending on the acid and salt concentrations [20]. Other researches are conducted to investigate the effect of pH value on the fretting wear behavior [21]. In addition, the third body effects on friction and wear during fretting of steel contacts seem to have significant influence [22], especially in such erosive and corrosive environment as in water slag trap of the industrial boilers.

As reported in [23], researchers are now trying to simulate the cooling of the slags to better understand the thermodynamics of this process. These simulations are next used for the design of complete recovery systems with dedicated to specific conditions special slag heat recuperators [24]. There are even attempts to use the slags as energy storage materials [25]. Various techniques of preparing the slag for heat energy extraction are taken into consideration, including slag breaking with rotary equipment like rotary drums [11], spiral blades and other methods [26].

A relatively new process for heat recovery from district heating grate-fired boiler slag is described in this paper. The aim of the work is to estimate the waste energy potential of the slag in the analyzed boiler and to propose the system for recovering this energy. Moreover, recovering the heat from the slag decreases the temperature of water in the deslagger. It is anticipated that lowering the water temperature will extend the operation time of the slag remover significantly. As an example, a typical grate-fired steam boiler with an installed capacity of 45 MW is examined for the purpose of heat recovery.

## 2. System Description

The investigation concerns assessment of available waste heat energy potential in a coal slag generated during hard coal combustion in a district heating grate boiler as well as proposals of its recovery. The research has been made on the basis of a steam boiler installed in the heat source of the municipal district heating system. The nominal capacity of the boiler fed with fine coal is 50 t/h of steam (45 MW thermal power). The considered plant is a heat source for high temperature municipal district heating system with the maximum demand for heat reaching 185 MW. The hot water of nominal pressure $p_n = 1.6$ MPa and nominal (maximum) supply temperature $t_{s,n} = 150$ °C is a heat carrier in the system. The plant consists of the following heat generating units, as shown in Figure 1:

- high efficient cogeneration unit based on a steam turbine STU supplied with hard coal stoker-fired steam boiler SSCB (labelled as OR50), examined for slag heat recovery,
- one pulverized hard coal-fired water boiler PCB (labelled as WP120) with the nominal heat capacity 125 MW,
- two hard coal stoker-fired water boilers SCB (labelled as WR25 no. 1 and no. 2) with the nominal heat capacity 33 MW each,
- a high-efficiency cogeneration unit based on a gas turbine unit GTU fired with natural gas of a nominal electric capacity 7.4 MW and integrated with a heat recovery boiler HRB of a nominal heat capacity 14.7 MW.

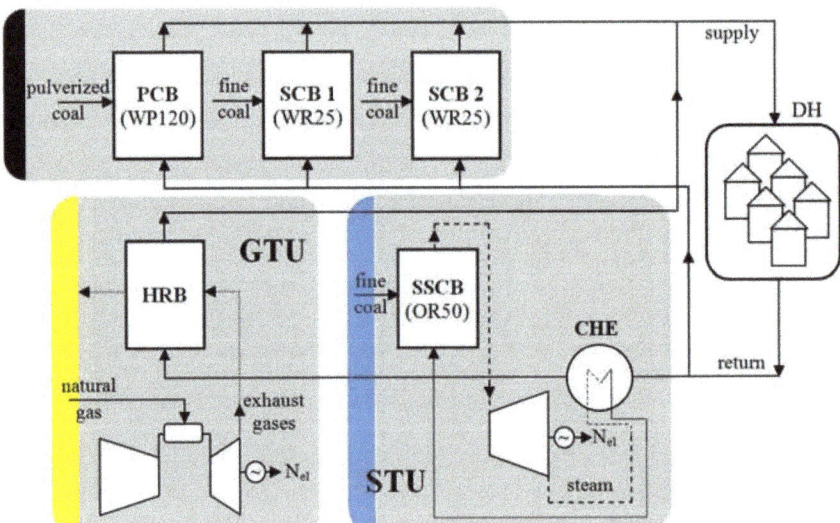

**Figure 1.** Schematic diagram of the current configuration of the analyzed district heating (DH) plant with heat and electricity generation.

The simplified layout of stoker-fired boiler considered as a slag source is presented in the Figure 2. The output slag is a blend of coal slag formed on the grate furnace during combustion and fly ashes captured during dedusting processes. Ashes are reversed to the water slag trap. Slag and ash mixture is being cooled in the water and then carried out by set of belt conveyors for further treatment or disposal by landfilling.

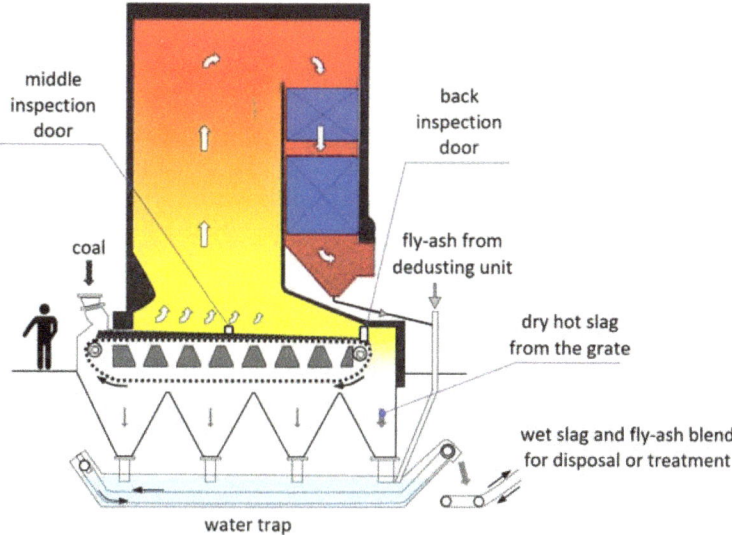

**Figure 2.** Schematic layout of the stoker-boiler under investigation with a deslagger with tub fueled with water (water trap at the bottom).

## 3. Problem Formulation, Methods and Input Data

The combustion process of hard coal in a stoker-boiler is deeply connected with a grate furnace as a key component of the boiler. The slag is formed in the furnace where the layer of fine coal changes smoothly into slag during combustion. The slag formation process takes place already from the beginning part of the furnace where fine coal is fed through the feeding hopper. The layer thickness of the coal results from expected boiler capacity during operation. The slag volume, temperature and distribution on the grate results from the boiler load. The temperature of the slag at the end part of the grate determines the waste energy potential.

The solid material distribution in the furnace was observed during the research and the exemplary situation is shown in the Figure 3. It presents combusting zones for the case of a 50% load boiler operation. The combustion zone, with the highest temperature reaching 1200 °C, ends up in the middle part of the grate. In this case, the back part of the grate is covered by the agglomerating slag with a temperature reduced to 120–110 °C.

The middle part of the boiler with combusting coal on a grate is presented in the Figure 4. The picture has been taken through the inspection door of the boiler indicated in the Figure 2 and the gas area with flames can be easily recognized as well as the layer of coal combusted on the grate. The length of that zone is a direct reflection of the boiler load. In case of district heating boilers, the load results from the operation mode which depends on what is the boiler configuration in the plant. For the optimal sized boilers operated in a base of a load duration curve, the output capacity is usually close to nominal during the whole year time. In other configuration cases the capacity can be modulated and thus the temperature and volume of the output slag can change significantly.

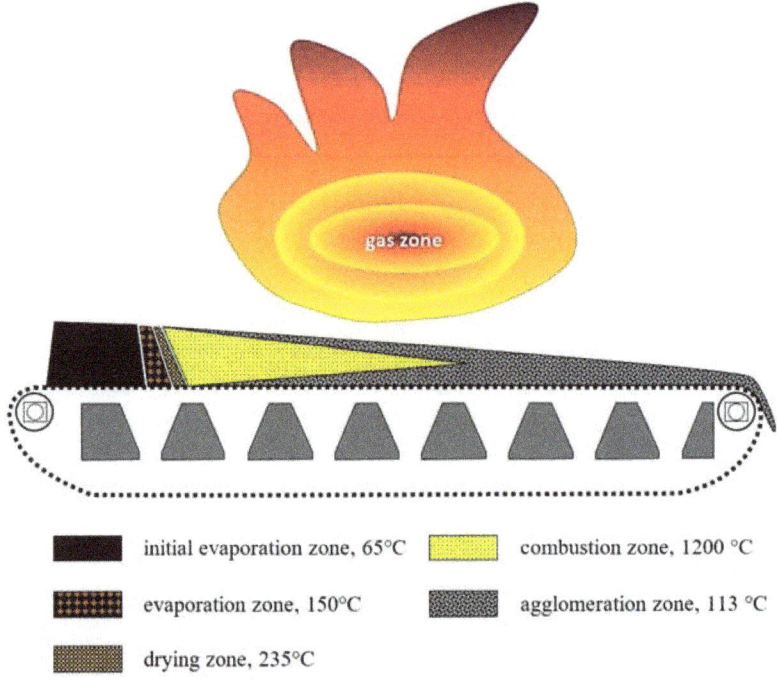

initial evaporation zone, 65°C          combustion zone, 1200 °C

evaporation zone, 150°C          agglomeration zone, 113 °C

drying zone, 235°C

**Figure 3.** The illustrative distribution of the temperature of coal and slag on the grate of a stoker-boiler.

**Figure 4.** The middle part of grate—view from the middle inspection door of the boiler.

The difference between the slag properties in case of two different loads of the same analyzed boiler can be noticed on the pictures presented in the Figure 5. Opening the back inspection door of the boiler operated with maximum load enabled to record combusting zone occurred even at the discharge part of the grate (Figure 5a). During the research observation, the combusting coal was even noticed falling down to the slag water trap. In case of decreased capacity, mainly during off-season operation, the combusting zone is shortening what can be confirmed by the Figure 5b.

a)             b)

**Figure 5.** The end section of the grate with discharge of the slag to the slag trap—view from the back inspection door of the boiler: (**a**) operation with full load, (**b**) operation with 30% of load during off-season period.

The idea of recovering waste heat from the slag is not only energy-saving in itself but also cools the water in the deslagger tub for the sake of minimizing tribocorrosion effects. In the analyzed case, triborrocion is the more intensive and problematic the higher temperatures of water in the tub. Figure 6 presents salt efflorences and corrosion traces on the walls of deslagger water trap after one month of operation in high temperature conditions.

**Figure 6.** Tribocorrosion traces on the walls of deslagger water trap occurred during operation with high temperature water in the tub.

In order to estimate waste energy potential of the slag leaving the boiler it was necessary to evaluate the volume of the slag as well as its temperature. The chosen physical and chemical properties of the slag and coal supplying the boiler also needed to be investigated.

In the beginning of the research procedure, the properties of hard coal were determined weekly. The results of the measurements are then used for further evaluation of waste energy potential of the coal slag. Chosen exemplary values are shown in the Table 1.

The slag available for energy recovery directed to the water trap is in fact a mixture of slag from the grate and fly ashes from the dedusting facility. Its properties have been measured and the exemplary data is presented in Table 2.

**Table 1.** Properties of hard coal used in the tested steam boiler.

| No. | Property | Unit | Value * | Testing Standard |
|-----|----------|------|---------|------------------|
| 1 | Moisture content | wt.% | 10.5 | PN—80/G-04511 |
| 2 | Ash content | wt.% | 14.9 | Accredited testing procedure |
| 3 | Carbon content | wt.% | 62.83 | PN-G-04571:1998 |
| 4 | Hydrogen content | wt.% | 3.31 | ISO/TS 12902:2007 |
| 5 | Nitrogen content | wt.% | 1.10 | ISO/TS 12902:2007 |
| 6 | Sulphur content | wt.% | 0.40 | PN-ISO 351:1999 |
| 7 | Lower Heating Value | MJ kg$^{-1}$ | 23.668 | Calculated |
| 8 | Higher Heating Value | MJ kg$^{-1}$ | 27.185 | PN-81/G-04513 |

* As received.

**Table 2.** Properties of the mixed slag and fly ash from the grate boiler.

| No. | Property | Unit | Value * | Testing Standard |
|-----|----------|------|---------|------------------|
| 1 | Moisture content | wt.% | 31.0 | Accredited testing procedure |
| 2 | TOC | wt.% | 23.33 | Accredited testing procedure |
| 3 | Carbon content | wt.% | 15.3 | Accredited testing procedure |
| 4 | Size distribution: | wt.% | - | - |
| 4a | >10 | wt.% | 45.8 | - |
| 4b | 10 ÷ 5 | wt.% | 17.6 | - |
| 4c | 5 ÷ 3 | wt.% | 8.7 | - |
| 4d | 3 ÷ 1 | wt.% | 13.6 | - |
| 4e | 1 ÷ 0.5 | wt.% | 6.2 | - |
| 4f | <0.5 | wt.% | 8.1 | - |

* As received.

The temperature of the slag leaving the boiler $t_s$ has been measured and recorded online by the measurement set equipped with K-type (NiCr-NiAl) thermocouples (TP-204K-1b-1200-2,0). In order to determine slag volume, it was necessary to measure the fuel flow supplying the boiler. The measurement has been done automatically by the weighting system installed in the feeding hoppers of the boiler and recorded by SCADA system Pro2000.

The slag volume has been evaluated in the function of the coal quantity supplying the furnace, according to Formula (1):

$$s_S = a \cdot A_r \cdot (1 - TOC_s) \tag{1}$$

where:

$s_s$—unitary slag volume, kg/kg of coal

$a$—coefficient of ash contraction in coal, assumed as 0.9

$A_r$—ash content in coal, wt.%

$TOC_s$—total organic carbon of slag-ash mixture, wt.%

Figure 7 presents results of $A_r$ and $TOC_s$ weekly measurements taken during the one year of boiler operation, while Figure 8 contains unitary slag volume $s_s$ determined from Formula (1).

It can be observed that total organic carbon in the output slag is significant higher during summer time, where boiler is operated with a lower load, which makes the combustion process less effective.

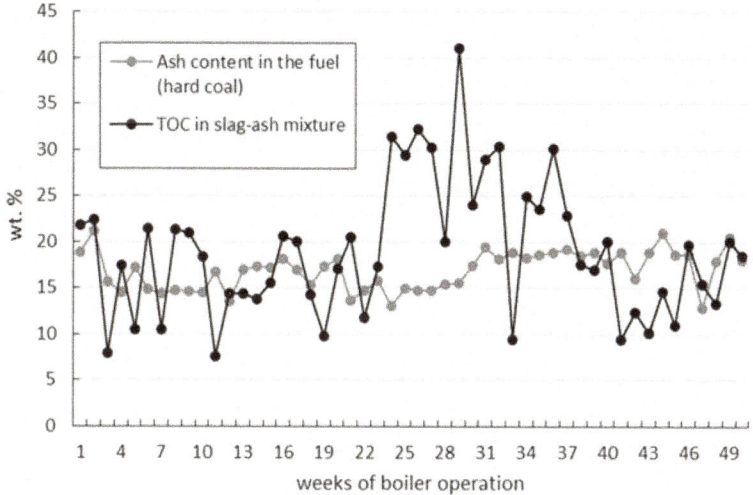

**Figure 7.** Ash content of the fuel and total organic carbon (TOC) in slag-ash mixture measured weekly.

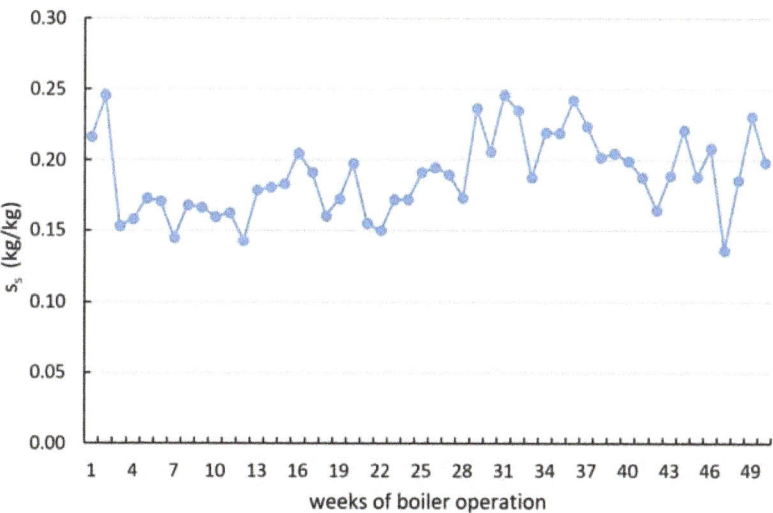

**Figure 8.** Unitary slag volume $s_s$.

The $\dot{Q}_s$ content in the slag is the matter of chemical and physical enthalpy of the slag, according to Formula (2):

$$\dot{Q}_s = \dot{H}_s = \dot{H}_{f,s} + \dot{H}_{ch,s} \qquad (2)$$

where:

$\dot{H}_s$—enthalpy flux of the slag, kW

$\dot{H}_{f,s}$—physical enthalpy flux of the slag, kW

$\dot{H}_{ch,s}$—chemical enthalpy flux of the slag, kW

For the purpose of the research, it has been assumed that only physical enthalpy of the slag is the subject of the recovery, thus:

$$\dot{Q}_s = \dot{H}_{f,s} = \dot{G}_s \cdot h_s = \dot{G}_s \cdot c_{p,s} \cdot t_s \tag{3}$$

where:

$\dot{G}_s$—slag flux, kg/s
$h_s$—specific physical enthalpy of the slag, kJ/kg
$c_{p,s}$—specific heat of the slag, assumed as 1 kJ/kgK [9,27]
$t_s$—slag temperature, °C.

The annual quantity of waste energy in the slag can be derived basing on Formula (4):

$$Q_s = \int_{\tau=0}^{\tau=8760} \dot{H}_{f,s}d\tau \tag{4}$$

It has been assumed that the energy that it is possible to recover is accumulated in the water inside the deslagger tub which can be calculated according to the energy balance presented in Figure 9 and in Formula (5).

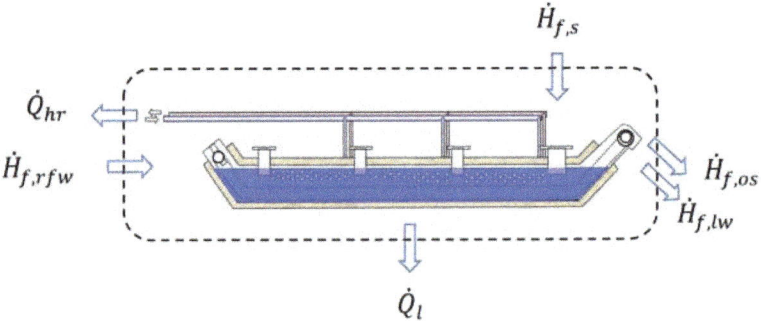

**Figure 9.** Energy balance of the deslagging device of the boiler.

$$\dot{Q}_{hr} = \dot{H}_{f,s} + \dot{H}_{f,rfw} - \dot{H}_{f,os} - \dot{H}_{f,lw} - \dot{Q}_l \tag{5}$$

where:

$\dot{Q}_{hr}$—heat flux available for recovery, kW
$\dot{H}_{f,s}$—physical enthalpy flux of the slag, kW
$\dot{H}_{f,os}$—physical enthalpy of the slag leaving deslagger, kW
$\dot{H}_{f,lw}$—physical enthalpy of the water lost with the slag, kW
$\dot{Q}_l$—heat energy losses, kW, assumed as 2% of $\dot{H}_{f,s}$
$\dot{H}_{f,rfw}$—physical enthalpy of the refill water, kW

The quantity of waste energy annually available for recovery can be derived from Formula (6):

$$Q_{hr} = \int_{\tau-0}^{\tau=8760} \dot{Q}_{hr}d\tau \tag{6}$$

## 4. Results and Discussion

As the result of assumed method and presented input data the slag waste energy flux $\dot{Q}_s$ and its part available for the recovery $\dot{Q}_{hr}$ have been derived. The output capacity distribution of the analyzed boiler $\dot{Q}_B$ and the temperature of the slag are presented in Figure 10.

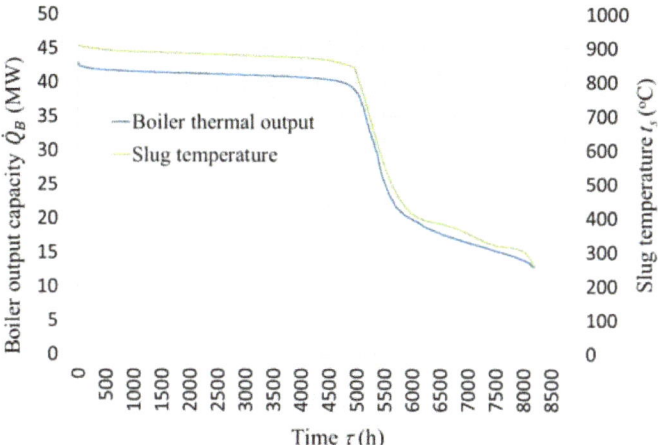

**Figure 10.** Annual distribution curve of boiler output capacity $\dot{Q}_B$ and slag temperature $t_s$ at the end section of the grate.

The decrease of the boiler thermal output after around 5000 h of operation, shown in the Figure 9, is the result of boiler operation at the base of the heat load duration curve. In the analyzed district heating plant, the heat demand decreases significantly during the summer time, which requires a reduction of the boiler thermal output. During the off season periods, the heat demand of district heating system is the result only of domestic hot water needs. It should also be stated here that the total annual time of boiler operation does not exceed 8200 h due to the maintenance works and other necessary standstills. The content of Figure 10 confirms strict correlation between load of the boiler and slag temperature which remains in line with previous discussion presented in the problem formulation section of the paper.

The annual distribution curve of enthalpy flux of the slag is presented in Figure 11—as the result of proposed evaluation procedure. It can be seen that the highest values of waste energy flux $\dot{Q}_s$ occurs during maximum load of the boiler and equals to 508 kW. The value is relatively small, compared to maximum thermal output of the boiler, which is around 45 MW. It needs to be noticed here that off-season period has much less potential of slag waste energy, due to the slower temperatures of the slag and decreased slag volume.

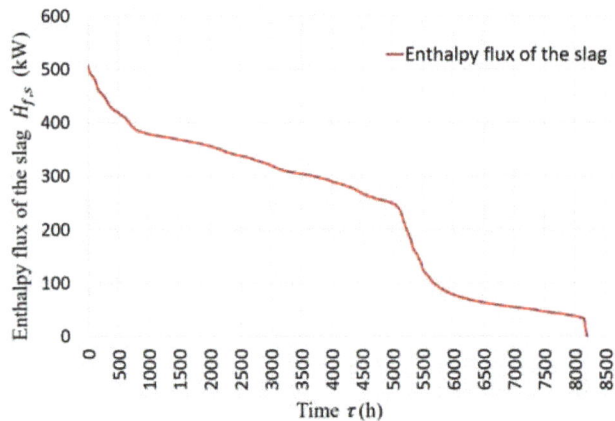

**Figure 11.** Annual distribution curve of enthalpy flux of the analyzed slag.

The annual quantity of waste energy available in the slag, derived from Formula (4) equals 1,969,451 kWh (7090 GJ). This amount can be gained by means of an appropriate recovery system. The authors propose the solution based on a high temperature heat pump used for transferring heat from water in the deslagger to district heating network. The heat sink of the heat pump is the return pipe of the DH system with temperatures from 55 °C to 70 °C, according to the ambient temperatures and DH heat demand.

The idea of using heat pumps for heat recovery purposes is well-known solution, according to the literature. High temperature heat pump applications are reported to be efficient solutions, particularly in industrial cases. During simulations, Yu at al. proved that high temperature heat pumps (HTHP) are capable to produce heat at the temperature of 120 °C with good performance [28]. Wu at al. explained that in some industrial cases the on-site testing the HTHP system could be reliably operated to heat the liquid temperature up to 95 °C with an average system coefficient of performance COP of 4.2 during the entire heating process [29]. Yu et al. showed that under the experimental conditions of the inlet water temperature of evaporator at 50–70 °C, the outlet water temperature of condenser could reach 80–110 °C [30]. Although Zhang et al. reported even 135 °C on condensing unit [31]. Experimental and numerical investigations of a new high temperature heat pump for industrial heat recovery using water as refrigerant was performed by Camoun at al. and their experimental investigations of this heat pump were carried out in the condensing temperature range of 130–140 °C. Economic analysis indicated that the HTHP system could save about 47% of the operating cost in comparison to the traditional steam heating [32].

## 5. Proposal of Technical Solution for Slag Energy Recovery

On the basis of the research and observations of current slag treatment system of the analyzed boiler, heat pump system was proposed for slag energy recovery. The heat source of the pump is in the deslagger tub filled with the water and the heat sink is the return pipe of the district heating network. The target configuration of the district heating plant equipped with heat pump (HP) is presented in Figure 12.

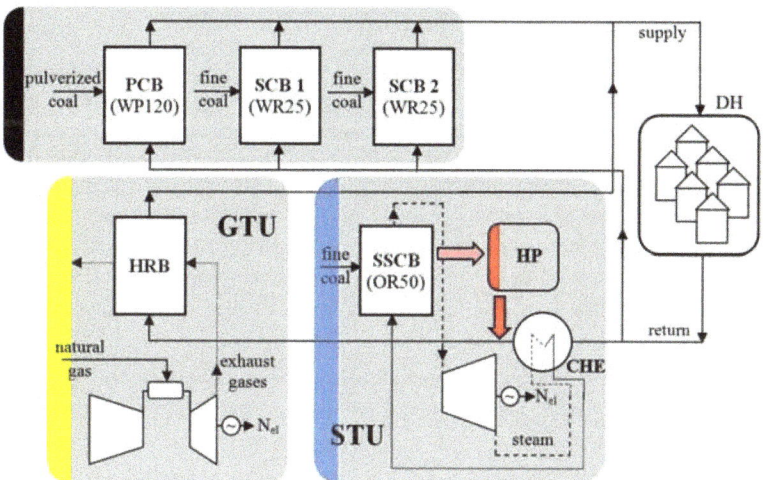

**Figure 12.** Schematic diagram of the proposed configuration of analyzed DH plant retrofitted with heat pump (HP) system recovering heat from coal stoker-fired steam boiler (SSCB) slag.

A detailed diagram of the proposed system is shown in Figure 13. It consists of a heat source equipped with heat exchanger HE submerged in the deslagger water tub and low loss header. The header separates the high temperature heat pump HP from polluted tub water. At the high temperature side, a heat pump is connected to the heat sink. The heat sink is the return pipe of the district heating system.

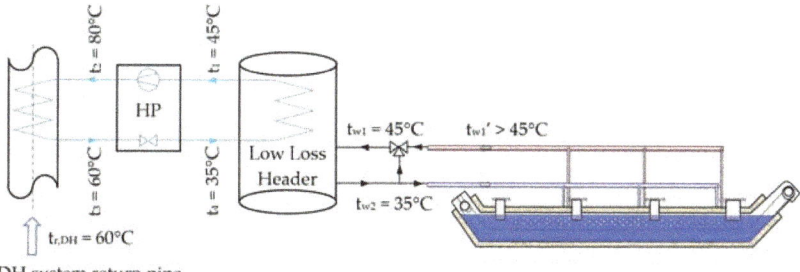

**Figure 13.** Detailed schematic diagram of the proposed slag heat recovery system based on a high temperature heat pump HP.

In order to enable the waste heat to be obtained from the deslagger water tub, the following two proposals of modification of the tub unit has been proposed and assumed to be technically possible to apply:

- direct heat recovery in the existing slag trap,
- first proposal extended by installing a worm conveyor tube covered by a heat exchanger coil.

Both ideas of heat recovery in the slag trap have been presented in Figures 14 and 15. The first proposal given by the author, based on a direct heat recovery from a slag trap, requires simple modification of slag trap water chute. The proposal assumes installing water-water heat exchanger inside the trap and thermal insulation of the trap. As it can be seen in Figure 14, heat exchanger pipes are placed in the upper liquid layer, where the water has highest temperature. Optionally, circulating pumps for mixing water in the top layer for better heat exchange can be applied.

The second idea, presented in Figure 15, contains the same improvements on the first proposal. However, it is extended and additionally results from an approach to solve the problem of low heat conductivity of the slag, which is usualyl in the form of different sized coarse solid rock-like particles (of equivalent diameter in the range of 0.01–0.5 m). The principle of this concepts is based on crushing and homogenizing of the slag by a worm conveyor tube covered by a heat exchanger coil. In this waste heat recovery system, hot water from the exchanger is introduced counter-current to the slag discharge direction, which increases water even more. The slag goes to the opposite end of the trap which extends its contact with the cooling water. The system can produce steam, but requires better construction materials and a pumping facility.

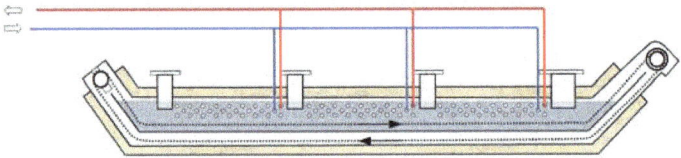

**Figure 14.** A direct slag waste heat recovery proposal.

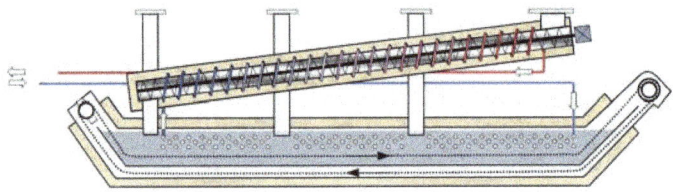

**Figure 15.** A direct slag waste heat recovery extended by crushing and homogenizing of the slag in a worm conveyor tube covered by the additional heat exchanger coil.

*5.1. Energy Performance of the Slag Waste Heat Recovery System*

Annual distribution curve of heat flux recovered from the slag due to the heat pump system is presented in Figure 16—as the result of proposed evaluation procedure.

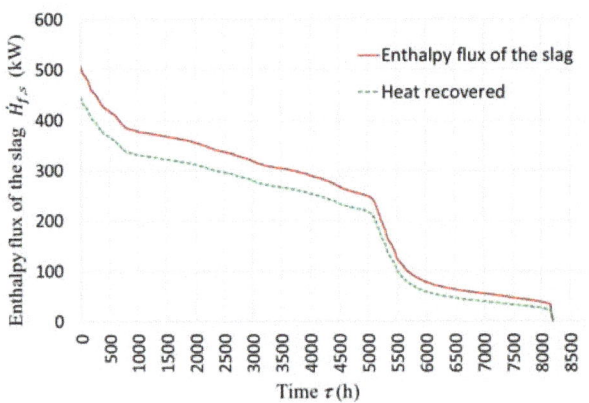

**Figure 16.** Annual distribution curve of enthalpy flux of the analyzed slag.

Derived heat flux is in the range of 58.8–88.0% of the waste energy potential of the slag. The annual quantity of waste energy recovered from the slag $Q_{hr}$ derived from Formula (6) equals to 1,700,600 kWh (6122 GJ).

It should be mentioned in this section of the paper that application of high temperature heat pump for slag waste energy recovery requires considering specific temperatures conditions in the deslagger on one side and heat pump operating cycle on the other side.

In the deslagger tub there are water temperature fluctuations depending on the efficiency of the boiler. The temperature may drop below 50 °C and also reach 100 °C. The temperature of the water in the tub has a big influence on the durability of the entire slag trap. From the point of view of technology, it is advantageous to keep the temperature in the bathtub as low as possible.

On the other hand, we have temperatures from 55 °C to 70 °C at the return pipe from district heating system. This means that it is not possible to recover the heat directly from water in the slag tub. In addition, in case of failure of such a direct heat exchanger, there is a danger of district heating water contamination. Therefore, a solution based on a heat pump is proposed, which on the one hand will lower the temperature in the slag tub and stabilize it, on the other hand it will be able to heat the district heating return water even when its temperature is 70 °C. For this purpose, the refrigerant in the heat pump should be able to reach temperatures above 70 °C during condensation. A good and economical solution is to use 1,1,1,2-Tetrafluoroetan (R134a, also known as HFC-R134a) with 1430 of global warming potentials (GWP). The R134a allows us to obtain 78 °C in a condenser, while maintaining the temperature of the bottom source at 45 °C (Figure 17). The heat pump operated with R134a according to the cycle presented in Figure 17 can reach a COP of up to 6.6 taking into account only thermodynamic cycle performance.

Due to the F-Gas Regulation (EU) No 517/2014 of April 16, which lowers the maximum GWP limits (to 2 500 by the year of 2020 and 150 by the year of 2022), the alternative for the proposed system may be a more expensive refrigerant 2,3,3,3-Tetrafluoropropene (R1234yf, also known as HFO-1234yf), with GWP less than 7.

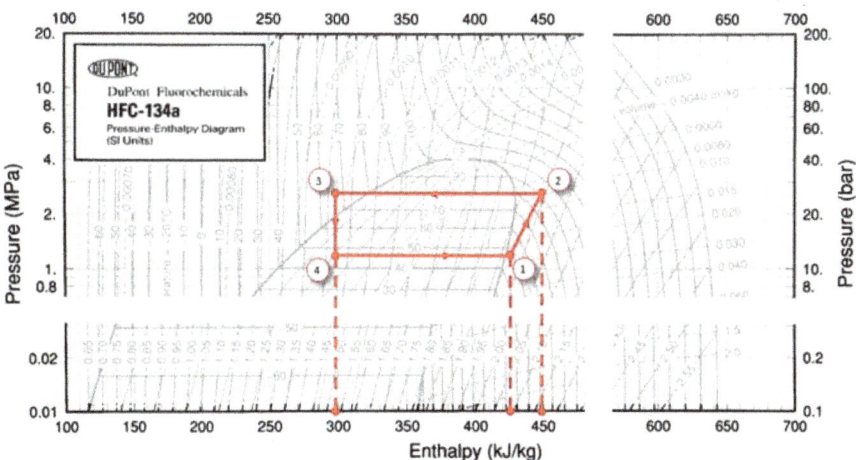

**Figure 17.** 1,1,1,2-Tetrafluoroetan (HFC-134a) cycle of proposed high temperature heat pump presented on p-H diagram.

## 5.2. Energy Performance of the Waste Heat Recovery System

On the basis of presented discussion the high temperature heat pump operated with R134a working medium was sized regarding the heat flux available for recovery. The main design parameters of the pump are presented in Table 3.

The annual operation parameters have been derived regarding annual distribution curve of recovered heat available for recovery $\dot{Q}_{hr}$ (Figure 16) and design parameters of the heat pump, as follows:

- annual operation time of the heat pump: $\tau_{HP}$ = 3115 h,
- heat recovered from the deslagger: $Q_{hr}$ = 1,700,600 kWh (6122 GJ),
- heat supplied into the return pipe of district heating system: $Q_{DH}$ = 2,270,582 kWh (8174 GJ),
- electricity consumed by the heat pump: $E_{el,DH}$ = 616,390 kWh.

**Table 3.** Main design parameters of the selected heat pump.

| Performance Information | Value |
|---|---|
| Heating capacity | 729 kW |
| Heating efficiency (coefficient of performance—COP) * | 3.69 kW/kW |
| Cooling capacity | 546 kW |
| Unit power input | 198 kW |
| Leaving temperature (evaporator) | 35 °C |
| Supply temperature (evaporator) | 45 °C |
| Leaving temperature (condenser) | 80 °C |
| Supply temperature (condenser) | 60 °C |

* Real COP, lower than theoretical COP = 6.6, due to the efficiencies of the particular components of designed heat pump.

### 5.3. Economic Evaluation

The simple economic pre-feasibility evaluation of the proposed waste heat recovery system was carried out on the basis on the annual cash flows generated by the recovery system [33]. The design parameters of the selected heat pump along with technical assumption of the recovery system allowed us to estimate total outlays of the project $J_0$. Next, derived energy quantities including effects and inputs were used to calculate annual cash flows. For the sake of rough calculations, the authors took into account the recovered heat supplied into the district heating system $Q_{DH}$ (positive effect) and electricity consumption of the heat pump $E_{el,DH}$ (negative effect). To calculate the annual profits $CF_{in}$ and costs $CF_c$, heat and electricity prices were also assumed as follows:

- price of heat generated in the analyzed plant: $p_h$ = 6 €/GJ,
- electricity price consumed by the heat pump: $p_{el}$ = 50 €/MWh.

It should be stated here that electricity used for driving the heat pump is generated onsite—in the cogeneration units installed in the DH plant so the price is attractive compared to price of electricity offered by distribution companies.

The summary economic effects of proposed solution for proposed heat recovery are shown in Table 4.

**Table 4.** Economic effects of heat recovery from slag

| Parameter | Value |
|---|---|
| Total investment cost $J_0$ | 130,000 € |
| Annual incomes $CF_{in}$ | 49,044 € |
| Annual costs $CF_c$ | 30,820 € |
| Simple payback time SPB | 7.1 years |

## 6. Conclusions

The direct result of the presented research is the estimation of a waste heat recovery from the coal slag generated in grate boiler supplied with hard coal. The waste heat available for recovering derived during the analysis is relatively low and does not exceed 1% of maximum thermal output of the boiler, which is reflected in the considerations presented in [9].

Regardless of the low share of energy contained in this slag in the total energy balance of the furnaces, the authors decided that it is still worth emphasizing the proposal of a technically possible and feasible application for waste heat recovery from the slag. The proposed modifications are focused on high temperature heat pump connected to the deslagging system as a strategic device within the heating plant infrastructure. The failure-free operation of the slag trap is critical for the system and the observations show that in the majority of cases, the customers are not satisfied with the long-term used deslagging solutions. The proposed modification may then be applied in case of retrofitting old, defective traps.

On the basis of the conducted analysis, the following final conclusions can be drawn.

1. High temperature of the slag generated in the boiler can be a challenge for developing heat recovery system.
2. Waste energy potential of the slag produced in stoker-boilers fired with hard coal has relatively low share in boiler energy balance.
3. Recovering the heat from the slag not only improves the energy efficiency of the district heating system but, due to the reduction of the temperature of water in the slag tub, limits the tribocorrosion phenomena in the deslagger.
4. Applying a heat pump for energy recovery from the slag can be more feasible if electricity is available onsite from a production facility (cogeneration units in the analyzed case).
5. The proposal of heat recovery systems from low temperature slag is a technical and economic challenge but can be an economically justified solution in certain circumstances.
6. Slag energy recovery facilities should be applied in line with old traps replacement to improve economic feasibility of the recovery.

**Acknowledgments:** The authors would like to thank the ECO SA—Heat Engineering Company engineers for the cooperation and for the access to operational data of heat generating district heating plant. The authors also gratefully acknowledge the possibility of using the design data of analyzed unites to the ECO SA company.

**Author Contributions:** Maciej Masiukiewicz and Mariusz Tańczuk conceived and designed the experiments and analysis; Maciej Masiukiewicz and Mariusz Tańczuk provided the analysis tools and analyzed the data; Robert Junga performed the measurements of the properties of the hard coal and slag; Stanisław Anweiler and Mariusz Tańczuk wrote the paper.

**Conflicts of Interest:** The authors declare no conflict of interest.

## Abbreviations

| | |
|---|---|
| $a$ | coefficient of ash contraction in coal |
| $A_r$ | ash content in coal, wt.% |
| $c_{p,s}$ | specific heat of the slag, kJ/kgK |
| $E_{el,DH}$ | electricity consumption of the heat pump, kWh |
| $\dot{G}_s$ | slag flux, kg/s |
| $h_s$ | specific physical enthalpy of the slag, kJ/kg |
| $\dot{H}_{ch,s}$ | chemical enthalpy flux of the slag, kW |
| $\dot{H}_{f,lw}$ | physical enthalpy of the water lost with the slag, kW |
| $\dot{H}_{f,os}$ | physical enthalpy of the slag leaving deslagger, kW |
| $\dot{H}_{f,rfw}$ | physical enthalpy of the refill water, kW |
| $\dot{H}_{f,s}$ | physical enthalpy flux of the slag, kW |
| $\dot{H}_s$ | enthalpy flux of the slag, kW |

| $k$ | thermal conductivity, W/mK |
|---|---|
| $N_{el}$ | electric power output of cogeneration units, kW |
| $p_n$ | hot water nominal pressure, MPa |
| $\dot{Q}_b$ | output capacity distribution of the boiler, MW |
| $Q_{DH}$ | annual quantity of waste energy supplied to the district heating DH system, GJ |
| $Q_{hr}$ | annual quantity of waste energy recovered from the slag, GJ |
| $\dot{Q}_{hr}$ | heat flux available for recovery, kW |
| $\dot{Q}_l$ | heat energy losses, kW |
| $Q_s$ | annual quantity of waste energy in the slag, GJ |
| $\dot{Q}_s$ | waste energy flux, kW |
| $\dot{Q}_t$ | heating capacity, kW |
| $s_s$ | unitary slag volume, kg/kg of coal |
| $\tau$ | time, h |
| $\tau_{DH}$ | annual operation time of heat pump, h |
| $TOC$ | total organic carbon, % |
| $TOC_s$ | total organic carbon of slag-ash mixture, wt.% |
| $t_{r,DH}$ | temperature of DH return water, °C |
| $t_s$ | slag temperature, °C |
| $t_{s,n}$ | nominal (maximum) supply temperature, °C |
| $t_w$ | water temperature in the deslagger tub, °C |
| CHE | condenser heat exchange unit |
| DH | district heating |
| GTU | gas turbine unit |
| HE | heat exchanger |
| HP | heat pump |
| HRB | heat recovery boiler |
| PCB | pulverized hard coal-fired water boiler |
| SCB | stocker coal boiler |
| SPB | simple payback time |
| SSCB | hard coal stoker-fired steam boiler |
| STU | steam turbine unit |

## References

1. Rose, L. (Ed.) *Energy: Modern Energy Storage, Conversion, and Transmission in the 21st Century*; Nova Science Publishers, Inc.: New York, NY, USA, 2013.
2. Klemeš, J.J. (Ed.) *Assessingand Measuring Environmental Impact and Sustainability*; Butterworth-Heinemann: Oxford, UK, 2015.
3. Duić, N.; Rosen, M.A. Sustainable development of energy systems. *Energy Convers. Manag.* **2014**, *87*, 1057–1062. [CrossRef]
4. Sadek, D.M. Effect of cooling technique of blast furnace slag on the thermal behavior of solid cement bricks. *J. Clean. Prod.* **2014**, *79*, 134–141. [CrossRef]
5. Wzorek, M.; Kozioł, M.; Ścierski, W. Emission characteristics of granulated fuel produced from sewage sludge and coal slime. *J. Air Waste Manag. Assoc.* **2010**, *60*, 1487–1493. [CrossRef] [PubMed]
6. Stijepovic, M.Z.; Linke, P. Optimal waste heat recovery and reuse in industrial zones. *Energy* **2011**, *36*, 4019–4031.
7. Chae, S.H.; Kim, S.H.; Yoon, S.G.; Park, S. Optimization of a waste heat utilization network in an eco-industrial park. *Appl. Energy* **2010**, *87*, 1978–1988. [CrossRef]
8. Nemet, A.; Klemeš, J.J.; Kravanja, Z. Optimising entire lifetime economy of heat exchanger networks. *Energy* **2013**, *57*, 222–235. [CrossRef]
9. Ma, G.Y.; Cai, J.J.; Zeng, W.W.; Dong, H. Analytical research on waste heat recovery and utilization of China's iron & steel industry. *Energy Procedia* **2012**, *14*, 1022–1028.
10. Szargut, J.; Ziębik, A.; Kozioł, J.; Kurpisz, K.; Majza, E. *Industrial Waste Energy*; WNT: Warsaw, Poland, 1993; (In Polish). ISBN 83-204-1626-4.
11. Barati, M.; Esfahani, S.; Utigard, T.A. Energy recovery from high temperature slags. *Energy* **2011**, *36*, 5440–5449. [CrossRef]

12. Bisio, G. Energy recovery from molten slag and exploitation of the recovered energy. *Energy* **1997**, *22*, 501–509. [CrossRef]

13. Mills, K.C. Heat Transfer and Thermal Conductivity of Coal Slags. In *Mineral Matter and Ash in Coal*; ACS Symposium Series 301; Vorres, K.S., Ed.; American Chemical Society: Washington, DC, USA, 1986; pp. 256–276.

14. Rezaei, H.R.; Gupta, R.P.; Bryant, G.W.; Hart, J.T.; Liu, G.S.; Bailey, C.W.; Wall, T.F.; Miyamae, S.; Makino, K.K.; Endo, Y. Thermal conductivity of coal ash and slags and models used. *Fuel* **2000**, *79*, 1697–1710. [CrossRef]

15. Henry, P.; Takadoum, J.; Berçot, P. Tribocorrosion of 316L stainless steel and TA6V4 alloy in $H_2SO_4$ media. *Corros. Sci.* **2009**, *51*, 1308–1314. [CrossRef]

16. Jiang, J.; Stack, M.M.; Neville, A. Modelling the tribo-corrosion interaction in aqueous sliding conditions. *Tribol. Int.* **2002**, *35*, 669–679. [CrossRef]

17. Mischler, S. Triboelectrochemical techniques and interpretation methods in tribocorrosion: A comparative evaluation. *Tribol. Int.* **2008**, *41*, 573–583. [CrossRef]

18. Mi, X.; Cai, Z.B.; Xiong, X.M.; Qian, H.; Tang, L.C.; Xie, Y.C.; Peng, J.F.; Zhu, M.H. Investigation on fretting wear behavior of 690 alloy in water under various temperatures. *Tribol. Int.* **2016**, *100*, 400–409. [CrossRef]

19. Diomidis, N.; Mischler, S. Third body effects on friction and wear during fretting of steel contacts. *Tribol. Int.* **2011**, *44*, 1452–1460. [CrossRef]

20. Hasan, B.O.; Sadek, S.A. The effect of temperature and hydrodynamics on carbon steel corrosion and its inhibition in oxygenated acid–salt solution. *J. Ind. Eng. Chem.* **2014**, *20*, 297–307. [CrossRef]

21. Wang, Z.H.; Lu, Y.H.; Li, J.; Shoji, T. Effect of pH value on the fretting wear behavior of Inconel 690 alloy. *Tribol. Int.* **2016**, *95*, 162–169. [CrossRef]

22. López-Ortega, A.; Bayón, R.; Arana, J.L.; Arredondo, A.; Igartua, A. Influence of temperature on the corrosion and tribocorrosion behaviour of High-Strength Low-Alloy steels used in offshore applications. *Tribol. Int.* **2018**, in press.

23. Sun, Y.; Shen, H.; Wang, H.; Wang, X.; Zhang, Z. Experimental investigation and modeling of cooling processes of high temperature slags. *Energy* **2014**, *76*, 761–767. [CrossRef]

24. Trashorras, A.J.G.; Álvarez, E.Á.; González, J.L.R.; Cuesta, J.M.S.; Bernat, J.X. Design and evaluation of a heat recuperator for steel slags. *Appl. Therm. Eng.* **2013**, *56*, 11–17. [CrossRef]

25. Ortega-Fernández, I.; Calvet, N.; Gil, A.; Rodríguez-Aseguinolaza, J.; Faik, A.; D'Aguanno, B. Thermophysical characterization of a by-product from the steel industry to be used as a sustainable and low-cost thermal energy storage material. *Energy* **2015**, *89*, 601–609. [CrossRef]

26. Zhang, H.; Wang, H.; Zhu, X.; Qiu, Y.J.; Li, K.; Chen, R.; Liao, Q. A review of waste heat recovery technologies towards molten slag in steel industry. *Appl. Energy* **2013**, *112*, 956–966. [CrossRef]

27. Mills, K.C. Estimation of physicochemical properties of coal slags. In *Mineral Matter and Ash in Coal*; ACS Symposium Series 301; Vorres, K.S., Ed.; American Chemical Society: Washington, DC, USA, 1986; pp. 195–214.

28. Yu, X.; Zhang, Y.; Kong, L.; Zhang, Y. Thermodynamic analysis and parameter estimation of a high-temperature industrial heat pump using a new binary mixture. *Appl. Therm. Eng.* **2018**, *131*, 715–723. [CrossRef]

29. Wu, X.; Xing, Z.; He, Z.; Wang, X.; Chen, W. Performance evaluation of a capacity-regulated high temperature heat pump for waste heat recovery in dyeing industry. *Appl. Therm. Eng.* **2016**, *93*, 1193–1201. [CrossRef]

30. Xiaohui, Y.; Yufeng, Z.; Na, D.; Chengmin, C.; Yan, Z. Experimental performance of high temperature heat pump with near-azeotropic refrigerant mixture. *Energy Build.* **2014**, *78*, 43–49. [CrossRef]

31. Zhang, Y.; Zhang, Y.; Yu, X.; Guo, J.; Deng, N.; Dong, S.; He, Z.; Ma, X. Analysis of a high temperature heat pump using BY-5 as refrigerant. *Appl. Therm. Eng.* **2017**, *127*, 1461–1468. [CrossRef]

32. Chamoun, M.; Rulliere, R.; Haberschill, P.; Peureux, J.L. Experimental and numerical investigations of a new high temperature heat pump for industrial heat recovery using water as refrigerant. *Int. J. Refrig.* **2014**, *44*, 177–188. [CrossRef]

33. Tańczuk, M.; Skorek, J.; Bargiel, P. Energy and economic optimization of the repowering of coal-fired municipal district heating source by a gas turbine. *Energy Convers. Manag.* **2017**, *149*, 885–895. [CrossRef]

*Article*

# Novel Concept of an Installation for Sustainable Thermal Utilization of Sewage Sludge

**Wilhelm Jan Tic** [1,2], **Joanna Guziałowska-Tic** [1], **Halina Pawlak-Kruczek** [3,*],
**Eugeniusz Woźnikowski** [2], **Adam Zadorożny** [2], **Łukasz Niedźwiecki** [3] ,
**Mateusz Wnukowski** [3], **Krystian Krochmalny** [3], **Michał Czerep** [3],
**Michał Ostrycharczyk** [3], **Marcin Baranowski** [3], **Jacek Zgóra** [3] **and Mateusz Kowal** [3]

[1] Department of Environmental Engineering, Opole University of Technology, 45-758 Opole, Poland;
    w.tic@po.opole.pl (W.J.T.); j.guzialowska@po.opole.pl (G.-T.)
[2] West Technology &Trading Polska Sp. z o.o., 45-641 Opole, Poland; ew@wttpolska.pl (E.W.);
    az@wttpolska.pl (A.Z.)
[3] Department of Boilers, Burners and Energy Systems, Wroclaw University of Science and Technology,
    50-370 Wrocław, Poland; lukasz.niedzwiecki@pwr.edu.pl (Ł.N.); mateusz.wnukowski@pwr.edu.pl (M.W.);
    krystian.krochmalny@pwr.edu.pl (K.K.); michal.czerep@pwr.edu.pl (M.C.);
    michal.ostrycharczyk@pwr.edu.pl (M.O.); marcin.baranowski@pwr.wroc.pl (M.B.);
    jacek.zgora@pwr.wroc.pl (J.Z.); mateusz.kowal@pwr.edu.pl (M.K.)
* Correspondence: halina.pawlak@pwr.edu.pl; Tel.: +48-71-320-3942

Received: 16 February 2018; Accepted: 22 March 2018; Published: 26 March 2018

**Abstract:** This study proposes an innovative installation concept for the sustainable utilization of sewage sludge. The aim of the study is to prove that existing devices and technologies allow construction of such an installation by integration of a dryer, torrefaction reactor and gasifier with engine, thus maximizing recovery of the waste heat by the installation. This study also presents the results of drying tests, performed at a commercial scale paddle dryer as well as detailed analysis of the torrefaction process of dried sewage sludge. Both tests aim to identify potential problems that could occur during the operation. The scarce literature studies published so far on the torrefaction of sewage sludge presents results from batch reactors, thus giving very limited data of the composition of the torgas. This study aims to cover that gap by presenting results from the torrefaction of sewage sludge in a continuously working, laboratory scale, isothermal rotary reactor. The study confirmed the feasibility of a self-sustaining installation of thermal utilization of sewage sludge using low quality heat. Performed study pointed out the most favorable way to use limited amounts of high temperature heat. Plasma gasification of the torrefied sewage sludge has been identified that requires further studies.

**Keywords:** sewage sludge; torrefaction; gasification; drying; sustainability

---

## 1. Introduction

### 1.1. Introduction

Sewage sludge is a waste that comes from the processing of wastewater. It is a biologically active mixture that consists predominantly of water and contains organic matter, dead or alive microorganisms, including pathogens, as well as toxic contaminants such as polycyclic aromatic hydrocarbons (PAHs) or heavy metals [1,2]. Currently management methods involving storage are being replaced in the EU by methods leading to waste stabilization and safe recycling [3]. Moreover, landfilling is deemed to be the most costly way to dispose of the sewage sludge, with average total costs ranging from 260 to 350 €/t of dry matter [4].

Land spreading is an alternative utilization route that allows the use of nutrients still present in the sewage sludge. Land spreading is the most cost effective way to dispose of sewage sludge, from an overall economic point of view, with an estimated total cost between 110 and 160 €/t of dry matter [4]. Land spreading of composted sewage sludge as well as land spreading in sylviculture and land reclamation practices bear slightly higher overall cost, ranging from 210 up to 250 €/t of dry matter [4]. Also in this case, properties and composition of the sewage sludge are of crucial importance, especially from the landowner's perspective, due to his potential liability [5].

Incineration and co-incineration are also feasible options. Besides influencing the technical aspects of a combustion installation, some of the fuel properties (moisture content and ash content) are detrimental from a point of view of the logistics, which prevents one from taking full advantage of the effect of scale in large power and combined heat and power (CHP) plants as in the cement industry, due to transportation costs. Total cost of disposal via incineration is deemed to be on average comparable with landfilling [4], although caution is advised since much depends on the overall supply chain.

Due to increasing amounts of environmental restrictions, not favorable to commonly used utilization pathways, such as landfilling [6], novel thermal processes are currently a subject of active investigation, in terms of their suitability for sewage sludge utilization as well as their applicability for other waste streams. It seems reasonable to pursue two distinctive paths in order to mitigate problems related with sewage sludge:

(A) Upgrade sewage sludge into a fuel with better fuel properties and increased energy density. Enabling the possibility of further removal of pathogens would be a significant gain from the health and safety perspective.
(B) Utilize sewage sludge using a thermal process that can potentially make inorganic residues easier to handle (turn it into a useful by-product or at least allow storage at a regular landfill).

*1.2. Novelty and Relevance Aspects of the Proposed Installation along with the Scope of Work*

Currently there is little literature data available on installations for thermal utilization of sewage sludge, capable of achieving full sustainability, i.e., the ability to sustain its work without any external source of heat and electricity. This paper presents the results of drying and torrefaction of sewage sludge and offers a holistic solution, based on the obtained results. The presented results allow one to draw meaningful conclusions in terms of the detailed choice of the individual elements of the system as well as suitable operational parameters for the proposed installation.

The main goal of the study is to prove existing devices and technologies allow the construction of an installation for sustainable utilization of sewage sludge, by integration of a dryer, torrefaction reactor and gasifier with engine, thus maximizing the recovery of the waste heat by the installation. The study aims to identify the most suitable unit operations, that should be included in the proposed installation. Justification of each of the proposed elements is based either on the performed tests or on the available literature data. Feasibility is assessed in conjunction with possible influence of the elements on the overall operation and maintenance of such unit. Due to scarce amount of literature studies published so far on the composition of the torgas from torrefaction of sewage sludge, results from the torrefaction of sewage sludge in a continuously working, laboratory scale, isothermal rotary reactor are presented. Moreover, torrefaction reactor of feasible scale is described in details, including brief description of the operational advantages of the proposed solution. Literature data on slagging gasifiers, used for gasification of low quality solid fuels, is discussed in the paper. Special emphasis is put on the plasma gasification as a potentially feasible technology from the operational point of view.

## 2. State of the Art—the Foundations of the Proposed Concept

### 2.1. Dewatering and Drying of Sewage Sludge

Water distribution in the sewage sludge, determined by the factors such as structure and origin, is crucial in terms of its dewatering and drying propensity. According to the literature, moisture in the sewage sludge takes forms of [7]:

- Free water that is not bound, in any way, by the particles of the sludge
- Interstitial water, trapped by the flocs of solids or existing in capillaries
- Surface water, held by adhesion and adsorption
- Intracellular and chemically bound water

Another classification is process oriented and distinguishes between free water, that can be removed by mechanical dewatering and bound water, that is left after mechanical dewatering [7]. Free water from the latter classification includes free moisture from the former classification as well as a part of interstitial and surface moisture [7].

Among conventional dewatering technologies one can distinguish belt presses, which according to some sources are capable of achieving 15% up to 25% of dry matter, from the activated sludge with initial dry solid concentration of 2% up to 5% [7]. Other available, conventional, technologies are screw presses, rotary presses and centrifuges [7,8]. Peeters described an innovative machine that combines mechanical dewatering and drying in one compact unit [9]. This technology requires natural gas to generate hot sweeping gas (flue gas), that sweeps the volume of the dryer at temperatures between 230 °C and 260 °C [9].

The review by Bennamoun et al. distinguishes three main drying methods: conductive drying, convective drying and solar drying [10]. Conductive dryers rely solely on indirect heating, whereas convective dryers rely on convection and to some extent on conduction (for example in the case of rotary dryers), while solar drying is based on solar radiation [10]. The aforementioned review also connects removal of a free moisture with constant drying rate, removal of interstitial water with the first falling rate (linear) and removal of surface water with the second falling rate (non-linear) [10]. During convective drying three phenomena can be observed: shrinkage, cracks and skin formation [10]. During conductive drying, torque variations undergo three phases: pasty, lumpy and granular [10], each of them exhibiting different rheological behavior [11]. Constant evaporation rate was observed by Arlabosse at al. pasty phase, whereas a linear decrease was observed during the granular period [12].

One of the problems with drying of the sewage sludge is related to its cohesive behavior, that becomes most intensive within certain levels of moisture content of the dried sewage sludge [13,14]. Changes in the rheological behavior are caused by the increased concentration of extra cellular polymers, mainly because of the increase of their respective concentrations, due to the loss of water [14]. These polymers originate both from microorganisms present in the sludge and flocculants added prior to dewatering [14,15]. Peeters et al. proposed using polyaluminium chloride as an additive to prevent the formation of the sticky phase [15].

Among convective dryers one can distinguish fluidized bed dryers, flash dryers, rotary dryers and belt dryers [10]. Among conductive dryers popular designs include disc dryer and thin film dryer [10] as well as paddle dryer [10,16].

Energy consumption per kg of removed water for convective dryers is reported to vary between 2520 and 5040 kJ/kg$_{water}$, whereas for conductive dryers the consumption is lower, varying between 2880 and 3438 kJ/kg$_{water}$ [10].

### 2.2. Dry and Wet Torrefaction and Pyrolysis of Sewage Sludge

As much as drying is of the highest significance, some potential in the area is presented by both dry and wet torrefaction (also known as hydrothermal carbonization—HTC). Torrefaction is a thermal treatment performed under anaerobic conditions and elevated temperature (typically 250 °C to 300 °C) [17]. Wet torrefaction is also performed under elevated temperature, in saturated water, and has been proven as a viable means to decrease the ash content of solid fuels [18,19].

Some amount of work on thermal treatment for utilization of the sewage sludge has been performed, including some fundamental work, using thermogravimetric analysis [20,21]. Huang et al. torrefied waste from pulp industries in a laboratory scale, batch reactor and determined energy densification ratios varying from 1.26 up to 1.5 depending on the process conditions [22]. Studies performed by Huang et al. on microwave co-torrefaction of sewage sludge with leucaena reported synergistic effect between the two feedstocks, during torrefaction [23,24]. An increase in higher heating value on dry, ash free basis, reaching 48 MJ/kg was reported [23]. Achieved ratios of O/C and H/C were similar to those of anthracite [23]. Huang et al. investigated kinetics of the torrefaction of the sewage sludge, using the simplified distributed activation energy model [25]. Lim et al. performed calculations of the performance of a hypothetical plant, that used fry-drying and torrefaction as unit operations along with a steam boiler using a part of the product [26]. Study concluded that the installation would be self-sufficient with an additional output of 33% of the dry solid mass originally fed to the dryer, converted to a solid biofuel [26].

Pulka et al. established that torrefaction increases the higher heating value (HHV) of the pretreated material on a dry ash free (daf) basis, although the change in HHV was not as significant due to the increased ash content of torrefied samples [27]. Increase in ash content and HHV was also observed by Poudel et al. [28]. Poudel et al. also observed increase in ash content of sewage sludge blends with waste wood [29]. Atienza-Martinez et al. successfully torrefied sewage sludge both in a fluidized bed reactor [30] and in an auger reactor [31]. In both cases a bit more reasonable residence times were investigated (13 to 35 min for auger reactor and 3.6 up to 10.2 min for fluidized bed), in comparison to other authors (~1 h). Decrease in the energy density was observed in both cases (dry basis).

Wet torrefaction of sewage sludge has been investigated by He et al. [32] and Denso-Boateng et al. [33]. Both groups observed increase in ash content of hydrochars in comparison with raw sewage sludge, which is most likely a consequence of a much more profound loss of organics, in comparison to inorganics. Denso-Boateng et al. [34] also successfully performed wet torrefaction using primary sewage sludge. In this case an increase of the ash content could also be observed. HTC of the slaughterhouse cake, performed by Oh and Yoon demonstrated increase in the heating value of that residue and found optimum temperature of the HTC process to be 180 °C for that type of feedstock [35].

Pyrolysis of sludge has been a subject of a significant amount of studies [36–40]. Baltrenaite and Peckyte studied the properties of pyrolysis product of various types of industrial sewage sludge (from paper and leather industries) [41]. Performed research indicated, that the form of biochar restrained leaching of heavy metals [41], despite that concentrations were considerable, when compared with the restrictions set by the regulations [42]. Assessment of environmental effects performed by Wang et al. concluded that carbonization of the sewage sludge has overall positive environmental impact in comparison with landfilling and incineration [43]. However, special care should be taken in the cases when composition of a particular sludge makes leaching a problem during a subsequent utilization of the product of pyrolysis.

### 2.3. Slagging Gasifier as a Feasible Tool for Gasification and Inertization of Solid Residues

Slagging gasifiers could be a feasible solution, for the cases when inert solid residues are needed to solve the leaching problem. There are many different thermal conversion technologies that produce gas from solid fuel and the interest in these technologies as a viable mean of thermal utilization of sewage sludge has been recently reviewed [44,45]. As it has been already mentioned, solid residues from utilization of sewage sludge are a subject of strict regulations due to negative impact that might be caused by the leaching. In that context technologies that have a potential to limit the leaching are preferable.

There is a wide variety of technologies designed for different types of waste materials that offer a feature of a vitrification of the predominantly inorganic residues. The Purox process, developed at the Linde division of Oxide Carbide, is an updraft, fixed bed, slagging gasifier that uses oxygen as a gasifying agent [46]. It was patented in 1973. An operating plant with designed capacity of 200 t/d was built in South Charleston and operated on municipal solid waste (MSW) between 1974 and 1978 (75 t/d was reached) [46]. Also twin 100 t/d units were built at Showa Denko (Chichibu, Japan) and operated between 1981 and 1997 [46]. The process consumed roughly 0.2 t of $O_2$ per t of waste, pressure swing absorption being used for production of oxygen [46].

Caliqua (the Heat and Power division of the French company Sofresid) developed a slagging fixed bed gasifier, working with preheated air in 1979 [46]. The throughput capacity of the gasifier was designed to be 8 t/h of MSW (LHV of MSW 7.92 MJ/kg) [46]. Gas was burned and heat was recovered in a Heat Recovery Steam Generator (HRSG), that allowed 1.5 MW electric output and 10.5 MW of heat output to the district heating network [46]. The calculated efficiency reached 68% [46].

The Twin Rec process, developed by Japanese company Ebara, splits fluidized bed gasification and melting of the inorganics into two distinct stages. Vitrification of the fly ash is performed in the cyclonic ash melting furnace. Flue gases are used to generate steam [47]. Technology is fully commercial. Within a portfolio of the company projects installation sizes range from 15.7 t/d of MSW (Joetsu, Japan; commissioned in 2000) up to two installations with capacity of 275 t/d (Tokyo Rinki, Japan; commissioned in 2006) [47]. Plants in Joetsu and in Aomori (2 × 225 t/d; commissioned in 2000) have operated using sewage sludge [47].

Plants are typically CHP facilities. The plant in Aomori, for example, has installed power of 17 $MW_{el}$ and 40 $MW_{th}$. One exception is the Kurobe plant, where gas is used for melting residues containing copper for the purpose of recovering that metal [47].

A significant part of the inorganic solids can be used as a by-product. For instance in the Kawaguchi plant (3 × 140 t/d; commissioned in 2002) by-products from one ton of utilized MSW are as follows [47]: 10 kg of recyclable metals (ferrous and aluminum), 95 kg of vitrified ash (aggregate), 20 kg of inert materials, 25 kg of fly ash.

Some processes use pyrolysis as a primary treatment. An example of such technology is the Thermoselect High Temperature Reactor (HTR) process developed in Switzerland. Pyrolysis is followed by high temperature, oxygen gasification in a slagging fixed bed. The first commercial plant (95 tpd) operated between 1992 and 1999 in Fondoce (Italy). More installations followed, mostly in Japan, but also in Karlsruhe, Germany [47].

Due to a high energy density and a possibility to obtain very high temperatures (couple thousand °C) plasma technologies are considered to be a viable option for utilization of various waste types [48]. High temperature is generated by a plasma torch, which requires electricity to operate.

There are existing plasma gasification technologies that have been successfully implemented in the field of thermal conversion of MSW. The most mature, fully commercialized, is a process developed by Westinghouse Plasma Corporation and currently owned by Canadian company Alter NRG. Plasma Gasification Vitrification Reactor (PGVR) uses heat generated by plasma torches to gasify the waste and vitrify solid residues. Fully commercial facility was built in Utashinai, Japan by the EcoValley consortium. Installation was commissioned in April 2003 and ceased operation in 2013 [47], due to problems with obtaining sufficient quantities of feedstock [47]. Installation was designed to process MSW and Automobile Shredding Residues (ASR) in 50/50 proportion, with nominal processing capacity of 165 t/d [49]. Plant was able to achieve 220 t/d working with 100% of MSW [49]. Installation was able to meet strict Japanese emission criteria and vitrified slag shown limited solubility, which made it suitable to be used as an aggregate [47].

Syngas from the PGVR was burned in the refractory-lined combustor after-burner. Heat obtained this way was recovered in HRSG to generate steam for a steam turbine. Some operational problems have been reported [22,24]: cold spots in the coke bed, due to inadequate penetration by plasma torches [49]; too short life span of the refractory of the reactor—caused by adding additional refractory to remediate the problem of the cold spots [49]; erosion of the refractory of the after-burner (which caused the operator to lower the exit temperature of syngas from 1200 °C to 750 °C) [49]. Reactor requires auxiliary substances to operate. For every ton of waste it requires (our own calculations based on data from [49]) 40 kg of coke and 74 kg of limestone.

## 2.4. Plasma Gasification of the Sewage Sludge

Considering sewage sludge plasma treatment there is a very limited information on the industrial installations and scarce amount of scientific papers on the topic. Montouris et al. determined, using the GasifEq equilibrium model, that plasma gasification of sewage sludge from Psittalia Island, can lead to a net production of electricity [50]. Assuming processing of 250 t/d of sewage sludge with moisture content of 68%, modelled installation has shown possibility to supply electric power of 2.85 MW [50]. Tar conversion is crucial in the context of gasification since their presence can cause technical problems—its removal is important if the gas is planned to be used in turbines, engines or subsequent synthesis [51]. Two research groups conducted successful experiments with a two-step plasma processing units [52,53]. In both cases applying plasma improved the gas quality. During the study performed by Striugas et al. the use of arc plasma allowed to achieve over 99% conversion of tar, while the raw syngas composition was not a subject of a significant change [53]. In the second work, a microwave plasma reactor was used to reform a raw syngas, derived from the sewage sludge gasification. In the research, it was shown that applying plasma resulted not only in the tar content decrease (with conversion form 70% to 100% depending on the analyzed compound) but also in a significant improvement in the gas composition [52].

Plasma gasification is more flexible in terms of the quality of the fuel, in comparison with typical slagging gasifiers, as it uses electricity to deliver a heat in the form of hot plasma. However, in terms of the sewage sludge, it is desired to remove as much moisture as possible, before feeding the sludge into the gasifier, in order to improve the heat balance of the installation.

## 2.5. State of the Art—Summary

Drying of the sewage sludge is a relatively well known process, with a significant amount of published investigations. Drying of the sewage sludge is a mature technology and knowledge on the operational issues, such as increased stickiness of the sludge during drying, is extensive. Special emphasis should be put on the suitability of the dryer for the specific parameters of a sewage sludge and a waste heat source.

There are some works available on the torrefaction of sewage sludge. However, little is known on the composition of torgas, which is often an operational issue for the torrefaction installation. Technology selected for the proposed installation should have a design features capable of minimizing operational problems related with condensation of tars.

Possibility to obtain inert solid by-product, not susceptible to leaching, is the main advantage of the plasma gasification. Moreover, high temperatures of the process are potentially good for utilization of torgas, without the need to build a separate burner.

## 3. Materials and Methods

Samples of feedstock were obtained at the sewage treatment plant in Brzeg Dolny. Samples of the sewage sludge were obtained after fermentation and mechanical dewatering stages of the sewage treatment. Detailed characterization of the raw material is presented in Table 1, below.

Standard proximate analysis and ultimate analysis of both raw sewage sludge and torrefied product, was performed according to European Standards. References of all the relevant standard procedures are presented in Tables 1 and 4 (required accuracies are stated in the respective standards).

**Table 1.** Proximate and ultimate analysis of the sample of sewage sludge from Brzeg Dolny water treatment plant (after mechanical dewatering).

| Test | Symbol | Value | Unit | Standard Procedure |
|------|--------|-------|------|--------------------|
| Moisture content [1] | MC | 83.90 | % | EN ISO 18134-2:2015 |
| Volatile matter content | VM [d] | 62.40 | % | EN 15148:2009 |
| Ash content | A [d] | 33.40 | % | EN ISO 1822:2015 |
| Higher heating value | HHV | 16,100 | kJ/kg | EN 14918:2009 |
| Lower heating value [2] | LHV | 661.1 | kJ/kg | EN 14918:2009 |
| Carbon content | C [d] | 38.05 | % | EN ISO 16948:2015 |
| Hydrogen content | H [d] | 3.85 | % | EN ISO 16948:2015 |
| Nitrogen content | N [d] | 5.55 | % | EN ISO 16948:2015 |
| Sulfur content | S [d] | 0.95 | % | EN ISO 16994:2016 |
| Oxygen content | O [d] | 18.20 | % | EN ISO 16993:2015 |

[1] Wet basis; [2] Calculated using the formula from the standard; [d] Dry basis.

Diagram of the installation with a paddle dryer, used for drying trials, is presented at Figure 1. Drier was operated in a batch mode. A suite of three consecutive drying tests was performed using the same batch of material for each test. Initial batch size was approximately 1 t. Sludge was recirculated after each test (see Table 2) in order to obtain a product with moisture content below 5%. Dryer used for the tests was 2.5 m long, 1 m wide and 1 m high (outside dimensions), with heat exchanging surfaces of approximately 10 m².

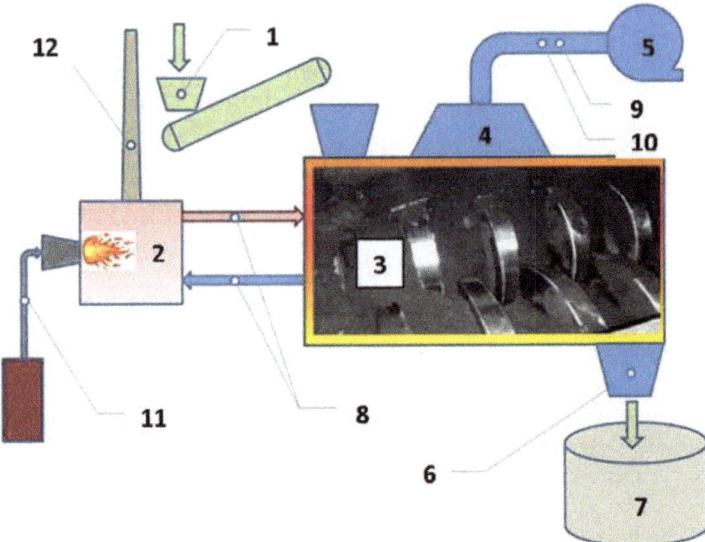

**Figure 1.** Diagram of the paddle dryer used for drying tests (1—feeding line of wet sewage sludge; 2—thermal oil boiler with oil burner; 3—paddle dryer; 4—outlet of the mixture of air and vapors; 5—extraction fan; 6—outlet of dried sewage sludge; 7—tank with dried sewage sludge; 8—temperature measurement for hot thermal oil supply and return; 9—sampling of the vapors with measurement of the velocity and temperature; 10—measurement of the relative humidity; 11—supply of the fuel oil to the burner; 12—stack).

**Table 2.** Results of the drying tests, performed with commercial scale paddle dryer.

| Test | Symbol | Test 1 | Test 2 | Test 3 | Unit |
|------|--------|--------|--------|--------|------|
| Moisture content of the feedstock [1] | $MC_{in}$ | 83.90 | 53.47 | 32.13 | % |
| Moisture content of the dried sewage sludge [1] | $MC_{out}$ | 53.47 | 32.13 | 4.02 | % |
| Residence time | $t_{res}$ | 75 | 55 | 60 | min |
| Average temperature of the thermal oil [2] | $T_{oil}$ | 178.7 | 177.2 | 183.3 | °C |
| Relative humidity of the air at the outlet of the dryer | RH | 60 | 62 | 58 | % |
| Mass of dried sludge | $m_{wet}$ | 966 | 329 | 227 | kg |
| Mass of removed water | $m_{vap}$ | 637 | 102 | 66 | kg |

[1] Wet basis; [2] Inside of the dryer.

Due to the size of the dryer and the sample itself, mass and energy balances of the dryer were both performed, using the indirect method. Volumetric flow rate of the mixture of vapors and air, out of the dryer, was measured, by measurement of the velocity of the gases in the duct. Sampling probe, used in the tests, had an integrated pitot probe, thus allowing measurement of the velocity of the gases in the duct. Due to relatively low diameter of the ducting (200 mm) and relatively high velocities in the ducting (between 10 and 20 m/s) velocity was measured in a single point of the cross section of the circular duct, and the velocity distribution was assumed to resemble the one of a plug flow. Uncertainty of the velocity measurement was ±2% of the measured velocity. Single sampling point is typical for the determination of low range mass concentration of dusts, for ducts with diameter smaller and equal to 350 mm [54]. Measurement of the relative humidity allowed to determine concentration of water vapors carried by air, using a Mollier diagram. Measurement was performed using a Testo 435 analyzer, with an accuracy of ±2% of the measuring range (0 to 100% Relative Humidity—RH). Energy balance was determined based on the flow rate of the thermal oil and the temperature difference between flow and return of the thermal oil. Flow rate of the thermal oil was performed using an ultrasonic flow meter FSD32 (Fuji Electric) with measuring range of 0.3 to 32 m/s, for temperatures of the fluid ranging from −40 °C up to 200 °C and accuracy of ±0.02 m/s. Flow meter was installed respectively more than 10 nominal diameters downstream and more than 5 nominal diameters upstream from the nearest obstacle. Distance from the nearest pump was higher than 50 nominal diameters. Temperature was measured using 1st class, K type thermocouples, with measurement uncertainty of 1.5 °C.

Torrefaction tests were carried out at a laboratory scale using isothermal rotary reactor presented at Figure 2. The core part of this test rig was a rotating pipe made of heat resistant steel and externally heated by a set of electric heaters. Test rig had its own temperature control system that could maintain temperature, set by the operator, up to maximum of 1000 °C. Temperature was measured at the outside surface of the pipe by three thermocouples, one in the middle and two on both ends of the heated pipe. Samples of the torgas were taken by the sampling probe introduced to the inside of the rotating pipe (at the far end of the pipe, along the central axis; No. 9 in Figure 2). All the solid products were collected at the bottom of the drop, out of the far end of the pipe (No. 8 in Figure 2).

Condensable gaseous compounds, present in torgas, were captured by the set of impinger bottles connected in series. Bottles were filled with isopropanol and cooled with thermostatic bath, keeping the temperature at −10 °C. Cold gas went through a conditioning unit in order to get rid of any residual moisture and filter all the remaining particulate impurities. Gaseous (non-condensable) products of torrefaction were measured using a Gasmet CX 4000 FTIR analyzer (Gasmet Technologies Oy, Company, Helsinki, Finland). Calibration of the analyzer was performed using nitrogen of 5.0 quality prior to measurements. The analyzer has a zero drift and linearity drift smaller than 2% of the range for each of the respective compounds.

Samples of the solutions from both impinger bottles were subsequently analyzed using GC-MS, which consisted of an Agilent 7820-A chromatograph (Agilent Technologies, Palo Alto, CA, USA) and an Agilent 5977B MSD spectrometer (Agilent Technologies, Palo Alto, CA, USA). In the chromatograph a Stabilwax-DA column (Restek) (Restek, Benner Circle, Bellefonte, PA, USA) was used. Helium was used as a carrier gas (1.5 mL/min). Heating up program was set to achieve 50 °C in 5 min and subsequently heat up the column with a ramp of 10 °C/min until the temperature of 200 °C was reached and hold for another 20 min. The data obtained with GC/MS was analyzed using the base peak chromatograms (BPC). Each sample was analyzed three times and the obtained results present an average value.

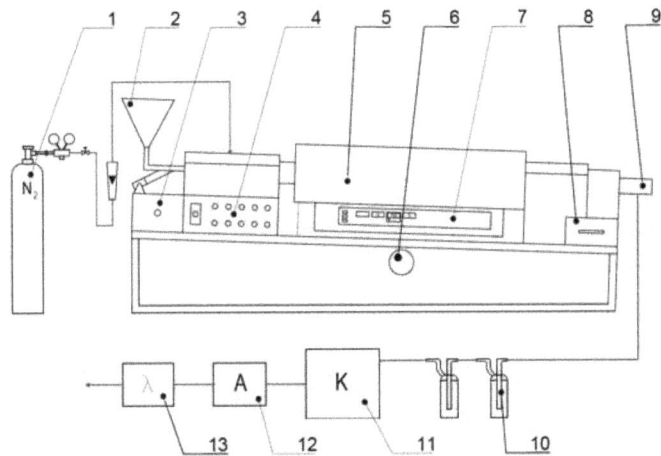

**Figure 2.** Diagram of the rotary isothermal torrefaction reactor (1—cylinder with nitrogen; 2—feedstock hopper; 3—electric motor; 4—control panel; 5—rotary drum; 6—regulation of the tilt; 7—temperature control panel; 8—container with torrefied product; 9—sampling of torgas; 10—impinger bottles; 11—conditioner; 12—FTIR gas analyzer; 13—Electrochemical $O_2$ analyzer).

## 4. Results

Results of the drying tests performed with the commercial scale paddle dryer are shown in the Tables 2 and 3 below. Tests from 1 to 3 were performed in a series, i.e., the material dried during test number 1 was subsequently recirculated back to the dryer and became the feedstock for test number 2, etc. This way cause it was possible to overcome limitations, such as length of the dryer.

The amount of energy needed to drive off 1 kg of water, for each of three performed tests is presented in Table 3, below. Consumption of the heat and consumption of the electricity were calculated separately, using methods described in paragraph 2.

Results of the proximate and ultimate analysis of the sewage sludge, torrefied at 300 °C are presented in Table 4, below. Initial moisture content of feedstock, prior to torrefaction was 10.5%.

**Table 3.** Energy consumed by the paddle dryer for the drying process for each of the tests.

| Result | Symbol | Test 1 | Test 2 | Test 3 | Unit |
|---|---|---|---|---|---|
| Average consumption of the heat | $E_{th}$ | 4141 | 6300 | 7203 | $kJ/kg_{H_2O}$ |
| Average consumption of the electricity | $E_{el}$ | 301 | 643 | 413 | $kJ/kg_{H_2O}$ |
| Average total energy consumption | $E_{tot}$ | 4441 | 6976 | 7615 | $kJ/kg_{H_2O}$ |

**Table 4.** Proximate and ultimate analysis of the torrefied sewage sludge.

| Test | Symbol | Value | Unit | Standard Procedure |
|---|---|---|---|---|
| Moisture content [1] | MC | 1.00 | % | EN ISO 18134-2:2015 |
| Volatile matter content | VM [d] | 41.50 | % | EN 15148:2009 |
| Ash content | A [d] | 48.00 | % | EN ISO 1822:2015 |
| Higher heating value | HHV | 15,700 | kJ/kg | EN 14918:2009 |
| Lower heating value [2] | LHV | 15,515 | kJ/kg | EN 14918:2009 |
| Carbon content | C [d] | 34.58 | % | EN ISO 16948:2015 |
| Hydrogen content | H [d] | 2.73 | % | EN ISO 16948:2015 |
| Nitrogen content | N [d] | 5.09 | % | EN ISO 16948:2015 |
| Sulfur content | S [d] | 0.76 | % | EN ISO 16994:2016 |
| Oxygen content | O [d] | 8.83 | % | EN ISO 16993:2015 |

[1] Wet basis; [2] Calculated using the formula from the standard; [d] Dry basis.

Major constituents of torgas are presented in Table 5 below. Except $N_2$ which was an inerting agent for the reactor, torgas consisted mainly of water, carbon dioxide and condensable hydrocarbons (tars). Compounds, that are considered condensable in this paper, are the compounds that have their boiling point high enough to condense in the ambient conditions.

**Table 5.** Composition of torgas—major constituents.

| Result | Symbol | Value | Unit |
|---|---|---|---|
| Concentration of $CO_2$ in torgas | $U_{CO_2}$ | 2.11 | $\%_{vol}$ |
| Concentration of condensable compounds in torgas | $D_{tars}$ | 63.83 | $g/m_{STP}^3$ |

Minor constituents, measured using the FTIR analyzer included simple, non-condensing (gaseous) hydrocarbons. The results of the measurements are shown at Figure 3. Their concentrations in torgas were insignificant, in comparison to the main components. However, this was caused partially by the dilution of the torgas by an inerting agent. Therefore constituents of torgas with concentrations sufficiently high (twice the detectability limit) are reported.

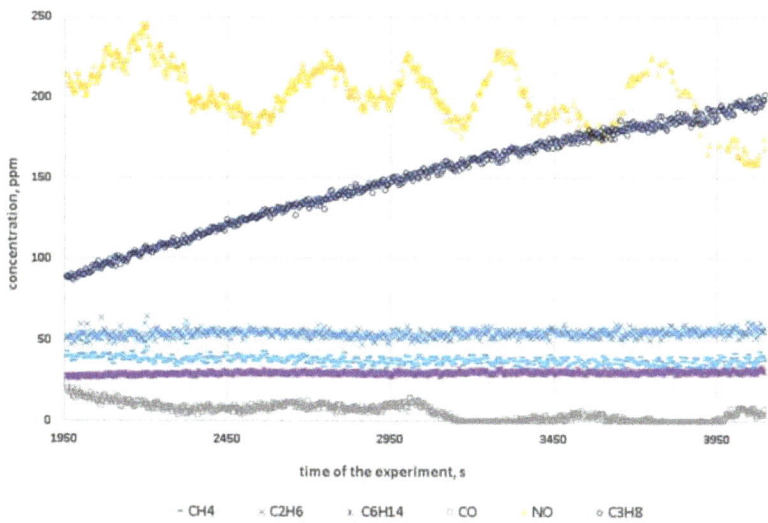

**Figure 3.** Composition of the torgas—minor constituents.

Condensable compounds detected in the samples of solvent (isopropanol) solution, are reported in Table 6. All of these compounds are heterorganic, i.e., they contain atoms of S, N or O within their particles. Water, which was also present (in abundance) in the solutions from impinger bottles was not included in the respective tables. Results presented in Table 6 are qualitative.

**Table 6.** Compounds identified in the first impinger bottle, using GC-MS analysis.

| Compound | Boiling Point | Molecular Mass | Area of the Peak | Relative Area of the Peak [2] |
|---|---|---|---|---|
| | °C | g/mol | a.u. $(10^{-6})$ [1] | % |
| Trimethylamine | 2.9 | 59.11 | 1.73 | 76.4% |
| Furan | 31.3 | 68.07 | 0.07 | 3.2% |
| Acetone | 56 | 58.08 | 1.14 | 50.5% |
| 2-Methylfuran | 63 | 82.1 | 0.47 | 20.6% |
| Methyl Alcohol | 64.7 | 32.04 | 2.19 | 100.0% |
| Acetonitrile | 82 | 41.05 | 0.54 | 23.7% |
| Dimethyl disulfide | 110 | 94.19 | 1.19 | 52.6% |
| *p*-Xylene | 138.4 | 106.16 | 0.20 | 9.0% |
| Pyridine | 115.6 | 79.1 | 0.57 | 25.2% |
| Pyrazine | 115 | 80.09 | 0.05 | 2.4% |
| Isocapronitrile | 155 | 97.16 | 0.01 | 0.6% |
| 2-Methylpyrazine | 135 | 94.11 | 0.25 | 11.2% |
| 4-Aminopyridine | 273 | 94.11 | 0.11 | 4.7% |
| Dimethyl trisulfide | 170 | 126.26 | 0.40 | 17.6% |
| Acetic acid | 118.1 | 60.05 | 1.66 | 73.3% |
| 1-(2-Furyl)ethanone | 173 | 110.11 | 0.05 | 2.3% |
| Pyrrole | 129 | 67.09 | 0.74 | 32.9% |
| 2-Methyl-1*H*-pyrrole | 148 | 81.12 | 0.07 | 3.0% |
| Acetamide | 221.2 | 59.07 | 1.29 | 57.1% |
| (*E*)-2-Butenoic acid | 185 | 86.09 | 0.08 | 3.4% |
| Phenol | 181.7 | 94.11 | 0.13 | 5.6% |
| Octanoic acid | 237 | 144.21 | 0.01 | 0.4% |
| *p*-Cresol | 201.8 | 108.13 | 0.03 | 1.2% |
| 3,4,5-Trimethylpyrazole | 170 | 110.16 | 0.11 | 4.9% |
| 3-Pyridinol | 180 | 95.1 | 0.60 | 26.5% |
| Indole | 254 | 117.15 | 0.35 | 15.4% |
| Succinimide | 288 | 99.09 | 0.04 | 1.7% |
| Dodecanoic acid | 298.9 | 200.32 | 0.06 | 2.8% |
| *N*-(Pyridin-3-yl)acetamide | 327 | 136.15 | 0.26 | 11.3% |
| *n*-Hexadecanoic acid | 351.5 | 256.43 | 0.05 | 2.3% |

[1] Arbitrary units; [2] Relative to the highest peak (Methyl Alcohol).

## 5. Discussion

### 5.1. Discussion of the Obtained Results

Sewage sludge is a non-Newtonian fluid, that poses some difficulties in a subsequent processing. Increase in the viscosity of the sludge, after partial drying can be clearly observed as an increased consumption of electricity by the installation. As parameters of the blowers and other auxiliaries had not been a subject of any significant change, this could be the only possible explanation of doubled consumption of electricity per kg of removed water during the second test. This is consistent with reports from various literature sources that state a profound change in the rheological properties of the sewage sludge after achieving certain level of moisture content [13–15,55]. In terms of the heat consumption for drying, specific drying heat was a subject of significant increase, with decreasing final moisture content of the dried material, which is by no means surprising. This could highlight potential problems with heat requirements, if deep drying was needed, which could significantly hamper the ambitions to obtain a self-sustaining installation. Latent heat of water could become a potential heat source. However, this would imply using another drying technology as high temperature heat sources are preferable for indirect-conductive units (paddle dryer), due to the constraints of heat transfer between hot surfaces and pasty dried material. Relative humidity of the gases at the outlet of the dryer was close to 60% in all of the cases, which gives some extra room for the optimization of the energy consumption by the dryer. Use of hybrid methods (convective-conductive) of drying might subsequently lead to significant decrease of energy/heat consumption, by enabling the possibility to recover heat from the vapors at the outlet of the dryer.

Torrefaction of the dried sewage sludge yielded the expected results, i.e., slight carbonization of the torrefied sludge and a significant decrease in the moisture content. The most profound change was the increase in ash content, as shown by the comparison of the results from Table 4 with the characteristics of the raw material, presented in the Table 1. This might not be a disadvantage in terms of a planned slagging gasifier and will be discussed further in more details. Moreover, relatively significant concentration of NO in the torgas is worth mentioning, as it clearly indicates the potential of the removal of nitrogen from the fuel by this technology.

Obtained torgas consisted mainly of condensable compounds and $CO_2$, as shown in Table 5. Other gaseous constituents were present, as shown at Figure 3. However, their respective concentrations were insignificant, which could be partly contributed to the dilution caused by using inerting agent during the experiment ($N_2$). Interesting was the fact that among permanent gases concentrations of methane, ethane and hexane, were much higher than the concentration of carbon monoxide, which is different in comparison to other types of biomass, with higher initial content of volatile matter (such as for example wood, palm kernel shells, olive mill residues).

Results of the GC-MS analysis of the solutions from the impinger bottles confirmed the presence of significant amounts of different heterorganic compounds in the torgas from torrefaction of sewage sludge. Their respective concentrations are presented in Table 6. Area of the peak for each substance, can be used as a simplified indicator of their respective concentrations. However, it should be noted that it depends not only on the concentration of the compound but also on its structure, as well as on the temperature of the compound that leaves the column of the gas chromatograph. Moreover, some of the compounds could not be detected either due to their concentrations being below the limit of detectability or due to their equivocal mass distribution, that did not allow proper identification by the mass spectrometer. Nevertheless, the most dominant compounds are clearly trimethylamine, acetone, methyl alcohol, dimethyl disulfide, acetic acid and acetamide.

The presence of compounds with high boiling point, such as *n*-hexadecanoic acid or *N*-(pyridin-3-yl) acetamide is particularly interesting, as in the case of this compound, the boiling point is higher than the temperature of the torrefaction. This might suggest the presence of such compounds on the surface of the torrefied material, which should be investigated further.

### 5.2. A Concept of the Sustainable Installation for Utilization of Sewage Sludge

The research group of Boilers, Combustion and Thermal Processes at Wroclaw University of Science and Technology, has developed and patented a flexible, multi stage tape reactor (Figure 4). This device depending on temperature regime can easily perform duties of dryer and/or torrefaction reactor. Successful trials of lignite drying and torrefaction of various types of biomass, have been performed in recent years [56,57].

Material fed into the reactor is being heated externally, through the surface of a metal plate. Material is being moved through the reactor by a chain conveying system and falls down from one plate into another, until it leaves the reactor.

Feedstock is being fed from the hopper, to the reactor also by a chain conveyor. On both ends of the reactor airlocks are being used to prevent the escape of torgas. Product is being cooled down in a screw conveyor with water jacket, at the bottom of the installation. Torgas is being burned in the combustion chamber, located beside the torrefier. An oil burner is used as a source of startup heat and as a pilot flame source during normal operation. When working in torrefaction mode installation is operating reasonably close to autothermal point, whereas the supply of auxiliary heat is determined by the moisture content of the feedstock. When working in drying mode, burner is the main source of heat.

The main advantage of such a solution is the fact that processed material is touching only hot surfaces of the reactor, thus significantly reducing the risk of the agglomeration of the particles, due to sticky tars condensing on the surfaces. This has a potential to reduce the operational cost, by reducing the risk of emergency shutdowns due to clogging.

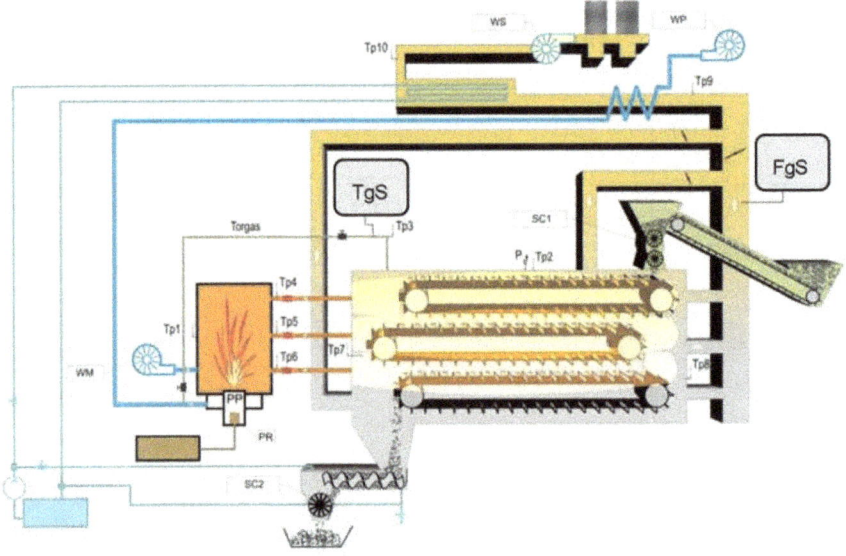

**Figure 4.** Multistage tape dryer/torrefier developed by Wroclaw University of Science and Technology (Tp—thermocouple; TgS—Torgas sampling port; FgS—Flue gas sampling port; WP—Preheated secondary air blower; WM—Primary air blower; WS—Flue gas extraction fan; SC1—airlock at the inlet; SC2—airlock at the outlet; PR—pressure regulator; PP—oil burner; P—pressure gauge).

Proposed installation for utilization of sewage sludge will use plasma gasification technology to produce syngas and vitrify inorganic by-products, as shown at the diagram, presented at Figure 5. Syngas shall be used by the engine + generator set to generate electricity. Heat from the exhaust gases will be recovered, and used for torrefaction and drying. Taking into consideration the available amount of the latent heat of vaporization it seems problematic to rely solely on the paddle dryer, despite its obvious merits. Among those one can count a relatively small consumption of electricity, in comparison with dryers using high flow rates of air, that consequently require higher amounts of electricity for the blowers. Relatively small footprint and consequently a supposedly smaller investment cost can also be considered an advantage of such a unit. However, the amount of low temperature heat, available for the process under the assumption of implementation of condensing heat exchanger, might suggest either using staged drying consisting of two units connected in series, or using a much bigger indirect dryer, using air as a drying agent.

It seems tempting to eliminate torrefaction as a unit operation in the proposed installation. However, the energy consumption of the subsequent unit operation (plasma gasification) should be taken into the account. Not using torrefaction, would cause that the heat for heating up the material from the temperature at the exit of the dryer to 300 °C, would effectively come from conversion of electricity into heat in plasma torch. Moreover, carbonization of the sewage sludge during torrefaction could have a positive influence on the overall quality of the syngas produced in the gasifier, through higher amounts of carbon available for Boudouard reaction. This could be highly significant in a plasma gasification, where temperatures are substantially higher in comparison with other gasification technologies, which would make reactions such as Boudouard generally faster. As the heat needed for decomposition of the material is also partially delivered during torrefaction stage, this could also be considered a way to save some amounts of precious electricity.

Introduction of the torgas into the gasifier can be seen as a way of simplification of the installation due to eliminating the need to build a separate torgas burner. Freeboard of the plasma gasifier seems to be a perfect place for a subsequent reforming of torgas, as refractory acts as a good insulation and temperatures in the chamber are generally high, due to the use of plasma. However, additional volume of the gases in the freeboard might require that section to be somewhat oversized, in comparison to regular plasma gasifiers. Further tests with plasma gasifier are needed to confirm merits of integration of the torrefaction unit into the installation.

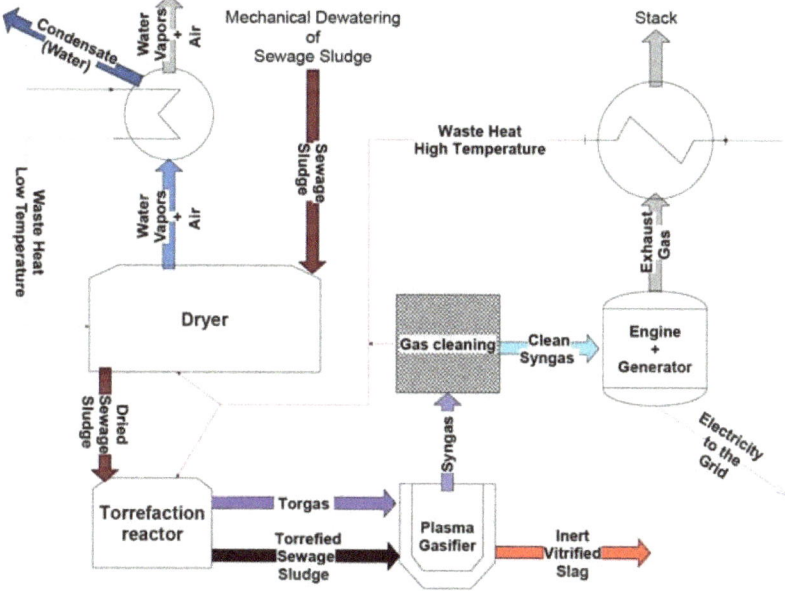

**Figure 5.** Diagram of the novel installation for utilization of sewage sludge.

## 6. Conclusions

A suite of experiments performed within the scope of this work, indicated the feasibility of the concept of proposed installation for sustainable, thermal utilization of sewage sludge. If self-sustainability of the installation is taken into consideration, minimalization of the footprint and the energy consumption is not the only factor that determines the choice of drying technology. Presence of relatively high amounts of low temperature heat, dictates the use of a staged, hybrid drying process or big unit, that would be able to effectively use a drying agent at relatively low temperatures that could be achieved by condensing heat exchangers. In that context maximalization of the dewatering capacity, during mechanical dewatering stage, seems to be crucial for achieving a favorable heat balance of the installation. Moreover, special attention should be directed at hybrid drying methods (convective-conductive) as they enable the possibility to recover heat from the vapors at the outlet of the dryer, thus potentially improving the efficiency of the process.

The use of torrefaction as an intermediate stage, between drying and plasma gasification, seems to be beneficial overall. Torrefaction of the sewage sludge leads to its further carbonization and increase of the ash content, which is not problematic in the proposed installation, as released volatiles are introduced into the gasifier anyway and inorganic fraction of the material will become a subject of a subsequent vitrification. Ability to remove a part of the nitrogen from the sewage sludge, during torrefaction, seems to be beneficial, due to a reducing conditions in the subsequent stage (gasification) which might lead to overall decrease of the $NO_x$ emissions during subsequent combustion of the gas in the engine.

Further tests with plasma gasifier are required to fully confirm all the benefits of combining torrefaction and plasma gasification into the complete system of thermal utilization of sewage sludge, in the way that has been proposed in this study. The proposed installation corresponds well with the goal 12 of the Sustainable Development Goals agenda, set by the United Nations for the years 2015–2030, in terms of reduction of waste, recovery of water, as well as energy savings.

**Acknowledgments:** This work was supported by Regional Operational Program for the Opole Voivodship 2014–2020, Priority Axis 01 Innovation in the Economy, Measure 1.1 Innovation in the Economy. Key Project No. RPOP.01.01.00-16-44/2016.

**Author Contributions:** Wilhelm Jan Tic, Joanna Guziałowska-Tic, Halina Pawlak-Kruczek, Eugeniusz Woźnikowski and Adam Zadorożny conceived the concept of the installation as well as the concept of the paper. They analyzed the results obtained from the measurements and took part in writing the paper. Moreover, Halina Pawlak-Kruczek played an advisory role, checked and approved the study at its present form. Łukasz Niedźwiecki wrote significant part of the paper and took part in the design of modernization of the torrefaction rig. Mateusz Wnukowski designed the sampling system for the measurement of the concentration of tarry compounds in torgas and performer GC-MS analysis. Krystian Krochmalny designed the sampling system for measuring of gaseous compounds of torgas. Michał Czerep and Michał Ostrycharczyk took part in the design of the measurement system for the drying tests. Marcin Baranowski took part in the drying tests and drawn the diagram of the dryer and the measurement system for drying tests. Jacek Zgóra analyzed the results obtained from the measurements of the flow rates and performed calculations of the heat delivered to the dryer. Mateusz Kowal performed proximate and ultimate analyses.

**Conflicts of Interest:** The authors declare no conflict of interest.

### References and Note

1. Kacprzak, M.; Neczaj, E.; Fijałkowski, K.; Grobelak, A.; Grosser, A.; Worwag, M.; Rorat, A.; Brattebo, H.; Almås, Å.; Singh, B.R. Sewage sludge disposal strategies for sustainable development. *Environ. Res.* **2017**, *156*, 39–46. [CrossRef] [PubMed]
2. Lee, L.H.; Wu, T.Y.; Shak, K.P.Y.; Lim, S.L.; Ng, K.Y.; Nguyen, M.N.; Teoh, W.H. Sustainable approach to biotransform industrial sludge into organic fertilizer via vermicomposting: A mini-review. *J. Chem. Technol. Biotechnol.* **2018**, *93*, 925–935. [CrossRef]
3. Cieślik, B.M.; Namieśnik, J.; Konieczka, P. Review of sewage sludge management: Standards, regulations and analytical methods. *J. Clean. Prod.* **2015**, *90*, 1–15. [CrossRef]
4. Andersen, A. *Disposal and Recycling Routes for Sewage Sludge Part 4: Economic Report*; Office for Official Publications of the European Communities: Luxembourg, 2002.
5. Andersen, A. *Disposal and Recycling Routes for Sewage Sludge: Part 1–Sludge use acceptance*; Office for Official Publications of the European Communities: Luxembourg, 2001.
6. Werle, S.; Wilk, R.K. A review of methods for the thermal utilization of sewage sludge: The Polish perspective. *Renew. Energy* **2010**, *35*, 1914–1919. [CrossRef]
7. Chen, G.; Yue, P.L.; Mujumdar, A.S. Sludge dewatering and drying. *Dry. Technol.* **2002**, *20*, 883–916. [CrossRef]
8. Tunçal, T.; Uslu, O. A Review of Dehydration of Various Industrial Sludges. *Dry. Technol.* **2014**, *32*, 1642–1654. [CrossRef]
9. Peeters, B. Mechanical dewatering and thermal drying of sludge in a single apparatus. *Dry. Technol.* **2010**, *28*, 454–459. [CrossRef]
10. Bennamoun, L.; Arlabosse, P.; Léonard, A. Review on fundamental aspect of application of drying process to wastewater sludge. *Renew. Sustain. Energy Rev.* **2013**, *28*, 29–43. [CrossRef]
11. Ferrasse, J.H.; Arlabosse, P.; Lecomte, D. Heat, momentum, and mass transfer measurements in indirect agitated sludge dryer. *Dry. Technol.* **2002**, *20*, 749–769. [CrossRef]
12. Arlabosse, P.; Chavez, S.; Lecomte, D. Method for thermal design of paddle dryers: Application to municipal sewage sludge. *Dry. Technol.* **2004**, *22*, 2375–2393. [CrossRef]
13. Peeters, B.; Vernimmen, L. Challenges of Handling Filamentous and Viscouis Wastewater Sludge. *Chem. Eng.* **2016**.

14. Peeters, B.; Dewil, R.; Van Impe, J.F.; Vernimmen, L.; Smets, I.Y. Using a Shear Test-Based Lab Protocol to Map the Sticky Phase of Activated Sludge. *Environ. Eng. Sci.* **2011**, *28*, 81–85. [CrossRef]
15. Peeters, B.; Dewil, R.; Vernimmen, L.; Van den Bogaert, B.; Smets, I.Y. Addition of polyaluminiumchloride (PACl) to waste activated sludge to mitigate the negative effects of its sticky phase in dewatering-drying operations. *Water Res.* **2013**, *47*, 3600–3609. [CrossRef] [PubMed]
16. Deng, W.; Su, Y.; Yu, W. Theoretical Calculation of Heat Transfer Coefficient When Sludge Drying in a Nara-Type Paddle Dryer Using Different Heat Carriers. *Procedia Environ. Sci.* **2013**, *18*, 709–715. [CrossRef]
17. Moscicki, K.J.; Niedzwiecki, L.; Owczarek, P.; Wnukowski, M. Commoditization of biomass: Dry torrefaction and pelletization-a review. *J. Power Technol.* **2014**, *94*, 233–249.
18. Moscicki, K.J.; Niedzwiecki, L.; Owczarek, P.; Wnukowski, M. Commoditization of wet and high ash biomass: Wet torrefaction—A review. *J. Power Technol.* **2017**, *97*, 354–369.
19. Wnukowski, M.; Owczarek, P.; Niedźwiecki, Ł. Wet Torrefaction of Miscanthus—Characterization of Hydrochars in View of Handling, Storage and Combustion Properties. *J. Ecol. Eng.* **2015**, *16*, 161–167. [CrossRef]
20. Magdziarz, A.; Wilk, M.; Kosturkiewicz, B. Investigation of sewage sludge preparation for combustion process. *Chem. Process Eng.—Inz. Chem. Proces.* **2011**, *32*, 299–309. [CrossRef]
21. Magdziarz, A.; Werle, S. Analysis of the combustion and pyrolysis of dried sewage sludge by TGA and MS. *Waste Manag.* **2014**, *34*, 174–179. [CrossRef] [PubMed]
22. Huang, M.; Chang, C.C.; Yuan, M.H.; Chang, C.Y.; Wu, C.H.; Shie, J.L.; Chen, Y.H.; Chen, Y.H.; Ho, C.; Chang, W.R.; et al. Production of torrefied solid bio-fuel from pulp industry waste. *Energies* **2017**, *10*, 910. [CrossRef]
23. Huang, Y.-F.; Sung, H.-T.; Chiueh, P.-T.; Lo, S.-L. Microwave torrefaction of sewage sludge and leucaena. *J. Taiwan Inst. Chem. Eng.* **2017**, *70*, 236–243. [CrossRef]
24. Huang, Y.-F.; Sung, H.-T.; Chiueh, P.-T.; Lo, S.-L. Co-torrefaction of sewage sludge and leucaena by using microwave heating. *Energy* **2016**, *116*, 1–7. [CrossRef]
25. Huang, Y.W.; Chen, M.Q.; Luo, H.F. Nonisothermal torrefaction kinetics of sewage sludge using the simplified distributed activation energy model. *Chem. Eng. J.* **2016**, *298*, 154–161. [CrossRef]
26. Do, T.X.; Lim, Y.; Cho, H.; Shim, J.; Yoo, J.; Rho, K.; Choi, S.-G.; Park, B.-Y. Process modeling and energy consumption of fry-drying and torrefaction of organic solid waste. *Dry. Technol.* **2017**, *35*, 754–765. [CrossRef]
27. Pulka, J.; Wiśniewski, D.; Gołaszewski, J.; Białowiec, A. Is the biochar produced from sewage sludge a good quality solid fuel? *Arch. Environ. Prot.* **2016**, *42*, 125–134. [CrossRef]
28. Poudel, J.; Ohm, T.I.; Lee, S.H.; Oh, S.C. A study on torrefaction of sewage sludge to enhance solid fuel qualities. *Waste Manag.* **2015**, *40*, 112–118. [CrossRef] [PubMed]
29. Poudel, J.; Karki, S.; Gu, J.H.; Lim, Y.; Oh, S.C. Effect of Co-Torrefaction on the Properties of Sewage Sludge and Waste Wood to Enhance Solid Fuel Qualities. *J. Residuals Sci. Technol.* **2017**, *14*, 23–36. [CrossRef]
30. Atienza-Martínez, M.; Fonts, I.; ábrego, J.; Ceamanos, J.; Gea, G. Sewage sludge torrefaction in a fluidized bed reactor. *Chem. Eng. J.* **2013**, *222*, 534–545. [CrossRef]
31. Atienza-Martínez, M.; Mastral, J.F.; Ábrego, J.; Ceamanos, J.; Gea, G. Sewage sludge torrefaction in an auger reactor. *Energy Fuels* **2015**, *29*, 160–170. [CrossRef]
32. He, C.; Giannis, A.; Wang, J.Y. Conversion of sewage sludge to clean solid fuel using hydrothermal carbonization: Hydrochar fuel characteristics and combustion behavior. *Appl. Energy* **2013**, *111*, 257–266. [CrossRef]
33. Danso-Boateng, E.; Shama, G.; Wheatley, A.D.; Martin, S.J.; Holdich, R.G. Hydrothermal carbonisation of sewage sludge: Effect of process conditions on product characteristics and methane production. *Bioresour. Technol.* **2015**, *177*, 318–327. [CrossRef] [PubMed]
34. Danso-Boateng, E.; Holdich, R.G.; Martin, S.J.; Shama, G.; Wheatley, A.D. Process energetics for the hydrothermal carbonisation of human faecal wastes. *Energy Convers. Manag.* **2015**, *105*, 1115–1124. [CrossRef]
35. Oh, S.Y.; Yoon, Y.M. Energy recovery efficiency of poultry slaughterhouse sludge cake by hydrothermal carbonization. *Energies* **2017**, *10*, 1876. [CrossRef]
36. Syed-Hassan, S.S.A.; Wang, Y.; Hu, S.; Su, S.; Xiang, J. Thermochemical processing of sewage sludge to energy and fuel: Fundamentals, challenges and considerations. *Renew. Sustain. Energy Rev.* **2017**, *80*, 888–913. [CrossRef]

37. Tomasi Morgano, M.; Leibold, H.; Richter, F.; Stapf, D.; Seifert, H. Screw pyrolysis technology for sewage sludge treatment. *Waste Manag.* **2018**, *73*, 487–495. [CrossRef] [PubMed]

38. Fonts, I.; Gea, G.; Azuara, M.; Ábrego, J.; Arauzo, J. Sewage sludge pyrolysis for liquid production: A review. *Renew. Sustain. Energy Rev.* **2012**, *16*, 2781–2805. [CrossRef]

39. Atienza-Martínez, M.; Ábrego, J.; Mastral, J.F.; Ceamanos, J.; Gea, G. Energy and exergy analyses of sewage sludge thermochemical treatment. *Energy* **2018**, *144*, 723–735. [CrossRef]

40. Manara, P.; Zabaniotou, A. Towards sewage sludge based biofuels via thermochemical conversion—A review. *Renew. Sustain. Energy Rev.* **2012**, *16*, 2566–2582. [CrossRef]

41. Pečkytė, J.; Baltrėnaitė, E. Assessment of heavy metals leaching from (bio) char obtained from industrial sewage sludge. *Environ. Prot. Eng.* **2015**, *7*, 399–406. [CrossRef]

42. Council of the European Union 2003/33/EC—Council Decision establishing criteria and procedures for the acceptance of waste at landfills pursuant to Article 16 of and Annex II to Directive 1999/31/EC. *Off. J. Eur. Communities* **2003**, *L 11/27*, 27–49.

43. Wang, N.Y.; Shih, C.H.; Chiueh, P.T.; Huang, Y.F. Environmental effects of sewage sludge carbonization and other treatment alternatives. *Energies* **2013**, *6*, 871–883. [CrossRef]

44. Werle, S.; Dudziak, M. Analysis of organic and inorganic contaminants in dried sewage sludge and by-products of dried sewage sludge gasification. *Energies* **2014**, *7*, 462–476. [CrossRef]

45. Werle, S. Gasification of a Dried Sewage Sludge in a Laboratory Scale Fixed Bed Reactor. *Phys. Procedia* **2015**, *66*, 253–256. [CrossRef]

46. Reed, T.B.; Gaur, S. *A Survey of Biomass Gasification 2001: Gasifier Projects and Manufacturers Around the World*; The Biomass Energy Foundation: Golden, CO, USA, 2001.

47. Pigneri, A.; Asbjerg, M.; Collin, C.; Dicks, A.; Sproule, G. *Gasification Technologies Review*; The council of the city of Sydney: Sydney, Australia, 2014.

48. Maczka, T.; Sliwka, E.; Wnukowski, M.; Niedzwiecki, L. Pilot installation for thermal plasma treatment of plastic wastes. In *Finnish—Swedish Flame Days 2013*; IFRF International Flame Research Foundation: Sheffield, UK, 2013.

49. Willis, K.P.; Osada, S.; Willerton, K.L. Plasma Gasification: Lessons Learned at Eco-Valley WTE Facility. In Proceedings of the 18th Annual North American Waste-to-Energy Conference, Orlando, FL, USA, 11–13 May 2010; pp. 133–140.

50. Mountouris, A.; Voutsas, E.; Tassios, D. Plasma gasification of sewage sludge: Process development and energy optimization. *Energy Convers. Manag.* **2008**, *49*, 2264–2271. [CrossRef]

51. Fabry, F.; Rehmet, C.; Rohani, V.; Fulcheri, L. Waste gasification by thermal plasma: A review. *Waste Biomass Valoriz.* **2013**, *4*, 421–439. [CrossRef]

52. Kordylewski, W.; Michalski, J.; Ociepa, M.; Wnukowski, M. A microwave plasma potential in producer gas cleaning—preliminary results with a gas derrived from a sewage sludge. In *VI Konferencja Naukowo-Techniczna Energetyka Gazowa 2016*; Silesian University of Technology: Gliwice, Poland, 2016.

53. Striūgas, N.; Valinčius, V.; Pedišius, N.; Poškas, R.; Zakarauskas, K. Investigation of sewage sludge treatment using air plasma assisted gasification. *Waste Manag.* **2017**. [CrossRef] [PubMed]

54. CEN (European Comitte for Standardisation) EN 13284-1:2001 Stationary source emissions—Determination of low range mass concentration of dust—Part 1: Manual gravimetric method 2001.

55. Peeters, B.; Dewil, R.; Smets, I. Challenges of drying sticky wastewater sludge. *Chem. Eng. (United States)* **2014**, *121*, 51–54.

56. Pawlak-Kruczek, H.; Krochmalny, K.K.; Niedźwiecki, Ł.; Mościcki, K.J. Slow pyrolysis of the sewage sludge with additives: Calcium oxide and lignite. In *The Clearwater Clean Coal Conference: Proceedings of the 42nd International Technical Conference on Clean Energy*; Sakkestad, B.A., Ed.; Coal Technologies Associates: Clearwater, FL, USA, 2017.

57. Pawlak-Kruczek, H.; Zgóra, J.M.; Krochmalny, K.K. Characterization of torrefied biomass depends on process condition. In Proceedings of the 40th International Technical Conference on Clean Coal & Fuel Systems, Clearwater, FL, USA, 31 May–4 June 2015.

*Article*

# Development and Validation of 3D-CFD Injection and Combustion Models for Dual Fuel Combustion in Diesel Ignited Large Gas Engines

Lucas Eder [1,*] , Marko Ban [2] , Gerhard Pirker [1], Milan Vujanovic [2], Peter Priesching [3] and Andreas Wimmer [4]

[1]   Large Engines Competence Center, 8010 Graz, Austria; gerhard.pirker@lec.tugraz.at
[2]   Faculty of Mechanical Engineering and Naval Architecture, University of Zagreb, 10002 Zagreb, Croatia; marko.ban@fsb.hr (M.B.); milan.vujanovic@fsb.hr (M.V.)
[3]   AVL List GmbH, Graz 8020, Austria; peter.priesching@avl.com
[4]   Institute of Internal Combustion Engine and Thermodynamics, Graz University of Technology, 8010 Graz, Austria; andreas.wimmer@lec.tugraz.at
*   Correspondence: lucas.eder@lec.tugraz.at; Tel.: +43-(0)-316-873-30141

Received: 27 February 2018; Accepted: 11 March 2018; Published: 14 March 2018

**Abstract:** This paper focuses on improving the 3D-Computational Fluid Dynamics (CFD) modeling of diesel ignited gas engines, with an emphasis on injection and combustion modeling. The challenges of modeling are stated and possible solutions are provided. A specific approach for modeling injection is proposed that improves the modeling of the ballistic region of the needle lift. Experimental results from an inert spray chamber are used for model validation. Two-stage ignition methods are described along with improvements in ignition delay modeling of the diesel ignited gas engine. The improved models are used in the Extended Coherent Flame Model with the 3 Zones approach (ECFM-3Z). The predictive capability of the models is investigated using data from single cylinder engine (SCE) tests conducted at the Large Engines Competence Center (LEC). The results are discussed and further steps for development are identified.

**Keywords:** dual fuel combustion; 3D-CFD modeling; ECFM-3Z; diesel ignited gas engine; ignition delay modeling; experimental validation

---

## 1. Introduction

### 1.1. Motivation

With the greater use of renewable energy sources (solar and wind), large engines are becoming more important for grid stabilization because of their very good transient behavior and the possibility of setting up modular systems [1]. In the transportation sector, which will triple by 2050 according to a recent study [2], large engines will also play a central role. While power generation will mainly rely upon gas engines [3], the diesel engine will still dominate the transportation (marine, rail) sector as the main source of propulsion. Accounting for 80% of transportation services worldwide, the marine sector in particular is coming under increased pressure and needs to catch up to land-based applications in terms of emission limits [4,5]. Low emission gas engines are favored as a solution to meet the very stringent emission requirements for marine applications. This requires further expansion of the gas infrastructure; in recent years a great amount of research into liquefied natural gas has been undertaken [6].

Besides standard gas and diesel combustion concepts, fully flexible dual fuel combustion concepts that can burn diesel and gas simultaneously have also become established in the large engine sector in recent years. Mobile applications such as marine or rail traction as well as stationary applications such as generator sets for power generation are available with dual fuel combustion systems [7–10]. However, today dual fuel technology is employed primarily in the marine sector as a consequence of the currently unfavorable development in the price of natural gas compared to the price of crude oil [11].

Besides producing lower emissions, dual fuel engines have a great advantage over diesel engines because they are able to generate the required power even in case of an interruption in the gas supply due to the redundancy of usable fuels. Especially in the marine sector, it is important to ensure robust and reliable ignition and combustion as well as high fuel flexibility.

## 1.2. Dual Fuel Technology

This paper focuses on one of these dual fuel concepts, the diesel ignited gas engine concept. It is also referred to as the diesel-gas engine, cf. [12]. The diesel ignited gas engine is operated using a homogeneous lean mixture in the combustion chamber. A small amount of diesel fuel provides the required ignition energy. It is injected into the lean mixture and starts the flame front propagation. Its application in large bore engines, which are mainly used in maritime transportation and energy production, is being researched in [13–16]. As mentioned previously, the advantages of the diesel ignited gas engine concept over the pure diesel engine are its fuel flexibility and lower $NO_x$ and soot emissions. Nevertheless, several disadvantages come hand in hand with greater fuel flexibility and have to be taken into account; these are outlined in [17]. The main disadvantage is that the efficiency is lower than with the pure gas engine. In addition, $NO_x$ emissions are higher than with a gas engine. However, the comparatively low engine-out $NO_x$ emissions are prescribed by the International Maritime Organization (IMO) Tier III [4] limit, which can be achieved by gas engines without making use of exhaust gas aftertreatment [18–20]. Therefore, it is critical to introduce dual fuel engines into the area of ship propulsion to meet the limits in Emission Controlled Areas (ECAs) in particular [18].

## 1.3. State-of-the-Art Simulation of Dual Fuel Combustion

Since the combustion principle in the diesel ignited gas engine includes more than one combustion regime, understanding the physical phenomena involved is a complex task. 3D-CFD simulation provides spatial and temporal information on the parameters for diesel ignited gas engine combustion. Besides experiments, it is therefore the most suitable tool for better understanding these phenomena. Numerical modeling of dual fuel combustion has proven to be challenging as all combustion regimes (i.e., autoignition, diffusion combustion and premixed flame front propagation) have to be modeled simultaneously. This issue can be addressed with detailed chemistry, where combustion is depicted by describing all the reactions involved in a reaction mechanism. Reduced chemical reaction mechanisms are used in 3D-CFD simulations to accurately depict the chemistry of the combustion process. The works of several authors [21–25] show that detailed chemistry to depict dual fuel combustion has to include all necessary reactions for both fuels used in the combustion. To describe turbulence, engine simulation with detailed chemistry is mostly limited to the Reynolds Averaged Navier Stokes (RANS) framework for engineering applications. Even with this approach to turbulence modeling, the computational time required to calculate combustion in internal combustion engines (ICE) with detailed chemistry is comparably large. Combustion models, on the other hand, can provide faster results but are usually only suitable for one type of combustion regime. A viable combination of different combustion models for autoignition and flame front propagation has been shown in [26]. An approach that can cover all regimes at the same time is the ECFM-3Z model. Developed by Colin et al. [27] and subsequently improved regarding ignition, mixing and post-oxidation phenomena [28–30], the ECFM-3Z is able to depict all necessary combustion regimes simultaneously. A feasible workflow with the ECFM-3Z for dual fuel combustion simulation is shown in [31], where the regime transition from autoignition to premixed flame front propagation is covered by an approach to modeling the initial flame surface density.

The present paper describes the numerical modeling of a diesel ignited gas engine using a 3D-CFD tool. The challenges of numerical simulation are identified with a special focus on the modeling of pilot injection and the interaction of the two fuels. The paper is divided into several subsections that describe the modified phases in the diesel ignited gas engine in sequence: injection, autoignition of the mixture and flame front propagation. The ECFM-3Z framework is only partly stated in this paper; cf. [11,16,32] for detailed information on the combustion model framework and mathematical background. N-heptane is used as the diesel surrogate fuel and methane as the natural gas, but the framework is not solely limited to these fuels. The influence of the fuels on ignition delay is considered by expanding the two-stage ignition delay tabulation method by an additional table dimension to include the fuel mixture fraction of both fuels. Furthermore, an interpolation function for the two-stage interpolation is introduced. An interpolation between the laminar flame speed of diesel and the laminar flame speed of natural gas is proposed to compare the flame front propagation of the two fuels. To obtain a better depiction of the diesel pilot injection, measurements of the injector were performed in a quiescent spray chamber and a Bosch Tube. The rate of injection with the diesel pilot injector was measured as well as the liquid and vapor penetration lengths, which were obtained using Schlieren and Mie scattering imaging methods [33]. The numerical and experimental results were compared and satisfactory agreement was found between the penetration lengths of the diesel masses injected. The parameter set found in these spray tests was later used in the engine simulations. The combustion modeling was then validated with engine measurements performed on a single cylinder engine at the test beds of the Large Engines Competence Center (LEC) at Graz University of Technology. The results are discussed for their validity and further ideas for model improvements are provided.

## 2. Challenges of Dual Fuel Combustion Modeling

Numerical simulation of a dual fuel engine has to depict several key phenomena, cf. Figure 1. The following subsection will refer to this figure as a general guideline for describing the individual challenges. Each challenge will be described along with recent research including experimental work and modeling approaches. As this paper focuses on the spatial and temporal resolution of injection, subsequent ignition and flame front propagation, a separate section is devoted to each of these challenges following this section.

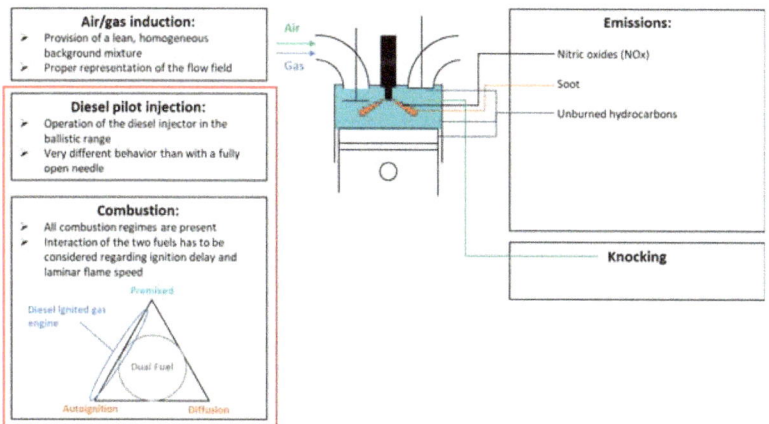

**Figure 1.** Numerical challenges for the 3D-CFD simulation of a dual fuel engine. The combustion regimes encountered are shown in the triangle. The red rectangle shows the focus of this paper.

The first challenge that arises is mixture preparation of the natural-gas air mixture. The air and gas induction into the cylinder has to be depicted correctly to determine the mixture fraction of this natural gas and air background mixture as well as the flow field and turbulence conditions at inlet valve closing conditions. This cold flow problem has already been thoroughly researched and described in the literature [34,35] and will not be discussed in this paper.

Diesel injection following the compression of the natural gas-air mixture is important for the ignition process. The penetration length of the diesel spray as well as its mixing with the surrounding media determines the ignition location(s), cf. [36,37]. The ignition process is also controlled by chemistry. The research in [38–41] has shown experimentally and numerically that ignition delay is dependent on the fuel mixture fraction of natural gas and diesel. The autoignition of the mixture then starts the flame front propagation throughout the lean background mixture.

The transition process between these two combustion regimes is the focus of ongoing scientific research. Based on direct numerical simulation, recent works [42,43] have identified key parameters that influence the time between autoignition and flame front propagation.

As soon as a premixed flame front has developed, the flame propagates through the homogeneous background mixture. An accurate description of the laminar flame speed is necessary to depict the heat release rate coming from flame front propagation. Once again, the laminar flame speed depends on the two fuels involved in the combustion process; with n-heptane methane mixtures, it was able to be shown experimentally [44] and numerically [41,45].

Emission modeling of nitric oxides ($NO_x$) is possible with the $NO_x$ mechanisms currently available such as the extended Zeldovich mechanism [46]. Donateo et al. have applied this mechanism for emission prediction in dual fuel engines in their recent work [21,47]. With soot formation as well, the literature on diesel and gasoline engines provides a framework that can be used for dual fuel engine combustion simulation [35]. However, the exact process of soot formation is still not understood perfectly and therefore its prediction is greatly limited. Recent research, therefore, relies on experimental work, cf. [15]. To predict unburned hydrocarbons, which is especially important in dual fuel engines due to the methane slip, Kuppa [48] has developed a numerical model.

Another topic of active research is knocking in dual fuel engines. Knock models of gasoline and diesel engines can serve as a basis for knocking in dual fuel engines. The knocking index proposed in [49] is the numerical approach used in this paper.

Numerical research was performed using AVL FIRE (Version 2014.2) (AVL List GmbH, Graz, Austria). Focusing on injection and ignition modeling and numerical description of flame front propagation, the following subsections briefly introduce the models used and emphasize the adjustments required for dual fuel simulation. Fundamental equations of the submodels are only mentioned if needed for the dual fuel adjustments.

### 2.1. Injection Modeling

In the injection modeling process, evaporation and breakup are the two most important factors for locating the space in which the diesel spray will ignite. The WAVE child breakup model based on the work of Reitz [50] was used for the breakup process of the droplets. Further information on the software implementation can be found in [51]. The Dukowicz evaporation model was used to model the evaporation process—basic model information can be found in [52]; for its implementation, see [51].

Especially the breakup process is influenced by the location of cavitation in the nozzle hole. Various authors have demonstrated these influences with a common rail type diesel injector [35,53,54]. As there was no information available on the cavitation locations to feed the simulation model with an accurate dataset, the penetration length was calibrated by modifying the nozzle hole diameter used in the blob model [50]. Based on a simple mass flow equation for a spherical opening, Equation (1) states how the droplet velocity is calculated at the beginning of the injection:

$$v = \frac{4\dot{m}}{\rho d^2 \pi} \tag{1}$$

$v$ represents the velocity of the droplets, $\dot{m}$ is the fuel mass flow, $\rho$ describes the fuel density and $d$ is the diameter of the nozzle hole. Since the mass flow measured equals zero at the beginning of the injection, the droplet velocity that is specified in the simulation model is also zero at the beginning. This means that the speed at which the droplets enter the simulation domain is too low. This does not have a significant impact when normal diesel engines are simulated since the needle stays open for a longer amount of time than with the diesel ignited gas engine concept, where the needle mostly operates within the ballistic region. To compensate for the problem of an inlet velocity initially too low, a non-constant nozzle hole diameter $d$ is specified. The shape of this non-constant diameter curve follows the shape of the rate of injection (ROI), meaning the diameter is smaller at the beginning of the injection and reaches a maximum when the needle is fully open. The minimum possible diameter is a modification parameter; the maximum diameter can be calculated by Equation (2), where the maximum possible diameter $d_{max}$ is a function of the nominal nozzle hole diameter $d_{nom}$ as well as drag coefficient $c_d$ and velocity coefficient $c_v$. Values for the coefficients were estimated using the information in [35]:

$$d_{max} = d_{nom}\sqrt{\frac{c_d}{c_v}} \tag{2}$$

This leads to a significant increase in droplet velocity at the beginning of the simulation as shown in Figure 2. One can clearly see that the velocity of the droplets increases significantly when compared to the standard calculation from Equation (1). It can also happen that velocities higher than the maximum possible velocity may occur. Therefore, the diameter is corrected again to allow only for velocities lower than the one shown in Equation (3):

$$v_{max} = c_v\sqrt{\frac{2(p_{inj} - p_c)}{\rho_{diesel}}} \tag{3}$$

$v_{max}$ is the maximum velocity used to limit and correct the diameter. $c_v$ is the velocity coefficient as previously shown in Equation (2). $p_{inj}$ is the injection pressure, $p_c$ is the back pressure of the medium and $\rho_{diesel}$ is the density of the diesel fuel. All of the variables were known for the injection measurements that are shown in the subsequent sections.

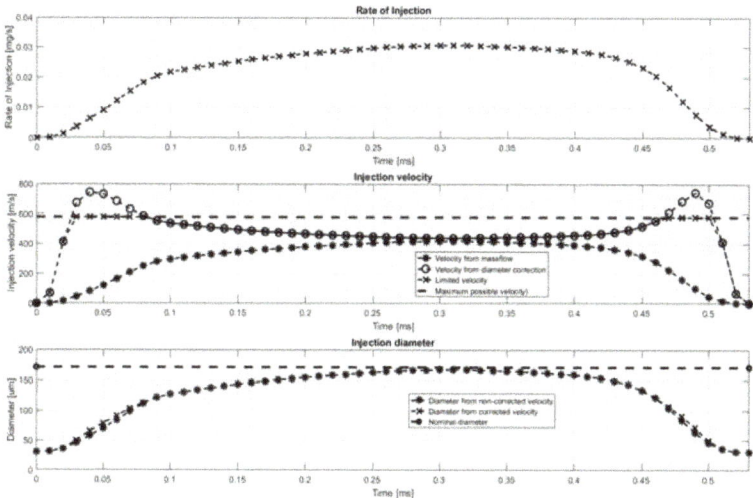

**Figure 2.** Specification of an inconstant nozzle diameter to modify the injection velocity of the droplets in the simulation model.

With this modification, the model can now consider the ballistic region of the needle lift and the resulting penetration lengths of vapor and liquid mass. Further calibration was achieved using the standard breakup and evaporation model constants as described in [51]. The results for the validation of this model approach can be found in Section 3 "Validation of injection modeling".

*2.2. Ignition Delay Modeling*

The first step in modeling ignition delay involves choosing surrogates for the fuels involved. As various compositions of the different species of diesel and natural gas are possible, a chemical surrogate must be selected for both the diesel fuel and the natural gas. One of the best characterized surrogates for diesel fuel is n-heptane—numerous studies show that this surrogate is chemically well understood [55–57]. As for natural gas, a combination of methane, ethane, propane and butane is adequate for the applications under investigation. In the following description of the chemical interaction between the two fuels during ignition, n-heptane is chosen as the diesel surrogate and pure methane as the natural gas surrogate.

The next step is to determine which kind of ignition delay modeling can be used. Since the ignition conditions in a diesel ignited gas engine allow cold flame combustion, a numerical modeling approach must be chosen that is able to capture cold flame heat release and the main ignition. An accurate numerical description of ignition delay timing can be achieved by using two-stage tabulation. In this numerical approach, the cold flame ignition delay and the main flame ignition delay are tabulated with respect to pressure, temperature, equivalence ratio and exhaust gas recirculation (EGR) rate. The descriptions provided below show the calculation of just one ignition delay time; this approach can be used for both cold flame and hot flame ignition delay.

After the fuel surrogates and numerical description are determined, the modeling approach considers the interaction of the fuels. As previously mentioned, experimental research [38,39] has revealed that there is a strong interaction between fuels used together in a Rapid Compression Expansion Machine (RCEM). The qualitative behavior of these experiments was also seen in the numerical approaches in [40,41]. 0D simulation carried out with the mechanism used below for dual fuel tabulation shows that there is strong non-linear behavior between the two fuels during ignition. 3 graphs ignition delay with selected pressure and global equivalence ratios versus temperature. The lines in the Figure 3 represent different fuel blends (note that the global equivalence ratio stays the same while the composition of the fuels changes; e.g., 95% CH4 indicates that if the global equivalence ratio is 1, 95 mole percent of the fuel is methane and 5 mole percent is n-heptane).

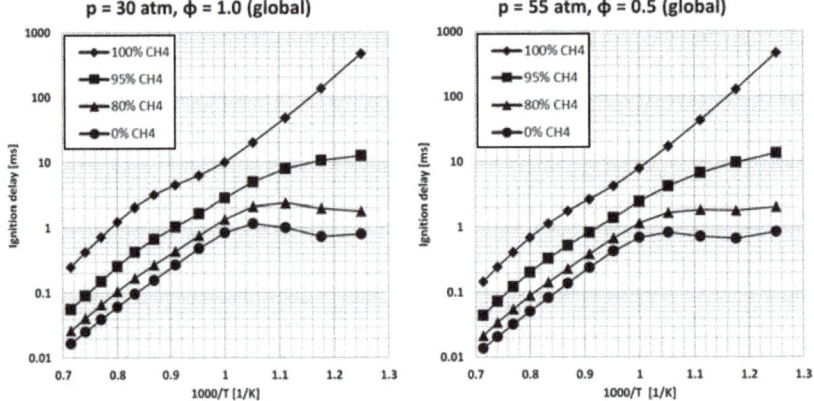

**Figure 3.** Ignition delay times with different fuel blends of n-heptane and methane at fixed global equivalence ratios and pressures.

The strong non-linear behavior of ignition delay versus temperature has to be taken into account in ignition delay modeling of the fuels. This behavior can be captured by one of the following two approaches.

The first approach applies an interpolation function if the ignition delay of each of the fuels has been tabulated with a different reaction mechanism. With a non-linear blending function, these two delays are combined to yield a dual fuel ignition delay.

A second and more accurate approach is to use a specific dual fuel mechanism as described in [45] and examined in more detail in the following subsections to tabulate the ignition delay with respect to temperature $T$, pressure $p$, global equivalence ratio $\Phi_{global}$ and EGR rate $x_{EGR}$ as well as the dual fuel fraction $\varphi_{DF}$. The schematic in Figure 4 shows these two approaches.

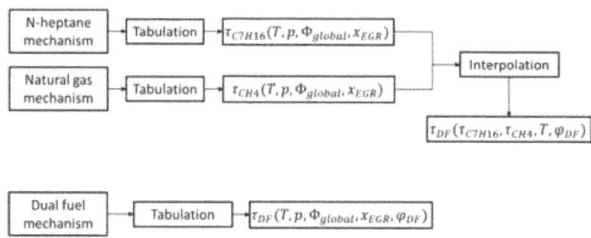

**Figure 4.** Schematic of the two proposed approaches for dual fuel ignition delay interpolation.

Both approaches are not limited to the fuel surrogates that have been described, yet adding a new surrogate (e.g., another natural gas component) requires a new tabulation. The following subsections describe each approach in detail, indicating their potential advantages and disadvantages.

*2.3. Two-Stage Ignition with an Interpolation Function*

As described in Figure 4, this approach combines two different tables with an interpolation function that is dependent on temperature and the dual fuel mixture fraction $\varphi_{DF}$ as described in Equation (4):

$$\varphi_{DF} = \frac{v_{C7H16}}{v_{C7H16} + v_{CH4}} \tag{4}$$

where $v_{C7H16}$ is the mole fraction of n-heptane and $v_{CH4}$ is the mole fraction of methane. Measurement data from a SCE shows that the injected diesel fuel usually ignites at a temperature below 1000 K. Therefore, it is not necessary that the interpolation function mentioned above covers the whole temperature range. The ignition delay of the individual fuels was tabulated using the LLNL-v3 mechanism for n-heptane [56] and the GRI 3.0 mechanism for natural gas [58]. Equation (5) shows the blending function used to obtain the dual fuel ignition delay with respect to the ignition delay of n-heptane $\tau_{C7H16}$, the ignition delay of methane $\tau_{CH4}$, temperature $T$ and dual fuel mixture fraction $\varphi_{DF}$:

$$\tau_{DF} = \tau_{C7H16} \times (1 - e^{-B(T) \times \varphi_{DF}}) + \tau_{CH4} \times e^{-B(T) \times \varphi_{DF}} \tag{5}$$

The parameter $B(T)$ is obtained by a mathematical equation based on 0D calculations at a given pressure of 55 atm and a global equivalence ratio of one. The selected case and the temperature range used for the fit is indicated by the black rectangle in Figure 5.

A temperature is then selected (dashed line in left graph) and the 0D calculation data for this point is shown along with the interpolation function (dashed line in the right graph). The interpolation function shows good agreement with the 0D data.

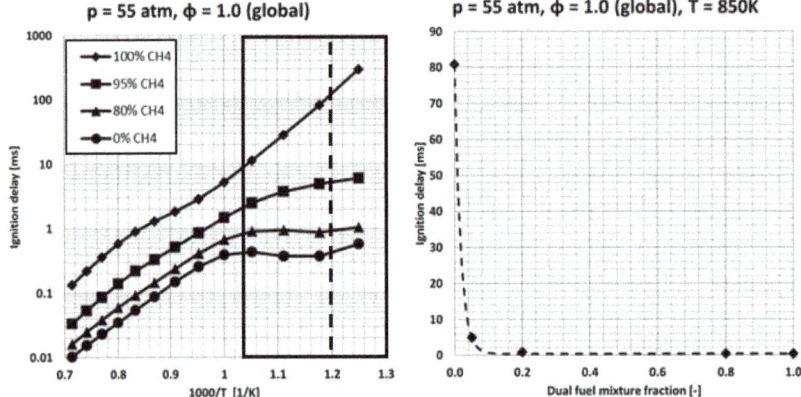

**Figure 5.** Selected temperature range (within the black rectangle) for the interpolation function at a given pressure and global equivalence ratio (**left**); selected interpolation temperature indicated by the dashed line. 0D results and interpolation function for the selected temperature (**right**).

This approach allows the combination of the arbitrary delay times of the two different fuels. One great disadvantage of this interpolation approach, however, is that only the influence of temperature is considered while changes due to pressure, equivalence ratio and EGR rate are not. This non-linear interaction can be depicted more accurately by using a specific dual fuel mechanism for the tabulation and introducing the dual fuel mixture fraction as an additional tabulation dimension.

*2.4. Two-Stage Ignition with a Specific Dual Fuel Mechanism*

For this approach, it is imperative to have a mechanism that is able to give accurate information on the ignition delay of the fuels involved, especially for the case in which both fuels interact with each other. Such a mechanism was developed by combining one mechanism for n-heptane [59] and one for natural gas [60]. The full scale mechanism was then reduced using numerical approaches as outlined in [37,61]. Further information on the selection of the separate mechanisms as well as the reduction process is provided in [45]. Figure 6 shows the ignition delay of the reduced mechanism compared to the full scale mechanism.

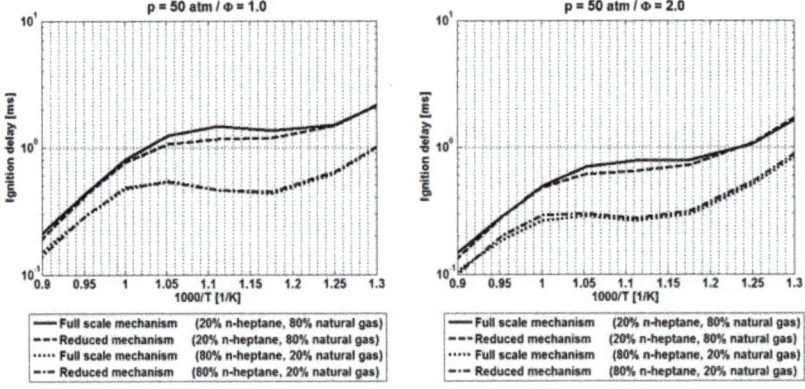

**Figure 6.** Comparison of ignition delay calculations for dual fuel mixtures. The full scale dual fuel mechanism is compared to the reduced mechanism.

To the best of the authors' knowledge, no experimental data on dual fuel ignition delays is currently available; only the specific operating conditions of the two mechanisms were compared. The results show that the reduced mechanism is also able to reproduce the ignition delay with satisfactory accuracy. Furthermore, the validation of the measurements of laminar flame speeds is shown. Figure 7 compares the full scale mechanism and the reduced mechanism, whereas Figure 8 provides the results of the reduced mechanism in relation to the experimental data found in [44]. As the laminar flame speed might also be the subject of tabulation in future research, this underlines the fact that the same mechanism can tabulate dual fuel ignition delay as well as laminar flame speed.

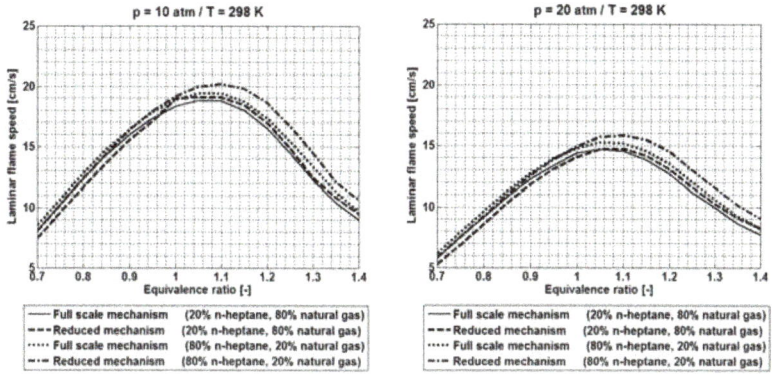

**Figure 7.** Comparison of laminar flame speed calculations for dual fuel mixtures. The full scale dual fuel mechanism is compared to the reduced mechanism.

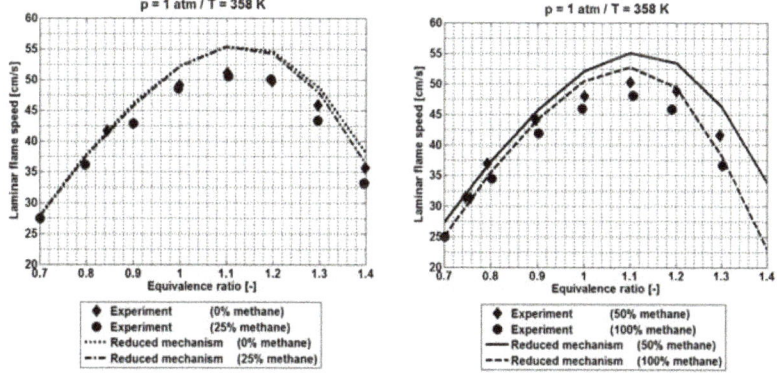

**Figure 8.** Comparison of laminar flame speed calculations for dual fuel mixtures. The reduced dual fuel mechanism is compared to the experimental data found in [44].

In order to improve the accuracy of this approach, a separate database was created with a comprehensive chemical mechanism that includes 695 species and 3037 reactions. A specific dual fuel tabulation procedure (cf. Figure 4) was applied. An additional tabulation parameter was introduced that was defined by applying Equation (4) from the previous section (dual fuel mixture fraction).

By varying the initial parameters, a total of 91,350 0D calculations were performed and post-processed in order to store the low temperature ignition delay time, the main ignition delay time, and respective heat release, which was used to determine the amount of fuel consumed during low temperature ignition (if applicable).

The introduction of a new parameter increased the order of the interpolation database from four to five dimensions, necessitating the adaptation of the previous interpolation routine. A new step was added to obtain the final value of each of the tabulated variables. Initially, the 4D interpolation routine returned a scalar for each of the values required by the combustion model. Increasing the order of the matrix caused the returned values to be vectors, whose size was determined by the number of fuel mixture fraction points. The vectors were processed by a separate interpolation function that returns the final scalar. It should be noted that a linear interpolation was used and proved to be adequately accurate. Since the dependence of the stored data on each of the parameters is far from linear, the distribution of data points was chosen so as to ensure that interpolation would not be one of the governing influences on the returned values. A simple option to improve the accuracy (if necessary) of the interpolation is to run the routine over a series transformed by a logarithmic function and the returned resulting scalar by an exponential function. The concept that makes use of the tabulated values is shown in Figure 9.

In the figure below, the temperature curve calculated with the complex chemistry mechanism (solid curve) and the one obtained with 3D-CFD calculation of a simple cube domain consisting of 27 cells (dashed line) show good agreement.

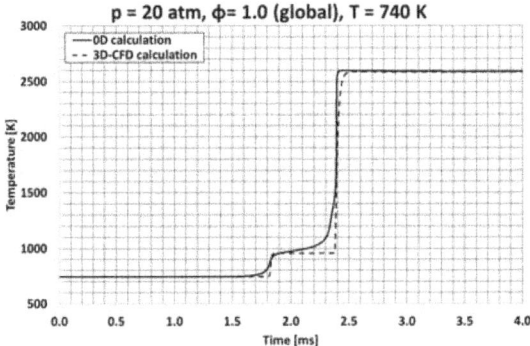

**Figure 9.** Comparison of 0D calculation with 3D-CFD calculation with tabulated values.

The two-stage ignition model includes a tracking ignition progress variable and a fuel tracer variable; fuel is consumed rapidly when the tracked values reach the tabulated ignition delays. At low temperature ignition, the amount of fuel consumed relates to the ratio of tabulated heat releases, while at main ignition, it is simply the remainder of the fuel that is consumed.

### 2.5. Premixed Flame Propagation Modeling

As soon as autoignition has occurred, a flame kernel develops and the premixed flame front propagation in the natural gas-air mixture can start. The heat release rate for this regime is calculated by the coherent flame model framework, which assumes decoupled chemistry and turbulence:

$$\tilde{\omega}_F^{CFM} = \bar{\rho}_u \times \tilde{Y}_{TF} \times s_L^O \times \Sigma \tag{6}$$

The Reynold's averaged density in the fresh gases is written as $\bar{\rho}_u$; $\tilde{Y}_{TF}$ is the tracer fuel mass fraction. The term $s_L^O$ corresponds to the unstretched laminar flame speed of the fuel. This variable represents the chemical influence of the heat release rate. Special treatment of this variable is required when the diesel ignited gas engine is simulated; it will be described in the next section. $\Sigma$ is the flame surface density, which represents the influence of turbulence on the combustion. The turbulence only wrinkles and therefore expands the laminar flame surface. This is described by the transport equation for the flame surface density, which is noted as follows:

$$\frac{\partial \Sigma}{\partial t} + \frac{\delta \tilde{u}_i \Sigma}{\delta x_i} = \frac{\partial}{\partial x_i}\left(\left(\frac{\mu_t}{Sc_t} + \frac{\mu}{Sc}\right)\frac{\partial}{\partial x_i}\left(\frac{\Sigma}{\bar{\rho}}\right)\right) + (P_1 + P_2 + P_3 - D) \times \Sigma \tag{7}$$

The equation considers convection, diffusion and temporal evolution of the flame surface density as well as several source and sink terms indicated by $P_1$, $P_2$, $P_3$ and $D$. As these terms are not investigated separately in this paper, they will not be explained in detail; cf. the original paper by Colin [32] for the ECFM framework and the description of the flame surface density balance equation. The ECFM requires starting values for the progress variable and for the initial flame surface density. The progress variable can be obtained either via the tracer fuels in the two-stage autoignition interpolation or directly from the tables in the case of the TKI tabulation [28]. The initial flame surface density needs additional considerations. Initially proposed by Colin [32] for knock modeling and later used for dual fuel applications by Belaid-Saleh et al. [31], the gradient of the progress variable is represented by:

$$\Sigma_{ini} = |\nabla \tilde{c}| \tag{8}$$

As Belaid-Saleh describes, this formula is insufficient to properly model the initial flame surface density. The influences of turbulent wrinkling as well as the regime transition from autoignition to flame front propagation have to be considered [31]. As a result, an additional constant is added to the formula as well as the turbulent intensity, which is composed of the turbulent kinetic energy $k$ and the average velocity $\bar{u}$ that can further fold the flame kernel:

$$\Sigma_{ini} = C \times |\nabla \tilde{c}| \times \left(1 + \frac{\sqrt{k}}{\bar{u}}\right) \tag{9}$$

The constant $C$ can take values between 0 and 1 just as Belaid-Saleh explains and can be seen as an adjustment for the regime transition.

As the flame now propagates throughout the mixture, the heat release is calculated according to $\tilde{\dot{\omega}}_F^{\sim CFM}$ as described in the previous section. The unstretched laminar flame speed $s_L^0$ must also be considered. If the ignition delay time after the diesel is injected is long as a result of early injection or a very lean background mixture, there is a long time during which the diesel can penetrate the combustion chamber. After autoignition has occurred, the flame front encounters zones of premixed diesel, natural gas and air. To take into account the influence of chemistry during flame front propagation as well, the laminar flame speed has to be modified. Figure 10 shows values for the laminar flame speed with respect to the volumetric methane fraction. The calculations were performed under various pressure, temperature and global equivalence ratio conditions.

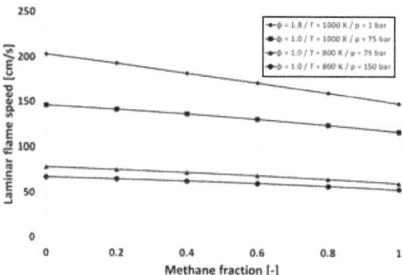

**Figure 10.** Laminar flame speed calculations for selected operating conditions. Calculations were performed with the full LLNL-v3 n-heptane mechanism. 0 on the x-axis indicates pure n-heptane, 1 indicates pure methane.

Unlike the ignition delay calculations, the results for the laminar flame speed of the fuel mixture exhibit linear behavior with respect the fuel mixture fraction. Recent experimental studies carried out by Li [44] support this assumption. The laminar flame speed for n-heptane was tabulated with the LLNL-v3 mechanism and that of methane was tabulated using the GRI 3.0 mechanism for natural gas. To consider the influence of the fuel mixture on the laminar flame speed, a linear interpolation between the two tabulated values can be used. The modified laminar flame speed for the dual fuel application, referred to as $s_{L_{df}}^0$, is written as follows:

$$s_{L_{df}}^0 = s_{L_{C7H16}}^0 \times \varphi_{df} + s_{L_{CH4}}^0 \times (1 - \varphi_{df}) \tag{10}$$

$s_{L_{C7H16}}^0$ and $s_{L_{CH4}}^0$ are the unstretched laminar flame speeds of n-heptane and methane. Both are determined from the tables described above. The tables consider the dependence of pressure, temperature, EGR ratio and global equivalence ratio. $\varphi_{df}$ is once again the dual fuel mixture fraction as described in Equation (4).

## 3. Validation of Injection Modeling

This section describes the validation of the injection modeling process discussed above. First, the experimental setup for rate of injection (ROI) measurement is introduced followed by the setup for non-reactive spray visualization. Next, the numerical setup is explained. Selected cases for validation are then presented in which the measured and simulated penetration lengths are compared. Finally, an optical comparison is provided for one selected test case.

### 3.1. Experimental Setup

A Bosch Tube was used for ROI measurements as described in [62]. The injector sprays the diesel mass into a measurement tube and the ROI is derived from the resulting pressure increase. The vapor and liquid penetration lengths are determined in a chamber filled with gaseous nitrogen. The absence of oxygen prevents combustion. The diesel is injected into a quiescent flow field. Schlieren and Mie scattering measurements are then performed to visualize the vapor and liquid phases of the injected diesel [33]. Further information on the experimental test setup can be found in [63]. A wide range of variations, including chamber temperatures and pressures as well as injection pressures and injected masses, has been tested and validated. Only selected test cases are discussed in this paper. Table 1 provides a short overview of the conditions in the spray chamber that were varied for the validation described in this paper.

**Table 1.** Hardware summary and selected test cases for validation of the injection modeling. The image on the right shows the measurement equipment.

| Hardware Summary | | |
|---|---|---|
| Chamber | Constant volume flow |  |
| Conditions | Preconditioned pressure and temperature | |
| Fluid | Nitrogen (N2) | |
| Turbulence | Low | |

| Selected Test Conditions | | | |
|---|---|---|---|
| Rail Pressure (bar) | Injected Mass (mg) | Chamber Pressure (bar) | Chamber Temperature (K) |
| 1600 | 12 <br> 8 <br> 4 | 60 | 780 |

## 3.2. Numerical Setup

The ROI measurements were used directly in the 3D-CFD simulation to accurately set up the spray model. As previously mentioned, the submodels of the spray model include the WAVE breakup model with the model improvements described in the previous section and a Dukowicz evaporation model. For further information on the models, cf. [51]. In addition to the spray model itself, mesh topology, cell size and simulation time step have a significant influence on the predicted spray penetration lengths. The settings for all three factors were similar to the engine mesh described in the following section. Table 2 provides an overview of the numerical setup.

**Table 2.** Summary of the numerical setup for validating the injection modeling. The image on the right shows the mesh.

| Meshing | | |
|---|---|---|
| Type | Sector mesh | |
| Cell size | 0.6 mm (average) | |
| Dimensions | 200 mm in a radial direction | |
| **General Setup** | | |
| Temporal discretization | 10 μs | |
| Turbulence modeling | RANS approach, k-ζ-f model | |
| Spray submodels | WAVE breakup<br>Dukowicz evaporation<br>O'Rourke turbulent dispersion | |

## 3.3. Validation Cases

Figure 11 shows the measured and simulated vapor and liquid penetration lengths. The vapor penetration length is plotted in triangles; the liquid penetration length is plotted in squares. In both cases, the bars indicate the standard deviation of the measurement fluctuations. The solid line indicates the simulated vapor penetration length, and the dashed line indicates the simulated liquid penetration length.

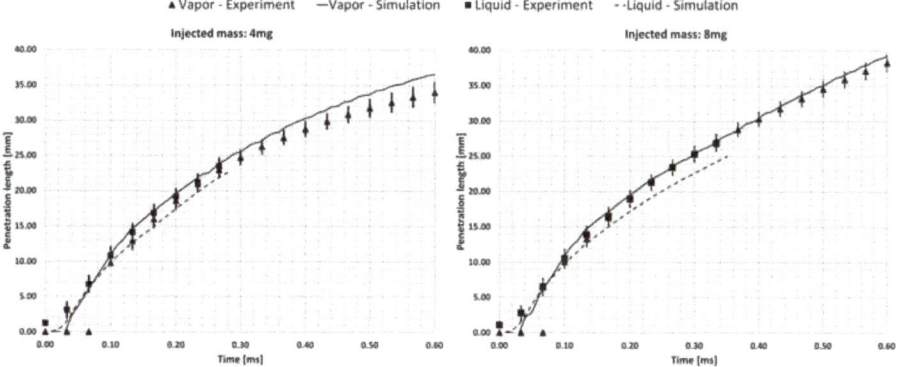

**Figure 11.** Comparison of the numerical and experimental results of selected test cases; 4 mg and 8 mg diesel masses injected.

Reasonable agreement of the penetration lengths is observed in one parameter set that was used in all the selected validation cases. Further validation was carried out using the visualization techniques of the measurement setup. The 3D results from the simulation of a diesel mass of 12 mg were used to recreate the images of the measurement. Figure 12 compares the penetration length once again as shown in the previous figure and also provides images of selected time steps. The top row shows the averaged images of the liquid phase of the measurement. The vertical, grey, dashed lines indicate the according time when the images of measurement and simulation were taken. The second row shows the particles in the simulation. The simulation images have been mirrored from the single spray to mimic the images of the measurement. Therefore, no hole to hole scattering can be seen in the simulation images. The rows of images below the penetration length graph provide the same information about the vapor phase. In the case of simulation, a cut plane is shown that runs through the centerline of the spray; the images show the diesel mass fraction.

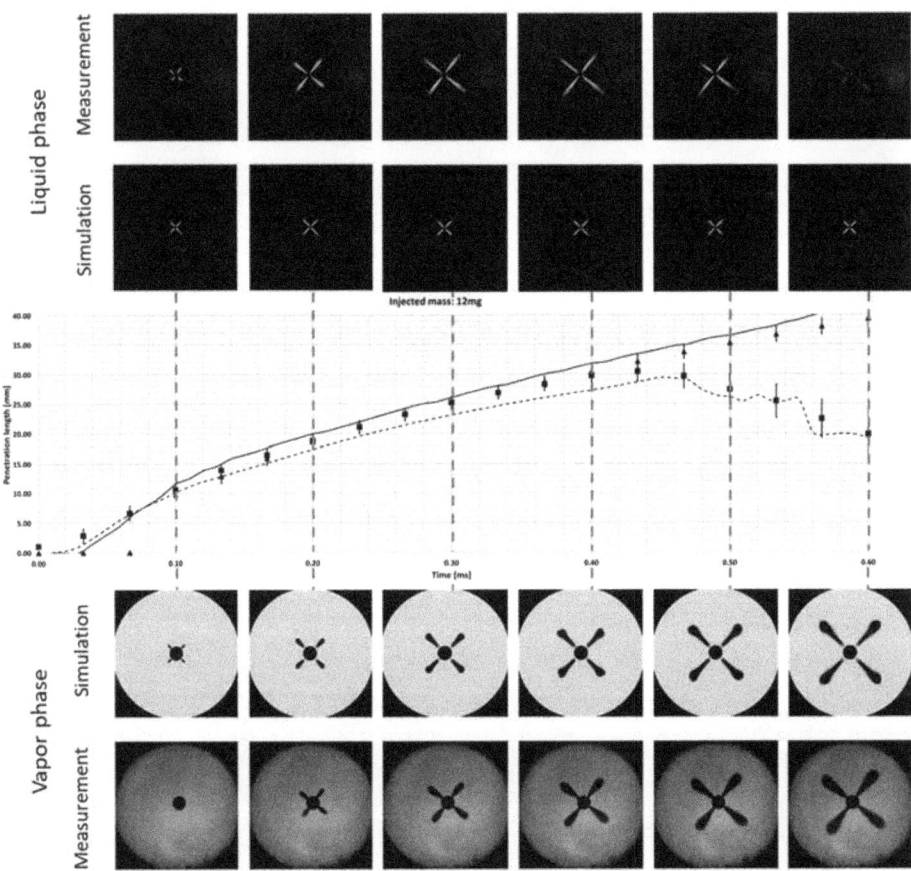

**Figure 12.** Visual comparison of simulation and measurement of a 12 mg diesel mass.

The visual comparison shows good agreement between measurement and simulation for the selected test case. One can notice a difference for the liquid phase, which is due to post-processing of the particle diameter in the simulation. The quantitative agreement of the penetration length can still be seen. With this validation process, the spatial resolution of the pilot diesel can be successfully predicted by the simulation. Further validation will now focus on the combustion modeling as described in the previous section, using experimental results from a single cylinder research engine (SCE).

## 4. Validation of Combustion Modeling

### 4.1. Experimental Setup

The approach to ignition delay modeling using a dual fuel mechanism as the basis for tabulation was validated using experimental data from a SCE at the LEC in Graz. Table 3 provides relevant engine data along with an image of the SCE. The natural gas-air mixture is prepared in a Venturi mixer several meters upstream of the intake ports, which substantiates the assumption of homogeneous distribution in the combustion chamber. The diesel fuel was injected with a 4-hole injector nozzle with a common rail system. Further information on the measurement setup can be found in [17]. All measurements have been carried out without the use of EGR.

**Table 3.** Technical specifications of the SCE and CAD mockup of the engine.

| Single Cylinder Engine—Technical Specifications | |
|---|---|
| Rated speed | 1500 rpm |
| Displacement | $\approx$6 dm$^3$ |
| Swirl/tumble | $\approx$0/0 |
| Charge air | Provided by external compressors with up to 10 bar boost pressure |
| Gas fuel supply | External mixture formation via Venturi mixer |
| Diesel fuel supply | Common rail system with up to 1600 bar rail pressure |

### 4.2. Numerical Setup

The 3D-CFD simulations only covered the high pressure phase of the working cycle. A 90 degree sector mesh was used for the simulation, cf. Table 4. The number of cells varied from 20,000 at top dead center (TDC) to 150,000 at bottom dead center (BDC), and the average cell size was 2 mm with a refinement in the spray region. Measured rates of injection (ROI) provided the input for the spray model. The spray submodel parameters were calibrated from optical measurements of injection sprays into inert atmospheres; for further information on the calibration, see [64]. The turbulence model was a k-$\zeta$-f model [65].

**Table 4.** Numerical setup for the engine simulations.

| Simulation Setup | |
|---|---|
| Turbulence modeling | RANS approach, k-$\zeta$-f model |
| Spray | Lagrangian particle tracking Dukowicz evaporation model WAVE breakup model |
| Combustion model | ECFM-3Z Two-stage ignition tabulation Initial flame surface density modeling Laminar flame speed interpolation from values |

*4.3. Validation Cases*

Two different engine operation condition variations were chosen for validation. The first variation is represented by the start of current (SoC). The SoC of the common rail injector was varied by ±10° CA related to a baseline timing. These cases correspond to the first three rows in Table 5. The second variation was in diesel mass, whereby the selected diesel shares of 1.5%, 1.0% and 0.5% correspond to the diesel masses injected shown in the section on the injection modeling variation. Further information on the measurement setup can be found in [65].

**Table 5.** Selected operating conditions for the combustion model validation.

| Engine Parameter Variations | | | |
|---|---|---|---|
| Variation | Diesel Share (Energetic) (%) | Global Lambda (–) | SOC |
| Start of Current (SoC) | 1.5 | 1.7 | Early |
| | 1.5 | 1.7 | Middle |
| | 1.5 | 1.7 | Late |
| Injected diesel mass | 1.5 | 1.7 | Late |
| | 1.0 | 1.7 | Late |
| | 0.5 | 1.7 | Late |

At first, the SoC variation is shown in Figure 13. Figure 13 below shows the pressure trace of the average pressure cycle of 100 experimental cycles as well as the maximum and minimum pressure cycles of these cycles. The black dashed line represents the pressure curve from 3D-CFD simulation. The graph on the right shows the normalized heat release rate (HRR) from measurement and simulation. It is clear that when injection timing is early, the ignition delay is well depicted. The predicted heat release rate due to flame front propagation through the natural gas air mixture is higher than the measured heat release rate.

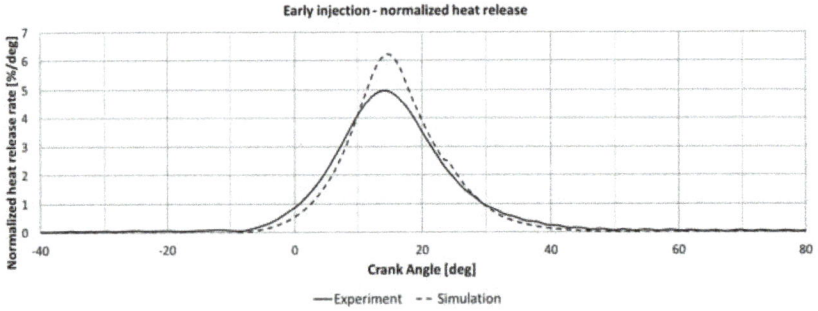

**Figure 13.** Measured and simulated heat release rates, early injection timing.

With medium injection timing, which can be seen as the baseline timing, the new ignition delay tables also correctly predict ignition delay and the subsequent flame front propagation is correctly captured. When the heat release rate of medium injection timing, as seen in Figure 14, is compared to the rate of early SoC timing, the premixed peak of diesel combustion becomes more significant when SoC timing is shifted to later crank angles.

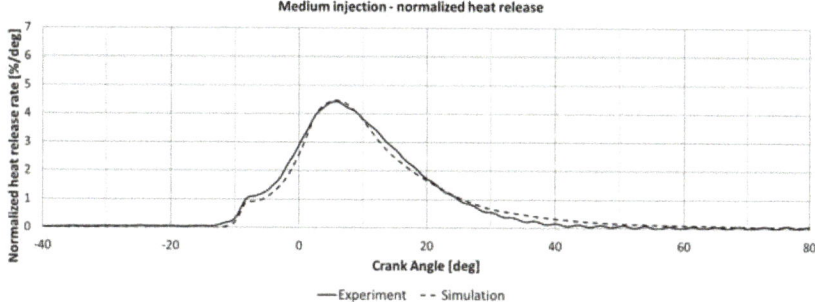

**Figure 14.** Measured and simulated rates, medium injection timing.

The effect of the diesel peak on the heat release rate is increased further by moving the SoC to a late crank angle, cf. Figure 15. The ignition delay prediction is also correct in this case.

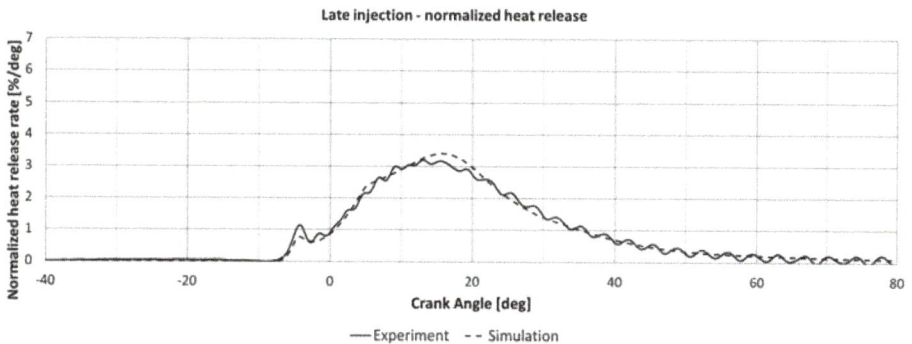

**Figure 15.** Measured and simulated heat release rates, late injection timing.

The diesel mass variation as described in Table 5 is shown for further validation. With the 1.5% diesel mass injected shown in Figure 16.

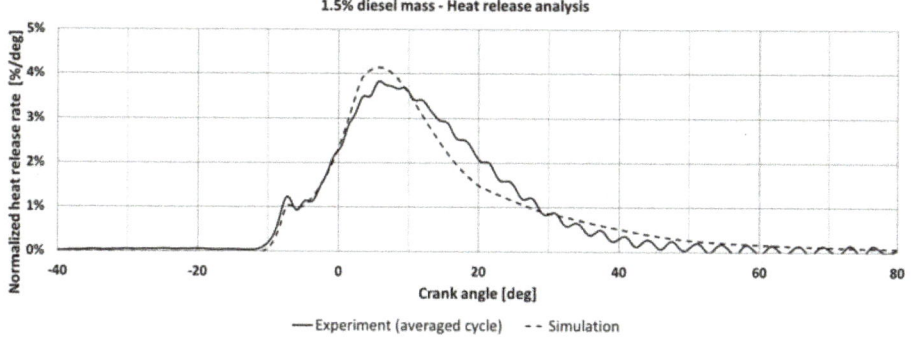

**Figure 16.** Measured and simulated heat release rates, 1.5% diesel mass injected.

As the diesel masses injected become smaller, a slightly less developed "diesel peak" is apparent, cf. Figure 17. The modeling of injection and combustion allows accurate reproduction of the heat release rates and pressure traces from the experiment.

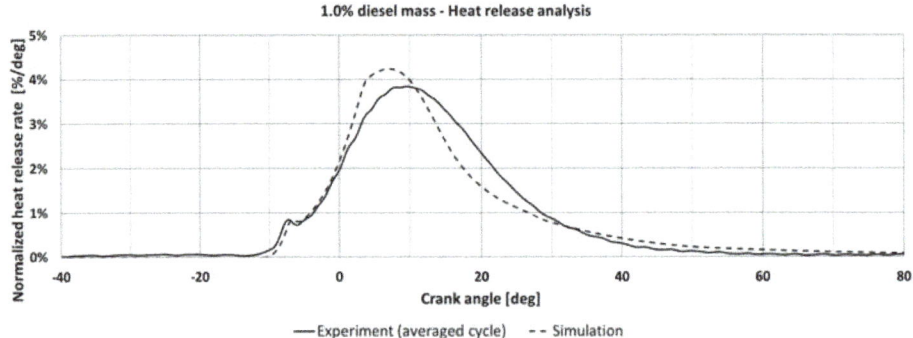

**Figure 17.** Measured and simulated heat release rates, 1.0% diesel mass injected.

The lower diesel peak becomes even more pronounced when the diesel mass injected is the smallest possible that still permits stable ignition at the specific engine conditions as shown in Figure 18.

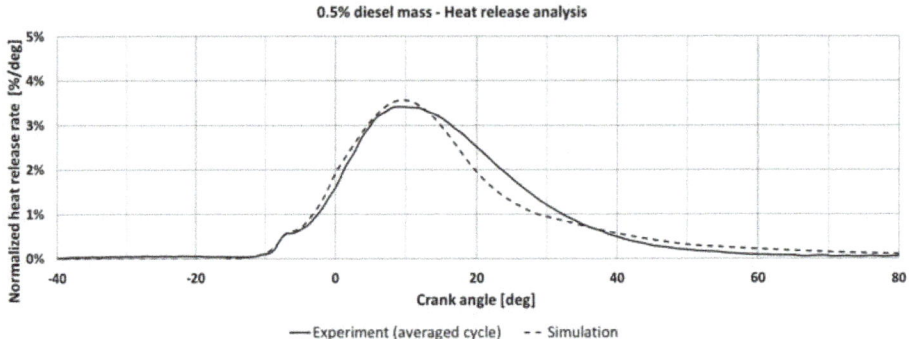

**Figure 18.** Measured and simulated heat release rates, 0.5% diesel mass injected.

## 5. Discussion of Results

### 5.1. Injection and Ignition Delay Modeling

The comparison of the simulated and measured vapor and liquid penetration lengths from the selected test cases in Section 3 show reasonable agreement. Furthermore, the qualitative comparison of the post-processed images for measurement and simulation yields satisfactory results. The calibration of the spray model with the diameter specification, therefore, is a valid approach to modeling the pilot injection with the given input data. It should be noted that this is only true when the injector is operated in the ballistic range. With larger diesel masses, where the injector needle reaches and maintains a fully open state for a longer period of time, the calibration approach may no longer be valid. The modeling of the ignition delay Section 2.2 showed reasonable agreement with the SCE results. Currently there are no basic experiments that solely focus on the ignition delay calculation being undertaken, thereby preventing a fundamental investigation of this process.

## 5.2. Combustion Modeling

The overall improvements to the modeling framework of injection, ignition and combustion have been tested on the diesel ignited gas engine concept using results from a single cylinder research engine. The model was able to reproduce the average pressure trace satisfactorily in six selected tests. The further interest of the authors in SoC variation led to a closer analysis of the ignition process in the simulation. Figure 19 shows a top down view into the piston from simulation. Since the sector mesh has been mirrored to mimic the full piston bowl, all spray cones appear to be the same and no hole to hole scattering is observed with this post-processing. The yellow surfaces are isosurfaces of the progress variable, which can be seen as an indicator of where the flame is currently located.

Though this is a qualitative comparison, several basic conclusions can be drawn. At the early injection timing, the diesel fuel has more time to penetrate throughout the combustion chamber and ignites in a rather spherical manner. The later the diesel is injected, the less time there is for mixture formation to occur, which causes the diesel to ignite in more of a conical shape closer to the injector nozzle as with the late injection timing. Although the interpretation seems plausible, it must be compared to the measurements before it can be regarded as a solid argument.

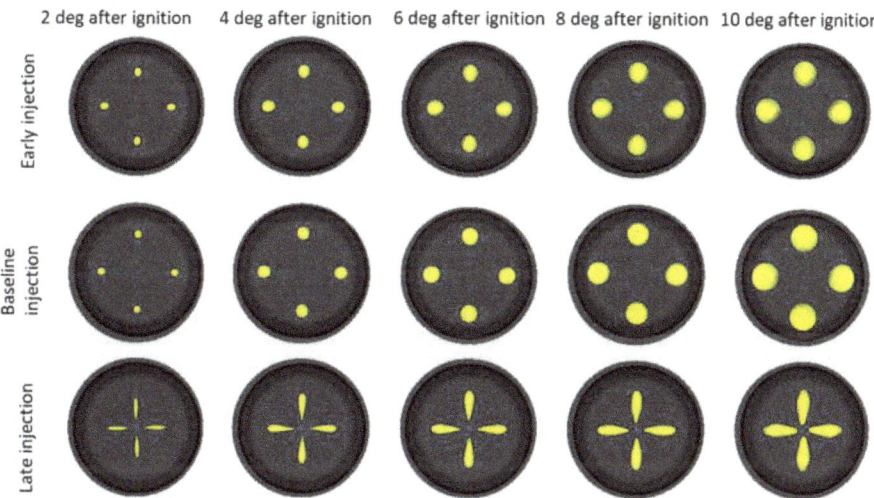

**Figure 19.** Isosurface of the progress variable for the SoC variation (top down view of the piston).

## 6. Conclusions and Future Work

A comprehensive collection of fundamental experiments regarding diesel pilot injection and single cylinder research engine measurements has led to the development and validation of injection and combustion models to simulate diesel ignited combustion. The results of the inert injection simulations show good agreement regarding liquid and vapor penetration lengths as well as qualitative agreement for visual post-processing. Further work on injection modeling will focus on simulations of a fully opened injector needle as well as reactive simulations. The diesel pilot quantities will be injected into synthetic air to characterize the ignition behavior within the spray box and to generate further data for the validation of the simulation model.

Based on the calibrated spray model for pilot injection, the ignition delay, initial flame surface density deposition and laminar flame speed of dual fuel combustion were able to be simulated with the improved ECFM-3Z model. The results of the simulation show good agreement between simulated and experimental heat release rates. A further analysis of the simulation results has also revealed the plausible behavior of the ignition process. Further work will focus on supporting the assumptions from the 3D-CFD simulation regarding ignition behavior described in Section 5 with endoscopic measurements on the SCE. Modeling efforts will focus on tabulation of the laminar flame speed based on the dual fuel mechanism developed in this paper in order to replace the linear interpolation currently in use. Furthermore, the emissions of the engine, with a special focus on nitric oxides ($NO_x$), soot and unburned hydrocarbons, will be subject of subsequent studies.

**Acknowledgments:** The authors would like to acknowledge the financial support of the "COMET—Competence Centres for Excellent Technologies Programme" of the Austrian Federal Ministry for Transport, Innovation and Technology (BMVIT), the Austrian Federal Ministry of Science, Research and Economy (BMWFW) and the Provinces of Styria, Tyrol and Vienna for the K1 Centre LEC EvoLET. The COMET Programme is managed by the Austrian Research Promotion Agency (FFG). The authors would like to express their gratitude for the measurement data provided by CMT-Motores Térmicos (Universitat Politècnica de València).

**Author Contributions:** Lucas Eder, Gerhard Pirker and Andreas Wimmer conceived and designed the experiments/simulations and analyzed the data; Marko Ban and Milan Vujanovic provided the tabulated ignition delays; Peter Priesching provided the software support; Lucas Eder and Gerhard Pirker wrote the paper.

**Conflicts of Interest:** The authors declare no conflict of interest.

## References

1. Verdolini, E.; Vona, F.; Popp, D. *Bridging the Gap: Do Fast Reacting Fossil Technologies Facilitate Renewable Energy Diffusion*; National Bureau of Economic Research (NBER): Cambridge, MA, USA, 2016.
2. International Transport Forum. *ITF Transport Outlook 2015*; International Transport Forum: Paris, France, 2015.
3. Pirker, G.; Wimmer, A. Sustainable power generation with large gas engines. *Energy Convers. Manag.* **2017**, *149*, 1048–1065. [CrossRef]
4. International Maritime Organization. Nitrogen Oxides ($NO_x$)—Regulation 13, 2015. Available online: http://www.imo.org/en/OurWork/Environment/PollutionPrevention/AirPollution/Pages/Nitrogen-oxides-(NOx)----Regulation-13.aspx (accessed on 12 September 2016).
5. Buchholz, B. Saubere Großmotoren für die Zukunft—Herausforderung für die Forschung. In *Rostock Large Engine Symposium*; FVTR: Rostock, Germany, 2014.
6. Zelenka, J.; Kammel, G. The Quality of Gaseous Fuels and Consequences for Gas Engines. In Proceedings of the 10th Internationale Energiewirtschaftstagung (IEWT 2017), Vienna, Austria, 15–17 February 2017; pp. 1–16.
7. Aaltonen, P.; Järvi, A.; Vaahtera, P.; Widell, K. Paper No. 251: New DF Engine Portfolio (Wärtsilä 4-Stroke). In Proceedings of the 28th CIMAC World Congress, Helsinki, Finland, 6–10 June 2016.
8. Dillen, E.; Yearce, D.; Trask, L.; Klingbeil, A. Paper No. 214: GE Transportation Dual Fuel Locomotive Development. In Proceedings of the 28th CIMAC World Congress, Helsinki, Finland, 6–10 June 2016.
9. Issei, O.; Nishida, K.; Hirose, K. Paper No. 049: New marine gas engine development in YANMAR. In Proceedings of the 28th CIMAC World Congress, Helsinki, Finland, 6–10 June 2016.
10. Yoon, W. Paper No. 201: Development of HiMSEN Dual Fuel Engine Line-up. In Proceedings of the 28th CIMAC World Congress, Helsinki, Finland, 6–10 June 2016.
11. Energy Information Administration (EIA). Available online: https://www.eia.gov/dnav/ng/NG_PRI_FUT_S1_M.htm (accessed on 2 February 2018).
12. Mooser, D. Brenngase und Gasmotoren. In *Handbuch Dieselmotoren*, 3rd ed.; Aufl, K., Mollenhauer, K., Tschöke, H., Eds.; Springer: Berlin/Heidelberg, Germany, 2007.
13. Redtenbacher, C.; Kiesling, C.; Wimmer, A.; Sprenger, F.; Fasching, P.; Eichelseder, H. Dual Fuel Brennverfahren—Ein zukunftsweisendes Konzept vom PKW-bis zum Großmotorenbereich? In Proceedings of the 37th International Vienna Motor Symposium, Vienna, Austria, 28–29 April 2016; pp. 403–428.
14. Krenn, M.; Redtenbacher, C.; Pirker, G.; Wimmer, A. A new approach for combustion modeling of large dual-fuel engines. In Proceedings of the Heavy-Duty, On- und Off-Highway Engines 2015—10th International MTZ Conference, Speyer, Germany, 24–25 November 2015; pp. 1–19.
15. Königsson, F. On Combustion in the CNG—Diesel Dual Fuel Engine. Ph.D. Thesis, KTH Royal Institute of Technology, Stockholm, Sweden, 2014.
16. Manns, H.J.; Brauer, M.; Dyja, H.; Beier, H.; Lasch, A. *Diesel CNG—The Potential of a Dual Fuel Combustion Concept for Lower $CO_2$ and Emissions*; SAE International: Warrendale, PA, USA, 2015.

17. Redtenbacher, C.; Kiesling, C.; Malin, M.; Wimmer, A.; Pastor, J.V.; Pinotti, M. Potential and Limitations of Dual Fuel Operation of High Speed Large Engines. In Proceedings of the ASME 2016 Internal Combustion Fall Technical Conference, San Diego, CA, USA, 4–7 November 2018.

18. Hanenkamp, A.; Böckhoff, N. The 51/60 DF and V32/40 PGI—Modern Gas engines from MAN Diesel SE. Their way from development to serial application. In Proceedings of the 6th Dessauer Gasmotoren-Konferenz, Dessau-Roßlau, Germany, 26–27 March 2009; pp. 129–142.

19. Troberg, M.; Portin, K.; Jarvi, A. Paper No. 406: Update on Wärtsilä 4-stroke Gas Product Development. In Proceedings of the 27th CIMAC World Congress, Shanghai, China, 13–16 May 2013.

20. Böckhoff, N.; Mondrzyk, D.; Terbeck, S. Continuous Development of the 51/60G to the 51/60G TS of the MAN Diesel and Turbo SE. In Proceedings of the 10th Dessauer Gasmotoren-Konferenz, Dessau-Roßlau, Germany, 6–7 April 2017; pp. 61–71.

21. Donateo, T.; Carlucci, A.P.; Strafella, L.; Laforgia, D. *Experimental Validation of a CFD Model and an Optimization Procedure for Dual Fuel Engines*; SAE International: Warrendale, PA, USA, 2014.

22. Hockett, A.; Hampson, G.; Marchese, A.J. Development and Validation of a Reduced Chemical Kinetic Mechanism for Computational Fluid Dynamics Simulations of Natural Gas/Diesel Dual-Fuel Engines. *Energy Fuels* **2016**, *30*, 2414–2427. [CrossRef]

23. Maghbouli, A.; Saray, R.K.; Shafee, S.; Ghafouri, J. Numerical study of combustion and emission characteristics of dual-fuel engines using 3D-CFD models coupled with chemical kinetics. *Fuel* **2013**, *106*, 98–105. [CrossRef]

24. Mousavi, S.M.; Saray, R.K.; Poorghasemi, K.; Maghbouli, A. A numerical investigation on combustion and emission characteristics of a dual fuel engine at part load condition. *Fuel* **2016**, *166*, 309–319. [CrossRef]

25. Li, Y.; Guo, H.; Li, H. *Evaluation of Kinetics Process in CFD Model and Its Application in Ignition Process Analysis of a Natural Gas-Diesel Dual Fuel Engine*; SAE International: Warrendale, PA, USA, 2017.

26. Maurya, R.K.; Mishra, P. Parametric investigation on combustion and emissions characteristics of a dual fuel (natural gas port injection and diesel pilot injection) engine using 0-D SRM and 3D CFD approach. *Fuel* **2017**, *210*, 900–913. [CrossRef]

27. Colin, O.; Benkenida, A. The 3-Zones Extended Coherent Flame Model (ECFM3Z) for Computing Premixed/Diffusion Combustion. *Oil Gas Sci. Technol.* **2004**, *59*, 593–609. [CrossRef]

28. Colin, O.; da Cruz, A.P.; Jay, S. Detailed chemistry-based auto-ignition model including low temperature phenomena applied to 3-D engine calculations. *Proc. Combust. Inst.* **2005**, *30*, 2649–2656. [CrossRef]

29. Subramanian, G.; da Cruz, A.P.; Colin, O.; Vervisch, L. *Modeling Engine Turbulent Auto-Ignition Using Tabulated Detailed Chemistry*; SAE International: Warrendale, PA, USA, 2007; pp. 776–790.

30. Subramanian, G.; Vervisch, L.; Ravet, F. *New Developments in Turbulent Combustion Modeling for Engine Design: ECFM-CLEH Combustion Model*; SAE International: Warrendale, PA, USA, 2007.

31. Belaid-Saleh, H.; Jay, S.; Kashdan, J.; Ternel, C.; Mounaim-Rousselle, C. *Numerical and Experimental Investigation of Combustion Regimes in a Dual Fuel Engine*; SAE International: Warrendale, PA, USA, 2013.

32. Colin, O.; Benkenida, A.; Angelberger, C. 3D Modeling of Mixing, Ignition and Combustion Phenomena in Highly Stratified Gasoline Engines. *Oil Gas Sci. Technol.* **2003**, *58*, 47–62. [CrossRef]

33. Pastor, J.V.; Payri, R.; Garcia-Oliver, J. Analysis of Transient Liquid and Vapor Phase Penetration for Diesel Sprays under Variable Injection Conditions. *At. Sprays* **2011**, *21*, 503–520. [CrossRef]

34. Malalasekera, W.; Versteeg, H.K. *An Introduction to Computational Fluid Dynamics—The Finite Volume Method*; Pearson Education Limited: London, UK, 1995.

35. Merker, G.P.; Schwarz, C. *Grundlagen Verbrennungsmotoren*; Vieweg Teubner: Wiesbaden, Germany, 2009.

36. Mastorakos, E. Ignition of turbulent non-premixed flames. *Prog. Energy Combust. Sci.* **2009**, *35*, 57–97. [CrossRef]

37. Kuo, K.K.; Acharya, R. *Fundamentals of Turbulent Multi-Phase Combustion*; John Wiley & Sons: New York, NY, USA, 2012.

38. Schlatter, S.; Schneider, B.; Wright, Y.; Boulouchos, K. *Experimental Study of Ignition and Combustion Characteristics of a Diesel Pilot Spray in a Lean Premixed Methane/Air Charge Using a Rapid Compression Expansion Machine*; SAE Technical Paper; SAE International: Warrendale, PA, USA, 2012.

39. Schlatter, S.; Schneider, B.; Wright, Y.M.; Boulouchos, K. *N*-heptane micro pilot assisted methane combustion in a Rapid Compression Expansion Machine. *Fuel* **2016**, *179*, 339–352. [CrossRef]

40. Aggarwal, S.K.; Awomolo, O.; Akber, K. Ignition characteristics of heptane–hydrogen and heptane–methane fuel blends at elevated pressures. *Int. J. Hydrogen Energy* **2011**, *36*, 15392–15402. [CrossRef]

41. Ban, M.; Vujanovic, M. Investigation of Dual-fuel Combustion Properties for CFD Simulation Purposes. In Proceedings of the 11th Conference on Sustainable Development of Energy, Water and Environment Systems, Lisbon, Portugal, 4–9 September 2016.

42. Demosthenous, E.; Borghesi, G.; Mastorakos, E.; Cant, R.S. Direct Numerical Simulations of premixed methane flame initiation by pilot n-heptane spray autoignition. *Combust. Flame* **2016**, *163*, 122–137. [CrossRef]

43. Wang, Z.; Abraham, J. Fundamental physics of flame development in an autoigniting dual fuel mixture. *Proc. Combust. Inst.* **2015**, *35*, 1041–1048. [CrossRef]

44. Li, G.; Liang, J.; Zhang, Z.; Tian, L.; Cai, Y.; Tian, L. Experimental investigation on laminar burning velocities and Markstein lengths of premixed methane-n-heptane-air mixtures. *Energy Fuels* **2015**, *29*, 4549–4556. [CrossRef]

45. Eder, L.; Kiesling, C.; Pirker, G.; Wimmer, A. Development and Validation of a Reduced Reaction Mechanism for CFD Simulation of Diesel Ignited Gas Engines. *SAE Int. J. Fuels Lubr.* **2009**, *1*, 675–702.

46. Heywood, J.B. *Internal Combustion Engine Fundementals*; McGraw-Hill Education: New York, NY, USA, 1988.

47. Donateo, T.; Strafella, L.; Laforgia, D. *Effect of the Shape of the Combustion Chamber on Dual Fuel Combustion*; SAE International: Warrendale, PA, USA, 2013.

48. Kuppa, K.; Butzbach, G.; Ratzke, A.; Dinkelacker, F. A Numerical Approach for the Prediction of Unburned Hydrocarbon Emissions in Gas Engines. In Proceedings of the 8th International Seminar on Flame Structure: Flame Structure, Berlin, Germany, 21–24 September 2014.

49. Chevillard, S.; Colin, O.; Bohbot, J.; Wang, M.; Pomraning, E.; Senecal, P.K. *Advanced Methodology to Investigate Knock for Downsized Gasoline Direct Injection Engine Using 3D RANS Simulations*; SAE International: Warrendale, PA, USA, 2017.

50. Reitz, R.D. Modeling Atomization Processes in High-Pressure Vaporizing Sprays. *At. Sprays Technol.* **1987**, *3*, 309–337.

51. AVL. *AVL FIRE Spray Module—Version Manual 2014.1*; AVL: Graz, Austria, 2014.

52. Dukowicz, J.K. *Quasi-Steady Droplet Phase Change in the Presence of Convection*; Los Alamos Scientific Laboratory: Los Alamos, NM, USA, 1979.

53. Roth, H.; Giannadakis, E.; Gavaises, M.; Arcoumanis, C.; Omae, K.; Sakata, I.; Nakamura, M.; Yanagihara, H. Effect of Multi-Injection Strategy on Cavitation Development in Diesel Injector Nozzle Holes. *SAE Trans.* **2005**, *114*, 1029–1045.

54. Arcoumanis, C.; Flora, H.; Gavaises, M.; Badami, M. *Cavitation in Real-Size Multi-Hole Diesel Injector Nozzle*; SAE International: Warrendale, PA, USA, 2000.

55. Curran, H.J. Rate constant estimation for C1 to C4 alkyl and alkoxyl radical decomposition. *Int. J. Chem. Kinet.* **2006**, *38*, 250–275. [CrossRef]

56. Conaire, M.Ó.; Curran, H.J.; Simmie, J.M.; Pitz, W.J.; Westbrook, C.K. A comprehensive modeling study of hydrogen oxidation. *Int. J. Chem. Kinet.* **2004**, *36*, 603–622. [CrossRef]

57. Mehl, M.; Pitz, W.J.; Westbrook, C.K.; Curran, H.J. Kinetic modeling of gasoline surrogate components and mixtures under engine conditions. *Proc. Combust. Inst.* **2011**, *33*, 193–200. [CrossRef]

58. Smith, G.; Bowman, T.; Frenklach, M. "GRI 3.0 Mechanism", Gas Research Institutue—GRI, 2000. Available online: http://combustion.berkeley.edu/gri-mech/index.html (accessed on 18 November 2016).

59. Mehl, M.; Chen, J.Y.; Pitz, W.J.; Sarathy, S.M.; Westbrook, C.K. An Approach for Formulating Surrogates for Gasoline with Application toward a Reduced Surrogate Mechanism for CFD Engine Modeling. *Energy Fuels* **2011**, *25*, 5215–5223. [CrossRef]

60. Petersen, E.L.; Kalitan, D.M.; Simmons, S.; Bourque, G.; Curran, H.J.; Simmie, J.M. Methane/propane oxidation at high pressures: Experimental and detailed chemical kinetic modeling. *Proc. Combust. Inst.* **2007**, *31*, 447–454. [CrossRef]

61. Niemeyer, K.E.; Sung, C.J.; Raju, M.P. Skeletal mechanism generation for surrogate fuels using directed relation graph with error propagation and sensitivity analysis. *Combust. Flame* **2010**, *157*, 1760–1770. [CrossRef]

62. Bosch, W. Der Einspritzgesetz-Indikator, ein neues Meßgerät zur direkten Bestimmung des Einspritzgesetzes von Einzeleinspritzungen. *Mot. Z.* **1964**, *25*, 268–282. (In German)

63. Kiesling, C.; Redtenbacher, C.; Kirsten, M.; Andreas, W.; Imhof, D.; Berger, J.M. IngmarGarcía-oliver, Detailed Assessment of an Advanced Wide Range Diesel Injector for Dual Fuel Operation of Large Engines. In Proceedings of the CIMAC Congress 2016, Helsinki, Finland, 6–10 June 2016.

64. Eder, L.; Kiesling, C.; Pirker, G.; Priesching, P.; Wimmer, A. *Multidimensional Modeling of Injection and Combustion Phenomena in a Diesel Ignited Gas Engine*; SAE International: Warrendale, PA, USA, 2017.

65. Hanjalić, K.; Popovac, M.; Hadžiabdić, M. A robust near-wall elliptic-relaxation eddy-viscosity turbulence model for CFD. *Int. J. Heat Fluid Flow* **2004**, *25*, 1047–1051. [CrossRef]

Article

# Co-Combustion of Low-Rank Coal with Woody Biomass and Miscanthus: An Experimental Study

**Anes Kazagic [1],\*, Nihad Hodzic [2] and Sadjit Metovic [2]**

[1]  Department for Strategic Development, JP Elektroprivreda BiH d.d.-Sarajevo—Power utility, Vilsonovo setaliste 15, 71000 Sarajevo, Bosnia and Herzegovina

[2]  Faculty of Mechanical Engineering, University of Sarajevo, Vilsonovo setaliste 9, 71000 Sarajevo, Bosnia and Herzegovina; hodzic@mef.unsa.ba (N.H.); metovic@mef.unsa.ba (S.M.)

\*  Correspondence: a.kazagic@epbih.ba; Tel.: +387-61-217-228

Received: 9 February 2018; Accepted: 23 February 2018; Published: 9 March 2018

**Abstract:** This paper presents a research on ash-related problems and emissions during co-firing low-rank Bosnian coals with different kinds of biomass; in this case woody sawdust and herbaceous energy crops Miscanthus. An entrained-flow drop tube furnace was used for the tests, varying fuel portions at a high co-firing ratio up to 30%wt woody sawdust and 10%wt Miscanthus in a fuel blend. The tests were supposed to optimize the process temperature, air distribution (including OFA) and fuel distributions (reburning) as function of $SO_2$ and $NO_x$ emissions as well as efficiency of combustion process estimated through the ash deposits behaviors, CO emissions and unburnt. The results for 12 co-firing fuel combinations impose a reasonable expectation that the coal/biomass/Miscanthus blends could be successfully run under certain conditions not producing any serious ash-related problems. $SO_2$ emissions were slightly higher when higher content of woody biomass was used. Oppositely, higher Miscanthus percentage in the fuel mix slightly decreases $SO_2$ emissions. $NO_x$ emissions generally decrease with an increase of biomass co-firing rate. The study suggests that co-firing Bosnian coals with woody sawdust and Miscanthus shows promise at higher co-firing ratios for pulverized combustion, giving some directions for further works in co-firing similar multi-fuel combinations.

**Keywords:** co-firing; coal; biomass; woody sawdust; Miscanthus

---

## 1. Introduction—Specific Objectives and State-of-the-Art

Transition of traditional fossil-fueled power stations into multi-fuel power plants is ongoing. Multi-fuel operation of coal-fired power stations, running different kinds of biomass in co-firing with coal, is nowadays aimed to provide fuel mix diversity to reduce $CO_2$ emissions, improve security of supply and reduce operational costs by fuel cost optimization. In the last decade, significant progress was made in biomass co-firing in coal-fired power stations, [1]. Over 250 units worldwide have either tested or demonstrated biomass co-firing or are currently co-firing on a commercial basis, as reported by KEMA (Arnhem, The Netherlands), [2]. Coal is often replaced with up to 30% of biomass by weight in pulverized coal (PC) fueled power stations. Most of these projects refer to co-firing of biomass with high-rank coal (both bituminous and anthracite), (e.g., [3]), while projects with biomass co-firing with low-rank sub-bituminous coal, brown coal or lignite are more scarce, like the project involving Greek lignite reported by Kakaras, [4]. Furthermore, progress is made in application of different types of biomass, [5], such as municipal solid waste (solid recovered fuel—SRF or refuse derived fuel—RDF, including its gasification), or even co-firing with sewage sludge, see [6,7]. Examples of biomass co-firing can be found in other industries as well, such as the cement industry, [8].

Nevertheless, research, development and demonstration projects and technologies which include combination different types of biomass co-fired with coal should be investigated, to achieve sustainable solution for future solid-based power plants. Over the last decade many research studies have been conducted to investigate the phenomenon of biomass co-firing, but only few of them included a multi-fuel concept. Wang et al. (2014) used a drop tube furnace to assess the combustion behavior and ash properties of several renewable fuels, like rice husk, straw, coffee husk, pine and waste derived fuels, [9]. Kupka et al. (2008) investigated the ash deposits formation during co-firing coal with sewage sludge, saw-dust and refuse derived fuels in a drop tube furnace, to optimize biomass co-firing blends, [10]. Williams et al. (2012) investigated the pollutants emissions from solid biomass fuel combustion, [11]. A combination of fuels can give rise to positive or negative synergy effects, of which the interactions among S, Cl, K, Al and Si are the best known, and may give rise to or prevent deposits on tubes, as reported in [12], or may have an influence on the formation of dioxins, see [13]. Furthermore, biomass can be successfully used for reburning in order to reduce $NO_x$ emissions [14].

Although different types of energy crops have been recognized as significant energy resource in future low carbon society, co-firing energy crops with other types of biomass or coal have not been widely, if at all, investigated so far.

This work presents an experimental study on co-firing Bosnian low-rank coal with woody biomass and herbaceous energy crop Miscanthus, by a multi-fuel pulverized combustion concept. Emissions and ash-related problems were already investigated in case of Bosnian low rank coal, also in co-firing Bosnian low rank coal with waste woody biomass, see [12,15], where some specific benefits and synergy effects of biomass co-firing with that coal type have been observed. The same coal type, used in co-firing with biomass and natural gas, was subject of research into reburning, see [14]. A novelty of the work can be assigned to the fact that paper, as pioneering, deals with investigation of possibility to use mix of low-rank coal, woody biomass and energy crops Miscantus, for direct PF co-firing, at different and high biomass co-firing rate. The results and findings in the next sections answer the question whether such a multi-fuel system can be beneficial, while giving directions for further research in the field.

## 2. Experimental

### 2.1. Test Methodology and Fuel Test Matrix

An electrically heated entrained PF flow Lab-scale furnace was used for the tests, varying fuel portions in a fuel blend with up to 30%wt woody sawdust and 10%wt of Miscanthus. The purpose of the tests is to optimize the process temperature, air distribution including OFA, and fuel portions and distributions including reburning, as functions of emissions of $NO_x$ and $SO_2$ as well as combustion process efficiency estimated through the ash deposits behaviors, CO emissions and unburnt. For all six basic fuels, namely four basic coal blends: TET5-100%, TET7-100%, TEK6-100% and TEK8-100%, and two biomass types: WB-100% (woody sawdust) and M-100% (Miscanthus), chemical analysis of fuel (F-CA), ash chemical composition (A-CA) and ash fusion test (AFT) were performed. For each of the 12 co-firing tests, analyses of ash deposit (D), slag (S) and fly ash (FA) have been carried out as specified in Table 1.

**Table 1.** Fuel test matrix, temperatures and air ratio used, scope of chemical analyses performed.

| | No. | Temperature, °C | 1250 | 1450 |
|---|---|---|---|---|
| | | $\lambda_1/\lambda$ | 0.95/1.20 | 0.95/1.20 |
| | | **Fuel Combinations** | 1 | 2 |
| **TET** | 1. | TET5:WB = 93:7%wt | F-CA, A-CA, AFT | |
| | 2. | TET5:WB = 85:15%wt | D-UBC | |
| | 3. | TET5:M = 93:7%wt | S-CA + AFT | |
| | | | FA-CA + AFT | |
| | 4. | TET7:WB = 85:15%wt | F-CA, A-CA, AFT | |
| | 5. | TET7:WB:M = 80:13:7%wt | D-CA | |
| | 6. | TET7:WB:M = 75:15:10%wt | S-UBC | |

**Table 1.** *Cont.*

| | No. | Temperature, °C | 1250 | 1450 |
|---|---|---|---|---|
| | | $\lambda_1/\lambda$ | 0.95/1.20 | 0.95/1.20 |
| | | Fuel Combinations | 1 | 2 |
| TEK | 7. | TEK6:WB = 85:15%wt | | F-CA, A-CA, AFT |
| | 8. | TEK6:WB = 75:25%wt | | D-UBC |
| | 9. | TEK6:M = 93:7%wt | | S-CA + AFT |
| | | | | FA-CA + AFT |
| | 10. | TEK8:WB = 75:25%wt | F-CA, A-CA, AFT | |
| | 11. | TEK8:WB:M = 85:8:7%wt | D-CA | |
| | 12. | TEK8:WB:M = 75:15:10%wt | S-UBC | |

*2.2. Fuel Types, Sampling and Preparation*

Coal sampling and transport. Coal samples have been prepared in Tuzla Thermal power plant, namely mix of lignite and brown coal—TET5 and mix of lignite—TET7 and in Kakanj Thermal power plant, namely brown coal mix TEK6 and brown coal mix TEK8. Approximately 30 kg of each coal are prepared for the lab-scale tests. The coal samples were taken in pulverized form, capturing the coal dust from the channel behind the mill, to have realistic test fuel size distribution which fully corresponding to real situation in large boiler.

Biomass sampling and transport. The forest residues (waste woody biomass—small size wood residue of low quality, which cannot be used for other purposes) have been collected in the forestry in the region of Middle Bosnia and East Bosnia, by potential supplier of the waste woody biomass. The wood chips have been made from these forest residues, beech and spruce, and after they have been mixed at approximate ratio 50:50%wt, chipped and then pulverized in the laboratory mill. Use of such a waste wood residues have double positive GHG effect: prevent formation of $CH_4$ while improve forest sink effect, in parallel with reducing $CO_2$ emission from coal-fueled power station which running in biomass co-firing.

Miscanthus. The Miscanthus test samples have been taken from the trial field of Miscanthus in Butmir (Bosnia) in November 2015. The chips from Miscanthus have been made from these samples in February 2016 and they have been pulverized in the laboratory mill. There is a plan of EPBiH power utility to grow Mischantus for carbon sequestration in post-mining area on its coalmines and then to use it in a mix of fuels in its Tuzla and Kakanj coal-based power stations, which would have double positive carbon sink effect.

*2.3. Fuel Analysis*

Before the lab scale tests, all analyses have been done for all fuels from the fuel matrix. The basic characteristics of the mixed coals from TPP Tuzla (TET5—Unit 5 of TPP Tuzla, TET7—new Unit 7 of TPP Tuzla), from TPP Kakanj (TEK6—Unit 6 of TPP Kakanj, and TEK8—new Unit 8 of TPP Kakanj) and from the different biomass are shown in the Table 2. Table 2 depicts trend for the respective characteristics to give an impression about possible changes in the fuel composition.

For the components of fuel matrix: TET5, TET7, TEK6, TEK8, WB (woody biomass, i.e., mix of beech and spruce at 50:50%wt) and M (Miscanthus), proximate and ultimate analyses, including particle size distribution, ash chemical analysis and ash fusion test have been done.

**Table 2.** The basic characteristics of the mixed coals from TPP Tuzla, TPP Kakanj and different biomass.

| Item | TET5 | TET7 | TEK6 | TEK8 | B100 | S100 | M100 |
|---|---|---|---|---|---|---|---|
| Mixed fuel (ratio by weight) | Šikulje:Dubrave:br. coal I 40:35:25 | Dubrave:Šikulje 50:50 | 100 | Kakanj:Breza:Zenica 70:20:10 | beech 100 | spruce 100 | miscanthus 100 |
| **Proximate analysis, %** | | | | | | | |
| Moisture ($w_{crude}$) | 0 | 0 | 9.4 | 8.4 | - | - | - |
| Moisture ($w_{hygro}$) | 3.39 | 3.27 | 2.64 | 2.89 | - | - | - |
| Moisture ($w_{total}$) | 3.39 | 3.27 | 12.04 | 11.29 | 39.57 | 44.07 | 12.33 |
| Ash | 41.67 | 38.28 | 36.98 | 41.38 | 0.53 | 0.24 | 4.28 |
| Combustible | 54.94 | 58.45 | 50.98 | 47.32 | 59.91 | 55.69 | 83.39 |
| Volatiles | 30.9 | 33.64 | 29.9 | 26.86 | 50.12 | 47.83 | 71.4 |
| Char | 65.71 | 63.09 | 58.06 | 61.84 | 10.31 | 8.1 | 16.28 |
| Fixed carbon ($C_{fix}$) | 23.9 | 24.79 | 20.94 | 20.38 | 9.79 | 7.86 | 11.99 |
| Sulfur ($S_{total}$) | 1.65 | 1.057 | 2.46 | 2.47 | 0.07 | 0.11 | 0.15 |
| Sulfur ($S_{ash}$) | 0.57 | 0.67 | 1.37 | 1.34 | 0.01 | 0.02 | 0.04 |
| Sulfur ($S_{combustible}$) | 1.08 | 0.9 | 1.1 | 1.14 | 0.06 | 0.1 | 0.11 |
| **Ultimate analysis, %** | | | | | | | |
| Carbon (C) | 36.69 | 38.72 | 35.42 | 31.89 | 29.65 | 27.92 | 42.6 |
| Hydrogen (H) | 3.07 | 2.95 | 2.64 | 2.71 | 3.68 | 3.39 | 4.79 |
| Sulfur ($S_{combustible}$) | 1.08 | 0.9 | 1.1 | 1.14 | 0.06 | 0.1 | 0.11 |
| Nitrogen (N) | 0.79 | 0.86 | 0.62 | 0.64 | 0.11 | 0.11 | 0.11 |
| Oxygen (O) | 13.31 | 15.02 | 11.21 | 10.95 | 26.4 | 24.17 | 35.78 |
| Ash | 41.67 | 38.28 | 36.98 | 41.38 | 0.53 | 0.24 | 4.28 |
| Moisture ($w_{total}$) | 3.39 | 3.27 | 12.04 | 11.29 | 39.57 | 44.07 | 12.33 |
| **Heating values fuel, kJ/kg** | | | | | | | |
| HHV ($H_h$) | 14,688 | 14,870 | 14,018 | 12,989 | 11,215 | 10,476 | 15,361 |
| LHV ($H_l$) | 13,977 | 14,187 | 13,198 | 12,170 | 9546 | 8763 | 14,090 |
| **Mineral analysis of ash, %** | | | | | | | |
| $SiO_2$ | 54 | 59.01 | 43.2 | 46.12 | 6.62 | 7.15 | 67.76 |
| $Fe_2O_3$ | 9.98 | 8.38 | 9.18 | 9.58 | 1.2 | 0.9 | 4 |
| $Al_2O_3$ | 20.08 | 17.53 | 19.58 | 22.63 | 0.61 | 0.23 | 0.1 |
| CaO | 6 | 6.2 | 11.7 | 8.95 | 33.2 | 34.08 | 8.3 |
| MgO | 4 | 3 | 4.3 | 4.35 | 11 | 17.5 | 6.4 |
| $SO_3$ | 3.44 | 4.36 | 9.26 | 6.05 | 2.57 | 2.99 | 6.34 |
| $TiO_2$ | 0.4 | 0.4 | 0.35 | 0.35 | <0.10 | <0.10 | <0.10 |
| $Na_2O$ | 0.429 | 0.258 | 1.609 | 1.211 | 8.15 | 10.02 | 1.13 |
| $K_2O$ | 0.976 | 0.804 | 0.809 | 0.659 | 16.25 | 17.56 | 2.12 |

**Table 2.** *Cont.*

| Item | TET5 | TET7 | TEK6 | TEK8 | B100 | S100 | M100 |
|---|---|---|---|---|---|---|---|
| Ash temperatures, °C | | | | | | | |
| $t_2$—softening. | 1300 | 1300 | 1240 | 1250 | 1300 | 1300 | 1150 |
| $t_3$—hemisphere | 1350 | 1360 | 1310 | 1320 | 1400 | 1370 | 1250 |
| $t_4$—flow | 1400 | 1400 | 1350 | 1360 | 1450 | 1420 | 1350 |
| Particle size distribution, % | | | | | | | |
| >1.00 mm | 0 | 0 | 0 | 8.35 | 0.59 | 1.57 | 0.83 |
| >0.50 mm | 14.3 | 15.88 | 0.08 | 21.74 | 16.2 | 44.82 | 30.24 |
| >0.20 mm | 30.4 | 43.91 | 8.46 | 26.95 | 60.73 | 30.13 | 43.31 |
| >0.09 mm | 28.7 | 21.57 | 41.05 | 17.13 | 17.4 | 18.5 | 12.58 |
| <0.09 mm | 26.5 | 18.64 | 50.41 | 25.83 | 5.08 | 4.98 | 13.04 |

## 2.4. Lab-Scale Furnace Used

The lab-scale furnace is comprised of a 3 m length alumina-silicate ceramic tube, with a diameter of 230/200 mm, where combustion takes place, surrounded by SiC stick-type electric heaters and three-layer insulation, Figure 1. The temperature of the reaction zone is controlled by a programmable logic controller (PLC) with thyristor units for each of the four heating zones, allowing the process temperature to be varied at will across the range from ambient to 1560 °C.

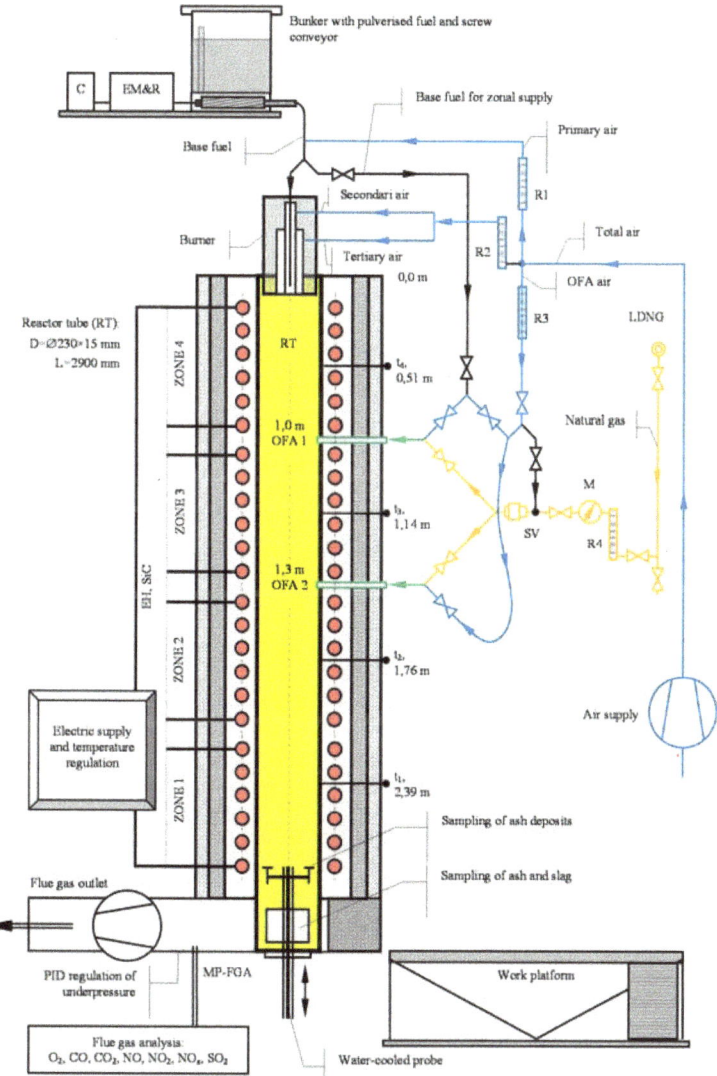

**Figure 1.** Principal scheme of the experimental furnace; C—speed controller of fuel supply, EM&R—electric motor and gearbox, RT—reaction tube, Ri—flow meters measuring the flow of air and natural gas, EH—SiC electric heaters arranged in four zones/levels, LDNG—laboratory distribution of natural gas, SV—safety valve with integrated flame arrester, M—manometer, MP-FGA—measuring point/flue gas analysis.

The maximum power of the electrical heaters used to maintain temperature in the reaction tube is 70 kW, while nominal or thermal power of the furnace is 20 kW. Pulverized fuel is fed into the furnace by means of a volumetric feeder, mounted above the reactor. The feeder is equipped with a speed controller, allowing mass flow in the range of 0.25–5 kg/h. Air for combustion, coming from the air blower, is divided into carrier air (primary air), secondary air, tertiary air, and over fire air (OFA). The first three air portions are fed into the furnace over the swirl burner settled on the top of the reactor, so the air-fuel particle mixture flows downward, [12,14,15]. The excess air ratio was adjusted by tuning the air flow in each airline, at a constant fuel flow. The ceramic probes for collecting ash deposit were attached to a water-cooled lance probe, which could be moved along the reaction tube axis and set at the desired position, Figure 1.

### 2.5. Components Test Demonstration

Different kind of fuels were used; coal, woody biomass and energy crops miscanthus—processed to appropriate particle size by a laboratory hammer mill. Particle size distribusion of the fules is given in Table 2.

Test parameters to be used are basically as follows:

- TET 5 based fuels (1–3): t = 1250 °C, OFA, (PFC dry),
- TET 7 based fuels (4–6): t = 1250 °C, OFA, (PFC dry),
- TEK 6 based fuels (7–9): t = 1450 °C, OFA, (PFC slag tap),
- TEK 8 based fuels (10–12): t = 1250 °C, OFA, (PFC dry).

Excess air ration is corresponding to the real situation in boilers of power units in Tuzla and Kakanj TPP. Total excess air ratio is set at 1.15–1.2 at furnace outlet, while in first stage of combustion (on burner) is set to be at 0.95. Ratio of primary and secondary/tertiary air is adjusted on swirl burner for optimal air staging in primary zone i.e., optimal $NO_x$ and CO ratio. Thus, in all tests, optimized $NO_x$ and CO emissions were provided if primary (carrier) air portion was set at 1.50 $m_n^3$/h providing primary air ratio to be between 0.30 and 0.33. For air staging in the furnace, OFA varied from 0 to 0.30 of the overall amount of combustion air. Depending on fuel and excess air ratio used, total airflow rate was between 4.2 and 4.5 $m_n^3$/h, while flue gas flow rate was between 4.5 and 5.0 $m_n^3$/h.

Fuel thermal load is kept at same thermal input (approximately 5 kWth) in all the runs to provide comparison of the results. Depending on the fuel and excess air used, the total airflow rate is identified. The primary (carrier) air flow rate is set at 1.50 $m_n^3$/h for all the runs, with the rest of the air is divided into secondary and tertiary portions, approximately at a ratio of 2.5:1. So, primary/secondary/tertiary/OFA ratio is comparable with real situation in Tuzla TPP and Kakanj TPP. During the tests, ceramic probes are set 2 m away from the top of the burner (see Figure 1). After about 120 min of test running and ash collecting, the ceramic probes with ash deposits are carefully removed for further analysis of the deposits.

Emissions of NO, $NO_2$, $SO_2$ and CO are measured by a 350XL instrument (TESTO, Lenzkirch, Germany).

### 2.6. Experimental Uncertainties—Assessment of Applicability and Reliability of the Results

Experimental study relies on lab-scale tests in pulverized fuel (PF) furnace, performed according to good-laboratory practice procedure, as well as some of standard techniques like SEM, AFT, Oxide determination (Chemical composition analysis) etc. To better understanding the combustion mechanism while mixing the fuels, basic experiments have to be done using the TG, FTIR or GCMS, which is already planned by authors.

With regard to performed lab-scale tests, the results presented here are valid for cross flow around the tubes of the heat exchangers and for refractory combustion chamber walls, in terms of the real situation in large boiler. For vertical walls with parallel particle flow, the ceramic probes for ash collection should be set at an appropriate angle to the particle flow. Furthermore, it should be noted that, as a result of combustion, the coal particle temperature can be higher than that of the furnace wall, with potential implications for the ash transformation and deposition processes, [12]. Specific particle size distribution was used during the tests, as given in Table 2, as well as specific air distribution, as described in previous section. Finally, regarding the run time of 120 min, it should be stressed that it was the initial stage of slagging that was being examined during these tests.

Concering the type of testing procedure used, it should be noted that such lab-scale tests are not yet considered standard testing procedure, although many laboratories worldwide do use them, including TU Clausthal [10], IVD Stuttgart [16] and the University of Newcastle [17]. This type of testing procedure is accompanied by standard ancillary techniques for sample preparation, see e.g., [10], or standard chemical techniques that provide useful information, e.g., AFT or Oxide determination. These chemical techniques are, however, imperfect slagging or fouling predictors, and testing by experimental facility is advisable, as it provides a more reliable evaluation of the slagging/fouling propensity of the given fuels [17].

With regard to measurement of emissions, the measurement error is estimated for the NO emissions and $SO_2$ emissions. Flue gas temperature is measured at the point in the partially insulated outlet tube where the gas sample is taken for emissions measurement. The processes in the flue gas line is frozen; there is no post combustion from the reactor to the TESTO 350XL instrument.

The aforementioned facts should be taken into account in considering the results presented here.

## 3. Results and Discussion

### 3.1. Combustion Efficiency

For biomass co-firing Tuzla coal blends TET5 and TET7 with woody biomass and Miscanthus, it can be noticed that the increase of the unburnt carbon content (UBC) in the ash deposits collected in Lab-scale furnace is only minor, from 0.0% to 0.04%. However, in the slag collected at the bottom of the furnace, the UBC is increased from 0.45% at 15% biomass co-firing, over 3.58% at 20% biomass co-firing to 6.92% at 25% biomass co-firing with TET7 coal. This implies that attention has to be paid to the combustion when co-firing above 15% of biomass with Tuzla coals at similar process conditions as used here, in order to keep combustion efficiency at an acceptable level.

For biomass co-firing with Kakanj coal TEK8, the increase of the unburnt carbon content (UBC) in the ash deposits collected in Lab-scale furnace is only minor, 0.0 to 0.02%. However, in the slag collected at the bottom of the furnace, the UBC is increased from 4.7% at 15% biomass co-firing, to over 8.48% at 25% biomass co-firing. This implies that attention has to be paid to the combustion when co-firing above 15% of biomass with coal at similar process conditions as used here, to keep combustion efficiency at an acceptable level.

### 3.2. Slagging

Evaluation of ash deposits is supported by visuals observation of the deposits (photographically and by microscope), as well as by chemical analysis of the deposits and deposition rate determination. So, multi-criteria assessment of slagging/deposition propensity has been applied, a method as reported in detail in [18]. Powdered, soft, hard and molten data points expressing different types of deposits can be identified, see for example in Figure 2 SEM images of ash deposits for different fuels. For coal alone ash deposits, see Figure 2a,b, one can notice melted ash spheres, indicating beginning of developed process of melting already at temperature of 1250 °C. It should be noted that melting is particularly expressed for ash deposit of mix of brown coal—Figure 2a, the coal which is more inclined to slagging than tested mix of lignite. This corresponded well with ash melting temperatures of the tested coal, see Table 2. From other side, one can notice that with adding woody biomass to mix of lignite, ash deposits becoming less melted, meaning the slagging propensity decreases, Figure 2c,d.

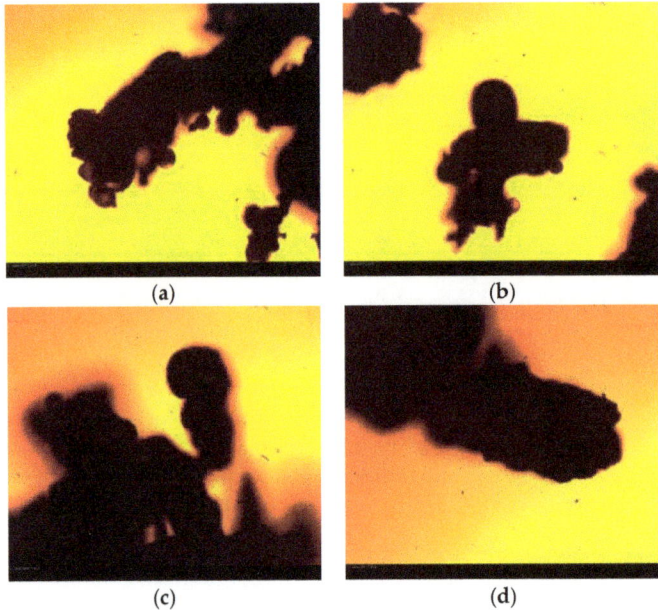

**Figure 2.** SEM images of ash deposits during co-firing of lignite mixture TET7 with woody biomass; (**a**) coal TEK6, (**b**) coal TET5, (**c**) TET5:WB = 93:7%wt, (**d**) TET5:WB = 85:15%wt.

For all of the tested fuels, ash deposition rate can be determined as a mass of the deposit divided by the deposition area and time of deposition. Furthermore, oxide determination and Ash Fusion Test (AFT) of the deposits, fly ash or slag are also done to provide information on alkali metals distribution as a function of the type of fuel and the process temperature, as well as to investigate ash fusion behaviors. Oxide determination of fly ash for selected co-firing mixtures are given in Figure 3.

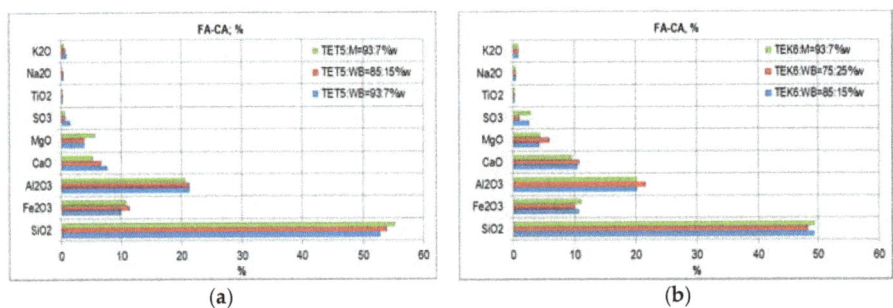

**Figure 3.** Fly ash—chemical analysis (FA-CA) for selected co-firing regimes (**a**) biomass co-firing with lignite TET5, (**b**) biomass co-firing with brown coal TEK6.

Based on the above presented multi-criteria assessment of ash deposits characteristics, final evaluation of slagging and fouling propensity for the given fuel and process conditions, is expressed in the following terms:

- Low slagging/fouling,
- Moderate slagging/fouling,

- Strong slagging/fouling,
- Very strong slagging/fouling.

Following aforementioned procedure, based on the presented results, evaluation of slagging for 12 co-firing regimes conducted is given in Table 3.

**Table 3.** Evaluation of slagging.

|  | No. | Temperature, °C | 1250 | 1450 |
|---|---|---|---|---|
|  |  | $\lambda_1/\lambda$ | 0.95/1.20 | 0.95/1.20 |
|  |  | Co-Firing Regime | Slagging Propensity | |
| **TET** | 1 | TET5:WB = 93:7%wt | Low | |
|  | 2 | TET5:WB = 85:15%wt | Low | |
|  | 3 | TET5:M = 93:7%wt | Low | |
|  | 4 | TET7:WB = 85:15%wt | Low to Medium | |
|  | 5 | TET7:WB:M = 80:13:7%wt | Low to Medium | |
|  | 6 | TET7:WB:M = 75:15:10%wt | Low to Medium | |
| **TEK** | 7 | TEK6:WB = 85:15%wt |  | Very Strong [1]/Medium [2] |
|  | 8 | TEK6:WB = 75:25%wt | Low to Medium | |
|  | 9 | TEK6:M = 93:7%wt | Low | |
|  | 10 | TEK8:WB = 75:25%wt | Low to Medium | |
|  | 11 | TEK8:WB:M = 85:8:7%wt | Low to Medium | |
|  | 12 | TEK8:WB:M = 75:15:10%wt | Low to Medium | |

[1] If considered to be used in Boiler with dry bottom furnace, [2] If considered to be used in Boiler with slag-tap furnace.

### 3.3. Emissions

In Figure 4a, results of $NO_x$ and $SO_2$ emissions during biomass co-firing with coal TET5 (lignite) are presented. Considering temperature of combustion of 1250 °C and sulphur content in the basic coal blend TET5 (S = 1.65%), $SO_2$ emissions are at expected level of 2500 mg/$m_n^3$ at 6% $O_2$ dry, which is comparable with $SO_2$ emissions on Tuzla power plant Unit 5. It can be noticed a slightly higher emissions of $SO_2$ when higher content of biomass is used, while Miscanthus slightly decreases $SO_2$ emissions as compared to woody biomass.

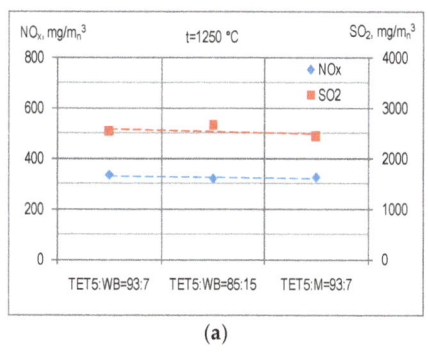 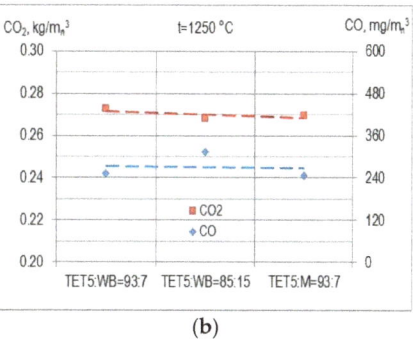

(a)  (b)

**Figure 4.** Emissions for different biomass co-firing with coal TET5, (**a**) $NO_x$ and $SO_2$ emissions (**b**) CO and $CO_2$ emissions.

Regarding $NO_x$ emissions, it can be noticed that those emissions are between 310 and 330 mg/$m_n^3$ at 6% $O_2$ dry, with a quite slight variation between three fuels tested. At this, a bit lower $NO_x$ emissions have been recorded for biomass co-firing blend TET5:WB = 85:15 as compared to other two fuels.

The same level of NO$_x$ emissions are on Tuzla power plant Unit 5 equipped with modern jet fuels in two rows and OFA system with ration between primary/secondary air against over fire air portion at 95:5—the same as used in these tests.

Higher CO emissions, as expected, have been recorded when co-firing coal TET5 with 15%wt of biomass as compared to 5% co-firing, see Figure 4b. This co-relates also with a bit lower NO$_x$ emissions in case of 15% co-firing, as shown in Figure 4a, which is phenomenon of NO$_x$ and CO correlation also reported by Wang [19], Li [20] and Rozendal [21]. During all co-firing test runs, lower NO$_x$ emissions were measured at the lower process temperatures, at the air staging and reburning, that is also reported in [20–22]. Despite of general decrease of NO$_x$ emissions recorded at higher biomass co-firing ratio, it was not, however, possible to identify clearly the correlation between the biomass content in the co-firing blend and NO$_x$ emissions during the tests.

In Figure 5a, results of NO$_x$ and SO$_2$ emissions during biomass co-firing with coal TET7 are presented. Considering temperature of combustion of 1250 °C and sulphur content in the basic coal blend TET7 (S = 1.57%), SO$_2$ emissions are at expected level of 2500 mg/m$_n^3$ at 6% O$_2$ dry, which is comparable with SO$_2$ emissions for biomass co-firing with coal blend TET5. It can be noticed that slightly higher emissions of SO$_2$ result when 25%wt of biomass is used as compared to 15% or 20% biomass co-firing. At this, it seems that miscanthus again slightly decreases SO$_2$ emissions as compared when woody biomass is used.

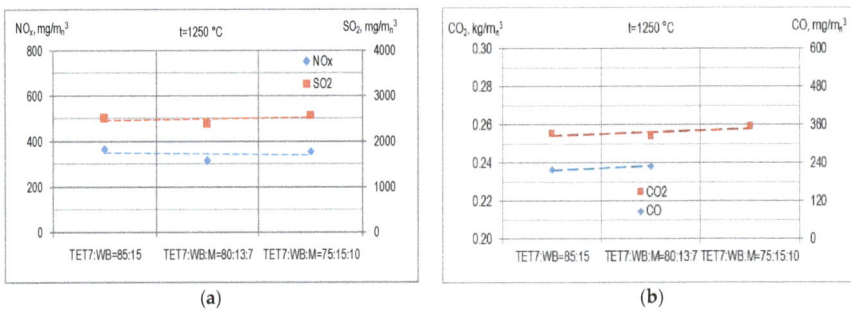

**Figure 5.** Emissions for different biomass co-firing with coal TET7, (**a**) NO$_x$ and SO$_2$ emissions (**b**) CO and CO$_2$ emissions.

Regarding NO$_x$ emissions, it can be noticed that emissions are now ranging between 310 and 375 mg/m$_n^3$ at 6% O$_2$ dry, with a bit wider variation between three fuels tested as compared to biomass co-firing with coal blend TET5. This can be explained with higher percentages of biomass being used in this case. At this, a bit lower NO$_x$ emissions have been recorded for 20% biomass co-firing as compared to other two fuels.

Higher CO emissions, as expected, have been recorded when co-firing coal TET7 with 20%wt of biomass as compared to 15% co-firing, see Figure 5b. This co-relates also with a bit lower NO$_x$ emissions in case of 20% co-firing, see Figure 5a. A bit lower CO$_2$ emissions are recorded, at level of 0.26 kg/m$_n^3$ as compared to case of TET5 based biomass co-firing (0.27 kg/m$_n^3$), see Figure 4b.

In Figure 6a, results of NO$_x$ and SO$_2$ emissions during biomass co-firing with coal TEK6 are presented, at temperature of combustion of 1450 °C. Considering that high process temperature and consequently lower bounding of sulphur to the alkalis from the ash, as well as pretty high Sulphur content in the basic coal blend TEK6 (S = 2.46%), SO$_2$ emissions are at expected level ranging from 6000 to 6400 mg/m$_n^3$ at 6% O$_2$ dry. This is some lower emissions of SO$_2$ as compared with SO$_2$ emissions in regular operation of Kakanj Unit 6 where SO$_2$ emissions are in the range of 7000–8000 mg/m$_n^3$ at 6% O$_2$ dry. At this, 25% biomass co-firing reduces SO$_2$ emissions as compared to 15% biomass co-firing, while it seems that Miscanthus again slightly decreases SO$_2$ emissions as compared when woody biomass is used.

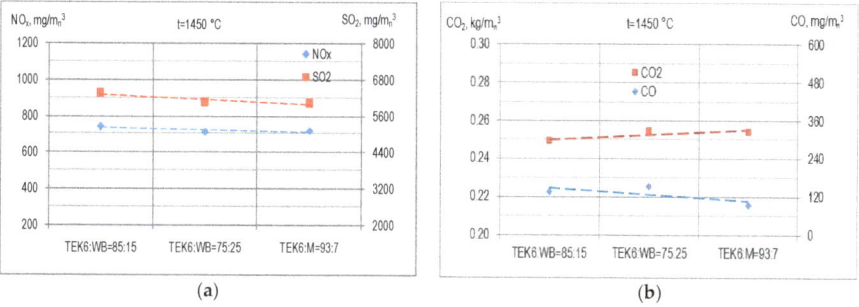

**Figure 6.** Emissions for different biomass co-firing with coal TEK6, (**a**) NO$_x$ and SO$_2$ emissions (**b**) CO and CO$_2$ emissions.

Regarding NO$_x$ emissions, it can be noticed that emissions are now in the range between 700 and 730 mg/m$_n$³ at 6% O$_2$ dry, which is slightly lower as compared to the NO$_x$ emission at Kakanj Unit 6 (750–850 mg/m$_n$³ at 6% O$_2$ dry) where the same swirl type burner as well 5% OFA ration is used as on Lab-scale furnace. This can suggest that biomass co-firing slightly decreased NO$_x$ emissions as compared to situation when coal alone is used.

Lower CO emissions (at 150 mg/m$_n$³ at 6% O$_2$ dry), as expected due to higher process temperature applied, have been recorded in this case as compared to biomass co-firing at lower temperature of 1250 °C. In the same time, lower CO emission is recorded when lower biomass percentage was used, see Figure 6b. This trend suits well to the trend of NO$_x$ emissions for the same fuel. A bit lower CO$_2$ emissions are recorded (0.25 kg/m$_n$³) as compared to TET5 and TET7 biomass co-firing, due to more convenient C/H ratio and higher calorific value of fuel in case of TEK6 biomass co-firing.

In Figure 7a, results of NO$_x$ and SO$_2$ emissions during biomass co-firing with coal TEK8 are presented, at temperature of combustion of 1250 °C. Considering specifically high content of the Sulphur in this fuel (S = 2.47%) and a lower process temperature with consequently better bounding of Sulphur to the alkalis from the ash, SO$_2$ emissions are at expected level ranging from 3500 to 4100 mg/m$_n$³ at 6% O$_2$ dry. This is much lower emissions of SO$_2$ as compared with SO$_2$ emissions in regular operation of Kakanj Unit 6 where SO$_2$ emissions are in the range of 7000–8000 mg/m$_n$³ at 6% O$_2$ dry, at a process temperature of 1450 °C.

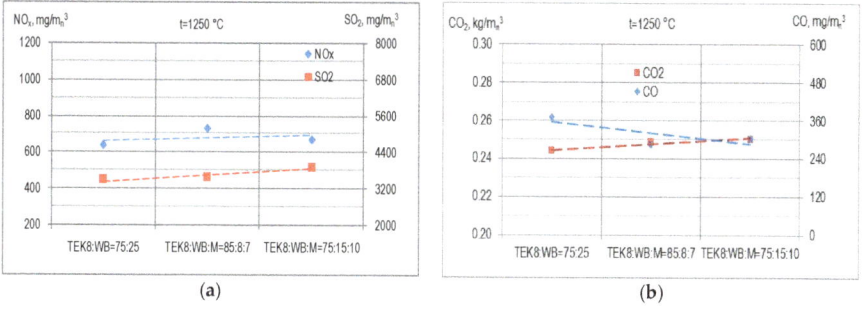

**Figure 7.** Emissions for different biomass co-firing with coal TEK8, (**a**) NO$_x$ and SO$_2$ emissions (**b**) CO and CO$_2$ emissions.

Regarding $NO_x$ emissions, it can be noticed that emissions are in the range between 630 and 710 mg/m$_n^3$ at 6% $O_2$ dry, which is lower as compared to the $NO_x$ emission at Kakanj Unit 6 (750–850 mg/m$_n^3$ at 6% $O_2$ dry) or in case of biomass co-firing with coal TEK8, due to lower process temperature used here. Additionally, it can be noticed that $NO_x$ emissions are lower for 25% biomass co-firing than in case 15% co-firing. This can suggest that an increase in biomass share produced slightly decreasing of $NO_x$ emissions in case of biomass co-firing with coal TEK8.

Double CO emissions, as expected due to lower process temperature applied, have been recorded in this case as compared to biomass co-firing for the similar coal and higher temperature of 1450 °C (case of biomass co-firing with TEK6). In the same time, lower CO emission is recorded when lower biomass percentage was used—case of 15% biomass co-firing, as compared to other two fuels with 25% biomass co-firing, see Figure 7b. This trend suits quite well to the trend of $NO_x$ emissions for the same fuel.

## 4. Conclusions

The given experimental results do show that there is reasonable expectation that tested coal/woody biomass/Miscanthus blends could be successfully co-fired in real operating conditions of PC Large Boiler, not producing any serious ash-related problem. Low or low-to-medium slagging propensity was noticed for all 12 co-firing case, at usual PC process temperature 1250 °C. The increase of the unburnt carbon content (UBC) in the ash deposits collected in Lab-scale furnace when increasing co-firing rate from 0.0 to 0.3 is only minor. However, in the slag collected at the bottom of the furnace, the UBC is increased for 0.15 and 0.2 biomass co-firing which implies that attention has to be paid to the combustion organization when co-firing above 15% of biomass at similar process conditions as used here, in order to keep combustion efficiency at an acceptable level.

In regard with emissions issue, it can be concluded that the results from performed biomass co-firing tests give some ground for optimism for achieving additional benefits, in particular for the co-firing with Miscanthus. While $SO_2$ emissions have been at expected level, it can be noticed a slightly higher emissions of $SO_2$ when higher content of biomass is used, while Miscanthus slightly decreases $SO_2$ emissions as compared to woody biomass. That is in good correlation with the gas temperature recorded during the tests, which related to phenomenon of Sulphur capture rate, considering an identified increasing of gas temperature when higher woody biomass was used, while higher Miscanthus co-firing rate decreased the gas temperature. Regarding $NO_x$ emissions, both for biomass co-firing with lignite and brown coal, it can be noticed slight variations of $NO_x$ emissions between the tested fuels, with generally lower emissions when biomass co-firing rate was increased. Higher CO emissions, as expected, have been recorded in co-firing coal with higher share of biomass, as compared to lower biomass co-firing rate. This co-relates also with a bit lower $NO_x$ emissions in case of higher biomass co-firing rate.

The results suggest that woody sawdust and Miscanthus co-firing with Bosnian low rank coal types shows promising effects at higher ratios for pulverized combustion.

**Acknowledgments:** Authors aknowladge with thanks to EPBiH power utility and VPC GmbH for their kind cooperation and support.

**Author Contributions:** Nihad Hodzic and Anes Kazagic conceived and designed the experiments; Nihad Hodzic and Sadjit Metovic performed the experiments; Nihad Hodzic, Sadjit Metovic and Anes Kazagic analyzed the data; Nihad Hodzic and Sadjit Metovic contributed reagents/materials/analysis tools; Anes Kazagic wrote the paper.

**Conflicts of Interest:** The authors declare no conflict of interest.

## References

1.    Kazagic, A.; Hodzic, N.; Metovic, S. Experimental Study on Co-Firing Low-Rank Coal and Woody Biomass with Harbacus Energy Crops Miscanthus in a Drop Tube Furnace. In Proceedings of the International Conference on Sustainable Development of Energy, Water and Environmental Systems—11th SDEWES, Dubrovnik, Croatia, 4–8 October 2017.

2.  KEMA. Technical status of biomass co-firing. In Proceedings of the IEA Bioenergy Task 32, Arnhem, The Netherlands, 11 August 2009.
3.  Lei, K.; Ye, B.; Cao, J.; Zhang, R.; Liu, D. Combustion Characteristics of Single Particles from Bituminous Coal and Pine Sawdust in $O_2/N_2$, $O_2/CO_2$, and $O_2/H_2O$ Atmospheres. *Energies* **2017**, *10*, 1695. [CrossRef]
4.  Kakaras, E. Low emission co-combustion of different waste wood species and lignite derived products in industrial power plants. In Proceedings of the XXXII Krafwerkstechnisches Colloquium: Nutzung Schwieriger Brennstoffe in Kraftwerken, Dresden, Germany, 24–25 October 2000; pp. 37–46.
5.  Aziz, M.; Budianto, D.; Oda, T. Computational Fluid Dynamic Analysis of Co-Firing of Palm Kernel Shell and Coal. *Energies* **2016**, *9*, 137. [CrossRef]
6.  Wischnewski, R.; Werther, J.; Heidenhof, N. Synergy Effects of the Co-combustion of Biomass and Sewage Sludge with Coal in the CFB Combustor of Stadtwerke Duisburg AG. *VGB PowerTech* **2006**, *86*, 63–70.
7.  Chen, W.; Wang, F.; Kanhar, A.H. Sludge Acts as a Catalyst for Coal during the Co-Combustion Process Investigated by Thermogravimetric Analysis. *Energies* **2017**, *10*, 1993. [CrossRef]
8.  Mikulcic, H.; von Berg, E.; Vujanovic, M.; Duic, N. Numerical Study of Co-firing Pulverized Coal and Biomass inside a Cement Calciner. *Waste Manag. Res.* **2014**, *32*, 661–669. [CrossRef] [PubMed]
9.  Wang, G.; Silva, R.B.; Azevedo, J.L.T.; Martins-Dias, S.; Costa, M. Evaluation of the combustion behaviour and ash characteristics of biomass waste derived fuels, pine and coal in a drop tube furnace. *Fuel* **2014**, *117*, 809–824. [CrossRef]
10. Kupka, T.; Mancini, M.; Irmer, M.; Weber, R. Investigation of ash deposit formation during co-firing of coal with sewage sludge, saw-dust and refuse derived fuel. *Fuel* **2008**, *87*, 2824–2837. [CrossRef]
11. Williams, A.; Jones, J.M.; Ma, L.; Pourkashanian, M. Pollutants from the combustion of solid biomass fuels. *Prog. Energy Combust. Sci.* **2012**, *38*, 113–137. [CrossRef]
12. Kazagic, A.; Smajevic, I. Synergy Effects of Co-firing of Woody Biomass with Bosnian Coal. *Energy* **2009**, *34*, 699–707. [CrossRef]
13. Leckner, B. Co-Combustion: A Summary of Technology. *Therm. Sci.* **2007**, *11*, 5–40. [CrossRef]
14. Hodzic, N.; Kazagic, A.; Smajevic, I. Influence of multiple air staging and reburning on $NO_x$ emissions during co-firing of low rank brown coal with woody biomass and natural gas. *Appl. Energy* **2016**, *168*, 38–47. [CrossRef]
15. Kazagic, A.; Smajevic, I. Experimental investigation of ash behaviour and emissions during combustion of Bosnian coal and biomass. *Energy* **2007**, *32*, 2006–2016. [CrossRef]
16. Maier, J.; Kluger, F.; Spliethoff, H.; Hein, K.R.G. Particle and emission behavior of raw and predried lignite in a 20 kW and in a 500 kW test facility. In Proceedings of the 23rd International Conference on Coal Utilization & Fuel Systems, Clearwater, FL, USA, 9–13 March 1998.
17. Wall, T.F.; Juniper, L.; Lowe, A. *State-of-the-Art Review of Ash Behaviour in Coal-Fired Furnaces*; ACARP Project C9055; The University of Newcastle: Callaghan, Australia, 2001.
18. Smajević, I.; Kazagić, A. Evaluation of Ash Deposits during Experimental Investigation of Co-firing of Bosnian Coal with Wooden Biomass. In Proceedings of the Tagungsband, Kunftiges Brennstoff- und Technologieportfolio in der Kraftwerkstechnik, 40. Kraftwerkstechnisches Kolloquium, Dresden, Germany, 14–15 October 2008; pp. 238–249.
19. Wang, X.; Tan, H.; Niu, Y.; Pourkashanian, M.; Ma, L.; Chen, E.; Liu, Y.; Liu, Z. Experimental investigation on biomass co-firing in a 300 MW pulverized coal-fired utility furnace in China. *Proc. Combust. Inst.* **2011**, *33*, 2725–2733. [CrossRef]
20. Li, S.; Xu, T.; Hui, S.; Wei, X. $NO_x$ emission and thermal efficiency of a 300 MWe utility boiler retrofitted by air staging. *Appl. Energy* **2009**, *86*, 1797–1803. [CrossRef]
21. Rozendaal, M. Impact of Coal Quality on NOx Emissions from Power Plants. Ph.D. Thesis, Delft University of Technology, Delft, The Netherlands, 1999.
22. Wang, Y.; Wang, X.; Hu, Z.; Li, Y.; Deng, S.; Niu, B.; Tan, H. NO Emissions and Combustion Efficiency during Biomass Co-firing and Air-staging. *BioResources* **2015**, *10*, 3987–3998.

Article

# Evaluation of Excess Heat Utilization in District Heating Systems by Implementing Levelized Cost of Excess Heat

**Borna Doračić \* [iD], Tomislav Novosel, Tomislav Pukšec and Neven Duić**

Faculty of Mechanical Engineering and Naval Architecture, Department of Energy,
Power and Environmental Engineering, University of Zagreb, Ivana Lučića 5, 10002 Zagreb, Croatia;
tomislav.novosel@fsb.hr (T.N.); tomislav.puksec@fsb.hr (T.P.); neven.duic@fsb.hr (N.D.)
\* Correspondence: borna.doracic@fsb.hr; Tel.: +385-1-6168-494

Received: 16 February 2018; Accepted: 5 March 2018; Published: 7 March 2018

**Abstract:** District heating plays a key role in achieving high primary energy savings and the reduction of the overall environmental impact of the energy sector. This was recently recognized by the European Commission, which emphasizes the importance of these systems, especially when integrated with renewable energy sources, like solar, biomass, geothermal, etc. On the other hand, high amounts of heat are currently being wasted in the industry sector, which causes low energy efficiency of these processes. This excess heat can be utilized and transported to the final customer by a distribution network. The main goal of this research was to calculate the potential for excess heat utilization in district heating systems by implementing the levelized cost of excess heat method. Additionally, this paper proves the economic and environmental benefits of switching from individual heating solutions to a district heating system. This was done by using the QGIS software. The variation of different relevant parameters was taken into account in the sensitivity analysis. Therefore, the final result was the determination of the maximum potential distance of the excess heat source from the demand, for different available heat supplies, costs of pipes, and excess heat prices.

**Keywords:** excess heat; levelized cost of excess heat; district heating; $CO_2$ emissions; heat demand mapping

## 1. Introduction

Security of energy supply and $CO_2$ emissions reduction have been recognized by the EU as the key topics that will define the development of its energy systems. For that reason, the utilization of highly efficient cogeneration with district heating systems should increase significantly, since these systems can greatly increase energy efficiency and reduce the $CO_2$ emissions of the energy sector. Currently, only 13% of the European heat supply is covered by district heating systems, which makes the potential for increasing this share significant, especially in urban areas which are characterised by high heat demand densities [1]. However, some northern countries, e.g., Sweden, already cover more than 50% of the residential and service sector heat demand with district heating [2], showing the way for the rest of the Europe. An analysis was conducted in Denmark as a case study, which examined the role of district heating systems in future renewable energy systems [3]. The primary conclusion was that the expansion of district heating to up to 70% of Danish net heat demand would be optimal. However, this could be limited by the uneven framework as shown in [4]. The expansion would result in significant fuel savings, reduction of $CO_2$ emissions, reduction of costs, as well as in better utilization of excess heat. Similarly, from the perspective of the consumers, the most important reasons for connecting to district heating are affordability, increased comfort, and the favourable environmental impact [5]. Prosumers, i.e., consumers who are at the same time producers of heat, will also have an important role in future district heating systems, as shown in [6] for Finland. This will facilitate the integration of renewable energy sources with these systems. The environmental benefit of district heating, combined with the implementation of renewables in other sectors, was shown in [7], providing detailed decarbonization scenarios by 2050 for Italy. These future district heating systems will be classified as fourth-generation district heating systems. They will incorporate low distribution temperatures, use of renewable energy sources and excess heat, use of large scale heat pumps and thermal storage, integration of the heat and electricity sectors, etc. [8]. Integrated with information and communication technologies, they will represent sustainable smart district heating systems as a part of the smart cities of the future [9]. The use of renewable energy sources in particular lowers both the environmental impact and the heat production costs in comparison to conventional district heating systems, as shown in [10]. Furthermore, low supply and return temperatures lower the losses in the distribution network, which are currently one of the biggest problems of the existing old systems, especially in Eastern Europe. This was presented in a study [11], which provided a comparative analysis between two district heating systems in Croatia and Denmark. It showed that because of the advanced age and high distribution temperatures in the Croatian system, heat losses are approximately three times higher than in the Danish system. However, the prerequisite for low temperature networks is the availability of adequate low temperature sources and their economic conditions, as shown for four cases in Austria [12]. Another way of reducing heat losses is the refurbishment of distribution pipes. Grid losses significantly influence the overall performance of district heating, as shown by data from several systems in Italy [13]. For that purpose, different designs of pipes can be considered, including twin pipes, asymmetrical insulation of twin pipes, double pipes, and triple pipes, which provide potential for energy savings [14]. Furthermore, an increase of insulation standards on pipes also facilitates heat savings. It was shown that the costs are still too high to implement the highest available standard, although it is expected to be feasible in the near future [15].

An interesting heat source for district heating systems is the excess heat from industrial facilities. A significant amount of energy used in industry is currently being wasted, as shown in the case of China, where these losses amount to at least 50% [16]. Moreover, research has shown that there is enough excess heat in the EU to cover the heat demands of all buildings from the service sector and households [17]. Furthermore, an analysis of these sources has been made for the EU-27 [18]. The main conclusion of this research is that the potential for implementation of excess heat in district heating systems is significant, but it is currently not being used. Similar studies have been carried out for various excess heat sources concerning different frameworks, for example analyses of excess heat utilization from thermal power plants in the EU-28 [19], industrial excess heat utilization in China [20], excess heat utilization from the petrochemical industry on the west coast of Sweden [21], and excess heat utilization in Japan [22]. Based on the methods from [18], authors in [23] made an analysis of various excess heat sources in Denmark. The focus was on their utilization in district heating systems, using heat pumps in order to increase the temperature level. Their results showed that these sources are often located far from potential consumers, i.e., heat demands, and therefore further research is required in this area. A similar conclusion was drawn in [24], where authors analyzed the potential for excess heat utilization in district heating systems in Great Britain. It is concluded that in the case of remote locations of these sources, it is not economically viable to utilize them in a district heating network. Some researchers are trying to tackle this problem. For example, the use of mobile heat storage units is proposed in [25]. These units are charged at the site of excess heat sources, transported by a train or a truck to the location of heat demand, and then discharged. Another concept that is being researched is the novel heat allocator concept, which is a combination of a heat engine, a heat exchanger, and a heat pump [26].

A number of studies regarding the economics of excess heat utilization in district heating have already been performed, showing its benefits. This was the focus of [27], where authors provided a system analysis of this source for a case in Sweden. The study implemented a model for minimizing system costs, and it was shown that excess heat is a feasible solution in all the investigated energy market scenarios. Similarly, other research [28] highlighted the economic and environmental advantages of utilizing excess heat in district heating, which also significantly increased the production of jointly operated cogeneration units. Moreover, the optimal contribution of excess heat from industrial facilities has been studied in [29], where the authors developed a method for determining the investment costs of its utilization from a cluster of industrial facilities. The impact of excess heat utilization in a district heating system on $CO_2$ emissions and the energy system as a whole has been studied for a region in Sweden [30]. The research showed that introducing excess heat into the energy system would reduce the use of fossil fuels and therefore the environmental impact of the energy system, although this is highly case-dependent.

A good criterion for the economic evaluation of energy production technologies is the levelized cost of energy, which takes into account all the cashflows during the lifetime of a plant. Numerous research studies have already been carried out by implementing the levelized cost of electricity calculations. Recently, it has been used in [31] and supplemented by including uncertainty and endogeneities in input parameters for analyzing the economic feasibility of gas and nuclear power plants, showing much higher feasibility for gas power plants. Furthermore, in [32] it has been used to analyze the feasibility of a solar chimney power plant, proving its competitiveness against other renewable power production technologies. However, a significantly smaller amount of research has been carried out in the heat sector by implementing the levelized cost of heat method, with most papers focusing on the calculation of the total costs, as shown for a building in [33]. The levelized cost of heat has been used for example in [34] for determining the feasible level of heat savings and heat production on the European level, in [35] for the Fresnel solar system, and in [36] for co-firing solid, liquid, and gaseous fuel in a heat-only boiler, but none of these papers include excess heat in the analyses. One of the main parameters in the calculation of the levelized cost of excess heat will be the procurement cost of excess heat, which has been analyzed in [37] for excess heat from data centers, while taking into account a scenario with the possibility of a heat market. The potential for heat market implementation, i.e., third-party access, has also been discussed in [38], giving some basic comments on its benefits for excess heat utilization in district heating systems.

This paper presents the continuation of the research conducted in [39], which provided the analysis of excess heat utilization in a district heating system in a small rural city. The concept proved to be feasible; however, the analysis considered only the potentially available excess heat supply, and no other parameters were taken into account. Therefore, this research has been expanded as described in the next few lines. In this paper, heat demand mapping has been utilized in order to provide the analysis of the feasibility of a natural gas district heating implementation for a small city. This way, both the environmental and economic advantages of this system over individual heating solutions are demonstrated. The analysis further includes potential excess heat utilization, taking into account its distance from the heat demand. The novelty of this study is the utilization of the levelized cost of excess heat method. The method is validated by performing a case study for the city of Ozalj, a small city in Croatia.

## 2. Materials and Methods

The method consists of two main steps: heat demand mapping and feasibility analysis by implementing the novel levelized cost of excess heat method. In the next sections, a more detailed description of the aforementioned steps will be provided.

### 2.1. Heat Demand Mapping

In order to assess the heat demand of the city of Ozalj and therefore provide the input for the scenario analysis, heat demand mapping was performed. A similar geographic information system analysis has already been done in [40], providing the potential for district heating expansion. However, mapping is not the focus of this paper but only provides the required input for further analysis.

For that purpose, Matlab [41] and QGIS [42] software were used. The data used in the process of mapping were mostly public in order to facilitate the replication of the method. The method was also complemented with the results from a survey carried out in Ozalj [43]. The questionnaire was developed as a part of the CoolHeating project [44], and the questions were specifically designed to collect good quality data from the citizens, in order to assess their heating needs and gather ideas, suggestions, and doubts for connecting to a district heating system. In order to get more precise energy consumption patterns, information was gathered both on the building stock (i.e., age of buildings, type of windows, insulation, net heating area, heating system, etc.) and on the annual fuel consumption. On the basis of this information, the heat demand of each surveyed household was calculated.

In order to better utilize these data for further analysis, the buildings were divided into eight categories with associated specific heat demands. The categories were determined by visually inspecting surveyed buildings and aggregating data from similar buildings into a specific category. This method is suited for smaller municipalities and can provide very detailed and more accurate heat demand maps, both on the building and on the aggregated level. However, when analyzing heat demands of larger areas, this method would not be appropriate since it would require too much time to carry out the survey. Specific heat demands for eight categories of buildings in the city of Ozalj are shown in Table 1. It has to be pointed out that the values for office building, public building, industry, and historic building have been taken from the city's Sustainable Energy Action Plan [45] because of the insufficient data for these categories. Specific heat demands of some categories deviate significantly from the mean values, as shown in Table 1. This is specifically the case for a house without insulation, since this category includes all the houses without any insulation on the outer walls. Therefore, the heat losses are the highest in this category. The survey was carried out in 391 households, which represents a share of 17% of the overall number of households in Ozalj. The results from Table 1 clearly show the status of energy consumption of building stock in the continental part of Croatia. These can also be applied to the whole region of southeastern Europe, because of the similar characteristics in this sector. Such high heat demands are the result of the relatively old age of buildings and low rates of refurbishment, with more than 50% of the surveyed households having no outer wall and roof insulation at all.

**Table 1.** Building categories and associated specific heat demands.

| Category | Number of Buildings Analyzed in the Survey | Specific Heat Demand (kWh/m$^2$) | Standard Deviation from the Mean Specific Heat Demand (159.2 kWh/m$^2$) |
|---|---|---|---|
| Old house | 241 | 177.75 | 18.56 |
| New house | 12 | 112.5 | −46.69 |
| House without insulation | 28 | 262.5 | 103.31 |
| Apartment building | 21 apartments | 161.25 | 2.06 |
| Office building | - | 135 | −24.19 |
| Public building | - | 270 | 110.81 |
| Historic building | - | 78.75 | −80.44 |
| Industry | - | 110 | −49.19 |

The heat demand mapping conducted for this research consisted of four main steps. These can be divided as follows:

1. The first step was to create a matrix in Matlab that contained information on the total gross area and locations of buildings from the Croatian online building census Geoportal [46].
2. In the second stage, the buildings were classified into eight categories according to their purpose and condition, in order to allocate their specific heat demands.
3. At the same time, data on the number of floors were collected by visually inspecting all the households in the analyzed area. This could be done by using free online tools like Google Earth, etc. Both the categories and the number of floors were added to the initial matrix by color coding.

4.   Afterwards, the final heat demand matrix was created by multiplying the total gross areas of the buildings with the associated specific heat demands. This final matrix was then transferred into a geographic information system interface using the QGIS tool.

The main steps are presented graphically in Figure 1, which gives an overview of the building locations map, category map, number of floors map, and, finally, the heat demand map on the 100 × 100 m level for the selected location. The final step, i.e., the GIS map is presented in the results section.

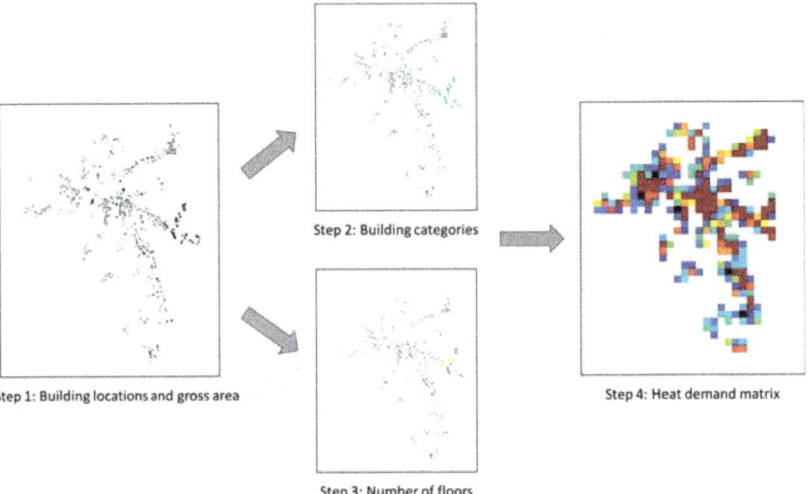

**Figure 1.** Graphical representation of the four main steps in the heat demand mapping method.

*2.2. Scenario Analysis*

In order to determine the feasibility and the environmental impact of district heating system implementation in a small rural city, different scenarios were developed. Microsoft Excel and QGIS software were used for the calculations. First, the implementation of a natural gas district heating system was analyzed in order to point out the advantages of such a system over individual heating solutions. The effect of excess heat utilization on the system costs was also researched by implementing the levelized cost of excess heat method, as described in more detail in the following paragraphs. Finally, a sensitivity analysis was implemented, taking into account various relevant parameters.

2.2.1. Implementation of a Natural Gas District Heating System

Feasibility calculations of the proposed scenarios were done on the level of aggregated 100 × 100 m heat demand areas. Furthermore, by using the cost data from [47], the levelized cost of heat was calculated for a potential natural gas district heating system, as shown in (1):

$$\text{LCOH} = \frac{I_c \cdot CRF \cdot (1 - TD_{pv})}{8760 \cdot i \cdot (1 - T)} + \frac{O_{total}}{8760 \cdot i} + c_{fuel} \quad [\text{€/kWh}] \qquad (1)$$

where $I_c$ is the capital cost of the production facility [€/kW], $CRF$ is the capital recovery factor which discounts the investment, $T$ is the tax rate, $D_{pv}$ is the present value of depreciation taken from [48], $i$ is the capacity factor of the production facility, $O_{total}$ are the total operation and maintenance costs [€/kW], and $c_{fuel}$ is the cost of the fuel being used [€/kWh].

Besides the heat production facility, the distribution network has also to be taken into account when calculating the feasibility of district heating implementation. The average specific network length could be calculated by dividing the total length of roads by the number of 100 × 100 m areas in the analyzed location. This could be used to calculate the cost of the district heating network installation in every 100 × 100 m area. The technical and cost data of the distribution network and the price of heat were taken from [49]. By integrating these data in the QGIS software (version 2.18.7) and using (2), the feasibility of the district heating system implementation was calculated for every 100 × 100 m heat demand area:

$$R = Q \cdot c_h - Q \cdot LCOH - 10000 \cdot l_s \cdot c_p \ [\text{€}] \tag{2}$$

where $R$ is the potential revenue for a district heating system in a single 100 × 100 m area [€], $Q$ is the heat demand of a single 100 × 100 m area [kWh], $c_h$ is the price of heat [€/kWh], $l_s$ is the average specific distribution network length [m/m²], while $c_p$ is the cost of the distribution network installation [€/m].

The price of heat for the final consumers is a crucial parameter in this kind of analysis since it determines the revenues from the district heating system, thus having a major influence on the feasibility of the whole system. It is accounted for in (2). In Croatia, the heat price for the final consumer is defined by every individual district heating system operator. It is then approved by the Croatian Energy Regulatory Agency. Since the analyzed city of Ozalj currently only uses individual heating systems, the price of the heat was assumed to be the same as for the district heating system in the nearby city of Karlovac, i.e., 66.6 €/MWh. However, this price also includes the connection fee.

Areas where $R > 0$ are feasible for district heating implementation. The calculation in QGIS provided the map with highlighted parts of the city which can be connected to a district heating system. The outputs of this analysis included total heat demand, total area of households, and number of 100 × 100 m areas for which it would be feasible to implement a district heating system.

These data were further used to examine the environmental impact of a natural gas district heating system compared with the existing individual heating systems. The analysis is based on the $CO_2$ emissions calculation, as well as local particulate matter (PM), CO, and $NO_x$ emissions calculations. The shares of different energy sources, which are currently used in the analyzed city, could be determined by using the data from the survey and the city's Sustainable Energy Action Plan. In the current situation, around 40% of the final heat demand is supplied by individual logwood furnaces, and another 40% by individual fuel oil boilers. The results of the survey showed that these are mostly old and inefficient boilers, causing a high environmental impact on the local level. Taking into account the low efficiency of old boilers and the high efficiency of district heating boilers, the emissions were calculated by multiplying the demand for each fuel with the respective emission factors.

### 2.2.2. Integrating Excess Heat into the District Heating System

When compared to conventional individual heating solutions, district heating already has significant advantages, both from the economic and the ecological point of view. However, integrating excess heat into a district heating system can provide further benefits, since there are no fuel costs for this source, and the environmental impact is even lower because this heat would otherwise be wasted, and the emissions related to its production would be existent anyway. Therefore, in the second scenario, a part of heat production from natural gas district heating was substituted by excess heat in order to analyze its effect on the overall system. Industrial and other facilities with high amounts of excess heat are often located outside cities, far from the heat demand. Consequently, a significant part of the investment into excess heat utilization is the distribution network which needs to be built in order to transport the heat from the source to the existing demand. The other, less capital-intensive investment is the cost of heat exchangers. It is assumed in this analysis that the temperature level of the available excess heat source is high enough for direct utilization. However, these sources often have low temperatures, especially if the heat is from the service sector. In these cases, heat pumps are needed in order to increase the temperature level of the heat. Low-grade excess heat has a particularly high potential in low-temperature fourth-generation district heating systems and should not be neglected.

In this scenario, the levelized cost of heat method was modified in order to serve as a criterion for investment into the excess heat utilization equipment. As mentioned above, in many cases, these sources are located far from the heat demand, and therefore this scenario includes a calculation of the maximum feasible distance of the potential excess heat source, taking into account different quantities of the available excess heat in the area. This way, both the investment into the heat exchangers and the distribution network are included in the analysis. The modified levelized cost of excess heat was calculated by using (3):

$$\text{LCOEH} = \frac{I_{HE} \cdot CRF \cdot (1 - TD_{pv})}{8760 \cdot i \cdot (1 - T)} + \frac{O_{HE,total}}{8760 \cdot i} + c_{excess\ heat} \quad [\text{€/kWh}] \tag{3}$$

where $I_{HE}$ is the investment cost for the heat exchangers [€/kW], $O_{HE,total}$ are the operation and maintenance costs for the heat exchangers [€/kW], and $c_{excess\ heat}$ is the cost of excess heat [€/kWh].

When calculating the levelized cost of excess heat, the cost for the installation of the distribution network is not included in the equation, since it is accounted for in Equation (4). This is done in order to calculate the maximum potential distance of the heat source from the demand, i.e., the extra revenue which can be used to finance the construction of the distribution network. Therefore, the investment and operation and maintenance costs in (3) only cover the heat exchangers, which are used to extract the excess heat from the source. The cost of excess heat includes the procurement costs, which are defined by the operator of the excess heat facility and agreed with the operator of a district heating system. Different values of excess heat price were analysed in the sensitivity analysis, as shown in Table 3.

Furthermore, the extra revenue was calculated for different values of available excess heat, by using (4). Then, this extra revenue was divided by the discounted cost of pipes in order to determine the maximum distance of the excess heat source from the heat demand:

$$R_{EH} = E_{total} \cdot r_{heat} - (E_{EH} \cdot LCOEH + E_{DH} \cdot LCOH) - l \cdot n \cdot c_{pipes} \quad [\text{€}] \tag{4}$$

where $R_{EH}$ is extra revenue, $E_{total}$ is the total heat demand of the area for which it would be feasible to establish a connection to a natural gas district heating system [kWh], $r_{heat}$ is the revenue from heat, i.e., the price of heat [€/kWh], $E_{EH}$ is the available excess heat [kWh], $LCOEH$ is the levelized cost of excess heat [€/kWh], $E_{DH}$ is the remaining heat demand being covered by the natural gas-based production facility of the district heating system [kWh], $LCOH$ is the levelized cost of heat for the natural gas district heating system [€/kWh], $l$ is the average length of the distribution network in a $100 \times 100$ m area [m], $n$ is the number of $100 \times 100$ m areas, and $c_{pipes}$ is the discounted cost of pipes [€/m].

Since numerous parameters affect the feasibility of excess heat utilization, a sensitivity analysis was made by changing the values of available excess heat, costs of pipes, and cost of excess heat.

## 3. Results

In this section, the main results of this paper, including heat demand mapping, feasibility analysis of switching from individual systems to natural gas district heating, and feasibility analysis of excess heat utilization in district heating systems, are presented and discussed.

The results of the first step, i.e., the heat demand mapping, can be seen in Figure 2. It shows that the areas with the highest heat demand densities are located around the city centre and the industrial zone, which is expected since most of the public and apartment buildings are situated in that part of the city. The final heat demand of the city amounts to 90.92 GWh. Apart from $100 \times 100$ m areas, the heat demand was mapped on the building level as well, therefore providing a more detailed insight into the current building stock of the city. This also showed the locations of the biggest heat consumers with the highest potential for connecting to a district heating system, thus providing important information in the planning process.

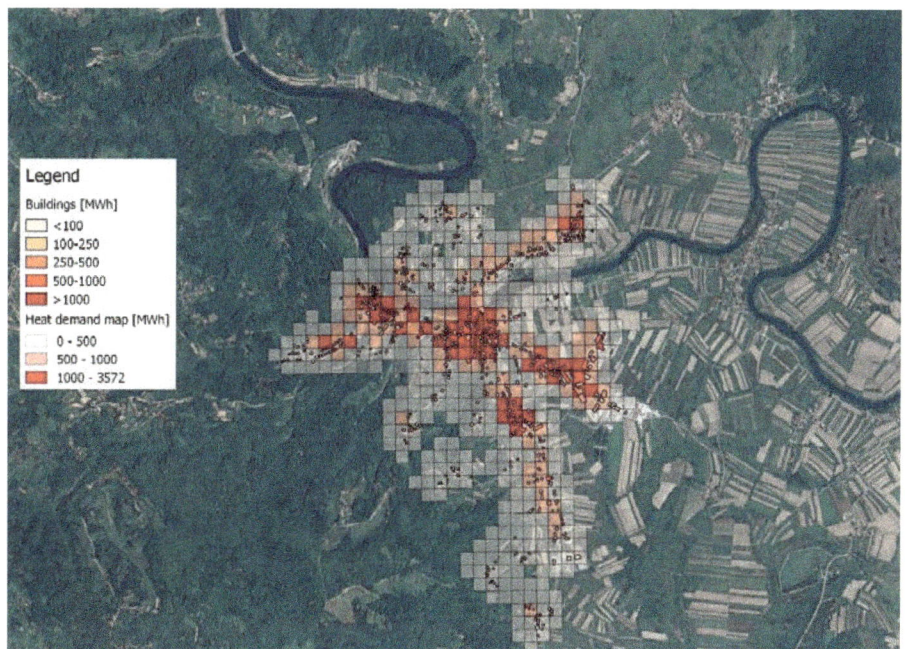

**Figure 2.** Heat demand map of the city of Ozalj, including the heat demands of each building.

By using the heat demand data from the aggregated 100 × 100 m areas, the share of demand which could be feasibly covered by a district heating system was calculated. Then, all the remote areas were excluded from further analysis, as shown in Figure 3. This was done because the real pipe length in these cases would be much higher, since the average specific distribution network length used in the calculations did not include the distance between the feasible 100 × 100 m areas. However, the average specific distribution network length could be applied in the final selected area from Figure 3, since most of the 100 × 100 m areas are connected or very near each other. The main results of this analysis can be seen in Table 2. They show that it would be feasible to cover 83.3% of the existing heat demand in the city by a natural gas district heating system, providing households with an inexpensive and comfortable way of heating.

On the basis of these results, the potential for excess heat utilization in the analysed system was calculated, as described in the Methods section. The main outcome of this analysis was the maximum distance of the excess heat source from the demand for different excess heat prices and costs of pipes. The latter is an important parameter since it presents the highest investment for a system utilizing a remote excess heat source. This cost also includes digging and the laying of pipes. The different costs of pipes and the prices of excess heat used in the analysis are shown in Table 3. All the variations of these parameters were analysed in the sensitivity analysis and presented in a form of a graph.

**Table 2.** Main results of the district heating implementation feasibility analysis.

| Heat Demand (MWh) | 75,383.00 |
|---|---|
| Gross household area (m$^2$) | 357,674.00 |
| Number of 100 × 100 m areas | 92 |

**Figure 3.** Parts of the city for which it is feasible to establish a connection to a district heating system (orange) and final area selection used in further analyses (red).

**Table 3.** Different excess heat prices, costs of pipes, and available excess heat supply used in the analysis of excess heat utilization.

| Excess Heat Price [€/MWh] | Cost of Distribution Pipes [€/m] | Available Excess Heat Supply (GWh) |
|---|---|---|
| 1 | 200 | 10 |
| 2 | 400 | 20 |
| 3 | 600 | 30 |
| 4 | 800 | 40 |

The results of the analysis can be seen in Figure 4. This figure shows that the maximum feasible distance of the excess heat source from the heat demand rose with the amount of available excess heat, as expected. However, when the excess heat price was increased, the maximum potential distance of the source decreased. This was also the case with the increasing costs of pipes. Nevertheless, all the variations of the important parameters resulted in a feasible integration of excess heat in a natural gas district heating system. The results showed that the levelized cost of excess heat method can be used as an efficient way of analyzing the feasibility of excess heat utilization in district heating systems, therefore serving as a criterion for the investment into excess heat utilization equipment.

This shows the great potential of this source, but also its limitations regarding the location of the source and its distance to the heat demand. The maximum potential distance varies significantly with different values of the relevant parameters. Therefore, it changed from 23.11 km in the case of 40 GWh available excess heat supply, at the price of 1 €/MWh and pipe cost of 200 €/m, to 2.7 km in the case of 10 GWh available excess heat supply, at the price of 4 €/MWh and pipe cost of 800 €/m. This showed that in the cases in which there is a high availability of excess heat, this excess heat could be utilized from various locations outside the analyzed city and even from larger cities in its vicinity.

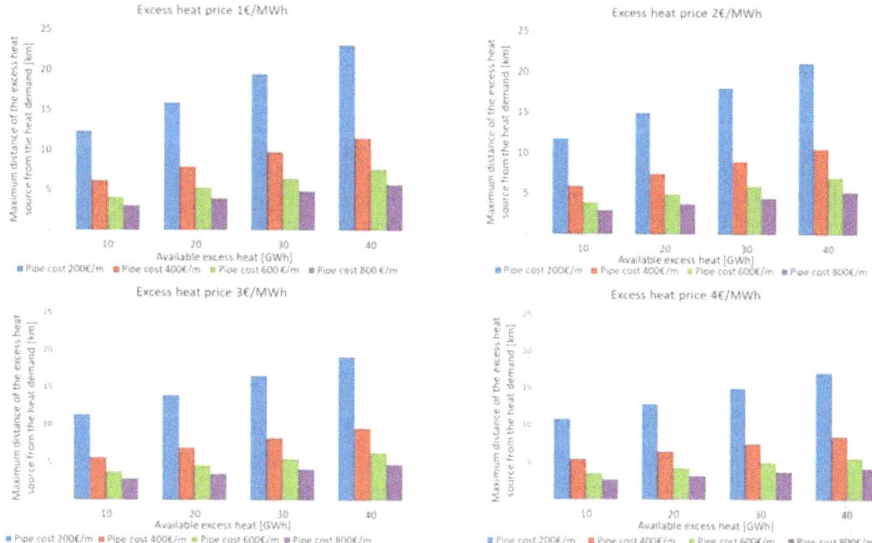

**Figure 4.** Maximum distance of the excess heat source from the heat demand for different values of available excess supply, excess heat price, and costs of pipes.

Finally, the district heating systems also provide significant environmental benefits due to the high efficiency of boilers and to the strict regulations regarding their pollutant emissions. The results of the $CO_2$ emissions analysis support this hypothesis by providing the $CO_2$ emissions savings achieved by switching from individual heating solutions to a natural gas district heating system, as shown in Figure 5. Even though the analysed district heating system uses natural gas as a fuel, its emissions were still lower than in the current situation, because of the aforementioned reasons. However, more significant benefits were achieved by reducing the PM, $NO_x$, and CO emissions, which are currently substantial because of a high share of old and inefficient logwood boilers without a filtration system. These have a much higher local impact on the environment. Their values were calculated and are presented in Table 4. The highest reductions were achieved for PM emissions, which were almost completely eliminated by introducing a natural gas district heating system. Furthermore, $NO_x$, and CO emissions were also substantially reduced, by 87% and 97%, respectively.

Additionally, when excess heat is integrated into the system, significantly higher $CO_2$ emission savings can be achieved. Figure 5 shows that in the case of 40 GWh of excess heat supply, the $CO_2$ emissions were around 50% of the emissions in the current situation. This is due to the fact that the emissions from the excess heat production facilities are already existent and are calculated in the industrial or service sectors, depending on the origin of excess heat. Therefore, the analysis included only the emissions from the part of the district heating system supplied by the natural gas boiler, significantly lowering the overall environmental impact of the system and the heating sector in general. This was also proven by analysing PM, $NO_x$, and CO emissions for the case of natural gas district heating plus 40 GWh excess heat, where all the emissions were reduced by more than 93% in comparison to the current situation.

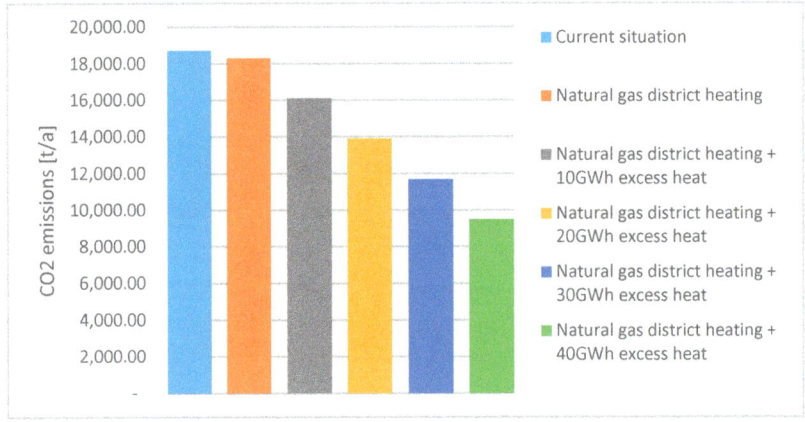

**Figure 5.** Results of $CO_2$ emission analysis for different cases.

**Table 4.** $NO_x$, PM, and CO emissions for different cases.

|  | Current Situation | Natural Gas District Heating | Natural Gas District Heating + 40 GWh Excess Heat |
|---|---|---|---|
| $NO_x$ emissions (kg/a) | 25,783.24 | 3292.07 | 1707.62 |
| PM emissions (kg/a) | 1,331,938.62 | 29.93 | 15.52 |
| CO emissions (kg/a) | 2,153,771.65 | 70,013.02 | 36,316.36 |

## 4. Discussion and Concluding Remarks

The idea for this paper was twofold. On the one hand, its purpose was to show the economic and environmental benefits of a district heating system implementation in a city which is currently using only individual heating solutions. On the other hand, the novel approach towards analyzing the feasibility of excess heat integration into a district heating system was proposed. The case study was the city of Ozalj, a small city with no existing district heating systems. The prerequisite step in the energy planning of district heating systems is the heat demand mapping of the focus area. This way, parts of the city in which it is feasible to implement a district heating system were determined. In this case, the final heat demand of an area which could be feasibly covered by a natural gas district heating system was 75,383 MWh. These results were then used for further analyses, taking into account the excess heat utilization in a natural gas district heating system.

First, the levelized cost of excess heat was calculated for this source, being significantly lower than for other heat production technologies because of its low investment costs and the lack of fuel costs. It has to be noted that the temperature level of the source was assumed to be high enough, and therefore no heat pumps were needed. Consequently, the only investment cost was for the heat exchangers. However, these sources are rarely located in the vicinity of the potential heat demand, and an investment into additional distribution pipes is necessary. In that case, these are the highest costs related to excess heat utilization. Therefore, by implementing the levelized cost of excess heat method, the maximum distance of the source from the heat demand was calculated, that way taking into account both the investment into heat exchangers and the distribution network. Three different parameters were varied in order to perform the sensitivity analysis: available excess heat supply, costs of pipes, and excess heat price. The maximum feasible distance of the excess heat source from the demand was 23.11 km in the case of 40 GWh available excess heat supply, at the price of 1 €/MWh and pipe cost of 200 €/m. On the other hand, the minimum feasible distance of the heat source from the demand was 2.7 km in the case of 10 GWh available excess heat supply, at the price of 4 €/MWh and pipe cost of 800 €/m. The results showed that excess heat is a feasible solution in all cases, but this is highly dependent on the available excess heat supply and on its distance from the heat demand. Furthermore, it can be concluded that the levelized cost of excess heat method can be used as a criterion for the investment into excess heat utilization equipment.

The final aspect analyzed in this paper was the environmental impact of a district heating system, in relation to individual heating systems. The analysis was conducted both for the $CO_2$ emissions and the $NO_x$, PM, and CO emissions, which have a higher influence on the local level. The analysis showed that a natural gas district heating system already has lower $CO_2$ emissions than individual solutions. Further benefits are achieved as the result of significantly lower local $NO_x$, PM, and CO emissions from highly efficient district heating boilers, while many individual biomass furnaces are old and do not have the necessary filtration system. Thus, reductions of more than 87% were achieved for all local emissions by switching to a natural gas district heating system. $CO_2$ emissions were drastically reduced if excess heat was additionally introduced into the system. This heat is already being wasted, and the emissions from its production can be allocated to the industrial or service sectors, depending on its origin. Therefore, when being utilized in a district heating system, this heat does not contribute to the emissions of a heating sector. If 40 GWh of excess heat supply is available, the $CO_2$ emissions of a district heating system are 50% lower than for individual heating solutions.

**Acknowledgments:** Financial support from the RESFLEX project funded by the Programme of the Government of Republic of Croatia for encouraging research and development activities in the area of Climate Change from 2015 to 2016, as well as support from the European Union's Horizon2020 project CoolHeating (grant agreement 691679) are gratefully acknowledged.

**Author Contributions:** Borna Doračić performed data collection and all the analysis regarding the district heating implementation and excess heat integration. He also performed the analysis of the $CO_2$ emissions for all the considered configurations. Tomislav Novosel performed the heat demand mapping in QGIS software. Tomislav Pukšec and Neven Duić supported and supervised these activities. All the authors contributed, read, and checked the paper.

**Conflicts of Interest:** The authors declare no conflict of interest.

## References

1. Connolly, D.; Lund, H.; Mathiesen, B.V.; Werner, S.; Möller, B.; Persson, U.; Boermans, T.; Trier, D.; Østergaard, P.A.; Nielsen, S. Heat roadmap Europe: Combining district heating with heat savings to decarbonise the EU energy system. *Energy Policy* **2014**, *65*, 475–489. [CrossRef]
2. Werner, S. District heating and cooling in Sweden. *Energy* **2017**, *126*, 419–429. [CrossRef]
3. Lund, H.; Möller, B.; Mathiesen, B.V.; Dyrelund, A. The role of district heating in future renewable energy systems. *Energy* **2010**, *35*, 1381–1390. [CrossRef]
4. Grundahl, L.; Nielsen, S.; Lund, H.; Möller, B. Comparison of district heating expansion potential based on consumer-economy or socio-economy. *Energy* **2016**, *115*, 1771–1778. [CrossRef]
5. Ahvenniemi, H.; Klobut, K. Future Services for District Heating Solutions in Residential Districts. *J. Sustain. Dev. Energy Water Environ. Syst.* **2014**, *2*, 127–138. [CrossRef]
6. Paiho, S.; Reda, F. Towards next generation district heating in Finland. *Renew. Sustain. Energy Rev.* **2016**, *65*, 915–924. [CrossRef]
7. Calise, F.; D'Accadia, M.D.; Barletta, C.; Battaglia, V.; Pfeifer, A.; Duic, N. Detailed Modelling of the Deep Decarbonisation Scenarios with Demand Response Technologies in the Heating and Cooling Sector: A Case Study for Italy. *Energies* **2017**, *10*, 1535. [CrossRef]
8. Lund, H.; Werner, S.; Wiltshire, R.; Svendsen, S.; Thorsen, J.; Hvelplund, F.; Mathiesen, B.V. 4th Generation District Heating (4GDH). *Energy* **2014**, *68*, 1–11. [CrossRef]
9. Sayegh, M.A.; Danielewicz, J.; Nannou, T.; Miniewicz, M.; Jadwiszczak, P.; Piekarska, K.; Jouhara, H. Trends of European research and development in district heating technologies. *Renew. Sustain. Energy Rev.* **2016**, *68*, 1183–1192. [CrossRef]
10. Mikulandrić, R.; Krajačič, G.; Duić, N.; Khavin, G.; Lund, H.; Mathiesen, B.V. Performance Analysis of a Hybrid District Heating System: A Case Study of a Small Town in Croatia. *J. Sustain. Dev. Energy Water Environ. Syst.* **2015**, *3*, 282–302. [CrossRef]

11. Čulig-Tokić, D.; Krajačić, G.; Doračić, B.; Mathiesen, B.V.; Krklec, R.; Larsen, J.M. Comparative analysis of the district heating systems of two towns in Croatia and Denmark. *Energy* **2015**, *92*, 435–443. [CrossRef]

12. Basciotti, D.; Schmidt, R.R.; Meissner, E.; Doczekal, C.; Giovannini, A. Low temperature district heating in Austria: Energetic, ecologic and economic comparison of four case studies. *Energy* **2016**, *110*, 95–104.

13. Noussan, M. Performance indicators of District Heating Systems in Italy—Insights from a data analysis. *Appl. Therm. Eng.* **2018**, *134*, 194–202. [CrossRef]

14. Rosa, A.D.; Boulter, R.; Church, K.; Svendsen, S. District heating (DH) network design and operation toward a system-wide methodology for optimizing renewable energy solutions (SMORES) in Canada: A case study. *Energy* **2012**, *45*, 960–974. [CrossRef]

15. Lund, R.; Mohammadi, S. Choice of insulation standard for pipe networks in 4th generation district heating systems. *Appl. Therm. Eng.* **2016**, *98*, 256–264. [CrossRef]

16. Lian, H.K.; Li, Y.; Shu, G.Y.Z.; Gu, C.W. An overview of domestic technologies for waste heat utilization. *Energy Conserv. Technol.* **2011**, *29*, 123–128.

17. Werner, S. Ecoheatcool: The European Heat Market. 2006. Available online: https://www.euroheat.org/wp-content/uploads/2016/02/Ecoheatcool_WP1_Web.pdf (accessed on 7 March 2018).

18. Persson, U.; Möller, B.; Werner, S. Heat Roadmap Europe: Identifying strategic heat synergy regions. *Energy Policy* **2014**, *74*, 663–681. [CrossRef]

19. Colmenar-santos, A.; Rosales-asensio, E.; Borge-diez, D.; Blanes-peiró, J. District heating and cogeneration in the EU-28: Current situation, potential and proposed energy strategy for its generalisation. *Renew. Sustain. Energy Rev.* **2016**, *62*, 621–639. [CrossRef]

20. Fang, H.; Xia, J.; Zhu, K.; Su, Y.; Jiang, Y. Industrial waste heat utilization for low temperature district heating. *Energy Policy* **2013**, *62*, 236–246. [CrossRef]

21. Morandin, M.; Hackl, R.; Harvey, S. Economic feasibility of district heating delivery from industrial excess heat: A case study of a Swedish petrochemical cluster. *Energy* **2014**, *65*, 209–220. [CrossRef]

22. Dou, Y.; Togawa, T.; Dong, L.; Fujii, M.; Ohnishi, S.; Tanikawa, H.; Fujita, T. Innovative planning and evaluation system for district heating using waste heat considering spatial configuration: A case in Fukushima, Japan. *Resour. Conserv. Recycl.* **2018**, *128*, 406–416. [CrossRef]

23. Lund, R.; Persson, U. Mapping of potential heat sources for heat pumps for district heating in Denmark. *Energy* **2016**, *110*, 129–138. [CrossRef]

24. Cooper, S.J.G.; Hammond, G.; Norman, J. Potential for use of heat rejected from industry in district heating networks, GB perspective. *J. Energy Inst.* **2016**, *1*, 57–69. [CrossRef]

25. Chiu, J.N.W.; Flores, J.C.; Martin, V.; Lacarriere, B. Industrial surplus heat transportation for use in district heating. *Energy* **2016**, *110*, 139–147. [CrossRef]

26. Zhang, Y.; Zhang, Y.; Shi, W.; Wang, X. Application of concept of heat adaptor: Determining an ideal central heating system using industrial waste heat. *Appl. Therm. Eng.* **2016**. [CrossRef]

27. Viklund, S.B.; Karlsson, M. Industrial excess heat use: Systems analysis and $CO_2$ emissions reduction. *Appl. Energy* **2015**, *152*, 189–197. [CrossRef]

28. Weinberger, G.; Amiri, S.; Moshfegh, B. On the benefit of integration of a district heating system with industrial excess heat: An economic and environmental analysis. *Appl. Energy* **2017**, *191*, 454–468. [CrossRef]

29. Eriksson, L.; Morandin, M.; Harvey, S. Targeting capital cost of excess heat collection systems in complex industrial sites for district heating applications. *Energy* **2015**, *91*, 465–478. [CrossRef]

30. Ekvall, T.; Ahlgren, E.O.; Fakhri, A.; Martin, B. Modelling environmental and energy system impacts of large-scale excess heat utilisation e A regional case study. *Energy* **2015**, *79*, 68–79.

31. Geissmann, T. A probabilistic approach to the computation of the levelized cost of electricity. *Energy* **2017**, *124*, 372–381. [CrossRef]

32. Guo, P.; Zhai, Y.; Xu, X.; Li, Y. Assessment of levelized cost of electricity for a 10-MW solar chimney power plant in Yinchuan China. *Energy Convers. Manag.* **2017**, *152*, 176–185. [CrossRef]

33. Picard, D.; Helsen, L. Economic Optimal HVAC Design for Hybrid GEOTABS Buildings and $CO_2$ Emissions Analysis. *Energies* **2018**, *11*, 314. [CrossRef]

34. Hansen, K.; Connolly, D.; Lund, H.; Drysdale, D.; Thellufsen, J.Z. Heat Roadmap Europe: Identifying the balance between saving heat and supplying heat. *Energy* **2016**, *115*, 1663–1671. [CrossRef]

35. Gabbrielli, R.; Castrataro, P.; del Medico, F.; di Palo, M.; Lenyo, P. Levelized Cost of Heat for Linear Fresnel Concentrated Solar Systems. *Energy Procedia* **2014**, *49*, 1340–1349. [CrossRef]

36. Fawzy, M.; Kazulis, V.; Veidenbergs, I.; Blumberga, D. Levelized cost of energy analysis of co-firing solid, liquid and gaseous fuel. *Energy Procedia* **2017**, *128*, 202–207. [CrossRef]

37. Wahlroos, M.; Matti, P.; Manner, J.; Syri, S. Utilizing data center waste heat in district heating—Impacts on energy efficiency and prospects for low-temperature district heating networks. *Energy* **2017**, *140*, 1228–1238. [CrossRef]

38. Broberg, S.; Backlund, S.; Karlsson, M.; Thollander, P. Industrial excess heat deliveries to Swedish district heating networks: Drop it like it's hot. *Energy Policy* **2012**, *51*, 332–339. [CrossRef]

39. Doračić, B.; Novosel, T.; Pukšec, T. Novel approach for the evaluation of excess heat utilization in small district heating systems. In Proceedings of the 12th Conference on Sustainable Development of Energy, Water and Environment Systems (SDEWES), Dubrovnik, Croatia, 4–8 October 2017.

40. Wyrwa, A.; Chen, Y. Mapping Urban Heat Demand with the Use of GIS-Based Tools. *Energies* **2017**, *10*, 5.

41. The MathWorks Inc. *MATLAB*; The MathWorks Inc.: Natick, MA, USA, 2016.

42. QGIS. A Free and Open Source Geographic Information System. 2018. Available online: https://www.qgis.org/en/site/ (accessed on 7 March 2018).

43. Pukšec, T.; Duić, N.; Sunko, R.; Mataradžija, M.; Fejzović, E.; Babić, A.; Gjorgievski, V.; Dimov, L.J.; Bozhikaliev, V.; Markovska, M.; et al. Survey on the Energy Consumption and Attitudes towards Renewable Heating and Cooling in the CoolHeating Target Communities. 2016. Available online: http://www.coolheating.eu/images/downloads/CoolHeating_Survey_3.4.pdf (accessed on 7 March 2018).

44. Rutz, D.; Rutz, D.; Janssen, R.; Ugalde, J.M.; Hofmeister, M.; Sorensen, P.A.; Jensen, L.L.; Doczekal, C.; Zweiler, R.; Pukšec, T.; et al. Small, modular and renewable district heating & cooling grids for communities in South-Eastern Europe. *Eur. Biomass Conf. Exhib. Proc.* **2016**, *2016*, 1654–1659.

45. Domac, J.; Kolega, V.; Djukić, S.; Horvat, I.; Lončar, I.; Maras, H.; Pržulj, I.; Šegon, V.; Cvijak, V. Akcijski Plan Energetski Održivog Razvitka Grada Ozlja. 2009. Available online: http://www.eko.zagreb.hr/UserDocsImages/dokumenti/seap-i%20hr%20gradova/SEAP_OZALJ_radna%20verzija_fin.pdf (accessed on 7 March 2018).

46. Geoportal—Državna Geodetska Uprava. 2018. Available online: https://geoportal.dgu.hr/ (accessed on 7 March 2018).

47. Technology Data for Energy Plants. 2012. Available online: https://energiatalgud.ee/img_auth.php/4/42/Energinet.dk._Technology_Data_for_Energy_Plants._2012.pdf (accessed on 7 March 2018).

48. Levelized Cost Calculations. Available online: http://en.openei.org/apps/TCDB/levelized_cost_calculations.html (accessed on 5 March 2018).

49. Doračić, B.; Pušić, T.; Novosel, T.; Pavičević, M.; Pukšec, T.; Duić, N. Techno-Economic Analysis of the Implementation of a Small Renewable District Heating System: Case Study for the City of Ozalj. In Proceedings of the 5th International Congress Mechanical Engineers Day, Amman, Jordan, 13–16 May 2017; pp. 27–32.

MDPI

St. Alban-Anlage 66

4052 Basel

Switzerland

Tel. +41 61 683 77 34

Fax +41 61 302 89 18

www.mdpi.com

*Energies* Editorial Office

E-mail: energies@mdpi.com

www.mdpi.com/journal/energies